SAFETY OF NANOMATERIALS ALONG THEIR LIFECYCLE

Release, Exposure, and Human Hazards

SAFETY OF NANOMATERIALS ALONG THEIR LIFECYCLE

Release, Exposure, and Human Hazards

EDITED BY

Wendel Wohlleben ▪ Thomas A. J. Kuhlbusch

Jürgen Schnekenburger ▪ Claus-Michael Lehr

CRC Press is an imprint of the
Taylor & Francis Group, an **informa** business

Cover images: (upper left) dust exposure measurement during production (IUTA), see Section I; (upper right) nanoparticles of TiO2 interacting with a human cell (WWU), see Section II; (lower left) nanocomposite with embedded SiO2 after outdoor use (BASF), see Section III; (lower right) plastics nanocomposites enabling automotive parts (BASF), see Editorial Perspective and Section IV.

CRC Press
Taylor & Francis Group
6000 Broken Sound Parkway NW, Suite 300
Boca Raton, FL 33487-2742

First issued in paperback 2021

Version Date: 20140912

ISBN 13: 978-1-03-223646-9 (pbk)
ISBN 13: 978-1-4665-6786-3 (hbk)

Publisher's Note
The publisher has gone to great lengths to ensure the quality of this reprint but points out that some imperfections in the original copies may be apparent.

Library of Congress Cataloging-in-Publication Data

Safety of nanomaterials along their lifecycle : release, exposure, and human hazards / edited by Wendel Wohlleben, Thomas A.J. Kuhlbusch, Jürgen Schnekenburger, and Claus-Michael Lehr.
pages cm
Includes bibliographical references and index.
ISBN 978-1-4665-6786-3 (hardback)
1. Nanostructured materials--Environmental aspects. 2. Nanostructured materials--Health aspects. 3. Nanoparticles--Toxicology. I. Wohlleben, Wendel.

TD196.N36S24 2014
363.17'9--dc23 2014032615

Visit the Taylor & Francis Web site at
http://www.taylorandfrancis.com

and the CRC Press Web site at
http://www.crcpress.com

Contents

Perspectives

NANOMATERIALS: AN EVOLVING TECHNOLOGY

Nanomaterials have pervaded our personal and professional everyday life during the last centuries, but the awareness of possible consequences for us and the environment was raised only during the last decades. At the end of the 19th century it was shown that nanomaterials, when added to rubber, make tires longer lasting and enhance their grip on the ground. Nanomaterials can also be used to tune the viscosity of fluids, influence optical properties, and make plastics more robust. The general interest in nanomaterials has gained momentum at the end of the 20th century when production methods became more controllable, larger volumes were produced, and nanomaterials with adjustable properties were generated, characterized, and imaged at atomic resolution. Nowadays, nanomaterials and systems based on nanomaterials are used everywhere: construction and building materials; energy production and storage; catalysis; microelectronics; plastics, coatings, and paints; and foods, cosmetics, and pharmaceuticals. The economic success of these nanomaterials is driven by the improvement of existing end-user products in terms of lower costs, easier maintenance, increased durability or enhanced performance, or because they enable products with novel functionalities. The focus of this book was set on economically viable nanomaterials—identified by significant production volumes—with a likeliness of release and human exposure.

Innovation never comes at zero risk. Having learned from the history of conventional nanomaterials and ubiquitous ultrafine particles in ambient air, the nanotechnology community recognizes the necessity to accompany technological development with research into possible negative consequences arising from the increased use of traditional nanomaterials and the targeted use of specially engineered nanomaterials. This was especially addressed in publicly funded, cross-disciplinary, and collaborative projects among industry, academia, and regulators. They have significantly advanced a common terminology, an understanding of mechanisms, and a toolbox of adapted methods. Even more, the collaboration and gained knowledge of the possible risk of using nanomaterials are an important contribution to the public acceptance of this technology and consequently to the use of its inherent capabilities to tackle societal and environmental challenges today and in the future.

SCOPE AND LIMITATIONS

This book presents the state of the art in nanosafety research from a lifecycle perspective. Although knowledge gaps will always remain, the editors focus on the valid options for pragmatic risk assessment and management. Project nanoGEM (2010–2013), a follow-up of NanoCare (2006–2009), both funded by the German Ministry for Education and Research, is the nucleus to this book. Results of nanoGEM account for about half of the content, complemented by essential contributions from international experts.

Basis for an exposure and hence a risk is the release of nanomaterials into the environment. Key scenarios possibly leading to nanomaterial release are here derived using the available knwoledge from existing products, because nanomaterials often replace or complement traditional materials with a higher level of control of microscopic structures. Some of the applications of nanotechnology with high societal impact, such as nanostructures buried in microelectronic devices, are not considered in this book because the likeliness of release and human exposure is viewed as negligible. Furthermore, the book does not cover the topic of environmental fate and ecotoxicology to gain space for in-depth discussions on release, exposure, and human hazards. Still, we acknowledge the research need of environmental effects for a completeness of understanding and integration into the decision-making process.

Our editorial perspective highlights those features that make nanomaterials different from chemicals at either molecular or bulk scales, describes the resulting challenges in exposure and toxicity testing, and sketches the risk assessment for humans along the materials lifecycle.

CORRELATION BETWEEN MATERIALS PROPERTIES AND TOXICOLOGICAL HAZARD ASSESSMENT

Nanomaterials share properties with macromolecules, such as the possibility to form stable dispersions and a small size, which allows the penetration of biological barriers. But they also share other properties with bulk materials, such as a defined physical boundary with a stable and possibly reactive surface. The obvious difference of nanomaterials compared to their bulk or micron-sized analogs is the highly increased mass-specific surface area that scales the surface reactivity and adsorption capacity inherent to any solid material. Further nanomaterial properties, which may be associated with biological activity, are ions leaching from the nanomaterial or the environment-dependent agglomeration that may modify uptake and distribution in the human body. However, only few materials show a threshold behavior in material properties when decreasing the size. Examples include the optical and electronic effects of spatially confined semiconductors and metals, such as quantum dots and surface plasmons, or the catalytic effects of gold observed exclusively on nanoparticles, but these are not economically viable large-volume products. With most materials, essential properties such as the crystalline defect density, redox activity, agglomeration tendency, absorption, or turbidity scale smoothly without threshold behavior.

The holy grail of nanomaterial toxicity testing is the quantitative structure–activity relationship (QSAR). This would be the definition of specific physico-chemical material properties that result in defined biological effects as stress response, cell death, DNA damage, or other toxicity endpoints. Although that grail remains to be discovered, models of structure–activity relationships have been proposed. Anyhow, they are presently predictive only for relatively narrow classes of materials. Some properties have repeatedly shown a strong correlation to the outcome of toxicological tests—such as a high-aspect ratio, leaching of toxic ions mediated by increased specific surface, oxidative surface reactions, or a positive charge—but we currently cannot systematize these parameters in a unifying model. Judiciously surface-functionalized nanomaterials are key to test structure–activity relationships via surface-induced transport and effects as the most probable mode-of-action of particulate materials on physiological systems. A lot can be learned here from the comparably young field of nanomedicine, because it advances the understanding and engineering of defined nanomaterials that can cross some relevant biological barriers.

The fact that nanomaterials with nominally identical properties and chemical composition were shown to differ in their toxicology potential has led to a critical demand for an extensive physico-chemical characterization of nanomaterials. Whether that *as-produced* surface, or, instead, the in situ surface after interaction with biological media, is the structure that elicits an activity, is an open hypothesis. Its verification requires detailed in situ characterizations of nanomaterials, including case studies on the most relevant scenarios of nanomaterial uptake by humans: inhalation of dusts or ingestion. The direct interplay between material modifications and toxicity testing could be intensified if in vitro assays achieved a closer match with the regulatorily approved in vivo tests. Stumblestones of previous tests—including the interferences of optical properties of nanomaterials with spectroscopic test readouts, or the adsorption of molecular markers and nutrient—occurred especially at nanomaterial in vitro doses that exceeded any in vivo situation. A relevant source for method development is dust toxicology and epidemiology. Generic dusts are related to airborne nanomaterial via their polydisperse aerosol-size distribution. This notion helps us to extrapolate long-term effects, which so far have not been investigated for engineered nanomaterials.

RELEASE AND EXPOSURE ASSESSMENT THROUGHOUT THE LIFECYCLE

Engineered nanomaterials in their pure form (powders) are not used as consumer products. With very few exceptions, consumers encounter nanomaterials only embedded in a composite material or formulated into a liquid, creamy, or solid product. The materials lifecycle contains therefore at least three phases: (1) professional handling—including production and packaging of pure nanomaterial, formulation of composites and their manufacturing into finished parts; (2) use as consumer products or in industrial applications; and finally (3) waste and recycling or degradation of the product. A reliable exposure measurement in all three phases is crucial for risk assessment, and it should determine the application of relevant test items and relevant concentrations in toxicity testing. The procedures for characterization, toxicity, and exposure assessment in phase 1 are well established, but the product lifecycle alters the properties of released pieces of matter. This calls for adapted techniques for exposure assessment as well as for hazard assessments of "aged" or "transformed" nanomaterials in phases 2 and 3. These phenomena are maybe the most fundamental difference between nanomaterial and molecular chemicals. Further, measurement procedures need to account for the high environmental background of both natural and involuntarily man-made ultrafine particles (<100 nm diameter) from which the released nanomaterial has to be differentiated.

Since 2011 many exposure-related studies have been reported, and some release test set-ups are close to harmonization and standardization, such as the simulation of powder handling by dustiness set-ups and the simulation of finished parts manufacturing by sanding set-ups. All lifecycle tests on consumer products report the release of composite fragments, where the properties are dominated by the matrix, in a few cases modulated by surface protrusions of the nanomaterial. Especially by chemical degradation of the polymer matrix, an additional nonzero fraction of free nanomaterials (estimated around 0.001 mass-%) is released into the environment over a 10-year use phase. Even lower release rates can be anticipated—and should be confirmed on exemplary cases—for transformed materials, for example, ceramic slurries that are sintered onto the surface of casting molds and cementitious materials that dissolve and recrystallize, thus losing their nanoparticulate identity but retaining internal structures that are important for product performance. It is a goal of the present book to put the recently very productive release and exposure research into a pragmatic perspective with toxicological advances, including case studies.

CONCLUSION AND INTERIM RISK ASSESSMENT

The research presented in this book highlights the recent results that are needed for the risk assessment of engineered nanomaterials. We conclude that size-dependent biological effects do exist, but no effects were observed that are not known from either molecular chemicals or larger particles or fibers, with the exception of soluble ion-releasing nanomaterials. The toxic potency can scale with the specific surface, but the nanomaterial's composition is the primary determinant of toxicity, modulated by size, and surface chemistry. It further emerges that surface functionalization can reduce effects, and this functionalization controls the in situ structures in a rational manner via colloidal mechanisms of charge and stericity. But these in situ structures are not (yet) more predictive for toxicity than *as-produced* properties such as charge and solubility themselves. Nanomaterials like silicon dioxide, zirconium dioxide, or aluminium oxide do not show a significant substance-specific toxicity over and beyond their particulate effect. The similarity of toxicological effects from these biopersistent nanomaterials without specific toxicological properties (granular biopersistent nanomaterials) as one group versus fiber-shaped materials as another group suggests risk groupings based on hazard, although knowledge of lifecycle releases and specific product use offers alternative grouping strategies. For instance, the relatively good convergence of release and exposure results substantiates that fragments released from consumer products are typically composite fragments, where the properties are dominated by the matrix.

Risk is always a function of exposure and hazard potential of a given material. In simple terms, when either of the two is not present, there is no risk for humans or the environment. The knowledge of both, exposure and hazard, builds the basis for risk assessment enabling decision-making for risk management. Risk management of new materials in Europe is founded on the precautionary principle, which ultimately means that new materials with unknown hazards can only be used while assuring no exposure during manufacturing and use. Nanomaterials, however, are not unknown materials anymore. The state-of-knowledge on hazard, release, and

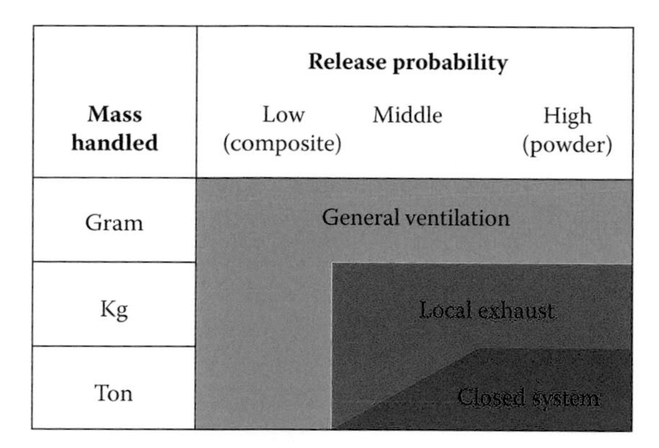

FIGURE P.1 Simple classification system for likeliness of exposure concentrations that need to be controlled by suitable measures as indicated. This scheme (BAuA, nanoGEM concluding conference, Berlin, June 2013) is applicable to GBP nanomaterials, for which an exposure limit of 0.5 mg/m³ is derived.

exposure, as summarized above, enables a risk assessment for professional workers and consumers. Different nanomaterials range from no, low-to-high exposure, and hazard potential. For GBP nanomaterials, the German Committee on Hazardous Substances (AGS announcement 527, 2013) derives an occupational exposure that should not be higher than 0.5 mg/m^3 (at a density of 2.5 g/cm^3), aiming at half of the currently valid and legally binding occupational limit value for the alveolar dust fraction. Considering that exposure can only occur if GBP nanomaterials are released into the environment, Figure P.1 links the probability of release, the mass handled, and advice on the risk management systems to be used at workplaces to limit exposure by ventilation, local exhaust, or closed systems.

This classification system can be extended for the hazard assessment of various nanomaterials, taking dose-response relationships and different health endpoints into account. One further scheme classifies nanomaterials into four groups (AGS announcement 527, 2013):

1. Biopersistent nanomaterials with a hazard potential that can be directly linked to its toxicological relevant chemical composition, for example, heavy metals, silver, and zinc oxides.
2. Biopersistent nanomaterials that have a hazard potential due to their morphology, more specifically fiber characteristics, like carbon nanotubes. High-aspect ratio serves as a primary identifier of concern with confirmed chronic pathologic effects.
3. Biopersistent nanomaterials without a specific chemical toxicity but possible effects related to their particulate nature, especially by dust inhalation. Examples include silica, carbon black, and titanium dioxide (GBP nanomaterials).
4. Soluble (not biopersistent) nanomaterials whose chemical composition supersedes the *as-produced* nanostructure for risk assessment.

Of the large-volume products mentioned at the beginning of this perspective, class 1 nanomaterials are predominantly found in energy production and storage, catalysis, and microelectronics, whereas applications in plastics, coatings, paints, foods, and cosmetics tend to belong to class 3 (GBP) nanomaterials. Certainly, in some cases no clear classification according to the above scheme is possible because no clear physically, chemically, or toxicologically driven cut points, for example, according to available traceable standards. The classifications are always based on comparisons to other components. It is also interesting to note that the suggested four classes, related mainly to hazard, are based on "classical" toxicological mechanisms. This comes as no surprise because no new nano-specific toxicological mechanisms have been identified so far. Within nanoGEM we came to the conclusion that often classical assessment methods can be applied to nanomaterials, but specific nanomaterial-related evaluations with regard to hazard potential and exposure have to be taken into account, although some are still lacking.

FUTURE DIRECTIONS OF RESEARCH

The steadily increasing commercialization of novel and engineered nanomaterials has created a sense of urgency for regulators. However, the European Commission recommendation for a regulatory definition of nanomaterials is very inclusive for conventional and natural materials. Accordingly, this nanodefinition is effectively a market screening of nanostructures and reports an enormous number of products in a large variety of materials classes, ranging from minerals, fillers, and pigments, over catalysts metals to cements and mortars to natural materials.

Across all stakeholders in nanoGEM, the need for simplified yet safe risk assessment approaches based on grouping/categorization was identified since (1) not all release/exposure scenarios can be tested for all nanomaterials, (2) the validation and integration of reliable in vitro assays in, for example, Organisation for Economic Co-operation and Development guidelines will take several years, and (3) it is impossible to determine the hazard potential for all nanomaterials by in vivo studies. This call for grouping to facilitate risk assessment is needed both for ethical (animal testing) and financial reasons. The above example for risk assessment was given for workers and workplaces, but similar simplified yet safe approaches are also needed for consumer and environmental risk assessments.

Grouping is key, but two major challenges for future risk assessment emerge:

1. The understanding of the underlying mechanisms of toxicity to further motivate grouping schemes and to substantiate measurable grouping criteria.
2. The understanding of nanomaterial release probability from composites and products, including environmental fates and effects.

The first challenge would avoid the full toxicity testing of each new nanomaterial, especially the currently essential in vivo testing even of closely related materials could be reduced. nanoGEM in vitro and in vivo nanomaterial toxicity testing has shown an excellent correlation of the applied in vitro test systems and the animal testing of the nanoGEM materials. Both testing strategies identified the identical grouping within the 16 nanomaterials studied. These encouraging results demonstrate the applicability of updated in vitro assays for nanomaterial grouping and prescreening of materials for further in vivo testing. The analysis of toxicity mechanisms would allow QSAR application and a classification of nanomaterials by abiotic or in vitro test systems.

The second challenge emerges from the increased use of nanomaterials within consumer products. How will the change of a materials environment along its complete lifecycle change essential material properties such as fate and transport, biomolecule corona and environmental deposition, and the resulting biological potency? These and many more questions require not only detailed research but also new technologies for the detection of nanomaterials. At present, we have good reasons to assume that the risk caused by nanomaterials is much reduced after their incorporation into solid matrices and products; however, the remaining risk potential especially for the environment is not completely clear. The high environmental background of nanomaterials of natural or man-made origin adds to the challenges that we face before we can reduce the uncertainties on the risks of engineered nanomaterials to acceptable levels not only for specific human activities, as presented in this book, but also for all relevant environmental aspects. I would suggest to add some free space after the text to better discriminate the text and the contributors.

Wendel Wohlleben
BASF SE, Advanced Materials and Systems Research
Ludwigshafen, Germany

Thomas A.J. Kuhlbusch
Institute of Energy and Environmental Technology e.V. (IUTA)
Duisburg, Germany
and
Centre for NanoIntegration University Duisburg-Essen (CENIDE)
Duisburg, Germany

Jürgen Schnekenburger
Biomedical Technology Center, Westfälische Wilhelms-Universität
Münster, Germany

Claus-Michael Lehr
Helmholtz-Institute for Pharmaceutical Research Saarland
Helmholtz Centre for Infection Research, and Saarland University
Saarbrücken, Germany

A Guide to the Reader

Because the editors have preferred to present their own perspective on the safety of nanomaterials as a leading contribution that summarizes the state of the art, this book does not contain a conventional editorial that would introduce structure and content. Instead, the following compilation of key statements from each chapter serves this purpose. The two elements of risk assessment—hazard and exposure—constitute Sections II and III. They are preceded by the characterization of the object of investigation in Section I and followed by case studies in Section IV, which are further brought to conclusions in an editorial perspective.

EDITORIAL PERSPECTIVE

Nanomaterials differ from bulk materials and from molecules—Advantages of nano-enabled products—In vitro test assay interferences add decision complexity to the cost and ethical issues of in vivo testing—Exposure background from natural and man-made origin is high—Four classes of nanomaterials—Risk assessment for granular biopersistent particles—Lifecycle induces changing properties, release of fragments—Grouping mechanisms and environmental fate as future challenges.

SECTION I: CHARACTERIZATION

Chapter 1 As-Produced: Intrinsic Physico-Chemical Properties and Appropriate Characterization Tools

Emilia Izak-Nau and Matthias Voetz

What does the material look like? What is the material made of? Which factors affect the material interactions?—ISO standards exist for relevant properties: particle size distribution—aggregation/agglomeration—shape—surface area—composition—surface chemistry—surface charge—solubility/dispersibility—control by surface functionalization is effective, but affects more than one property.

Chapter 2 Characterization Methods for the Determination of Inhalation Exposure to Airborne Nanomaterials

Christof Asbach

Four categories of methods: integrated/resolved on size and time, respectively—Nanomaterial aerosols are characterized by their sphere-equivalent diameters: electrical mobility diameter versus aerodynamic diameter versus optical diameter—Size-integrated, time-resolved handheld instruments are reliable for exposure screening—Scanning Mobility Particle Sizer delivers number metrics and is the most versatile size-and-time-resolving instrument.

Chapter 3 Classification Strategies for Regulatory Nanodefinitions
Wendel Wohlleben and Philipp Müller
Definitions of nanomaterials for regulatory purposes differ in thresholds, metrics, relevance of agglomerates, and relevance of novel properties—Methods to measure the size distribution in number metrics: for powders and for suspensions—Sample preparation is essential for the validity of fractionating and counting techniques—Tiered testing scheme for cost-effective screening and confirmation.

Chapter 4 Analyzing the Biological Entity of Nanomaterials: Characterization of Nanomaterial Properties in Biological Matrices
Christian A. Ruge, Marc D. Driessen, Andrea Haase, Ulrich F. Schäfer, Andreas Luch, and Claus-Michael Lehr
Nanomaterial characteristics change over their entire lifecycle, including toxicological testing—Varied, dynamically changing in situ properties—Colloidal and biochemical methods to characterize affinity, composition, and size of adsorbed proteins—Charge and hydrophobicity, size, and shape determine interactions—Case study on functionalized nanoparticles in lipids and surfactant proteins of the respiratory tract—Case study on nanoparticles in gastric juices.

SECTION II: HAZARD ASSESSMENT FOR HUMANS

Chapter 5 Lessons Learned from Unintentional Aerosols
Joseph Brain
Ultrafine fractions are familiar in sea spray, volcanoes, forest fires, tobacco smoke, motor vehicle emissions, coal dust, and welding fume—Same methods for hazard assessment, but need to aerodynamically select ultrafine particles from polydisperse aerosol—Ultrafine particles may be more toxic than larger particles, but as larger particles dissolve, they will inevitably go through a nanoparticle phase—Contribution of ultrafine fraction to toxicity of unintentional aerosols.

Chapter 6 Lessons Learned from Pharmaceutical Nanomaterials
Emad Malaeksefat, Sarah Barthold, Brigitta Loretz, and Claus-Michael Lehr
Nanoformulated drug delivery systems can enhance stability and bioavailability, can reduce side effects or target tissues—Group materials by biodegradability and potential for endocytosis—Quality by design must consider hydrophobicity, shape, protein-binding—Pulmonary route succeeds only for protein delivery, not for intracellular delivery—Case studies of drugs reformulated with nanocrystals, albumins, and liposomes.

Chapter 7 Measurement of Nanoparticle Uptake by Alveolar Macrophages: A New Approach Based on Quantitative Image Analysis
Darius Schippritt, Hans-Gerd Lipinski, and Martin Wiemann
Establish dose metrics: administered versus extracellular contact versus intracellular uptake—Nanoparticles in cell culture medium agglomerate, then sediment onto cells—Visualize sediment by phase contrast microscopy and tracking algorithm—Locate motile macrophages and their borderline—Model-based description of phagocytosis—Quantify particle number uptake by macrophages.

Chapter 8 Toxicological Effects of Metal Oxide Nanomaterials
Daniela Hahn, Martin Wiemann, Andrea Haase, Rainer Ossig, Francesca Alessandrini, Lan Ma-Hock, Robert Landsiedel, Marlies Nern, Antje Vennemann, Marc D. Driessen, Andreas Luch, Elke Dopp, and Jürgen Schnekenburger
Biological effects are mainly determined by chemical composition and solubility, modulated by size, surface functionalization, crystallinity—Oxidative stress varies on low level, genotoxicity is rare—Surface coatings can mitigate inhalation effects (less when positively charged) and adjuvant effects in allergic airway inflammation—Match in vitro to in vivo effect levels via mass dose per alveolar macrophage—Majority of metal oxides display no or only moderate toxicity.

Chapter 9 Toxicological Effects of Metal Nanomaterials
Rainer Ossig, Daniela Hahn, Martin Wiemann, Marc D. Driessen, Andrea Haase, Andreas Luch, Antje Vennemann, Elke Dopp, Marlies Nern, and Jürgen Schnekenburger
Effects of silver and copper nanomaterials depend on surface and size: oxidative stress, protein carbonylation, genotoxicity, lung inflammation, and liver injury—Released ions at least partially explain effects for silver, less for copper nanomaterials—Low cytotoxicity of gold or platinum nanoparticles, but uptake depends on surface modification and charge—Size, surface coating, and solubility act in concert on biodistribution and secondary biological responses.

Chapter 10 Uptake and Effects of Carbon Nanotubes
James C. Bonner
Rigidity, length, and the state of agglomeration determine macrophage uptake and clearance—Catalyst impurities elicit further acute cellular responses—Chronic pathologic effects modulated by state of agglomeration: fibrosis, granuloma formation, and pleural disease—Carcinogenicity of CNTs remains controversial—Effects beyond the lung—Susceptibility factors.

SECTION III: EMISSION AND EXPOSURE ALONG THE LIFECYCLE

Chapter 11 Measurement and Monitoring Strategy for Assessing Workplace Exposure to Airborne Nanomaterials
Christof Asbach, Thomas A.J. Kuhlbusch, Burkhard Stahlmecke, Heinz Kaminski, Heinz J. Kiesling, Matthias Voetz, Dirk Dahmann, Uwe Götz, Nico Dziurowitz, and Sabine Plitzko
Decision criteria to escalate tier by tier until uncertainty decreases—Tier 1: information gathering—Tier 2: simple measurement considering background and sources of emissions—Tier 3: extensive measurement campaign with size distributions and sampling—Risk management steps—Aerosol devices do not differentiate: need particle sampling with morphology and composition analysis to identify engineered nanomaterial against ubiquitous background particles.

Chapter 12 Release from Composites by Mechanical and Thermal Treatment: Test Methods
Thomas A.J. Kuhlbusch and Heinz Kaminski
Release studies are relevant to identify and model possible exposures—Dedicated test methods: mixing and sonication, abrasion, grinding, cutting, and sawing—High speed processes (sanding, drilling, and milling) may induce release via heating and degradation—Elevated release probability by combustion versus controlled oxidation during incineration—Dustiness, abrasion, sanding, and drilling are ready for standardization.

Chapter 13 Field and Laboratory Measurements Related to Occupational and Consumer Exposures
Derk Brouwer, Eelco Kuijpers, Cindy Bekker, Christof Asbach, and Thomas A.J. Kuhlbusch
Regulations are in place with exposure limits—Since 2011 many exposure related studies have been reported—Differentiation of four source domains: release during synthesis of nanomaterials, low energy handling and transfer of nanomaterials, high energy dispersion of powders and suspensions, actions resulting in fracturing or abrasion—Exposure is possible but often limited to matrix bound nano-objects—Meta analysis and models are needed for quantitative exposure assessment.

Chapter 14 Mechanisms of Aging and Release from Weathered Nanocomposites
Tinh Nguyen, Wendel Wohlleben, and Lipiin Sung
ISO weathering methods assess resilience of coatings and plastics: we need to collect fragments released—Aging of polymer nanocomposites by chain scission—The relevance of shear to induce release: particles release spontaneously, nanotubes don't—Detection, quantification, identification show polydisperse mixture of fragments, not excluding free nanofillers—Extrapolation to lifecycle emissions.

Chapter 15 Emissions from Consumer Products Containing Engineered Nanomaterials over Their Lifecycle
Bernd Nowack
Nanoparticles released from products are mostly matrix-bound, only some fraction of single, dispersed nanoparticles, for example, from pressurized sprays—Real-world predictivity, but also background issues increase from mechanistic degradation to release collection during use simulations to monitoring during normal use—Transformation products of nano-Ag from textile washing—Nano-release occurs from pigment-grade paints—Very limited evidence of migration from food containers—Significant fraction of modified nano-TiO_2 released from sunscreens into water—Lifecycle-based material flow modeling.

SECTION IV: INTEGRATING CASE STUDIES ON METHODS AND MATERIALS

Chapter 16 Concern-Driven Safety Assessment of Nanomaterials: An Integrated Approach Using Material Properties, Hazard, Biokinetic, and Exposure Data and Considerations on Grouping and Read-Across
Agnes Oomen, Peter Bos, and Robert Landsiedel
Integrated approaches for testing require grouping of nanomaterials. Derived from source-to-adverse-outcome-pathway, concerns are driven by release, physico-chemical characteristics, biokinetics, and early/apical biological effects—Depending on assessment purpose and concern, choose from these the appropriate parameters for grouping—Targeted testing enables a decision to group or waive, to proceed for next tier, or to apply specific risk management measures.

Chapter 17 Case Study: Paints and Lacquers with Silica Nanoparticles
Keld A. Jensen and Anne T. Saber
Silica affects hardness, gloss, adhesion, drying time, and the sanding properties of lacquers - Silica has inflammatory properties—Acute high dustiness levels during powder pouring and wet mixing, but silica contribution hidden by dust from talc and pigment TiO_2—Toxicity of total emitted sanding dust of paints and lacquers is controlled by the paint matrix rather than the added fillers and nanomaterials.

Chapter 18 Case Study: The Lifecycle of Conductive Plastics Based on Carbon Nanotubes
Richard Canady and Thomas A.J. Kuhlbusch
Conductive plastics are applied as light-weight replacement for metals, as antistatic packaging and electromagnetic interference shielding for electronics, or spark-resistant fuel lines in automotive and aerospace applications—Consumers often lack access, but releases are potentially non-zero to technicians—Across laboratories, release of free CNTs is

rarely observed, at least in initial release processes—Toxicity evaluation of CNT fibers greatly overestimates "fiber-related" health risk: No evidence that the released material would be more hazardous than material released from the same polymer without added CNTs.—Evaluation needed of polymer matrix degradation by weathering or heat, of subsequent environmental transport.

Chapter 19 Case Study: Challenges in Human Health Hazard and Risk Assessment of Nanoscale Silver

Christian Riebeling and Carsten Kneuer

Risks to human health can, in principle, be characterized using the established risk assessment paradigm—Substance identifiers and exposure assessment are challenged by transformation reactions of silver particles and ions along the product lifecycle—Uncertain relevance of data for specific nanosilver product versus the need for read-across—Data-deficient hazard characterization can result in restriction of nanosilver to essential applications and/or labelling: authorization in the U.S.A. occurred under condition of further data generation.

Acknowledgments

Part of the book chapters and results presented therein stem from the nanoGEM project funded by the Federal Ministry of Education and Research (BMBF, Germany) and Industry (03X0105A).

Contributors

Francesca Alessandrini
Focus Network Nanoparticles and Health (NanoHealth)
Division of Environmental Dermatology and Allergy
Technical University of Munich
Munich, Germany

Christof Asbach
Air Quality and Filtration Unit
Institute of Energy and Environmental Technology (IUTA) e.V
Duisburg, Germany

Sarah Barthold
Helmholtz-Institute for Pharmaceutical Research Saarland (HIPS)
Helmholtz Center for Infection Research (HZI)
Saarbrücken, Germany

Cindy Bekker
Netherlands Organization for Applied Scientific Research (TNO)
HE Zeist, The Netherlands

James C. Bonner
Department of Biological Sciences
North Carolina State University
Raleigh, North Carolina

Peter Bos
National Institute for Public Health and the Environment (RIVM)
Bilthoven, The Netherlands

Joseph Brain
Department of Environmental Health
Harvard School of Public Health
Boston, Massachusetts

Derk Brouwer
Netherlands Organization for Applied Scientific Research (TNO)
Zeist, The Netherlands

Richard Canady
ILSI—Center for Risk Science Innovation and Application
Washington, DC

Dirk Dahmann
Institute for Research on Hazardous Substances (IGF)
Bochum, Germany

Elke Dopp
Bayer Material Sciences
Leverkusen, Germany

Marc D. Driessen
Department of Chemicals and Product Safety
Federal Institute for Risk Assessment
Berlin, Germany

Nico Dziurowitz
Federal Institute of Occupational Safety and Health (BAuA)
Berlin, Germany

Uwe Götz
Department of Occupational Safety
BASF SE
Ludwigshafen, Germany

Andrea Haase
Department of Chemicals and Product Safety
Federal Institute for Risk Assessment
Berlin, Germany

Daniela Hahn
Biomedical Technology Center
Westfälische Wilhelms-Universität
Münster, Germany

Emilia Izak-Nau
Bayer Technology Services
Leverkusen, Germany

Keld A. Jensen
The Danish Center for NanoSafety
National Research Center for the Working Environment (NRCWE)
Copenhagen, Denmark

Heinz Kaminski
Air Quality and Sustainable Nanotechnology Unit
Institute of Energy and Environmental Technology (IUTA) e.V
Duisburg, Germany

Heinz J. Kiesling
Bayer Technology Services
Leverkusen, Germany

Carsten Kneuer
Department of Pesticide Safety
Federal Institute for Risk Assessment
Berlin, Germany

Thomas A.J. Kuhlbusch
Air Quality and Sustainable Nanotechnology Unit
Institute of Energy and Environmental Technology (IUTA) e.V
Duisburg, Germany

Eelco Kuijpers
Netherlands Organization for Applied Scientific Research (TNO)
Zeist, The Netherlands

Robert Landsiedel
Department of Experimental Toxicology and Ecology
BASF SE
Ludwigshafen, Germany

Claus-Michael Lehr
Helmholtz-Institute for Pharmaceutical Research Saarland (HIPS)
Helmholtz Center for Infection Research (HZI)
and
Department of Biopharmaceutics and Pharmaceutical Technology
Saarland University
Saarbrücken, Germany

Hans-Gerd Lipinski
Biomedical Imaging Group
Department of Computer Science
University of Applied Sciences and Arts
Dortmund, Germany

Brigitta Loretz
Helmholtz-Institute for Pharmaceutical Research Saarland (HIPS)
Helmholtz Center for Infection Research (HZI)
Saarbrücken, Germany

Andreas Luch
Department of Chemicals and Product Safety
Federal Institute for Risk Assessment
Berlin, Germany

Lan Ma-Hock
Department of Experimental Toxicology and Ecology
BASF SE
Ludwigshafen, Germany

Emad Malaeksefat
Helmholtz-Institute for Pharmaceutical Research Saarland (HIPS)
Helmholtz Center for Infection Research (HZI)
Saarbrücken, Germany

Philipp Müller
Advanced Materials and Systems Research
Department of Material Physics
BASF SE
Ludwigshafen, Germany

Marlies Nern
Bayer HealthCare
Wuppertal, Germany

Tinh Nguyen
Polymeric Materials Group
National Institute of Standards and Technology (NIST)
Gaithersburg, Maryland

Bernd Nowack
EMPA–Swiss Federal Laboratories for Materials Science and Technology
St. Gallen, Switzerland

Agnes Oomen
National Institute for Public Health and the Environment (RIVM)
Bilthoven, The Netherlands

Rainer Ossig
Biomedical Technology Center
Westfälische Wilhelms-Universität
Münster, Germany

Sabine Plitzko
Federal Institute of Occupational Safety and Health (BAuA)
Berlin, Germany

Christian Riebeling
Department of Chemicals and Product Safety
Federal Institute for Risk Assessment
Berlin, Germany

Christian A. Ruge
Department of Biopharmaceutics and Pharmaceutical Technology
Saarland University
Saarbrücken, Germany

Anne T. Saber
The Danish Center for NanoSafety
National Research Center for the Working Environment (NRCWE)
Copenhagen, Denmark

Ulrich F. Schäfer
Department of Biopharmaceutics and Pharmaceutical Technology
Saarland University
Saarbrücken, Germany

Darius Schippritt
Biomedical Imaging Group
Department of Computer Science
University of Applied Sciences and Arts
Dortmund, Germany

Jürgen Schnekenburger
Biomedical Technology Center
Westfälische Wilhelms-Universität
Münster, Germany

Burkhard Stahlmecke
Air Quality and Sustainable Nanotechnology Unit
Institute of Energy and Environmental Technology (IUTA) e.V
Duisburg, Germany

Lipiin Sung
Polymeric Materials Group
National Institute of Standards and Technology (NIST)
Gaithersburg, Maryland

Antje Vennemann
IBE Research and Development Institute for Lung Health gGmbH
Münster, Germany

Matthias Voetz
Bayer Technology Services
Leverkusen, Germany

Martin Wiemann
IBE Research and Development Institute for Lung Health gGmbH
Münster, Germany

Wendel Wohlleben
Advanced Materials and Systems Research
Department of Material Physics
BASF SE
Ludwigshafen, Germany

Section I

Characterization

1 As-Produced
Intrinsic Physico-Chemical Properties and Appropriate Characterization Tools

Emilia Izak-Nau and Matthias Voetz

CONTENTS

1.1 INTRODUCTION

It is well known that nanomaterials possess different chemical and physical properties than bulk materials of identical composition. When the particle size decreases to nanoscale, the fundamental characteristics of the material often changes, resulting in completely new properties. For instance, previously colorful materials become transparent, insulators become conductive, or the melting point of nanoparticles evidently drops when their size is below 100 nm (Nalwa 2004; Wang 2000). When the particle size decreases, the proportion of atoms found at the surface relative to the atoms in the interior of the particle increases. Thus, nano-objects have a much larger surface area per unit mass compared with bigger particles. The increase in the surface-to-volume ratio results in an increase in the particle surface energy, which may render them more reactive.

One of the major challenges in determining risks emerging from nanomaterials is their extremely wide diversity, in terms of not only different sizes but also a variety of shapes and compositions, or different surface chemistry and surface charge. An additional issue is the possible presence of impurities. When conducting any biological evaluation of a nanomaterial, it is important to know with completeness what substance is being tested. The main physico-chemical properties of nanomaterials can be organized into three general groups that can answer fundamental questions concerning nanomaterial characterization:

- What does the material look like?
  - Size/size distribution
  - Agglomeration/aggregation state
  - Shape
  - Specific surface area
- What is the material made of?
  - Composition (including chemical composition, crystal structure, and impurities)
  - Surface chemistry
- What factors affect the material interaction with its surroundings?
  - Surface charge
  - Solubility
  - Dispersibility

Different methods are available for the characterization of the nanomaterial properties. They are listed in Table 1.1 with relevant characterization standards. However, the majority of those methods have not necessarily been validated for different types of nano-objects. Thus, there is an urgent need for the development of additional measuring techniques. The standardization of the new methods is currently being undertaken by the International Organization for Standardization (ISO) technical committees (ISO/TC 20, ISO/TC 24, and ISO/TC 229). A variety of tools and methods for nanomaterial characterization are also being recommended by some international, as well as host nation organizations such as the Organization for Economic Co-operation and Development (OECD), the American Society for Testing and Materials (ASTM International), the U.S. Environmental Protection Agency (EPA), or the U.S. National Cancer Institute-Nanotechnology Characterization Laboratory.

Relevant methods are listed in the Registration, Evaluation, Authorisation and Restriction of Chemicals guidance for nanospecific characterization (ECHA 2012), in accord with the OECD sponsorship program (OECD 2010), and the methods were selected here by their proven reliability for nanomaterial characterization in industrial research and development.

TABLE 1.1
Measurement Methods and Relevant Standards Based on ISO/TR 13014:2012

Parameter	Measurement Methods	Relevant Standards
Particle size	Dynamic light scattering Small angle X-ray scattering Size exclusion chromatography Analysis of scanning electron microscopy (SEM), transmission electron microscopy (TEM) or scanning probe microscopy (SPM) images Centrifugal liquid sedimentation Raman spectroscopy Laser-induced incandescence	ISO 9276-1:1998; ISO 9276-1:1998/ Cor 1:2004; ISO 9276-2:2001; ISO 9276-3:2008; ISO 9276-4:2001; ISO 9276-5:2005; ISO 9276-6:2008; ISO 9277:2010; ISO 13318-1:2001; ISO 13318-2:2007; ISO 13318-3:2004; ISO 13320:2009; ISO 13321:1996; ISO 13322-1:2004; ISO 13322-2; ISO/TS 13762:2001; ISO 14488:2007; ISO 14887:2000; ISO 15900:2009; ISO 20998-1:2006; ISO 21501-1:2009; ISO 21501-2:2007; ISO 22412:2008; ASTM E2490-09; ISO 16700:2004
Aggregation/ agglomeration state	Analysis of cryo-SEM or cryo-TEM images Angle dependent scattering at different wavelengths Static light scattering Small angle X-ray scattering X-ray diffraction Small angle neutron scattering Rheology methods Centrifugal liquid sedimentation Laser diffraction Nanoparticle tracking analysis	ISO 9276-1:1998; ISO 9276-1:1998/ Cor 1:2004; ISO 9276-2:2001; ISO 9276-3:2008; ISO 9276-4:2001; ISO 9276-5:2005; ISO 9276-6:2008; ISO 9277:2010; ISO 13318-1:2001; ISO 13318-2:2007; ISO 13318-3:2004; ISO 13320:2009; ISO 13321:1996; ISO 13322-1:2004; ISO 13322-2; ISO/TS 13762:2001; ISO 14488:2007; ISO 14887:2000; ISO 15900:2009; ISO 20998-1:2006; ISO 21501-1:2009; ISO 21501-2:2007; ISO 22412:2008; ASTM E2490-09; ISO 16700:2004; ISO 13322-1:2004
Shape	Analysis of SEM or TEM or SPM images	ISO 16700:2004
	Scattering techniques	ISO 13322-1:2004
Surface area	Methods based on gas or liquid adsorption isotherms Liquid porosimetry Image analysis Laser-induced incandescence	ISO 15901-1:2005; ISO 15901-2:2005; ISO 15901-3:2005; ISO 18757:2003; ISO 13322-1:2004

(*Continued*)

TABLE 1.1 (*Continued*)
Measurement Methods and Relevant Standards Based on ISO/TR 13014:2012

Parameter	Measurement Methods	Relevant Standards
Composition	X-ray photoelectron spectroscopy X-ray fluorescence Auger electron spectroscopy X-ray diffraction Raman spectroscopy ThermoGravimetric Analysis Ultraviolet/visible spectrometry Nuclear magnetic resonance Inductively coupled plasma- optical emission spectrometer (ICP-OES) Inductively coupled plasma-mass spectrometer (ICP-MS) SEM	ISO 22309:2006; ISO 22489:2006; ISO 24173:2009; ISO 13084:2011; ISO 18114:2003
Surface chemistry	Auger electron spectroscopy Scanning Auger electron microscopy X-ray photoelectron spectroscopy Secondary ion mass spectrometry 3D atom probe tomography Energy dispersive X-Ray spectrometry Electron energy loss spectroscopy Low energy ion spectroscopy Raman and other molecular spectroscopies	Under development: ISO/DTR 14187; ISO 18115:2001; ISO 24236:2005; ISO 15471:2004; ISO/TR 19319:2003; ISO 17973:2002; ISO 18118:2004; ISO 20903:2006; ISO/TR 18394:2006; ISO 23830:2008; ISO 17560:2002; ISO 18114:2003; ISO 20341:2003; ISO 15472:2002; ISO 21270:2004; ISO 24237:2005; ISO 15470:2004; ISO 19318:2004; ISO/TR 18392:2005; ISO 18516:2006; ISO 18117:2009; ISO 23812:2009
Surface charge	Isoelectric point Electrophoretic light scattering Electrophoresis Electro-osmosis Electric sonic amplitude Colloidal vibration current	ISO 20998-1:2006
Solubility	Possible measurement methods: ICP-OES and ICP-MS	—
Dispersibility	Methods based on particle size measurements (see above)	ISO/TC 24 new work item on dispersion; ISO/TR 13097

1.2 PARTICLE SIZE/SIZE DISTRIBUTION

The particle size plays a crucial role in nanomaterial properties, thus determining the dimension of nano-objects, and their agglomerates and aggregates (NOAAs), is extremely important for science and technology. It is necessary to estimate not only particle size but also particle size distribution. Both of these parameters have a significant effect on different properties like mechanical strength, density, chemical reactivity, and electrical and thermal properties of NOAA.

The size of particles and the distribution of their size can be determined using numerous techniques and a variety of commercially available instruments. Microscopy techniques can be applied to study a wide range of materials with a broad distribution of sizes. Instruments used for the microscopy techniques include optical light microscopes, scanning electron microscopes (SEMs), and transmission electron microscopes (TEMs). The choice of the equipment is mainly determined by the size range of particles, magnification, and resolution that is desired. The optical microscopes are easier to operate than electron microscopes, but are more limited in magnification and resolution. The optical microscope analyzes particles of all kinds, including fibers, in the size range from 0.2 μm to 5 mm. The SEM enables an analysis at higher magnification and resolution, and it is suitable for particles in the size range of about 0.01–1000 μm (the lower limit depends upon the quality of the instrument being used). The TEM enables examination in the size range of 0.001–10 μm with a very high local resolution and is capable of imaging crystal lattice distances. However, most electron microscopic techniques do not provide good statistical assessment. In comparison to the other size measurement techniques, they are slow, expensive, and only a small amount of particles can be examined at the same time. Nevertheless, if the analyzed NOAAs can be demonstrated to be representative of a whole sample, these methods can provide useful information. Moreover, sample preparation for microscopy analysis (e.g., sample dilution or the way of sample drying) may cause some artifacts, which consequently affect the results of the measurements (see Figure 1.1).

The other method of defining a particle size/size distribution is based on the relationship between particle behavior and its size. For nano-objects that can be treated as spherical, dynamic light scattering (DLS) is mainly applied. The DLS enables rapid and simple estimation of average particle size and of the width of the size distribution in the measurement range up to 1000 nm, but it cannot reliably extract the actual shape of the distribution from the raw data (see Figure 1.2).

DLS characterizes particles previously dispersed in a liquid, such that the dependence on suitable dispersion protocols is the main limitation of this technique. Moreover, DLS lacks the ability to distinguish between primary particles and agglomerates, thus the method should be applied with caution for polydisperse nanomaterials (Calzolai et al. 2011).

Another technique to estimate the particle size distribution of a material dispersed in a liquid is centrifugal sedimentation. It enables measurement of particles in the size range of 0.1–5 μm by fractionation. Centrifugal sedimentation methods are based on the rate of NOAAs velocity under a centrifugal field, and the calculation

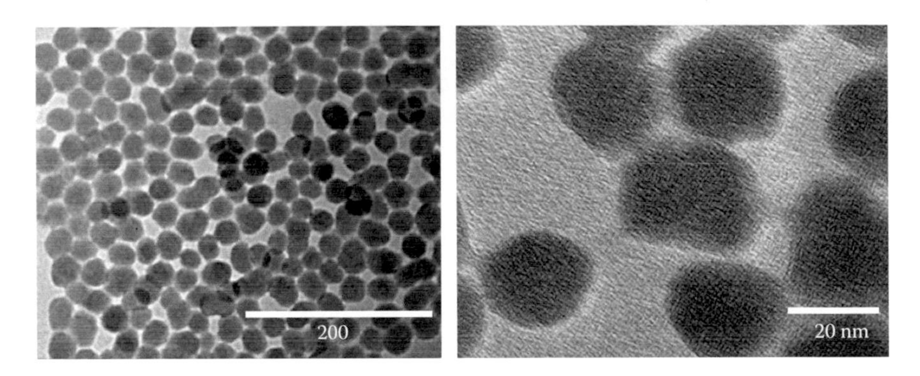

FIGURE 1.1 Transmission electron microscope (TEM) images of silica nanoparticles (NanoGEM.SiO_2.FITC).

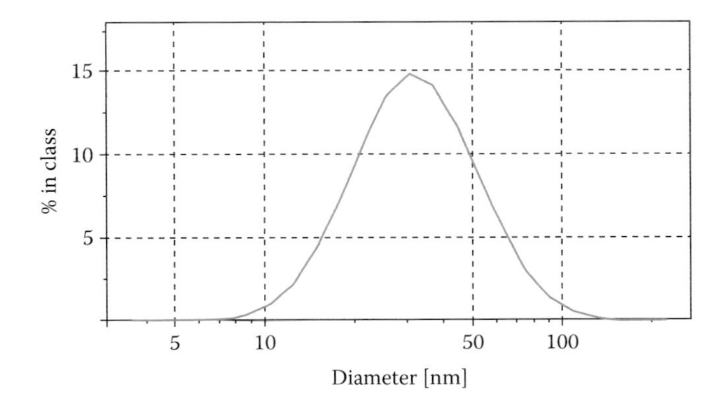

FIGURE 1.2 Intensity size distribution of silica nanoparticles (NanoGEM.SiO_2.FITC) analyzed by dynamic light scattering (DLS).

of particle size depends on Stokes' law (Batchelor 1967, 233). This method involves fewer artifacts and possible errors than integral methods such as DLS. However, the measuring concentration is very low and therefore significant dilution is necessary (see Figure 1.3).

It has to be noted that when comparing the particle size of the same sample measured by different techniques, it is important to report what type of distribution is being measured. Different types give different results, which is illustrated in the DLS example (Figure 1.4).

For a mixture of particles with sizes of 5 and 50 nm the number weighted distribution gives equal values to both of them, emphasizing the presence of the 5 nm particles, whereas the intensity weighted distribution has a signal one million times higher for the 50 nm particles. This is caused by the phenomenon that large particles scatter more light than small particles (the intensity of scattering of a particle is proportional to the sixth power of its diameter). The volume-weighted

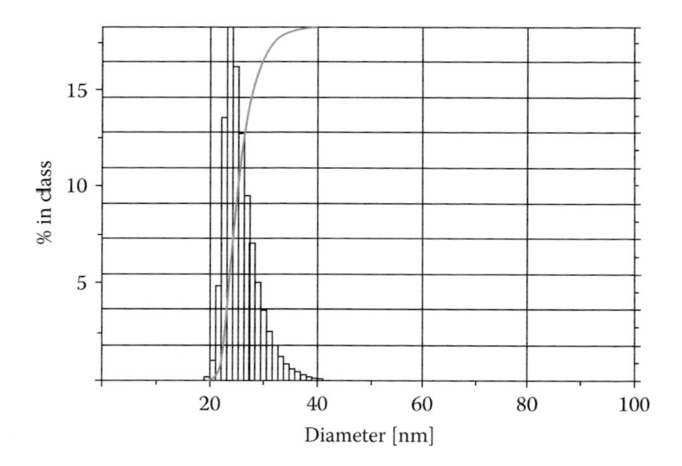

FIGURE 1.3 Analytical ultracentrifugation analysis of silica nanoparticles (NanoGEM. SiO_2.FITC).

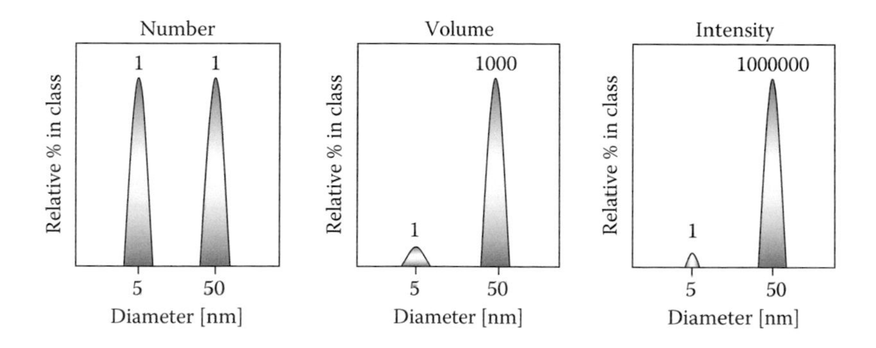

FIGURE 1.4 Number, volume, and intensity size distribution of nanoparticles determined by dynamic light scattering (DLS). (Based on Malvern Instruments (2012). A basic guide to particle characterization. Malvern Instruments Limited, UK. Retrieved from http://www.atascientific.com.au/publications/wp-content/uploads/2012/07/MRK1806-01-basic-guide-to-particle-characterisation.pdf.)

distribution is intermediate between the two. The area of the peak for the 50 nm particles is 1000 times larger than the area of the peak for the 5 nm particles (the volume of the 50 nm particle is 1000 times larger than the volume of the 5 nm particle).

In many cases, the measurements of particle size and size distribution, directly in a stock dispersion, may not be possible to perform (e.g., too high concentration of particles). When other solvents are used to prepare working suspensions, the exact composition of the solvents, pH, and temperature should be reported, as this may affect the obtained results. Methods of sample dispersion, such as stirring or sonication, could also have an impact on particle size. Thus, sample preparation is widely recognized as one of the most critical steps toward a successful characterization of nanomaterial size and size distribution.

1.3 AGGREGATION AND AGGLOMERATION STATE

According to ISO/TS 27687:2008, aggregates are strongly bonded or fused particles for which the resulting external surface area may be significantly smaller than the sum of calculated surface areas of the individual components. The forces holding an aggregate together are strong forces, for example, covalent bondings. Agglomerates, however, are weakly or loosely bonded particles or aggregates or mixtures of the two, and the forces holding the agglomerates together are weak forces, for example, van der Waals forces. Aggregates and agglomerates are also termed "secondary particles" and single particles are termed "primary particles." The state of aggregation/agglomeration refers to the number of secondary particles in comparison to the total number of primary particles in the suspending medium, as well as the number of primary particles in the agglomerate/aggregate.

Whether a particle remains as a single-dispersed particle or forms an aggregate/agglomerate is mainly dependent on the probability of nanoparticle–nanoparticle collisions and the relative mechanical or thermal energy possessed by the particles when the interaction occurs. Dispersion of nano-objects is controlled by not only the distance between the particles but also the balance between the particles and their surroundings. Thus, slight perturbations in the properties of the dispersing medium, such as temperature, pH, ionic strength, or the presence of biological molecules, can significantly modify the whole system. Therefore, conditions applied during sample preparation are crucial to obtain reliable results.

Methods used to determine the state of agglomeration and aggregation are mainly the techniques used for measurement of primary particles size (see Particle size/size

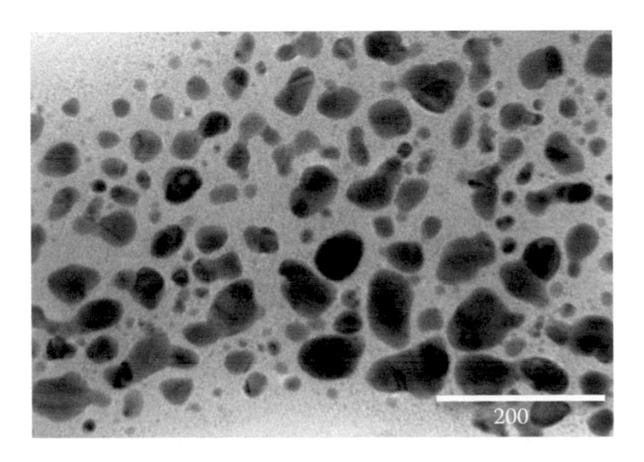

FIGURE 1.5 Aggregates of silver nanoparticles (NanoGEM.Ag_50.EO).

distribution). Special care must be taken during sample preparation and analysis to prevent artifacts, particularly when the method only provides two-dimensional images (e.g., SEM) (Figure 1.5).

For measurement in liquid-free states, additional care must be taken to avoid drying the artifacts.

It is worth noting that the absence of information about the state of dispersion may lead to misleading interpretations of experimental findings; the aggregation/agglomeration state may cause significant variations in an effective dose of nano-objects and modulate their fate and transport in vivo (Albanese and Chan 2011).

1.4 SHAPE

Engineered nano-objects of identical composition can be produced in a variety of shapes including spheres, fibers, and plates, and each of these shapes may possess different physical and chemical properties. The shape of a nano-object is mainly determined by the symmetry of its constituting crystallites, if any, and by the minimization of the bulk and surface free energy.

Many properties of NOAA, such as dissolution rate, aggregation behavior, or availability of reactive sites among others are strongly linked to the shape of a nano-object. The characterization of a nano-object shape mainly includes analysis of SEM, TEM, or SPM images.

However, care must be taken during sample preparation and analysis in order to prevent misleading results due to orientation effects and artifacts coming from contaminants and aggregates.

It has been reported by Oberdörster et al. (2005) and Powers et al. (2007) that differently shaped nanomaterials of identical composition would cause different biological responses. It has also been shown by Sun et al. (2011) that the toxicity and biodistribution of NOAAs in vivo are strongly shape dependent.

1.5 SURFACE AREA

The surface area of NOAAs can be described as the quantity of accessible surface of a sample when exposed to either gaseous or liquid adsorbate phase. Surface area is expressed as the mass specific surface area (m^2/g) or as the volume specific surface area (m^2/cm^3), where the total quantity of the area has been normalized to either the sample's mass or volume (ISO/TR 13014:2012).

The surface area of a nanomaterial is mostly measured through physical adsorption of an inert gas (typically nitrogen) using the method of Brunauer, Emmett, and Teller (1938) (BET). However, BET measurements of the surface area can only be performed on dry powders, which is the main limitation of this method. Moreover, the analysis of the surface area of many nanomaterials (e.g., hydrated polymeric nanoparticle systems, micro- and nanoemulsion, nanoliposomes, or dendrimers) cannot be sufficiently performed due to artifacts that arise from either sample drying or difficult sample recovery. In this case, the surface area can be estimated through DLS, cryo-TEM/SEM, or atomic force microscopy (AFM).

Chemical reactions mostly take place at surfaces, and thus a material with a high surface area can be expected to have a higher reactivity than the same material with a low surface area to volume ratio. This may have a direct impact on nano-object bioavailability and mechanisms of toxicity (Hassellöv and Kaegi 2009; Lison et al. 1997; Oberdörster et al. 1994; Oberdörster et al. 2005; Sager et al. 2008; Tran et al. 2000). The toxicity of some NOAAs has been reported in many studies to be highly dependent on their surface area (e.g., Duffin et al. 2002, 2007).

1.6 COMPOSITION

Information about the composition of an analyzed sample should include its chemical composition, crystalline structure, and impurities, if any.

The chemical composition refers to entities of which the material is composed. The function of a nanomaterial can be defined by its chemical composition. The chemical composition gives specific properties and consequently a unique application of the nanomaterial. The most common methods to analyze the nanomaterial composition are Raman spectroscopy, energy dispersive X-ray spectroscopy (EDAX), X-ray photoelectron spectroscopy (XPS), secondary ion mass spectrometry (SIMS), inductively coupled plasma-mass spectrometer (ICP-MS), and nuclear magnetic resonance (NMR). The majority of these methods give information about not only the composition of an NOAA but also the chemical composition of the suspending medium (see Figure 1.6).

In addition to the information about the chemical composition of an analyzed sample, crystallinity of a nano-object has to be reported. The differences in the crystal state of nanomaterials of identical chemical compositions can result in different properties of those nanomaterials. For instance, amorphous form of a silica particle has a totally different function than its crystalline form, and consequently, the crystal structure has direct impact on nanomaterial toxicity (the amorphous

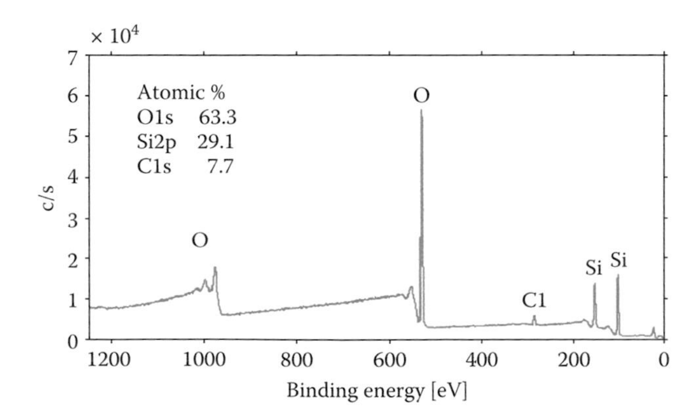

FIGURE 1.6 X-ray photoelectron spectroscopy (XPS) survey spectrum of silica nanoparticles (NanoGEM.SiO_2.FITC).

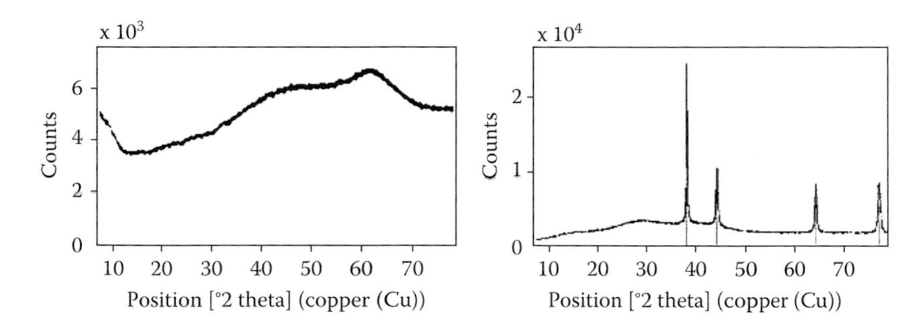

FIGURE 1.7 X-ray diffraction analysis (XRD) spectra of silica nanoparticles (NPs) (NanoGEM.SiO_2.FITC) amorphous form and silver NPs (NanoGEM.Ag_50.EO) crystalline form.

silica is benign, while its crystalline form is cytotoxic and carcinogenic) (Borm et al. 2001). Thus, the crystallinity of a nanomaterial should always be evaluated. The crystal structure of an NOAA can be determined by X-ray diffraction analysis (XRD) (see Figure 1.7).

The XRD is useful for distinguishing different crystal phases of materials of the same chemical composition and can be used to determine the average crystallite size and shape. Microscopy methods, such as the AFM, SEM, and TEM, can also be used to assess the crystallite size, which for monocrystalline primary particles also means average primary particle size.

An analysis of nanomaterial composition should be undertaken with special attention to possible impurities. Contamination can appear at any stage of nanomaterial production, handling, or sampling. Impurities can be delivered from reagents, not sufficiently cleaned laboratory equipment, or handling vessels. Methods mostly used for the detection of impurities are the Fourier transform infrared spectroscopy (FTIR), XPS, and SIMS. For biological evaluation of a nanomaterial, impurities may have an influence if they are present in sufficient quantity. Thus, before any studies on a nanomaterial, not only the composition of the possible impurities but also their quantity should be detected.

1.7 SURFACE CHEMISTRY

Chemical composition of the outermost layers of a nanomaterial highly defines its energy and reactivity. Surface chemistry is critically important with respect to nanomaterial agglomeration, aggregation, or interaction with its surroundings. Tailoring surface chemistry of nanomaterials with well-controlled surface coatings/functionalities enables their broad application. It is highly promising that nano-objects may have more and more adaptations in therapy, medicine delivery, or diagnostics. However, identification and quantification of the functional groups linked to the surface of NOAAs are very challenging due to the fact that they are coated with only a little amount of small molecules. Techniques applicable for identifying the surface composition and atomic arrangement on the surface of

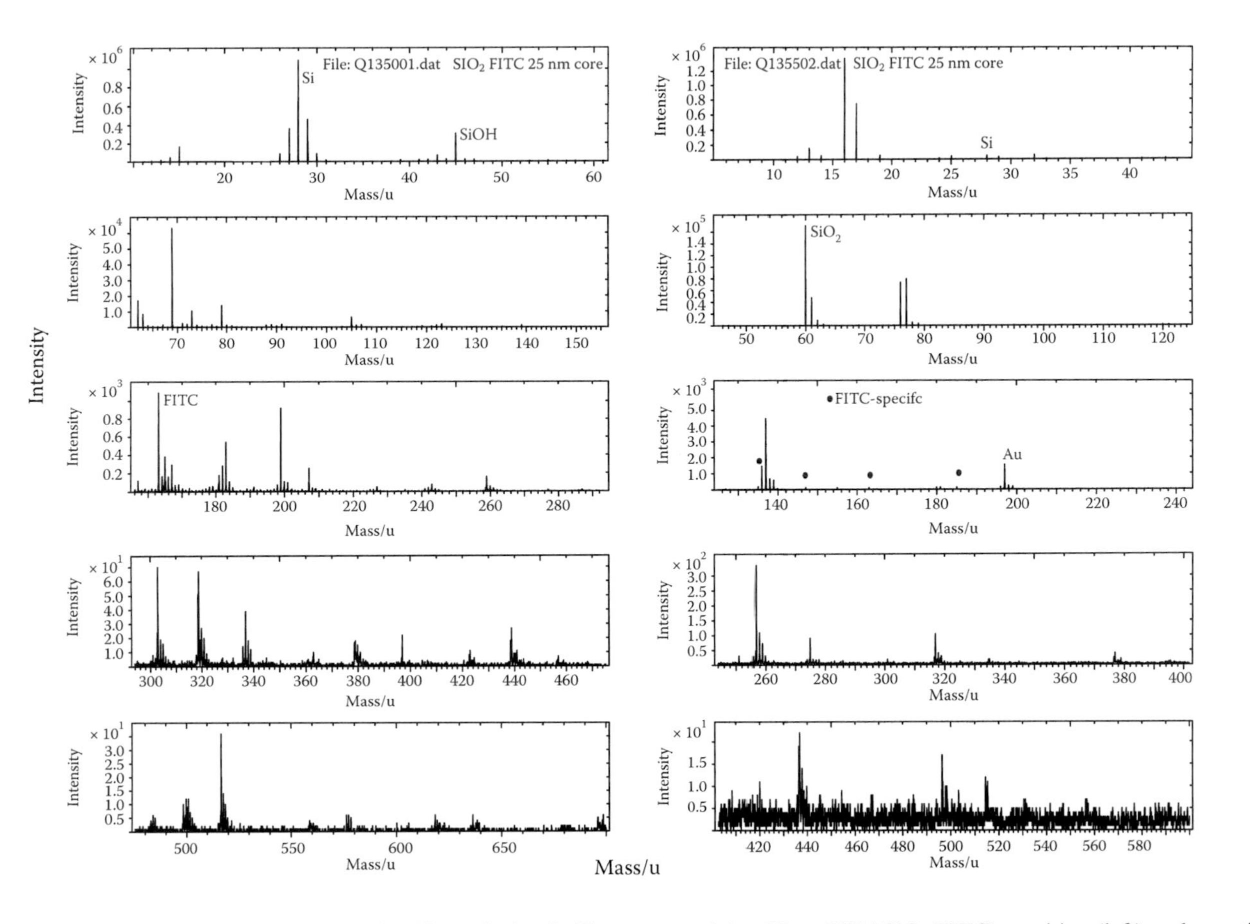

FIGURE 1.8 Secondary ion mass spectrometry (SIMS) analysis of silica nanoparticles (NanoGEM.SiO_2.FITC); positive (left) and negative (right) polarity.

NOAAs are mainly Auger electron spectroscopy, XPS, and SIMS (described in Figure 1.8).

It has been widely reported that surface chemistry can play one of the key roles in determining the hazard of any given NOAA. Several studies have shown that NOAAs of the same material but with different surface chemistry have different properties with respect to cell uptake and toxicological profile (Goodman et al. 2004; Chithrani et al. 2006; Walkey et al. 2011; Suresh et al. 2012).

1.8 SURFACE CHARGE

Surface charge influences the state of nanomaterial agglomeration and the adsorption of ions and biomolecules onto a nano-object's surface. Consequently, it can affect cell uptake of the nano-object and its biodistribution (e.g., Limbach et al. 2007; Limbach et al. 2009; Lockman et al. 2004). The toxicological role of surface charge has been deeply discussed by Oberdörster et al. (2005) and Nel et al. (2006).

Surface charge of a nano-object can be determined by zeta potential measurements. These measurements involve quantifying the velocity of particle movement in an applied electric field. The zeta potential is a function of the surface charge of the particle, adsorbed species on its surface, and composition and ionic strength of the surrounding solution. Hence, the zeta potential values alone do not provide much information and it is important to report additionally the other abiotic factors such as temperature, pH, and ionic strength of an analyzed sample. Zeta potential of an NOAA is normally measured by light-scattering electrophoresis or electroacoustophoresis methods. The theory used for zeta potential calculation should be noted (e.g., Smoluchowski, Huckel, Henry, and others) and the limitations of these theories should be recognized. The isoelectric point (IEP) (pH where the zeta potential is equal to zero) of a material under controlled conditions should be reported together with the zeta potential values (Figure 1.9).

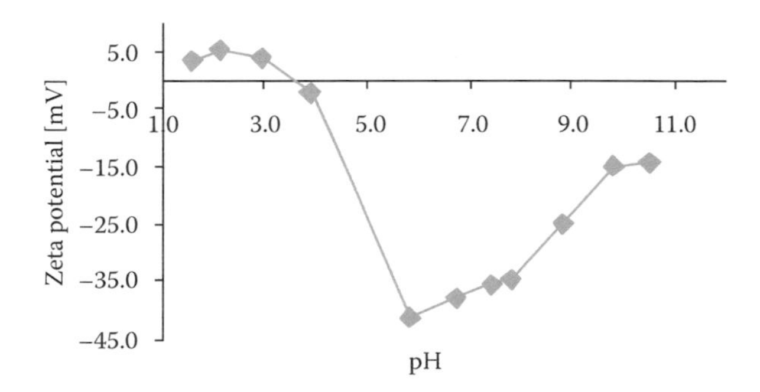

FIGURE 1.9 Zeta potential and isoelectric point of silica nanoparticles (NanoGEM.SiO_2.FITC).

Zeta potential values can be related to the stability of NOAAs dispersion. High zeta potential values (positive or negative) usually confirm dispersion stability and small values (positive or negative) indicate potential agglomeration/aggregation of the system.

1.9 SOLUBILITY/DISPERSIBILITY

According to ISO 7579, solubility is the maximum mass of a nanomaterial that is soluble in a given volume of a particular solvent under specified conditions. A nano-object is expected to exhibit greater solubility and faster dissolution than a bigger material of the same composition. Solubility mainly depends on temperature, pressure, and pH of a solvent. Dissolution of a NOAA can have a great effect on its behavior, which definitely affects the biological and environmental distribution of the material (Oyabu et al. 2008). Although changes in particle size can provide some indications of dissolution, the use of mass spectroscopies, such as ICP-MS, that monitor the concentration of dissolved NOAAs is preferred.

For a nanomaterial, it can be difficult to distinguish between when it is dispersed and when it is dissolved because of the small particle size. It is crucial to recognize that solubility and dispersibility are two different phenomena, and it is important to differentiate between them.

Dispersibility is described as the degree to which a solid material is uniformly distributed in another material (a dispersing medium), and the resulting dispersion remains stable. In general, the main difference between solubility and dispersibility is that solubility requires the molecules of a solid phase to be strongly disassociated by the process, while there is no significant disassociation involved in a phase that has been dispersed into another. The influence of dispersibility on toxicity results is not yet fully understood (Powers et al. 2009).

1.10 CONTROL OF PROPERTIES BY SURFACE FUNCTIONALIZATION

Two materials, nanoGEM.SiO_2.FITC and nanoGEM.Ag_50.EO, were used to exemplify characterization methods in the previous sections. Another 16 nanomaterials, all of roughly spherical shape, were characterized by the same methods of the nanoGEM project. Three compositions (SiO_2, ZrO_2, and Ag) with primary particle sizes from 7 nm to 134 nm (determined by TEM) were functionalized by organic molecules to control the surface chemistry. Relatively short molecules mainly aimed at surface charge reversal: trioxadecanoic acid, amino, phosphate, or citrate end groups. Relatively uncharged polymers aimed to impart steric functionalization: polyethylene glycols (PEG), polyvinylpyrrolidones (PVP), and ethylene oxides (EO). A combination of both mechanisms was tested using strongly charged acrylic polymer on ZrO_2. The characterization substantiated the successful control of size, composition, surface charge, and surface reactivity (Table 1.2). Full characterization results including the generation of particle-free reference samples, assays of reactivity, and approaches to the spectroscopy of impurities are accessible online (Hellack et al. 2013).

TABLE 1.2
Control of Physico-Chemical Properties by Surface Functionalization

Property, Method/Unit	SiO_2.naked	SiO_2.PEG	SiO_2.Amino	SiO_2.Phosphat	SiO_2.FITC	ZrO_2.Acryl	ZrO_2.PEG	ZrO_2.Amino	ZrO_2.TODacid	AG50.EO	AG50.PVP	AG200.PVP	AG50.citrat	TiO_2@Eu	TiO_2 NM105	ZnO NM110	$BaSO_4$ NM220	AlOOH I NanoCare
Primary particle size, TEM/nm	15	15	15	15	25	9	9	10	9	7	97	134	20	8	21	80	32	37
Specific surface, BET/m²/g	200	200	200	200	178	117	117	105	117	86	6.2	4.5	30	208	51	12	41	47
Average size in H_2O, DLS/nm	40	50	42	40	33	9	27	315	9	40	123	408	35	600	478		350	262
Average size in H_2O, AUC/nm	19	21	20	20	25	27	13	38	11	34	77	95	38	1093	980	1000	350	340
Dispersibility in H_2O//FCS/AAN	1//28	1//213	1//90	1//2	1//	3//12	2//65	4//50	1//86	1//1	1//1	1//1	1//2	130//	25//32	12//10	11//9	10//19
Crystallinity, XRD/qualitative	Amorph	Amorph	Amorph	Amorph	Amorph	Amorph, and tetragonal	Amorph, monocline and tetragonal	Amorph, monocline and tetragonal	Monocline and tetragonal	Cubic	Cubic	Cubic	Cubic	Rutile, anatase	Rutile, anatase	Hexagonal	Orthorhombic	Orthorhombic
Surface chemistry, XPS/element%	O 66 Si 29 C 4 Na 1	PEG identified (SIMS)	Amino identified (SIMS)	O 66 Si 29 C 5 Na 0.5 PO_2, PO_3 fragments	O 63 Si 29 C 8	O 58 Zr 23 C 19 Acrylic acid	PEG identified (SIMS)	Amino identified (SIMS)	O 63 Zr 24 C 11 N 0.7 S 0.2	C 63 O 24 Ag 14	C 59 O 18 Ag 16 Na 8	C 77 O 10 Ag 12 Na 1 N 1	C 21 O 15 Ag 62 Na 2	O 63 Ti 23 C 13 Eu 0.6	O 61 Ti 26 C 13	O 38 Zn 35 C 20 Cl 3 Na 3	O 52 Ba 13 C 17 S 11 Cl, P 3 N 1	O 66 Al 23 C 11

(Continued)

TABLE 1.2 (*Continued*)
Control of Physico-Chemical Properties by Surface Functionalization

Property, Method/Unit	SiO_2.naked	SiO_2.PEG	SiO_2.Amino	SiO_2.Phosphat	SiO_2.FITC	ZrO_2.Acryl	ZrO_2.PEG	ZrO_2.Amino	ZrO_2.TODacid	AG50.EO	AG50.PVP	AG200.PVP	AG50.citrat	TiO_2@Eu	TiO_2 NM105	ZnO NM110	$BaSO_4$ NM220	AlOOH I NanoCare
Iso-electric point, electrophoretic mobility/pH	<1	4	7.2	<1	3.5	<1	7	>10	7.1	2.5	3.6	4.2	2	6.5	6.7	9.6	3.3	8.2
Zeta-potential at pH 7.4/mV	−39	−26	0	−42.9	−39	−39	−7.8	3.9	−0.5	−20	−7	−7	−45	−20	−17	20	−39	5
Surface reactivity, ESR/relative to D_2O	4/p-f s: 0.88	1/p-f s: 3.4	1/p-f s: 1.1	2.2/p-f s: 1.2	n.d.	1/p-f s: 2	1.5/	1/	0.54/p-f s: 5.7	8.0/	45/p-f s: 1	72/	n.m.	2.7	0.82	22	2	2.3
Surface ROS generation, ESR/relative to D_2O	11/p-f s: 6.3	11/p-f s: 13	21/p-f s: 5.2	19/p-f s: 5	n.d.	3.6/p-f s: 1.5	1.7/	3.5/	0.94/p-f s: 1.3	0.5/	0.4/p-f s: 1	0.48/	n.m.	1.9	3	12	2	1.3

Dispersibility indicated by AAN, average agglomeration number (ECHA 2012) = ratio of particle size in medium to BET-derived primary particle size. AUC, area under the curve; FCS, DMEM with 10% FCS; DLS, dynamic light scattering; ESR, electron spin resonance; ROS, reactive oxygen species; SEM, scanning electron microscopy; TEM, transmission electron microscopy; TOD, trioxadecanoic acid; p-f s, particle-free supernatant; XRD, X-ray diffraction analysis; XPS, X-ray photoelectron spectroscopy (Hellack et al. 2013).

ACKNOWLEDGMENTS

This contribution represents the materials synthesis and characterization by Stefanie Eiden, Bryan Hellack, Tim Hülser, Frank Meyer, Matthias Voetz, Hartmut Wiggers, Wendel Wohlleben, and Roman Zieba.

REFERENCES

Albanese, A., & Chan, W. C. W. (2011). Effect of gold nanoparticle aggregation on cell uptake and toxicity. *ACS Nano*, 5: 5478–5489.

ASTM (2009), ASTM E2490-09, Standard Guide for Measurement of Particle Size Distribution of Nanomaterials in Suspension by Photon Correlation Spectroscopy (PCS), ASTM, West Conshohocken, PA.

Batchelor, G. K. (1967). *An Introduction to Fluid Dynamics*. Cambridge Mathematical Library. Cambridge, UK.

Borm, P. J. A., Clouter, A., Schins, R. P. F., & Donaldson, K. (2001). The quartz hazard revisited: the role of matrix and surface. *Gefahrstoffe Reinhaltung Der Luft*, 61: 359–363.

Brunauer, S., Emmett, P. H., & Teller, E. (1938). Adsorption of gases in multimolecular layers. *J Am Chem Soc*, 60: 309–319.

Calzolai, L., Gilliland, D., Garcia, C. P., & Rossi, F. (2011). Separation and characterization of gold nanoparticle mixtures by flow-field-flow fractionation. *J Chromatogr A*, 1218: 4234–4239.

Chithrani, B. D., Ghazani, A. A., & Chan, W. C. W. (2006). Determining the size and shape dependence of gold nanoparticle uptake into mammalian cells. *Nano Lett*, 6: 662–668.

Duffin, R., Tran, C. L., Clouter, A., Brown, D. M., MacNee, W., Stone, V., & Donaldson, K. (2002). The importance of surface area and specific reactivity in the acute pulmonary inflammatory response to particles. *Ann Occup Hyg*, 46: 242–245.

Duffin, R., Tran, L., Brown, D., Stone V., & Donaldson, K. (2007). Proinflammogenic effects of low-toxicity and metal nanoparticles in vivo and in vitro: Highlighting the role of particle surface area and surface reactivity. *Inhal Toxicol*, 19: 849–856.

ECHA 2012. Guidance on information requirements and chemical safety assessment: Appendix R7-1 Recommendations for nanomaterials applicable to Chapter R7a - Endpoint specific guidance - ECHA-12-G-03-EN. (accessed April 30, 2012).

Goodman, C. M., McCusker, C. D., Yilmaz, T., & Rotello, V. M. (2004). Toxicity of gold nanoparticles functionalized with cationic and anionic side chains. *Bioconjug Chem*, 15: 897–900.

Hassellöv, M., & Kaegi, R. (2009). Analysis and characterization of manufactured nanoparticles in aquatic environments. In J. R. Lead & E. Smith (Eds.), *Nanoscience Nanotechnology: Environmental and Human Health Implications* (pp. 211–266). Wiley, UK.

Hellack, B., Hülser, T., Izak, E., Kuhlbusch, T. A. J., Meyer, F., Spree, M., Voetz, M., Wiggers, H., & Wohlleben, W. (2013). Characterization Report for all nanoGEM Materials. Retrieved from www.nanogem.de/cms/nanogem/upload/Veroeffentlichungen/nanoGEM_Del1.3.1_Characterization_Materials_2013_04_24.pdf (accessed November 3, 2013).

ISO (1996), ISO 13321:1996, Particle size analysis–Photon correlation spectroscopy, ISO, Geneva.

ISO (1998), ISO 9276-1:1998, Representation of results of particle size analysis–Part 1: Graphical representation, ISO, Geneva.

ISO (2000), ISO 14887:2000, Sample preparation–Dispersing procedures for powders in liquids, ISO, Geneva.

ISO (2001a), ISO 13318-1:2001, Determination of particle size distribution by centrifugal liquid sedimentation methods–Part 1: General principles and guidelines, ISO, Geneva.

ISO (2001b), ISO 18115:2001, Surface chemical analysis–Vocabulary, ISO, Geneva.

ISO (2001c), ISO 9276-2:2001, Representation of results of particle size analysis–Part 2: Calculation of average particle sizes/diameters and moments from particle size distributions, ISO, Geneva.

ISO (2001d), ISO 9276-4:2001, Representation of results of particle size analysis–Part 4: Characterization of a classification process, ISO, Geneva.

ISO (2001e), ISO/TS 13762:2001, Particle size analysis–Small angle X-ray scattering method, ISO, Geneva.

ISO (2002a), ISO 15472: 2002, Surface chemical analysis–X-ray photoelectron spectrometers–Calibration of energy scales, ISO, Geneva.

ISO (2002b), ISO 17560:2002, Surface chemical analysis–Secondary-ion mass spectrometry–Method for depth profiling of boron in silicon, ISO, Geneva.

ISO (2002c), ISO 17973:2002, Surface chemical analysis–Medium resolution Auger electron spectrometers–Calibration of energy scales for elemental analysis, ISO, Geneva.

ISO (2003a), ISO 18114:2003, Surface chemical analysis–Secondary ion mass spectrometry–Determination of relative sensitivity factors from ion implanted reference materials, ISO, Geneva.

ISO (2003b), ISO 20341: 2003, Surface chemical analysis–Secondary-ion mass spectrometry–Method for estimating depth resolution parameters with multiple delta-layer reference materials, ISO, Geneva.

ISO (2003c), ISO/TR 19319:2003, Surface chemical analysis–Auger electron spectroscopy and X-ray photoelectron spectroscopy–Determination of lateral resolution, analysis area and sample area viewed by the analyser, ISO, Geneva.

ISO (2004a), ISO 13318-3:2004, Determination of particle size distribution by centrifugal liquid sedimentation methods–Part 3: Centrifugal X-ray method, ISO, Geneva.

ISO (2004b), ISO 13322-1:2004, Particle size analysis–Image analysis method–Part 1: Static image analysis methods, ISO, Geneva.

ISO (2004c), ISO 15470:2004, Surface chemical analysis–X-ray photoelectron spectroscopy–Description of selected instrumental performance parameters, ISO, Geneva.

ISO (2004d), ISO 15471:2004, Surface chemical analysis–Auger electron spectroscopy–Description of selected instrumental performance parameters, ISO, Geneva.

ISO (2004e), ISO 16700:2004, Microbeam analysis–Scanning electron microscopy–Guidelines for calibrating image magnification, ISO, Geneva.

ISO (2004f), ISO 18118:2004, Surface chemical analysis–Auger electron spectroscopy and X-ray photoelectron spectroscopy–Guide to the use of experimentally determined relative sensitivity factors for the quantitative analysis of homogeneous materials, ISO, Geneva.

ISO (2004g), ISO 19318:2004, Surface chemical analysis–X-ray photoelectron spectroscopy–Reporting of methods used for charge control and charge correction, ISO, Geneva.

ISO (2004h), ISO 21270:2004, Surface chemical analysis–X-ray photoelectron and Auger electron spectrometers–Linearity of intensity scale, ISO, Geneva.

ISO (2004i), ISO 9276-1:1998/Cor 1:2004, Representation of results of particle size analysis–Part 1: Graphical representation–Technical Corrigendum 1, ISO, Geneva.

ISO (2005a), ISO 24236:2005, Surface chemical analysis–Auger electrons spectroscopy–Repeatability and constancy of intensity scale, ISO, Geneva.

ISO (2005b), ISO 24237:2005, Surface chemical analysis–X-ray photoelectron spectroscopy–Repeatability and constancy of intensity scale, ISO, Geneva.

ISO (2005c), ISO 9276-5:2005, Representation of results of particle size analysis–Part 5: Methods of calculation relating to particle size analyses using logarithmic normal probability distribution, ISO, Geneva.

ISO (2005d), ISO/TR 18392:2005, Surface chemical analysis–X-ray photoelectron spectroscopy–Procedures for determining backgrounds, ISO, Geneva.
ISO (2006a), ISO 13322-2, Particle size analysis–Image analysis methods–Part 2: Dynamic image analysis methods, ISO, Geneva.
ISO (2006b), ISO 18516:2006, Surface chemical analysis–Auger electron spectroscopy and X-ray photoelectron spectroscopy–Determination of lateral resolution, ISO, Geneva.
ISO (2006c), ISO 20903:2006, Surface chemical analysis–Auger electron spectroscopy and X-ray photoelectron spectroscopy–Methods used to determine peak intensities and information required when reporting results, ISO, Geneva.
ISO (2006d), ISO 20998-1:2006, Measurement and characterization of particles by acoustic methods–Part 1: Concepts and procedures in ultrasonic attenuation spectroscopy, ISO, Geneva.
ISO (2006e), ISO/TR 18394:2006, Surface chemical analysis–Auger electron spectroscopy–Derivation of chemical information, ISO, Geneva.
ISO (2007a), ISO 13318-2:2007, Determination of particle size distribution by centrifugal liquid sedimentation methods–Part 2: Photocentrifuge method, ISO, Geneva.
ISO (2007b), ISO 14488:2007, Particulate materials–Sampling and sample splitting for the determination of particulate properties, ISO, Geneva.
ISO (2007c), ISO 21501-2:2007, Determination of particle size distribution–Single particle light interaction methods–Part 2: Light scattering liquid-borne particle counter, ISO, Geneva.
ISO (2008a), ISO 22412:2008, Particle size analysis–Dynamic light scattering (DLS), ISO, Geneva.
ISO (2008b), ISO 23830:2008, Surface chemical analysis–Secondary ion mass spectrometry–Repeatability and constancy of the relative intensity scale in static secondary ion mass spectrometry, ISO, Geneva.
ISO (2008c), ISO 9276-3:2008, Representation of results of particle size analysis–Part 3: Adjustment of an experimental curve to a reference model, ISO, Geneva.
ISO (2008d), ISO 9276-6:2008, Representation of results of particle size analysis–Part 6: Descriptive and quantitative representation of particle shape and morphology, ISO, Geneva.
ISO (2009a), ISO 13320:2009, Particle size analysis–Laser diffraction methods, ISO, Geneva.
ISO (2009b), ISO 15900:2009, Determination of particle size distribution–Differential electrical mobility analysis for aerosol particles, ISO, Geneva.
ISO (2009c), ISO 18117:2009, Surface chemical analysis–Handling of specimens prior to analysis, ISO, Geneva.
ISO (2009d), ISO 21501-1:2009, Determination of particle size distribution–Single particle light interaction methods–Part 1: Light scattering aerosol spectrometer, ISO, Geneva.
ISO (2009e), ISO 23812:2009, Surface chemical analysis–Secondary-ion mass spectrometry–Method for depth calibration for silicon using multiple delta-layer reference materials, ISO, Geneva.
ISO (2009f), ISO 7579:2009, Dyestuffs–Determination of solubility in organic solvents–Gravimetric and photometric methods, ISO, Geneva.
ISO (2010), ISO 9277:2010, Determination of the specific surface area of solids by gas adsorption–BET method, ISO, Geneva.
ISO (2012), ISO/TR 13014:2012, Nanotechnologies–Guidance on physico-chemical characterization of engineered nanoscale materials for toxicological assessment, ISO, Geneva.
ISO (2013), ISO/TR 13097–Guide for the characterization of dispersion stability, ISO, Geneva.

Limbach, L. K., Grass, R. N., & Stark, W. J. (2009). Physico-chemical differences between particle- and molecule-derived toxicity: Can we make inherently safe nanoparticles? *CHIMIA Int J Chem*, 63: 38–43.

Limbach, L. K., Wick, P., Manser, P., Grass, R. N., Bruinink, A., & Stark, W. J. (2007). Exposure of engineered nanoparticles to human lung epithelial cells: Influence of chemical composition and catalytic activity on oxidative stress. *Environ Sci Technol*, 41: 4158–4163.

Lison, D., Lardot, C., Huaux, F., Zanetti, G., & Fubini, B. (1997). Influence of particle surface area on the toxicity of insoluble manganese dioxide dusts. *Arch Toxicol*, 71: 725–729.

Lockman, P. R., Koziara, J. M., Mumper, R. J., & Allen, D. D. (2004). Nanoparticle surface charges alter blood–brain barrier integrity and permeability. *J Drug Target*, 12: 635–641.

Malvern Instruments (2012), A basic guide to particle characterization. Malvern Instruments Limited, UK. Retrived from http://www.atascientific.com.au/publications/wp-content/uploads/2012/07/MRK1806-01-basic-guide-to-particle-characterisation.pdf

Nalwa, H. S. (2004). *Encyclopedia for Nanoscience and Nanotechnology*. American Scientific Publishers, Valencia, CA.

Nel, A., Xia, T., Madler, L., & Li, N. (2006). Toxic potential of materials at the nanolevel. *Science*, 311: 622–627.

Oberdörster, G., Ferin, J., & Lehnert, B. D. (1994). Correlation between particle size, in vivo particle persistence and lung injury. *Environ Health Perspect*, 102: 173–179.

Oberdörster, G., Maynard, A., Donaldson, K., Castranova, V., Fitzpatrick, J., Ausman, K., Carter, J., Karn, B., Kreyling, W., Lai, D., Olin, S., Monteiro-Riviere, N., Warheit, D., & Yang, H. (2005). Principles for characterising the potential human health effects from exposure to nanomaterials: Elements of a screening strategy. *Part Fibre Toxicol*, 2: 8.

Oberdorster, G., Oberdorster, E., & Oberdorster, J. (2005). Nanotoxicology: an emerging discipline evolving from studies of ultrafine particles. *Environ Health Perspect*, 113: 823–839.

OECD (2010), Guidance Manual for the testing of manufactured nanomaterials: OECD's sponsorship programme; first revision. OECD Environment, Health and Safety Publications Series on the Safety of Manufactured Nanomaterials 25: ENV/JM/MONO(2009)20/REV.

Oyabu, T., Ogami, A., Morimoto, Y., Myojo, T., Murakami, M., Yamato, H., & Tanaka, I. (2008). Simple flow-through solubility measurement apparatus and its effectiveness for hazard assessment of particles/fibers. *J Occup Health*, 50: 279–282.

Powers, K. W., Palazuelos, M., Moudgil, B. M., & Roberts, S. M. (2007). Characterization of the size, shape, and state of dispersion of nanoparticles for toxicological studies. *Nanotoxicology*, 1: 42–51.

Sager, T. M., Kommineni, C., & Castranova, V. (2008). Pulmonary response to intratracheal instillation of ultrafine versus fine titanium dioxide: role of particle surface area. *Part Fibre Toxicol*, 5: 17.

Sun, Y. N., Wang, C. D., Zhang, X. M., Ren, L., & Tian, X. H. (2011). Shape dependence of gold nanoparticles on in vivo acute toxicological effects and biodistribution. *J Nanosci Nanotechnol*, 11: 1210–1216.

Suresh, A. K., Pelletier, D. A., Wang, W., Morrell-Falvey, J. L., Gu, B., & Doktycz, M. J. (2012). Cytotoxicity induced by engineered silver nanocrystallites is dependent on surface coatings and cell types. *Langmuir*, 28: 2727–2735.

Tran, C. L., Buchanan, D., Cullen, R. T., Searl, A., Jones, A. D., & Donaldson, K. (2000). Inhalation of poorly soluble particles. II. Influence of particle surface area on inflammation and clearance. *Inhal Toxicol*, 12: 1113–1126.

Walkey, C. D., Olsen, J. B., Guo, H., Emili, A. & Chan, W. C. W. (2011). A combination of nanoparticle size and surface chemistry determines serum protein adsorption and macrophage uptake. *J Am Chem Soc*, 134: 2139–2147.

Wang, Z. L. (2000). *Characterization of Nanophase Materials*. Wiley-VCH Verlag GmbH. Weinheim, Germany.

2 Characterization Methods for the Determination of Inhalation Exposure to Airborne Nanomaterials

Christof Asbach

CONTENTS

2.1 INTRODUCTION

Assessing the inhalation exposure to airborne nanomaterials requires characterization of the workplace aerosols. These aerosols can be characterized using various metrics, for example, the particle number, surface area, or mass concentrations. These concentrations can be measured either as particle size-resolved distributions or as size-integrated total concentrations. Depending on the type of measurement or sampling device used, these data are provided as either time-resolved data or time-integrated over the sampling time. In this chapter, the available measurement techniques are classified according to a scheme presented by Kuhlbusch et al. (2011), that is,

- Size integrated, time resolved
- Size resolved, time resolved
- Size resolved, time integrated
- Size integrated, time integrated

The aim of this chapter is to give an overview of the commercially available techniques for assessing exposure to airborne nanomaterials. Although nanomaterials are defined to be in dimensions at least smaller than 100 nm, measurement techniques for the micron particle size range are also described here, because nanoparticles tend to agglomerate quickly and can therefore form rather large agglomerates.

These instruments use a variety of measurement principles, and in order to be able to interpret the data obtained from particle sizing instruments, some knowledge on the measurement technique is required. The reason for this is that the measured particles are assumed to be spheres so that the particle size can be expressed as their diameters. If the particles are nonspherical, the particle sizes are given as equivalent diameters, where the equivalency describes that the particle considered behaves in the measuring instrument like a spherical particle of this particular size. The most common equivalent particle diameters delivered by aerosol measurement techniques are the electrical mobility diameter d_m (also referred to as Stokes diameter), the aerodynamic diameter d_{ae}, and the optical or polystyrene latex (PSL) equivalent diameter d_{PSL}. The electrical mobility diameter describes that the particle under consideration behaves in the electric field of a differential mobility analyzer (DMA) like a singly charged sphere of this particular diameter. For a spherical particle, the electrical mobility diameter is equal to the geometric diameter. The mobility equivalent diameter also describes the diffusional particle motion well. The aerodynamic diameter is used to describe the particle size in an inertial separators. The inertial particle motion not only depends on the particle geometry but also on the (effective) particle density $\rho_{p,eff}$:

$$d_{ae} = d_m \cdot \sqrt{\frac{\rho_{p,eff}}{\rho_0}} \tag{2.1}$$

where ρ_0 is the unit density, that is, 1 g/cm³ or 1000 kg/m³. The effective density equals the bulk density in the case of compact particles, but can be significantly lower in the case of (loose) agglomerates (McMurry et al. 2002; Ristimäki et al. 2002).

Optical particle sizing instruments measure the light scattered by the particles. The light scattering of particles depends not only on particle size but also on their geometry and refractive index. In some cases optical spectrometers are calibrated for a particular aerosol, taking into account the particles' morphologies and refractive indices. It is, however, more common to calibrate optical particle counters with PSL spheres of known properties. The determined particle size is then expressed as PSL equivalent size, that is, the particle under consideration scatters the same amount of light as a PSL particle of this size.

2.2 SIZE-INTEGRATED AND TIME-RESOLVED MEASUREMENTS

Size-integrating and time-resolving instruments deliver information on the total particle concentration (e.g., number, surface area, or mass) within a given particle size range with a high time resolution. The size range over which the concentration is integrated is commonly defined by the measurement technique involved. Additionally, the upper end of the size range is often, deliberately, limited by the use of an inertial preseparator, that is, a cyclone or an impactor.

2.2.1 Condensation Particle Counters

The most commonly used instrument for measuring nanoparticle number concentrations is the condensation particle counter (CPC, sometimes also referred to as condensation nucleus counter). In a CPC, the incoming particles are first exposed to a supersaturated atmosphere of a working fluid, commonly butanol, isopropyl alcohol, or water. The principle of the CPC dates back to the late 19th century, when Aitken (1888) designed the apparatus to quantify dust in the air. A schematic of a modern CPC is shown in Figure 2.1.

The particles, which otherwise would be too small to scatter light, grow to optically detectable sizes due to the condensation of the working fluid vapor onto their surfaces. The resulting droplets are counted in a measurement cell by light scattering to determine the number concentration with high time resolution of commonly

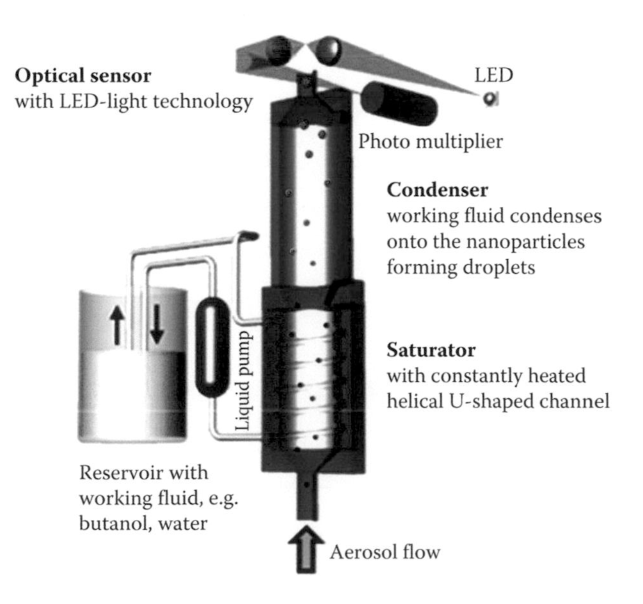

FIGURE 2.1 Schematic of a condensation particle counter (CPC). (Courtesy of Palas GmbH, Karlsruhe, Germany.)

1 s. In the lower concentration range, single light scattering events are counted. When the concentration increases, there is an increasing chance that more than a single particle is present in the measurement cell such that the signals can no longer be differentiated. This so-called coincidence error causes too low concentrations to be reported back by the instrument. Depending on the CPC model, the upper number concentration level that can still accurately be detected in the single particle count mode is between 10,000 #/cm³ for simple and/or old CPCs and 1,000,000 #/cm³ for the most sophisticated models. The lower detection limit of common CPCs is below 1 #/cm³. A review of the history of CPCs can be found in McMurry (2000). Figure 2.2 shows an example for the comparison of two CPCs that sampled an aerosol of varying number concentrations up to approximately 200,000 #/cm³. While CPC 1 is specified to measure accurately up to 300,000 #/cm³, CPC 2 only measures accurately up to 50,000 #/cm³. The figure shows that both CPC readings agree well up to a concentration of approximately 50,000 #/cm³ (range 1). When the number concentration exceeded 50,000 #/cm³ (range 2), concentrations measured with CPC 2 were increasingly below those measured by CPC 1, because two or more particles were simultaneously present in the measurement cell of CPC 2 and counted as one. Once the concentration exceeded approximately 180,000 #/cm³ (range 3), the number concentration provided by CPC 2 was constant, because the signals could no longer be differentiated.

Some CPC models switch to a photometric mode, once the coincidence limit is reached (i.e., in range 2). In the photometric mode, the CPC measures the total light intensity scattered by the particle ensemble in the measurement cell to calculate the particle number concentration from it. In the photometric mode, CPCs can commonly measure number concentrations up to 10^6–10^7 #/cm³. Determining the number concentration from the total intensity of scattered light requires the mean droplet

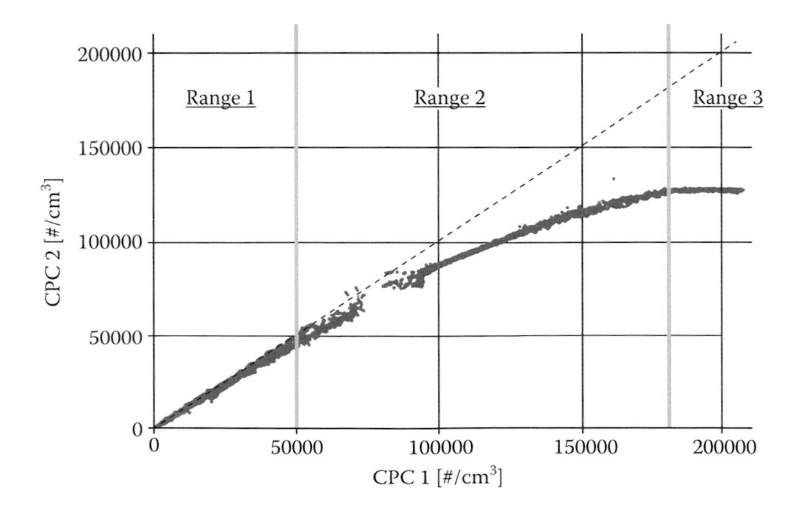

FIGURE 2.2 Comparison of two condensation particle counters (CPCs); CPC 1 can measure accurately up to 300,000 #/cm³ whereas CPC 2 only measures accurately up to 50,000 #/cm³ and shows coincidence errors at higher concentrations.

size to be known, constant, and reproducible. It was found that in the case of water-based CPCs the eventual droplet size is significantly smaller for hydrophobic than for hydrophilic particles. The photometric mode was therefore dropped in the latest versions of water CPCs.

Depending on the model, CPCs can measure particles with sizes down to 2.5 nm. Particularly in the case of water-based CPC, the lower size limit is affected by the hygroscopicity of the particle material. While Hering et al. (2005) and Petäjä et al. (2006) reported the cut-off diameter of model 3785 (TSI) water CPC to fluctuate by only about ±1 nm with solid hydrophilic and hydrophobic particles, Keller, Tritscher and Burtscher (2013) found that the newer water CPC model 3788 (TSI) under-counted even 70 nm soot particles by approximately 50%.

An upper particle size limit of 1 μm is often specified by the CPC manufacturers. The actual upper limit is, however, usually unknown and mainly defined by particle losses inside the CPC. It has been shown in the past that CPCs are capable of measuring concentrations of particles with sizes up to several microns (unpublished data). It should be noted that the number concentrations are usually dominated by particles that are much smaller than 1 μm, whereas the number concentration of particles >1 μm is commonly negligibly small.

Although most CPCs are rather large and mains operated, a small number of handheld CPCs are also available. These handheld CPCs are battery driven and operated with an alcohol cartridge to provide the working fluid. The battery lifetime and alcohol reservoir allow for an independent operation for approximately 6–8 hours. It has been shown that handheld CPCs can be accurate to within ±5% if well maintained and used within their specification limits (Hämeri et al. 2002; Asbach et al. 2012).

2.2.2 Diffusion Charger-Based Instruments

Another group of instruments that determine particle size-integrated and time-resolved particle concentrations have entered the market in recent years. These instruments use unipolar diffusion charging followed by particle charge measurement. It was shown that the charge level obtained by particles in unipolar diffusion charging is proportional to the fraction of the particle surface area that would deposit in the alveolar region of the human lung (Shin et al. 2007; Fissan et al. 2007). Since several studies have shown that the adverse biological response of inhaled particles seems to correlate best with the total surface area dose (Oberdörster, Oberdörster, and Oberdörster 2005) of the lung-deposited particles, possibility of measuring the lung-deposited surface area (LDSA) concentration has raised increased attention. Another study revealed that unipolar diffusion charging can only mimic the LDSA concentration in the range from 20 to 400 nm (Asbach et al. 2009a). Instruments that determine the LDSA concentration are the Nanoparticle Surface Area Monitor (NSAM, model 3550, TSI Inc., USA), Aerotrak 9000 (TSI), miniDiSC (Fierz et al. 2011, identical with DiSCmini, Matter Aerosol, Switzerland), nanoTracer (Marra, Voetz, and Kiesling 2010; Philips, The Netherlands, discontinued), and Partector (naneos, Switzerland). The NSAM was the first instrument of this kind and is still a rather bulky and mains-operated device. Aerotrak 9000 is basically the same instrument, but in a smaller packaging and means for operation independent of mains power and an external computer. miniDiSC/DiSCmini and nanoTracer are significantly smaller, handheld, and independently operating instruments that are particularly suited for tier 2 screening measurements (see Chapter 11), and the Partector is the first true personal monitor to measure the nanoparticle exposure in the breathing zone of a worker. All aforementioned diffusion charger-based instruments use a very similar set up as shown in Figure 2.3. The particles pass through a unipolar charger, where they acquire a known and predictable number of elementary charges (Kaminski et al. 2012). These unipolar chargers use corona discharge, caused by a highly inhomogeneous electric field near a corona electrode (usually a thin wire or a sharp tip), to ionize the air. Brownian motion causes these produced ions to collide with the particles and transfer their charge. In the next stage, the charged aerosol enters an ion trap, which is an electrostatic precipitator operated at a rather low voltage, sufficient to trap ions, but ideally leaving charged particles unaffected. In an NSAM and Aerotrak 9000, the ion trap can be set to two different voltages (100 or 200 V), which are both higher than required for the removal of ions only (20 V). As a consequence, a certain fraction of very small (<20 nm) charged particles are also captured in the ion trap. This is done to adjust the response function of the instrument to the functions required to mimic particle deposition in the alveolar or tracheobronchial region of the human lung (Fissan et al. 2007; Shin et al. 2007). It was, however, found that the measured differences are minute and that a single ion trap voltage can be used to mimic both deposition rates as well as the deposition in the total respiratory tract (Asbach et al. 2009a). The ion trap and electrical manipulation step, shown in Figure 2.3, are hence combined in an NSAM. The particles are eventually captured on a filter in a Faraday cup electrometer to measure the particle induced current. The LDSA is then derived from the current by the application of a simple calibration factor.

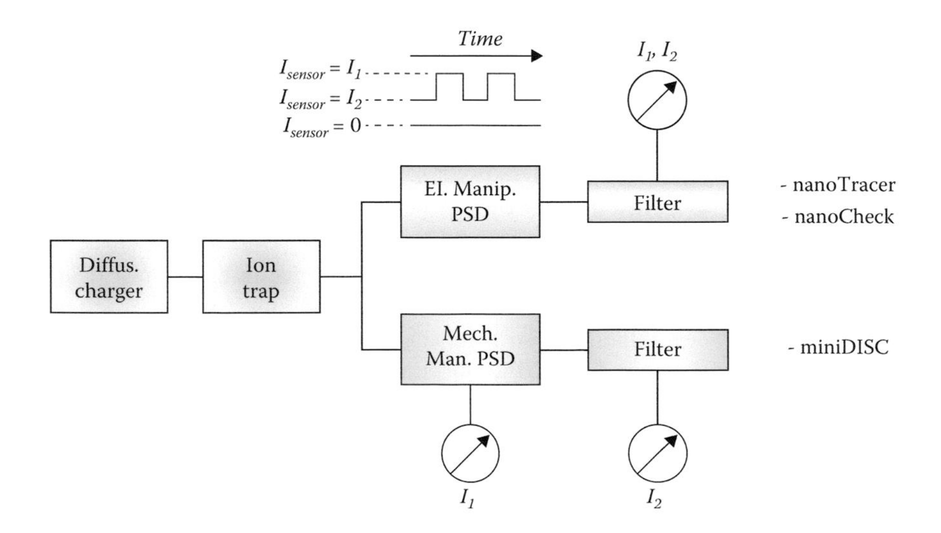

FIGURE 2.3 Schematic setup of diffusion charger-based monitors.

The nanoTracer (Marra et al. 2010) uses a slightly different principle, originally invented by Burtscher and Schmidt-Ott (2009). The charger is followed by an electrical manipulation stage, which consists of an electrostatic precipitator, with a square wave voltage applied to it. During times when the electrostatic precipitator (ESP) voltage is low, only ions are removed, whereas when the voltage is high, a certain fraction of small particles is also collected in the ESP. The current stemming from the particle-borne charges is continuously measured downstream of the ESP, resulting in two independent current levels being measured during high and low voltage periods. When a low voltage is applied to the ESP the measured current I_1 is again proportional to the LDSA concentration. Due to the size dependence of the particle charging and hence the particle collection in the ESP, the ratio of the currents I_1/I_2 measured with the different ESP voltages is a function of particle size and is therefore used to determine the mean particle diameter. The mean particle diameter and the LDSA concentration are then used to determine the total particle number concentration. Results are delivered with a time resolution of 16 s. According to the manufacturer's specifications, the modal diameters of the measured aerosols need to be in the range from 20 to 120 nm for the nanoTracer to measure accurately.

The miniDiSC (Fierz et al. 2011) and its commercial counterpart DiSCmini (Matter Aerosol, Switzerland) use a similar approach to the nanoTracer, but the manipulation of the particle size distribution is done by mechanical instead of electrostatic means, that is, it follows the lower branch in Figure 2.3. The instrument uses a dual stage particle deposition system. Small particles are preferentially deposited on a stack of diffusion screens in the first stage and all remaining particles are collected on a high efficiency filter in the second stage. Both stages are connected to separate electrometers such that the resulting two currents are measured simultaneously, resulting in a high time resolution of 1 s. The overall size range of the miniDiSC is limited to 10–300 nm.

The aforementioned diffusion charger-based instruments have been subject to several round robin and comparability tests (Asbach et al. 2012; Meier, Clark, and Riediker 2013; Mills, Park, and Peters 2013; Kaminski et al. 2013). The main outcome is that, although some limitations apply, this type of instrument can be reliably used for the determination of airborne particle concentrations. However, the accuracy of the results should be expected to be only around ±30%.

The Partector (naneos, Switzerland) is the latest and smallest diffusion charger-based instrument on the market. It is approximately the size of a cigarette box and can hence be used as a personal monitor. It follows the same overall principle as shown in Figure 2.3, but lacks a manipulation step. The particles are charged in a unipolar corona diffusion charger and the ions are removed in an ion trap. The particle charge is detected without particle deposition through induced currents in an induction stage. Since induction only occurs in the presence of a charge gradient, the particle charger is operated in a pulsed mode (Fierz et al. 2014). As previously discussed, the charge level acquired by particles in a unipolar diffusion charger is proportional to the alveolar LDSA concentration, which is hence the metric measured by the Partector. The manufacturer specifies the particle size range to be from 10 nm to 10 μm. Since the particles are not deposited for the measurement of the particle charge, they are available for electrostatic collection for further analyses. Depending on the model the Partector comes with an in-built sampler for sampling the particles onto transmission electron microscopy (TEM) grids.

2.3 SIZE-RESOLVED AND TIME-RESOLVED MEASUREMENTS

Size-resolved and time-resolved measurements provide information on the particle size distribution over time. This type of instrument not only provides the highest amount of physical information but also requires the highest efforts for data evaluation. Due to the nature of the particles, instruments for the submicron and micron size range require different measurement techniques. While micron particles are large enough to scatter sufficient light to be sized optically or provide a sufficient mass to be classified based on their inertia, particles below 300–500 nm can only be classified by their electrical mobility and, with some limitations, by inertial impaction. Since the measured particles are usually not spherical, the instruments can only provide particle size information based on equivalent diameters. In the case of electrical mobility analysis, the instruments' output is based on the electrical mobility equivalent diameter, which describes that the respective particle behaves in the electrostatic classifier like a spherical particle with this particular diameter. If the particles are classified according to their inertia, besides the particle size, the particle density also affects the classification. Results are hence reported as aerodynamic equivalent diameters. Particle sizing by light scattering is not only affected by the particle size but also by the surface properties of the particles as well as the refractive index. Particle sizes are therefore commonly reported as PSL-equivalent diameter, that is, the particle under consideration scatters the same amount of light as a spherical PSL particle of this diameter.

2.3.1 Instruments Based on Electrical Mobility Analysis

Instruments for measuring particle size distributions based on electrical mobility analysis commonly comprise of four main components, (a) a preseparator to remove too large particles, (b) a particle charger to establish a known particle charge distribution, (c) a differential electrical mobility classifier (DEMC, terminology according to ISO 15900), in the literature often found as DMA, and (d) an instrument for the quantification of the mobility-classified particle concentrations. Electrical mobility analysis is useful for measuring the number size distribution of submicron particles with time resolutions between $\leq$1 s and several minutes. Electrical mobility analysis is described in detail in ISO 15900:2009.

The most commonly used type of instrument is the scanning mobility particle sizer (SMPS, Wang, and Flagan 1990), shown in Figure 2.4. Incoming particles that are too large are removed in an inertial preseparator, usually an impactor. Particles are then brought to charge equilibrium (Fuchs 1963), that is, the particles acquire a bipolar charge distribution, which can be easily predicted (Wiedensohler 1988). Since the effective global charge level of the particles is nearly zero, the charger is usually referred to as neutralizer. In the past, the neutralizers all contained radioactive sources, mostly ^{241}Am or ^{85}Kr, to produce the required bipolar ion atmosphere. More recently, most manufacturers also offer the possibility of replacing the radioactive

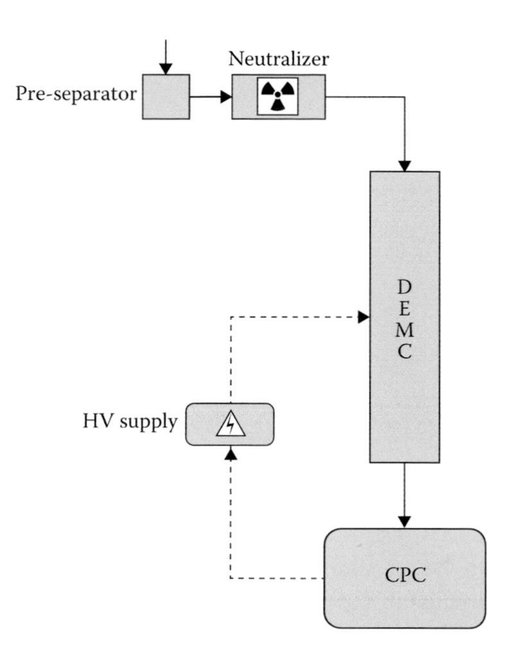

FIGURE 2.4 Typical setup of a scanning mobility particle sizer (SMPS).

neutralizer by soft X-ray sources, which provide comparable charge distributions (Shimada et al. 2002; Lee et al. 2005), but have lower regulatory and administrative requirements. Downstream of the neutralizer, the charged particles are introduced into a DEMC (Liu and Pui 1974; Winklmayr et al. 1991). A DEMC is essentially a coaxial arrangement with inner and outer electrodes as shown in Figure 2.5.

Particles are introduced near the outer electrode, whereas the space near the inner electrode is flushed with a particle-free sheath air flow. Depending on manufacturer and setting, the aerosol flow rate is commonly between 0.1 and 1.0 L/min and the ratio of sheath air to aerosol flow rate is usually 10:1. Toward the exit, the inner electrode contains a small slit, where the monodisperse aerosol flow leaves the classifier at the same flow rate as the aerosol inlet flow. If no voltage is applied to the DEMC, all particles leave the DEMC with the sheath air outlet and the monodisperse aerosol outlet is particle free. If a voltage is applied between inner and outer electrodes, particles of one polarity migrate toward the inner rod at a velocity that is defined by the classifier voltage and the electrical mobility Z_p of the particle (Hinds 1999):

$$Z_p = \frac{n \cdot e \cdot d_m}{3 \cdot \pi \cdot \eta \cdot C_c(d_m)} \tag{2.2}$$

In Equation 2.2, n is the number of particle-borne elementary charges e, d_m is the electrical mobility diameter, η is the gas viscosity, and C_c is the Cunningham slip correction factor (Cunningham 1910; Kim et al. 2005) that corrects for the fact that molecular reflection from particle surfaces in the submicron size range can no

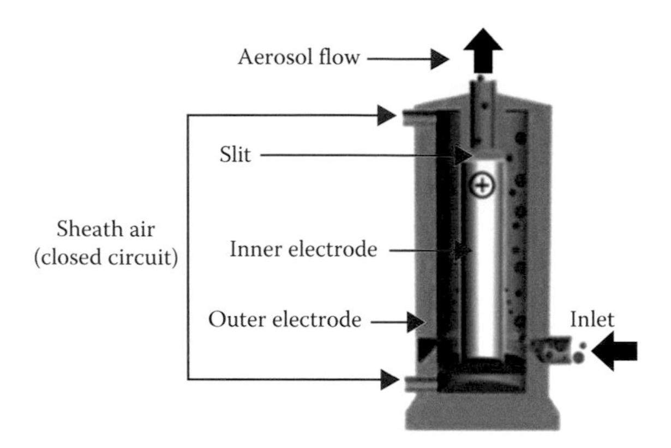

FIGURE 2.5 Schematic of a differential electrical mobility classifier (DEMC). (Courtesy of Palas GmbH, Karlsruhe, Germany.)

longer be assumed to be specular. The classified particles are then counted in a CPC. By changing the voltage in the classifier, the full range of electrical mobility can be covered. In a SMPS, the voltage is continuously or sequentially ramped, which requires approximately between 2 and 6 minutes. Only very recently much faster scanning SMPS systems have been introduced (Farnsworth et al. 2013) that accomplish a full scan in down to 10 s. As previously described, the primary output of an SMPS is the distribution of the particle number concentration downstream of the classifier as a function of the particle electrical mobility. A complex data deconvolution scheme is then used to obtain the particle number size distribution from these data (Hoppel 1978; Fissan, Helsper, and Thielen 1983). This data deconvolution requires the appropriate use of a preseparator. Otherwise, if particles larger than the ones to be measured in the largest particle size bin enter the DEMC, the multiple charge correction will overcompensate for multiply charged particles in a lower size bin, resulting in more or less strong dips in the determined particle size distribution, which require additional correction (He and Dhaniyala 2013). SMPS systems can cover particle sizes ranging from 2.5 nm up to around 1000 nm (Winklmayr et al. 1991; Chen et al. 1998), depending on the DEMC and CPC used and the SMPS settings and model. The sizes of particles and molecular clusters smaller than 2.5 nm can also be classified using a DEMC. The downstream quantification of the particle number concentration is only possible with an aerosol electrometer, because all commercially available CPCs are only usable down to 2.5 nm. In workplace aerosols, however, such small particles are usually irrelevant.

A possibility of measuring airborne particle number size distributions with a higher time resolution of up to 1 s was introduced by Tammet, Mirme, and Tamm (2002). Their instrument was commercialized as Fast Mobility Particle Sizer (FMPS model 3091, TSI) and covers a particle size ranging from 5.6 to 560 nm. It essentially follows a very similar principle set up to an SMPS. Initially, too large incoming particles are removed in a cyclone. The remaining particles pass a unipolar particle charger before being classified based on their electrical mobility. The classifier has a similar design to the DEMC shown in Figure 2.5, but particles enter near the inner electrode and are deflected toward the outer wall. The outer electrode consists of an array of electrometer rings. Since the location where the particles hit the outer electrode depends on the electrical mobility of the particles, each one of the currents from the electrometer electrodes along the outer electrode represents the concentration of particles in a certain electrical mobility range. The measured currents are then deconvoluted into a number size distribution by making use of the known particle charge distribution. Since concentrations are measured continuously over the entire size range covered, the FMPS can provide a much higher time resolution and thus also allows for the study of dynamically changing aerosols. The high measurement speed, however, comes at the price of a decreased accuracy, especially for agglomerated particles and for particles larger than 100 nm (Asbach et al. 2009b; Jeong and Evans 2009; Kaminski et al. 2013). In these cases, large deviations were found for both sizing and concentration measurements, whereas much higher accuracies were obtained in the same studies among SMPS systems of various manufacturers.

2.3.2 Instruments Based on Inertial Classification

Particles can be classified based on their inertia, which is defined by the particles' masses and velocities. Inertial classification is usually done by using impactors. In an impactor, the incoming flow is accelerated in a nozzle before facing a perpendicular impaction plate to form a stagnation point flow (see Figure 2.6). While small particles with low inertia follow the flow streamlines, particles with higher inertia divert from the streamlines and are deposited on the impaction plate. The aerodynamic particle size at which 50% of the particles are deposited, the cut-off diameter, can be easily predicted by the nozzle diameter, gas flow rate, and the particle density.

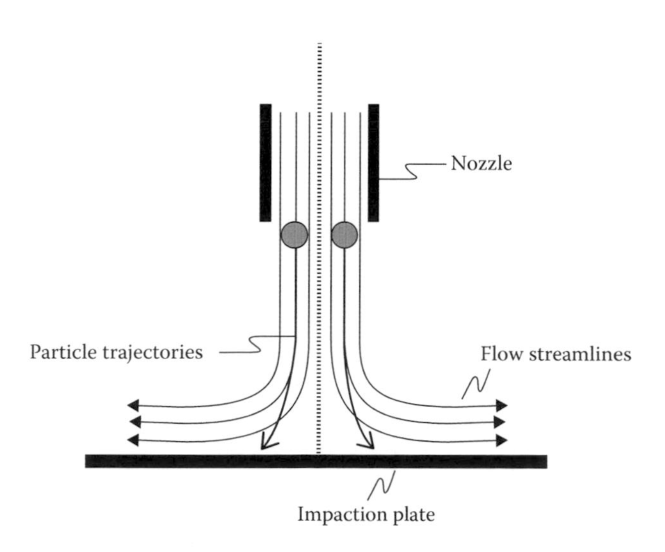

FIGURE 2.6 Principle of an impactor.

Impactors can be used as a cascade with cut-off diameters decreasing from stage to stage to obtain size-resolved samples. An Electrical Low Pressure Impactor (ELPI or ELPI+; Dekati, Finland; Keskinen, Pietarinen, and Lehtimäki 1992) uses a total of 14 stages (ELPI+) to sample particles in the size range from 6 nm to 10 μm. Since the particle inertia scales with the third power of the particle size, nanoscale particles can only be sampled by inertia if the gas pressure is significantly reduced. At reduced pressure, the drag force, which would otherwise cause the particle to be transported with the flow, decreases and therefore also particles of lower inertia deviate from the streamlines. ELPI uses a pressure of below 100 mbar in the last stage to collect such small particles. Before being introduced into the cascade impactor, particles are getting electrically charged to a known charge level in a unipolar diffusion charger. The current induced by the deposition of charged particles is measured from each impaction stage by the use of electrometers to obtain the particle concentration in the corresponding size bin. As a result, the particle size distribution is measured with a time resolution of up to 0.1 s. A schematic of ELPI is shown in Figure 2.7. ELPI is the only instrument on the market that covers the size range from a few nanometers up to 10 μm based on the same measuring principle and therefore using the same equivalent particle size, that is, the aerodynamic diameter. Furthermore, it is the only instrument that not only provides online information on the particle size distribution but also provides the particle samples from each impaction stage for consecutive analyses, for example, by electron microscopy or chemical analysis. However, ELPI

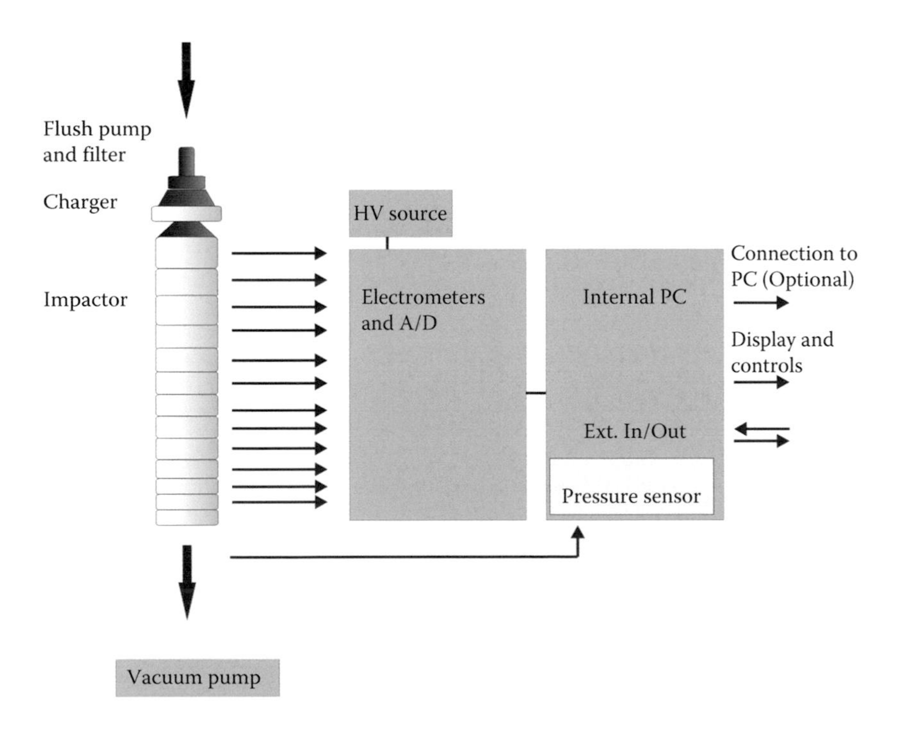

FIGURE 2.7 Schematic of the electrical low pressure impactor (ELPI). (Courtesy of Dekati, Kangasala, Finland.)

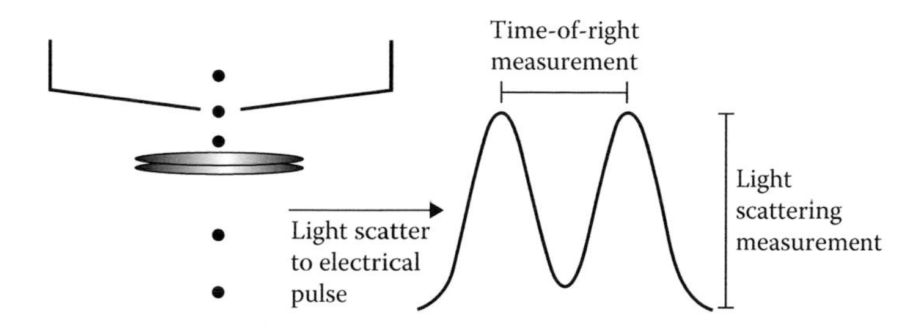

FIGURE 2.8 Schematic of the time of flight particle sizing in an aerodynamic particle sizer (APS). (Courtesy of TSI Inc., Shoreview, MN.)

and ELPI+ are rather bulky instruments and require the use of a large pump, which make them suitable only as stationary measurement instruments and the size information is limited to a resolution of 14 size channels.

Another instrument that determines the particle size distribution based on the aerodynamic particle size is the Aerodynamic Particle Sizer (APS, TSI Inc., USA; Baron 1986). An APS is a mains-operated desktop instrument, in which the aerosol is accelerated in a nozzle. The velocity of particles leaving the nozzle is a function of particle size. The particles pass through two laser light sheets and the particle size is determined by the time of flight between the two sheets (see Figure 2.8). The number of particles is then stored into the corresponding particle size bin to eventually determine the particle number size distribution. An APS nominally covers the size range from 0.5 to 20 μm and provides size distributions with a resolution of up to 32 size channels per size decade with a time resolution of up to 1 s. An APS is often used in combination with an SMPS to cover the full size range from a few nanometers up to 20 μm. The joint data evaluation from the two instruments requires conversion of either the aerodynamic diameters from the APS to electrical mobility diameters or vice versa by using Equation 2.1, which sometimes can be problematic if the (effective) density of the measured particles is unknown (Khlystov, Stanier, and Pandis 2004).

2.3.3 Instruments Based on Optical Classification

Optical particle spectrometers measure the light scattered by particles in an aerosol to measure the particle number size distribution. As a rule of thumb, the minimum detectable particle size is approximately half the wavelength of the light source used, that is, commonly around 250 nm. The upper size limit is usually defined by particle losses in the system and is typically between 10 and 40 μm, approximately. A schematic of an optical particle spectrometer is shown in Figure 2.9.

The aerosol is focused into a narrow particle stream, which passes the light beam from a light source. Each particle scatters light, which is detected in a photodetector, where it produces a voltage spike. The height of the spike depends on the amount of scattered light, which depends on the particle size and refractive index. If the refractive index is (assumed to be) known, the particle size can be determined. Since in many cases the particle material is unknown, the refractive index of PSL particles

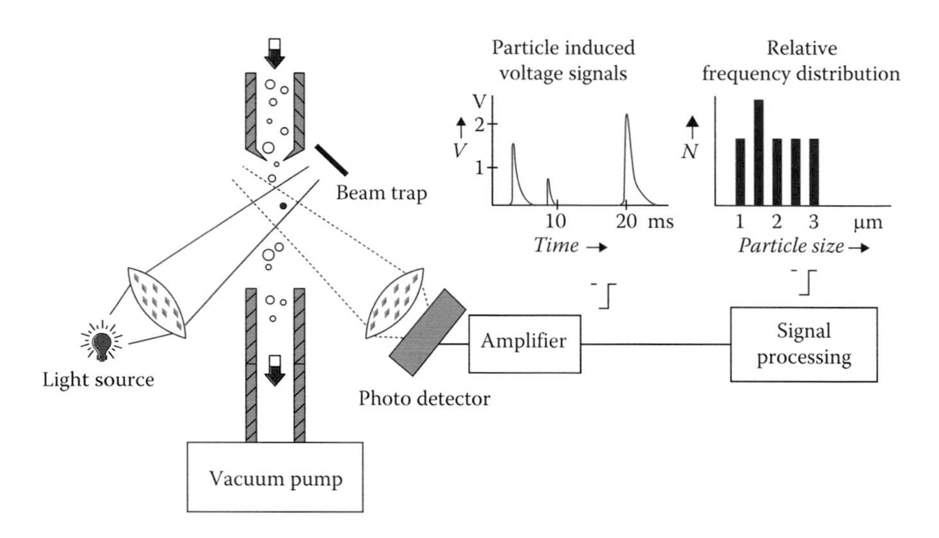

FIGURE 2.9 Principle of an optical particle spectrometer. (Borggräfe, P. (2000): Erweiterung des Meßbereichs von optischen Partikelzählern durch gezielte Reduzierung der Störquellen und mit digitalen Signalverarbeitungsmethoden, Ph.D. Thesis at Gerhard Mercator University Duisburg, available online: http://duepublico.uni-duisburg-essen.de/servlets/DerivateServlet/Derivate-5033/bodiss.pdf [accessed November 23rd, 2013].)

is used and the particle sizes delivered as "PSL-equivalent" sizes. By counting the number of particles of different sizes, the particle number size distribution is determined. Some of the commercially available optical particle spectrometers calculate the mass concentration fractions in specific size ranges (e.g., PM_{10} or $PM_{2.5}$), based on assumed particle size-dependent mean refractive indices and particle densities. These instruments are usually calibrated using empirically derived mean values of atmospheric particles. Hence, care needs to be taken when such spectrometers are used for the determination of particle mass concentrations in workplaces, where refractive indices and particle densities may differ significantly from those in the atmosphere.

Several manufacturers couple an optical spectrometer for the characterization of mainly micron size particles with SMPS measurements for submicron particles (e.g., Wide Range Aerosol Spectrometer WRAS model EDM 665, Grimm Aerosol, Germany, or Wide Range Particle Spectrometer model M1000XP, MSP, USA). Grimm Aerosol additionally introduced two different instruments that couple a spectrometer coupled with diffusion charger-based monitors. In the case of the nanoCheck, the diffusion charger instruments follow the identical principle to the nanoTracer (see the aforementioned) to provide the total number concentration and mean size of particles below 300 nm. The diffusion charger-based instrument in the miniWRAS (model 1.371) is more sophisticated and uses a simplified electrical mobility classifier to provide particle size distributions below 300 nm with a 10 channel resolution. Similar to the combination of APS with SMPS, merging of the data from the two measuring techniques requires the conversion from electrical mobility equivalent diameter to the optical equivalent diameter or vice versa.

2.4 SIZE-RESOLVED AND TIME-INTEGRATED MEASUREMENTS

Size-resolved and time-integrated measurements provide particle size-classified information such as the mass or number concentration or the chemical composition as a mean value over the sampling time, which is often a full working shift, that is, eight hours. Usually, the airborne particles are sampled onto a substrate, which is then used for downstream analyses. Size discrimination may be done already during sampling by a size classifier or by size determination during the subsequent analysis.

2.4.1 Inertial Particle Samplers

Size-resolved and time-integrated measurements can be achieved by using size-classified particle sampling along with subsequent analyses of the particles sampled on the different stages. Typically, cascade impactors are used for this purpose, that is, a cascade of impactor stages (see Figure 2.6) with cut-off diameters decreasing from stage to stage. Depending on the model, cascade impactors can sample particles between approximately 10 nm and 10 µm. While some cascade impactors use only three stages for sampling particles >10 µm, 2.5–10 µm, and 1–2.5 µm, others use many more stages to provide samples with higher size distribution. In a cascade impactor, the particles are sampled commonly onto flat substrates, which can later be used for weighing to determine the mass size distribution or electron microscopic analysis coupled with electron dispersive X-ray (EDX) spectroscopy to determine the morphology and chemical composition of the particles. Other cascade impactors have been designed to sample onto glassy carbon substrates for analyses by total X-ray reflection fluorescence spectroscopy (John et al. 2001) or to homogenously deposit the particles across the substrate by using a large number of very small parallel orifices per impactor stage (Micro Orifice Uniform Deposit Impactor, MOUDI, MSP Corp., USA; Marple et al. 1986). To further homogenize the sample, the substrates may rotate during sampling. MOUDI is available in different models. The stationary models operate at a flow rate of 30 L/min and can have between 3 and 14 stages to sample particles in an overall size range from 10 nm to 18 µm. Figure 2.10 shows the collection efficiency of the different stages of a MOUDI-II. Mini-MOUDI is a miniaturized MOUDI, which operates at a flow rate of only 2 L/min and can host six, seven, eight, or ten impaction stages, where the sixth and eighth stage versions can be used as personal samplers.

2.4.2 Thermal Particle Samplers

Thermal precipitators have been used as personal samplers for many decades to assess exposure to airborne dust. The sampling principle makes use of the thermophoretic motion of particles, which arises from inhomogeneous momentum transfer from colliding molecules in a temperature gradient. The molecules on the "warm side" of the particles transfer a higher momentum than their counterparts on the "cold side," resulting in a net particle motion from warm to cold regions. If a temperature gradient is established near a cold plate, particles are deposited onto this plate and are available for further analyses, for example, of the particle size distribution by (electron) microscopy.

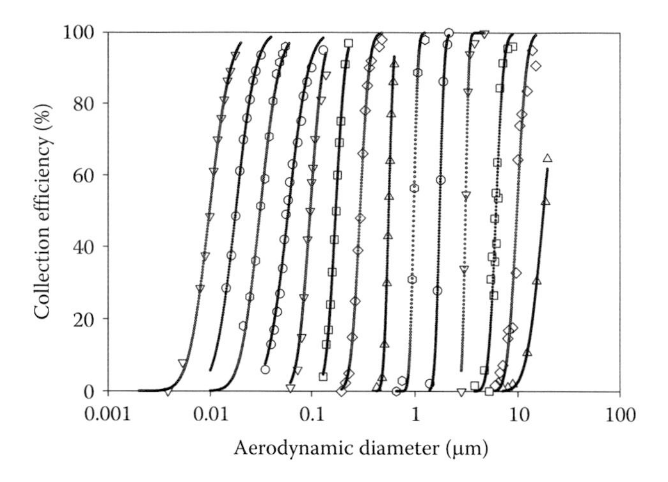

FIGURE 2.10 Collection efficiencies of a MOUDI-II with 14 impaction stages. (Courtesy of MSP Corp., Shoreview, MN.)

The originally designed thermal precipitators used a heated wire in the center between two parallel plates. The aerosol flow was perpendicular to the wires and parallel to the plates, such that the particles were deposited onto the plates (Roach 1959). The downside of this design was that particles were deposited spatially inhomogeneously, due to the inhomogeneous thermal gradient. The inhomogeneity of the deposit required high efforts to evaluate the particle size distribution. More recent developments (Azong-Wara et al. 2009, 2013; Thayer et al. 2011) use a homogenous thermal gradient between two parallel plates, which are maintained at different temperature levels to establish the required gradient. Since up to approximately 300 nm the thermophoretic velocity is nearly independent of particle size, these samplers deposit particles <300 nm with very high spatial homogeneity, which allows for a much simplified electron microscopic analysis of the particle size distribution, because only a low number of electron microscopic images need to be evaluated.

2.5 SIZE-INTEGRATED AND TIME-INTEGRATED MEASUREMENTS

Size-integrated and time-integrated measurements include particle sampling over the averaging time and subsequent analysis. Sampling and analysis do not include particle size discrimination. The sampling may, however, include a step to remove particles above a specified size limit, for example, an impactor as preseparator to sample only the alveolar fraction <4 μm. Due to its simplicity, this type of measurement can hence be considered as the simplest type of measurement.

2.5.1 Filtration Methods

Filters can be employed to collect all particles from an aerosol that is drawn through the filter without particle size segregation. Filter media can generally be divided into surface filtration and depth filtration media. Membrane filters are often used as surface filters, whereas fiber filters are used for depth filtration. In a membrane filter, the air is drawn through small well defined holes. Particles deposit on the surface by impaction, interception, or diffusion. A fiber filter contains many zigzagging fibers. Particles are deposited on the fibers inside the media based on the same deposition mechanisms as previously described for the membrane filter. The pressure drop of fiber filters is for a given efficiency lower than in membrane filters. The choice of surface or depth filtration media depends on the purpose of the sampling. Particles collected on a filter surface can much more easily be retained and then, for example, be used in toxicological in vitro or in vivo experiments or dispersed in a liquid to analyze their potential to generate reactive oxygen species (Hellack 2012). Depth filters are more commonly used for determining the particle mass of the sampled particles or for digesting wet chemical analyses.

Single stage impactors with subsequent particle filtration are often used to sample particles in certain size fraction, such as the thoracic (<10 μm) or the alveolar (<4 μm) fraction. The obtained filters can be weighed to determine the particle mass concentration and used for wet chemical analyses. These systems are available as large and bulky versions with high flow rates for sampling large amounts of particles, which may become necessary, depending on the objective of the sampling. Other such systems are small and usable as personal samplers (e.g., Tsai et al. 2012)

2.5.2 Electrostatic Particle Samplers

Electrostatic precipitators (ESPs) are used to deposit electrically charged particles by exposing them to a Coulomb force in an electric field. ESPs are used in a wide range of applications, including sampling of particles for consecutive analysis. In commercial ESPs, suitable for workplace aerosol analysis, the incoming aerosol flow faces a perpendicular substrate, which is maintained at a high electric potential, whereas the housing of the ESP is kept at ground potential (Dixkens and Fissan 1999). Consequently, particles of opposite polarity as the collecting electrode are collected on the substrate. These ESPs can be used to sample airborne particles by their naturally occuring charge, without additional charging. Aged particles in the atmosphere carry a bipolar equilibrium charge distribution and hence a nonnegligible fraction of the particles is charged and will be collected. To enhance the sampling efficiency, particles may be charged unipolarly upstream of the ESP. However, currently there are no commercially available particle chargers that could be used for this purpose and hence all studies that report on the use of a particle charger upstream of a commercial ESP used home-built unipolar chargers (e.g., Van Landuyt et al. 2014). Another option is to use an ESP downstream of a DEMC to sample mobility-classified particles. Particles leaving a DEMC are always charged, however, most of them bear only a single elementary charge, resulting in a rather low collection efficiency especially for larger particles.

More recently, a novel, commercial handheld ESP was introduced (ESPnano model 100, Miller et al. 2010). The sampler has a small size and is battery operated and can hence easily be taken to sample particles at a certain location, where an increased particle concentration or a particle release is expected. A schematic of the ESP is shown in Figure 2.11. This ESP includes a unipolar charger to generate

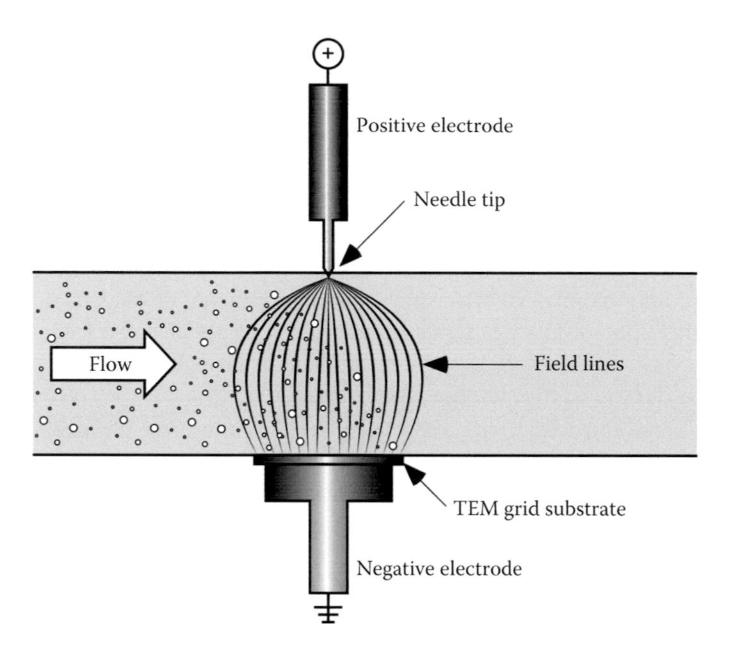

FIGURE 2.11 Principle of a handheld electrostatic precipitator to collect airborne (nano-) particles. (From Miller, A. et al. 2010. A handheld electrostatic precipitator for sampling airborne particles and nanoparticles, *Aerosol Sci Technol*, 44: 417–427.)

ions near a tip electrode. When a positive high voltage is applied to the tip, a corona is formed that ionizes the air. The tip electrode faces the sampling electrode. Consequently, the generated ions follow the electric field lines into the perpendicular aerosol flow, where they collide with the particles to charge them. The charged particles are then deposited onto the collection substrate within the same electric field that is used to generate the ions and charge the particles. A TEM grid is used as a sampling electrode and is available for analysis of the particle size and morphology by TEM and chemical composition by EDX.

2.6 SUMMARY

A large variety of methods exist to characterize airborne particles. The choice of instruments for a specific task depends strongly on the aerosol and particle characteristics (sizes, concentrations, etc.) and the specific requirements, that is, need for particle size-, time-, or spatial resolution. In general, all highly accurate instruments are rather bulky and heavy (especially for the characterization of nanoparticles). Smaller, portable, or personal instruments are available, but the wealth of information from these as well as the accuracy is limited. Chemical speciation and morphological characterization of the airborne particles require sampling for subsequent analyses, which is hence rather time consuming. In conclusion, there is not a single instrument that can fulfill all requirements of exposure assessment in workplaces. Instead, the choice of an instrument is always a compromise concerning instrument mobility, particle size and concentration range, measured metric, and accuracy.

Consequently, while single instruments can deliver valuable first insight into whether exposure to nanoparticles may occur in a workplace, a well thought-through suite of instruments tailored for the specific task is needed for a full workplace aerosol characterization.

REFERENCES

Asbach, C., Fissan, H., Stahlmecke, B., Kuhlbusch, T.A.J., Fissan, H. (2009a): Conceptual imitations and extensions of lung-deposited nanoparticle surface area monitor (NSAM), *J Nanopart Res*, 11: 101–109.

Asbach, C., Kaminski, H., Fissan, H., Monz, C., Dahmann, D., Mülhopt, S., Paur, H.R. et al. (2009b): Comparison of four mobility particle sizers with different time resolution for stationary exposure measurements, *J Nanopart Res*, 11: 1593–1609.

Asbach, C., Kaminski, H., Von Barany, D., Kuhlbusch, T.A.J., Monz, C., Dziurowitz, N., Pelzer, J. et al. (2012): Comparability of portable nanoparticle exposure monitors, *Ann Occup Hyg*, 56: 606–621.

Azong-Wara, N., Asbach, C., Stahlmecke, B., Fissan, H., Kaminski, H., Plitzko, S., Bathen, D., Kuhlbusch, T.A.J. (2013): Design and experimental evaluation of a new nanoparticle thermophoretic sampler, *J Nanopart Res*, 15: 1–12.

Azong-Wara, N., Asbach, C., Stahlmecke, B., Fissan, H., Kaminski, H., Plitzko, S., Kuhlbusch, T.A.J. (2009): Optimisation of a thermophoretic personal sampler for nanoparticle exposure studies, *J Nanopart Res*, 11: 1611–1624.

Baron, P.A. (1986): Calibration and use of the aerodynamic particle sizer (APS 3300), *Aerosol Sci Technol*, 5: 55–67.

Borggräfe, P. (2000): Erweiterung des Meßbereichs von optischen Partikelzählern durch gezielte Reduzierung der Störquellen und mit digitalen Signalverarbeitungsmethoden, *Ph.D. Thesis at Gerhard Mercator University Duisburg*, available online: http://duepublico.uni-duisburg-essen.de/servlets/DerivateServlet/Derivate-5033/bodiss.pdf (accessed November 23rd, 2013).

Burtscher, H., Schmidt-Ott, A. (2009): Method and device for measuring number concentration and mean diameter of particles suspended in a carrier gas, European Patent EP 1655595 B1, November 3rd, 2004.

Chen, D.R., Pui, D.Y.H., Hummes, D., Fissan, H., Quant, F.R., Sem, G.J. (1998): Design and evaluation of a nanometer aerosol differential mobility analyzer (Nano-DMA), *J Aerosol Sci*, 29: 497–509.

Cunningham, E. (1910): On the velocity of steady fall of spherical particles through fluid medium, *Proc Royal Soc Series A*, 83: 357–365.

Dixkens, J., Fissan, H. (1999): Development of an electrostatic precipitator for off-line particle analysis, *Aerosol Sci Technol*, 30: 438–453.

Farnsworth, J., Quant, F., Horn, H.-G., Osmondson, B., Caldow, R. (2013): Fast scanning mobility particle sizing system and classifier, Oral Presentation at the *European Aerosol Conference*, September 1st, 2013, Prague, Czech Republic.

Fierz, M., Houle, C., Steigmeier, P., Burtscher, H. (2011): Design, calibration, and field performance of a miniature diffusion size classifier, *Aerosol Sci Technol*, 45: 1–10.

Fierz, M., Meier, D., Steigmeier, P., Burtscher, H. (2014): Aerosol measurement by induced currents, *Aerosol Sci Technol*, 48: 350–357.

Fissan, H., Helsper, C., Thielen, H.J. (1983): Determination of particle size distribution by means of an electrostatic classifier, *J Aerosol Sci*, 14: 354–357.

Fissan, H., Neumann, S., Trampe, A., Pui, D.Y.H., Shin, W.G. (2007): Rationale and principle of an instrument measuring lung deposited nanoparticle surface area, *J Nanopart Res*, 9: 53–59.

Fuchs, N.A. (1963): On the stationary charge distribution on aerosol particles in bipolar ionic atmosphere, *Geofisica pura e applicata*, 56: 185–193.

Hämeri, K., Koponen, I.K., Aalto, P.P., Kulmala, M., (2002): Technical note: the particle detection efficiency of the TSI-3007 condensation particle counter, *J Aerosol Sci*, 33: 1463–1469.

He, M., Dhaniyala, S. (2013): A multiple charging correction algorithm for scanning electrical mobility spectrometer data, *J Aerosol Sci*, 61: 13–26.

Hellack, B. (2012): Räumliche, zeitliche und quellenbedingte Variation Feinstaub induzierter Hydroxylradikalgenerierung, *Ph.D. Thesis at the University Duisburg-Essen*, available online: http://duepublico.uni-duisburg-essen.de/servlets/DerivateServlet/Derivate-31869/Hellack_Diss.pdf (accessed November 23rd, 2013)

Hering, S.V., Stolzenburg, M.R., Quant, F.R., Oberreit, D.R., Keady, P.B. (2005): A laminar-flow, water-based condensation particle counter (WCPC), *Aerosol Sci Technol*, 39: 659–672.

Hinds, W.C. (1999): *Aerosol Technology—Properties, Behavior, and Measurement of Airborne Particles*, John Wiley & Sons, Inc., New York.

Hoppel, W.A. (1978): Determination of the aerosol size distribution from the mobility distribution of the charged fraction of aerosols, *J Aerosol Sci*, 9: 41–54.

ISO 15900:2009 (2009): Determination of particle size distribution—Differential electrical mobility analysis for aerosol particles.

Jeong, C.H., Evans, G.J. (2009): Inter-comparison of a fast mobility particle sizer and a scanning mobility particle sizer incorporating an ultrafine water-based condensation particle counter, *Aerosol Sci Technol*, 43: 364–373.

John, A.C., Kuhlbusch, T.A.J., Fissan, H., Schmidt, K.G. (2001): Size-fractionated sampling and chemical analysis be total-reflection x-ray fluorescence spectrometry of PMx in ambient air and emissions, *Spectrochim Acta B*, 56: 2137–2146.

Kaminski, H., Kuhlbusch, T.A.J., Fissan, H., Ravi, L., Horn, H.-G., Han, H.S., Caldow, R., Asbach, C. (2012): Experimental and mathematical evaluation of an improved unipolar diffusion charger for size distribution measurement, *Aerosol Sci Technol*, 46: 708–716.

Kaminski, H., Rath, S., Götz, U., Sprenger, M., Wels, D., Polloczek, J., Bachmann, V. et al. (2013): Comparability of mobility particle sizers and diffusion chargers, *J Aerosol Sci*, 57: 156–178.

Keller, A., Tritscher, T., Burtscher, H. (2013): Performance of a water-based CPC 3788 for particles from a propane-flame soot-generator operated with rich fuel/air mixtures, *J Aerosol Sci*, 60: 67–72.

Keskinen, J., Pietarinen, K., Lehtimäki, M. (1992): Electrical low pressure impactor, *J Aerosol Sci*, 23: 353–360.

Khlystov, A., Stanier, C., Pandis, S.N. (2004): An algorithm for combining electrical mobility and aerodynamic size distributions data when measuring ambient aerosol, *Aerosol Sci Technol*, 38: 229–238.

Kim, J.H., Mulholland, G.W., Kuckuck, S.R., Pui, D.Y.H. (2005): Slip correction measurements of certified PSL nanoparticles using a nanometer differential mobility analyzer (Nano-DMA) for Knudsen number from 0.5 to 83, *J Res NIST*, 110: 31–54.

Kuhlbusch, T.A.J., Asbach, C., Fissan, H., Göhler, D., Stintz, M. (2011): Nanoparticle exposure at nanotechnology workplaces: A review, *Part Fibre Tox*, 8: 1–18.

Lee, H.M., Kim, C.S., Shimada, M., Okuyama, K. (2005): Bipolar diffusion charging for aerosol nanoparticle measurement using a soft X-ray charger, *J Aerosol Sci*, 36: 813–829.

Marple, V., Rubow, K., Ananth, G., Fissan, H.J. (1986): Micro orifice uniform deposit impactor, *J Aerosol Sci*, 17: 489–494.

Marra, J., Voetz, M., Kiesling, H.J. (2010): Monitor for detecting and assessing exposure to airborne nanoparticles, *J Nanopart Res*, 12: 21–37.

McMurry, P.H. (2000): The history of condensation nucleus counters, *Aerosol Sci Technol*, 33: 297–322.

McMurry, P.H., Wang, X., Park, K., Ehara, K. (2002): The relationship between mass and mobility for atmospheric particles: A new technique for measuring particle density, *Aerosol Sci Technol*, 36: 227–238.

Meier, R., Clark, K., Riediker, M. (2013): Comparative testing of a miniature diffusion size classifier to assess airborne ultrafine particles under field conditions, *Aerosol Sci Technol*, 47: 22–28.

Miller, A., Frey, G., King, G., Sunderman, C. (2010): A handheld electrostatic precipitator for sampling airborne particles and nanoparticles, *Aerosol Sci Technol*, 44: 417–427.

Mills, J.B., Park, J.H., Peters, T.M. (2013): Comparison of the DiSCmini aerosol monitor to a handheld condensation particle counter and a scanning mobility particle sizer for submicrometer sodium chloride and metal aerosols, *Ann Occup Hyg*, 10: 250–258.

Oberdörster, G., Oberdörster, E., Oberdörster, J. (2005): Nanotoxicology: an emerging discipline evolving from studies of ultrafine particles, *Environ Health Perspect*, 113: 823–839.

Petäjä, T., Mordas, G., Manninen, H. Aalto, P.P., Hämeri, K., Kulmala, M. (2006): Detection efficiency of a water-based condensation particle counter 3785, *Aerosol Sci Technol*, 40:1090–1097.

Ristimäki, J., Virtanen, A., Marjamäki, M., Keskinen, J. (2002): On-line measurement of size distribution and effective density of submicron aerosol particles, *J Aerosol Sci*, 33: 1541–1557.

Roach, S.A. (1959): Measuring dust exposure using the thermal precipitator in colleries and foundries, *Br J Ind Med*, 16: 104–122.

Shimada, M., Han, B., Okuyama, K., Otani, Y. (2002): Bipolar charging of aerosol nanoparticles by a soft x-ray photoionizer, *J Chem Eng Jpn*, 35: 786–793.

Shin, W.G., Pui, D.Y.H., Fissan, H., Neumann, S., Trampe, A. (2007): Calibration and numerical simulation of nanoparticle surface area monitor (TSI model 3550 NSAM), *J Nanopart Res*, 9: 61–69.

Tammet, H., Mirme, A., Tamm, E. (2002): Electrical aerosol spectrometer of Tartu University. *Atmos. Res*, 62: 315–324.

Thayer, D., Koehler, K.A., Marchese, A., Volckens, J. (2011): A personal thermophoretic sampler for airborne nanoparticles, *Aerosol Sci Technol*, 45: 744–750.

Tsai, C.J., Liu, C.N., Hung, S.M., Chen, S.C., Uang, S.N., Cheng, Y.S., Zhou, Y. (2012): Novel active personal nanoparticle sampler for the exposure assessment of nanoparticles in workplaces, *Environ Sci Technol*, 46: 4546–4552.

Van Landuyt, K., Hellack, B., Van Meerbeek, B., Peumans, M., Hoet, P., Wiemann, M., Kuhlbusch, T.A.J., Asbach, C. (2014): Nanoparticle release from dental composites, *Acta Biomater*, 10: 365–374.

Wang, S.C., Flagan, R.C. (1990): Scanning electrical mobility spectrometer, *Aerosol Sci Technol*, 13: 230–240.

Wiedensohler, A. (1988): An approximation of the bipolar charge distribution for particles in the submicron size range, *J Aerosol Sci*, 19: 387–389.

Winklmayr, W., Reischl, G.P., Lindner, A.O., Berner, A. (1991): A new electromobility spectrometer for the measurement of aerosol size distributions in the size range from 1 to 1000 nm, *J Aerosol Sci*, 22: 289–296.

3 Classification Strategies for Regulatory Nanodefinitions

Wendel Wohlleben and Philipp Müller

CONTENTS

3.1 DEFINITIONS OF NANOMATERIALS FOR REGULATORY PURPOSES

What is a nanomaterial? Focusing on materials science, only materials with size-specific performance would be considered as nanomaterials; in a strict sense, one may even require that these properties do not simply scale with size, but that they exhibit a threshold behavior. Consider QDots as positive examples (luminescence vanishes for particles larger than a few nm), and ultraviolet (UV)-absorbing pigments as negative examples (UV absorption is less effective with increasing size, but does not vanish). Focusing on innovation processes, the intention at the time of production may be another criterion. On the contrary, the ISO terminology (ISO/TS 27687:2008, nanoparticle, nanofiber, and nanoplate) focuses on geometrical structures with the criterion that any external dimension of the material is in the nanoscale or the material having internal structure in the nanoscale.

The *practical* value of a nanodefinition for regulatory purposes is a filtered market screening that identifies materials that may require special attention and adapted methods to assess hazard, exposure, and risk—such as the methods presented in this book. Regulators around the world have combined several elements: size ranges, cut-offs, intention, size-specific properties, and/or properties that guide the hazard-assessment such as solubility (Table 3.1). The European Cosmetics Directive, the

TABLE 3.1
Definitions of Nanomaterials for Regulatory Purposes

Organization	Restricted to Size Range	Distribution Threshold and Metrics	Is "Solubility" an Element of the Definition?	Aggregates and Agglomerates Included?	Restricted to "Intentionally Manufactured/Engineered" Materials?	Are "Novel Properties" an Element of the Definition?
European Commission Cosmetics Directive	1–100 nm	No specific mention	Yes	Yes	Yes	No
European Commission Biocides Directive	1–100 nm	50% by number	No	Yes	No	No
European Commission (recommendation for a definition)	1–100 nm	50% by number	No	Yes	No	No
Australian Government Department of Health and Ageing	1–100 nm	10% by number	No	Yes	Yes	Yes
Health Canada	1–100 nm and larger	No specific mention	No	Yes	Yes	Yes
United States Food and Drug Administration	1–100 nm and larger	No specific mention	No	No specific mention	Yes	Yes
United States Environmental Protection Agency	1–100 nm	10% by weight	No	Yes	Yes	Yes
French Ministry of Ecology, Sustainable Development, Transport and Housing	1–100 nm	50% by number	No	Yes	Yes	No
Taiwan Council of Labor Affairs	1–100 nm and larger	No specific mention	No	No specific mention	Yes	Yes

European Biocides Directive, and the French Decret have regulatory status, whereas the other definitions have advisory status. The French Decret refers to the European Commission (EC) nanodefinition recommendation (Table 3.1) and narrows the scope by restriction to "intentionally produced" materials, but widens the scope by sub-kg notification thresholds. Particulate materials exhibit a distribution of particle sizes, are often polymorph, and in most cases several particles are sintered together to aggregates, which further agglomerate. Particles "bound" in agglomerates or aggregates are explicitly included—this is the first major analytical challenge. The other major analytical challenge is the number metrics employed for size distributions and cut-off criteria. Only the United States Environmental Protection Agency has stuck to the well-established mass metrics, for which standardized methods exist. All definitions refer to the ISO size range of 1–100 nm. To exclude molecular structures, "particle" structures are mentioned, and additionally macromolecules such as polymers and proteins are excluded. The upper size limit is softened in the Health Canada definition, which considers any manufactured substance or product and any component material, ingredient, device, or structure to be a nanomaterial if it is at or within the nanoscale in at least one external dimension, or has an internal or surface structure at the nanoscale, or if it is smaller or larger than the nanoscale in all dimensions and exhibits one or more nanoscale properties/phenomena.

The EC recommendation for a definition has possibly the broadest scope because it refers to number metrics and includes natural, incidental, and manufactured materials containing particles (Table 3.1) (EC 2011), and provides the framework for sector-specific definitions (Bleeker et al. 2013). However, it clearly expresses that classification as a nanomaterial does not imply that the material has a specific risk or new hazardous properties (EC 2011). The number metrics is the game-changing decision that inflates the scope. The first estimates of the impact of the EC nanodefinition recommendation by BiPRO and Öko-Institut expect about 2000–5000 substances in scope; 80,000–160,000 preparations and 800,000–1,300,000 articles alone in Belgium. About 35,000–45,000 enterprises (15%–20% of all enterprises in Belgium) would be affected from the following sectors: cosmetics, healthcare, food and feed, coatings and inks, cleaning and disinfection, tires and rubber products, plastic products, building and construction, textiles, paper and wood products, sporting goods, electronics, and so on (BiPRO 2013). Although the EC nanodefinition recommendation does not specify material composition, an understanding of correlated nano-object size and composition will be necessary, especially for formulated products with several particulate substances mixed (Brown et al. 2013). In the absence of a technical guidance, the following sections review the established and emerging methods that determine size distributions in number metrics, and propose practical elements of a technical guidance.

3.2 METHODS TO MEASURE THE SIZE DISTRIBUTION IN NUMBER METRICS: FOR POWDERS

Electron microscopy is generally accepted as the reference counting method (Figure 3.1) for the size distribution of particulate materials. For reasonably well-dispersible powders, transmission electron microscopy (TEM) can be performed on monolayer preparations, such that the remaining uncertainty is the attribution of an

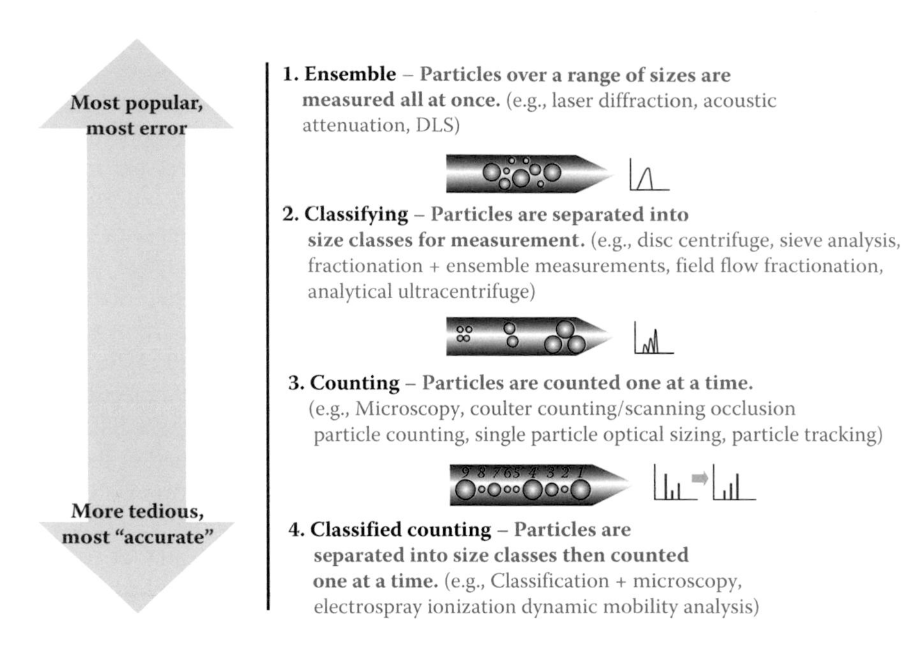

FIGURE 3.1 Systematization of methods to assess the size distribution of particulate materials in number metrics. (From Brown, S.C. et al., *Environ Health Perspect*, 121, 1282–1291, 2013.)

external diameter to irregularly shaped primary particles (Linsinger et al. 2012) and the counting of small agglomerates of nonseparated primary particles. Allen (1997a, 140ff) gives a comprehensive overview of the sample preparation techniques for TEM. The current emerging techniques for TEM preparation include electrospray ionization, ultrasonic dispersion, or ultrasonic evaporation (Lenggoro et al. 2006; Li et al. 2011; Pease et al. 2010; Taurozzi et al. 2010). For indispersible materials (e.g., those consisting completely of aggregates, or those designed for matrices incompatible with TEM), scanning electron microscopy (SEM) is applicable but often requires a multiple of time for image evaluation (compare ISO 13322-1). Currently, there are only exploratory methods to determine the distribution of the thickness of platelets; AFM (e.g., through ASTM E2859-11) could be an option, but has not been validated (Baalousha and Lead 2013). Further method development needs for electron microscopy are discussed by Brown et al. (2013).

The specific case of a rod-shaped iron oxide hydrate with the color index Pigment Yellow 42 highlights the difficulties in identifying the "boundary" of individual rods (Figure 3.2a). However, the same TEM grid displays also larger agglomerates (Figure 3.2b), which easily attain nonidentifiable structure or simply lose transparency for the electron beam. Hence these larger agglomerates must be excluded from automated evaluation, possibly introducing a sampling bias. Selecting only the small agglomerates, a semiautomated image analysis of 282 primary particles was performed, resulting in the distribution shown in Figure 3.2c. In order to enable comparisons to other methods (see the following), in this case the longer axis of the

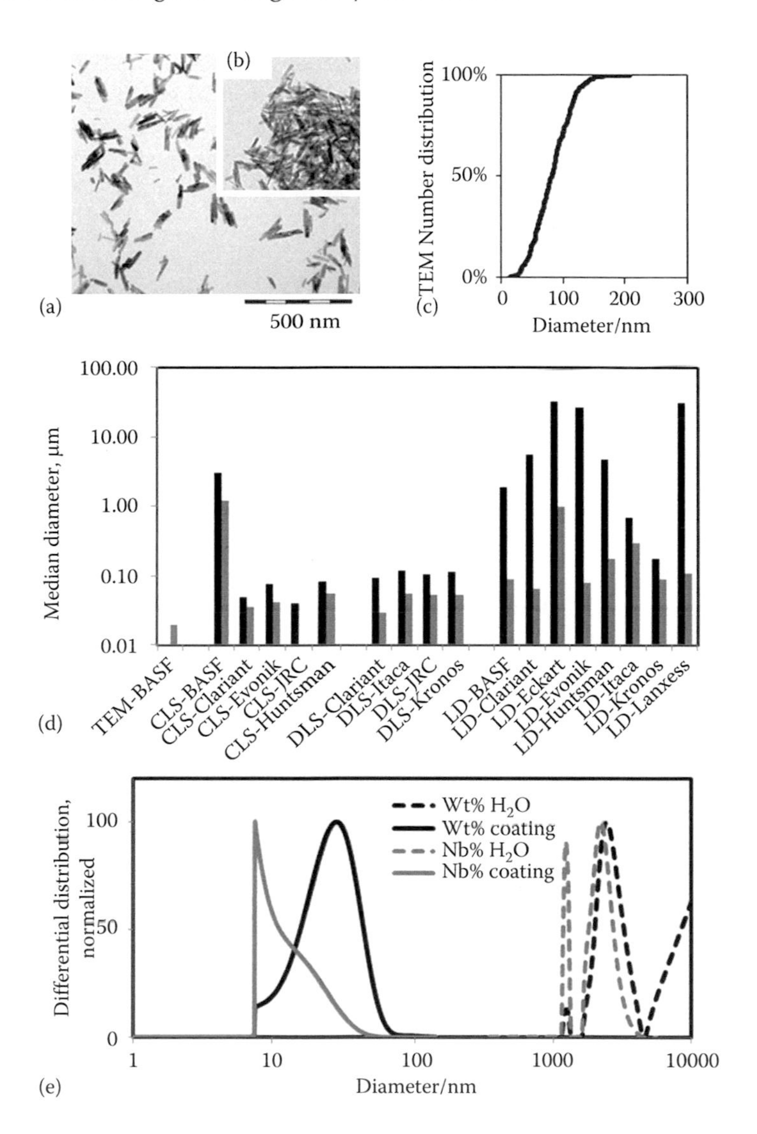

FIGURE 3.2 Classification results from multiple labs and several methods on an identical batch of iron oxide hydrate Pigment Yellow 42. (a) Monolayer-transmission electron microscopy (TEM) images after dipping the grid in the sample that was wetted by ethanol; (b) another sector on the same grid, which had to be excluded from evaluation; (c) TEM size distribution resulting from semiautomated evaluation of 282 primary particles; (d) median diameter in the originally measured metrics (black) and after conversion to number metrics (grey). (Replotted from Gilliland D, Hempelmann U (2013) Inter-laboratory comparison of particle size distribution measurements applied to industrial pigments and fillers. Presented to ECHA (European Chemicals Agency), October 2, 2013.) The values of TEM-BASF originate from plot (c), and the values of CLS-BASF originate from plot (e) dashed lines; (e) full size distributions from centrifugal liquid sedimentation (CLS) in mass metrics (black) converted to number metrics (grey), with powder dispersed by probe sonication in H_2O with a dispersant (dashed lines) or dispersed by application-specific shaking in organic coating (solid lines).

rods was determined, finding a median (D50) of 82.7 nm in number metrics, with a standard deviation of 30.5 nm, and min/max values of 13 nm/208 nm. In summary 58% of the primary particles are below 100 nm. For the smaller axis even a smaller D50 of 20 nm and even higher percentages below 100 nm result as the decisive criterion. Without doubt, this is a nanomaterial in the sense of the EC nanodefinition recommendation.

But a drastically simpler method also performs well for powder materials: Volume specific surface area (VSSA) was acknowledged as an agglomeration-tolerant ensemble method (Figure 3.1) with low cost and wide availability to identify nanomaterials (Allen 1997b; Kreyling, Semmler-Behnke, and Chaudhry 2010). VSSA has the important advantage over classifying and counting techniques (including TEM) that it does not involve dispersion protocols and achieves few-percent precision (Hackley and Stefaniak 2013). In the specific case of the iron oxide pigment, the mass-specific surface area is 80 m^2/g, as determined by the BET method; multiplied by the mass density of 3.9 g/cm^3, one obtains the VSSA of 312 m^2/cm^3 and a sphere-equivalent diameter of 20 nm, in agreement with the TEM evaluation. In a pilot round robin, the VSSA measurements from six labs were reproducible within a scatter of 10% on this substance (Gilliland and Hempelmann 2013). There are concerns that surface modifications on particles may dominate the BET and thus induce false positive classifications, but in such cases the number size distributions are to prevail (EC 2011).

Further, false negative classifications by VSSA occur only for platelet-shaped particles (Gilliland and Hempelmann 2013), suggesting that VSSA could also be a reliable method to screen for non-nanomaterials in a full testing strategy.

3.3 METHODS TO MEASURE THE SIZE DISTRIBUTION IN NUMBER METRICS: FOR SUSPENSIONS

The vast majority of nonmicroscopy based techniques for particle size measurement (Figure 3.1)

1. Provide size distributions in terms of an equivalent spherical diameter that is an average and not a minimum dimension measurement.
2. Interpret agglomerates and aggregates as individual particles.

Round-robin exercises (Lamberty et al. 2011) have typically employed relatively monodisperse materials that may not convey the complications associated with many real industrial materials, and have not substantiated the interexchangeability of number and volume distributions. Fractionating, counting, and microscopic techniques can still quantify smaller particles in the presence of agglomerates: Methodical comparisons on deliberately mixed multimodal distributions have evidenced that fractionating methods (Figure 3.1) successfully identified nanomaterials whereas ensemble methods failed and the innovative counting methods were too operator-dependent (Anderson et al. 2013; Wohlleben 2012). However, validation on well-dispersed spherical particles neglects irregular shapes and incomplete dispersion. A pilot round robin has compared the ensemble techniques (Figure 3.1) of Dynamic Light Scattering (DLS), Laser Diffraction (LD), and Centrifugal Liquid

Sedimentation (CLS, acc. to ISO13318) on eight pigments and fillers with a standardized dispersion protocol based by probe sonication in water with a dispersant (Gilliland and Hempelmann 2013). Focusing on the same iron oxide Pigment Yellow 42 as previously noted, Figure 3.2d presents the TEM, CLS, DLS, and LD results. With LD, the resulting D50 in number metrics originates from less than 5% of the originally measured LD distribution in volume metrics, which is fully within the error margin. There is hence little confidence on LD results for partially agglomerated pigments. With one exception, the results obtained with CLS are rather uniform, even though different manufacturers with different optics (X-ray and turbidity) (Planken and Cölfen 2010; Wohlleben 2012) were used. The BASF_CLS measurement was dominated by agglomerates, and is displayed in Figure 3.2e as dashed lines. One identifies several systematic sources of error: deviations in true size distribution after dispersion +, mathematical smoothing between raw data and size distribution +, sphere-equivalent conversion from volume to numbers. Error propagation and amplification of noise are major concerns since a 1% error in the ability of a method to accurately describe a mass or volume distribution at the nanoscale could translate into more than 50% error in a number distribution. An example of such an error propagation is displayed in Figure 3.2e, where a kink in the originally measured black solid line (mass metrics) introduces a seemingly dominant peak of the grey solid line (number metrics) at the lowest diameters, which is neither plausible for material chemistry nor compatible with TEM images in Figure 3.2a.

In summary from the experiments discussed here, VSSA and CLS are the preferred methods due to the low scatter between laboratories and due to inherently agglomerate-tolerant principles, but for CLS the dispersion protocol and metrics conversion must be further enhanced, as discussed in the following section.

Apart from the aforementioned techniques introduced, many alternatives (Figure 3.1) have been discussed and ranked elsewhere (Allen 1997a; Anderson et al. 2013; Brown et al. 2013; Calzolai, Gilliland, and Rossi 2012; Linsinger et al. 2012). A realistic roadmap from 2013 to 2017 will plan for only incremental improvements for techniques that have been established since decades with several commercial suppliers. The dynamic range of Field-Flow-Fractionation (FFF) per run condition will likely remain limited and multiple spacers, membranes, and run conditions would be required to access a dynamic range from a few nm to a few µm (Baalousha and Lead 2012; von der Kammer et al. 2011). Also the uncertainties of the optical detection in CLS will remain limiting at least for the low-cost commercial equipment, whereas Analytical Ultracentrifugation (AUC) with interference detection does resolve the issue but is still too expensive as an investment (Cölfen et al. 2009; Schäfer et al. 2012; Wohlleben 2012). The coupling of innovative, even counting detectors such as single-particle inductively coupled plama mass spectrometry (sp-ICP-MS) or condensation particle counter (CPC) to a fractionation channel does not resolve the necessity for dispersion. Major improvements in the applicability of FFF and CLS are expected for software tools that perform validity checks (see the following section) and circumvent the commercial conversion from raw data to smoothed volume distributions. Instead, these software tools would perform a direct error propagation to number metrics.

Especially the counting techniques are still in their infancy, often with only one commercial supplier and no standardization but several research projects; this situation is promising, but not more: Both Microchannel Resonators (e.g., Archimedes from Affinity) (Lee et al. 2010) and Scanning Ion Occlusion Spectroscopy (or Transient Resistive Pulse Sensing [TRPS], e.g., qnano from Izon) (Anderson et al. 2013) need to reduce the lower detection limits from current 70 nm to 10 nm, which—by their measurement principle—will further limit the upper diameter range, so that the entire distribution of unknown samples may not be accessible. Nanoparticle Tracking Analysis (NTA, e.g., from NanoSight or MicroTrac, see also ASTM E2834-12) is fundamentally based on localization of scattered light, whereas highly relevant particles such as 10 nm SiO_2 may never be detectable in a realistic environment. Electrospray-Dynamic Mobility Analyzer-Condensation particle counter (ES-DMA-CPC, several suppliers, see also ISO/DIN 27891:2013) has not been tested for the purpose and may suffer from salt impurities present in realistic products. sp-ICP-MS, like Microchannel resonators, measures mass and infers size assuming a density and shape.

But most fundamentally, all techniques other than VSSA and SEM require a predispersion via liquid or aerosol.

3.4 SAMPLE PREPARATION FOR FRACTIONATING AND COUNTING TECHNIQUES

The size distribution of suspensions or aerosols must be dominated in number metrics by individualized primary particles. Note that this validity criterion may still hold if primary particles do not dominate in volume or mass metrics. A content of agglomerates is acceptable if their presence does not compromise the quantification of the individualized primary particle by a specific technique. Pilot round robins with standardized dispersion protocols resulted generally in a high content of large agglomerates (cf. the LD data in Figure 3.2d) despite high ultrasonication energies. Some of the labs obtained significant contents of small agglomerates or aggregates (cf. the CLS data in Figure 3.2d), which allowed superficially correct identification as nanomaterials, but the disagreement with TEM or VSSA diameters (Figure 3.2c vs. Figure 3.2e dashed lines) remained significant (Gilliland and Hempelmann 2013).

However, the specific product is marketed by its performance as a "transparent" pigment, hence devoid of scattering agglomerates, and how to obtain the optimal dispersion is well known. DIN 53238-13 and DIN EN ISO 8781-1 specify the dispersion in low-viscous media by shaking with milling balls to determine the color strength of pigments. An iron oxide pigment identical to that mentioned previously was added (about 4.9%) into the EL2 organic coating and shaken for 1 h, then diluted in xylol to 0.05% and measured by CLS according to ISO13318. The size distribution in number metrics has a D50 of 14 nm (Figure 3.2e solid lines), in excellent agreement with the TEM value of the smaller axis, and the specific surface of 75 m^2/g derived from the CLS measurement is in excellent agreement with the BET value of 80 m^2/g. These values can only coincide if the size distribution is dominated by individualized primary particles, and provide a strong validity proof.

Considering that the EC nanodefinition differentiates between aggregates (nano) and nanoporous materials (non-nano) without a structural delimitation, some Associations have proposed a delimitation inspired by the lifecycle perspective

(Bleeker et al. 2013): The "best possible dispersion" achievable during the intended manufacture and use could be the basis of the measurement. For example, pigments would be dispersed in the coating where they achieve performance, but indispersible aggregates would be treated as constituent particles in the sense of the ISO working item of TC 24/SC 4 which defines "Constituent particles of agglomerates [...] are aggregates of fused together or elsewise combined former primary particles." However, an application of this approach to organic and inorganic pigments reveals that the successful dispersion in Figure 3.2c is an exception, whereas most engineered nanomaterials remain aggregated to diameters factors 2 to 10 above the primary particle size, and are no longer classified as nanomaterials. In summary, the "best possible dispersion" approach appears like a pragmatic technical guidance, but is in fact a major restriction of the scope of the nanodefinition. The best possible dispersion approach as very first step of risk screening (Chapter 16) is defendable if a low degree of dispersability is sufficient to suppress other effects due to the nanoscale structure, such as those from solubility, surface, or reactivity.

3.5 TIERED TESTING FOR COST-EFFECTIVE SCREENING AND CONFIRMATION

Calzolai et al. have recognized that "At the moment there is no single technique that can by itself provide a robust analytical method" (Calzolai, Gilliland, and Rossi 2012). Guidance on sample preparation, method selection, and evaluation is required. A rudimentary scheme of a tiered measurement strategy has been proposed recently (Figure 3.3) (Brown et al. 2013), and a slightly refined variant has been introduced in ISO as a new working item (Hayashi 2013). The key to validity is a quick validation of dispersion quality before measurement, benchmarked on expectations from VSSA or simple EM. If the size distribution in number metrics is dominated by constituent particles, an agglomeration-tolerant measurement technique will give the correct classification even if agglomerates are a significant fraction of the total mass. Alternative to the validation iteration, a future guidance could refer to the lowest degree of dispersion that is obtainable in the commercial application, as demonstrated by the example of pigments in coatings (Figure 3.2) and could establish standardized, sector-specific dispersion protocols.

Note that these schemes aim at the EC nanodefinition recommendation, and hence do not allow a direct path from VSSA <60 m^2/cm^3 to "not nano." Given that for a wide range of materials the VSSA criterion performs well to identify non-nano substances, it deserves a more prominent place as a refining criterion to make the EC nanodefinition practical. Another refining criterion could specify the volume% below 10 μm or any other cut-off (Brown et al. 2013). At the very least, these refining criteria can be used to group materials. As an example, one could group all $CaCO_3$ fillers with less than 50% in volume metrics below 1 μm diameter; only the $CaCO_3$ substance with the lowest average diameter would be measured by EM to confirm that it is not a nanomaterial, and this classification applies to the entire group then. Alternatively, also the VSSA will serve as an excellent grouping criterion since it has a proven sensitivity to the primary particle size (Gilliland and Hempelmann 2013) and proven precision (Hackley and Stefaniak 2013). A much more refined decision

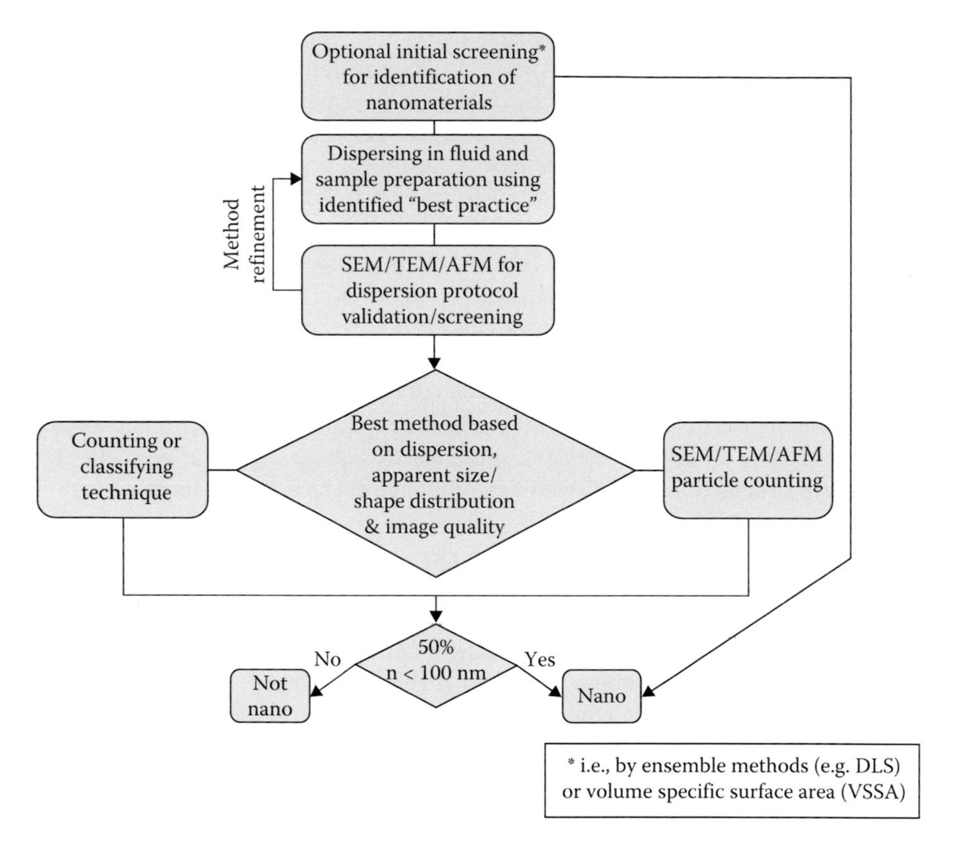

FIGURE 3.3 Draft of a tiered testing strategy that guides the user through a choice of techniques to assess the size distribution by microscopic or nonmicroscopic techniques. (From Brown, S.C. et al., *Environ Health Perspect*, 121, 1282–1291, 2013.)

tree, including the dispersion protocol measurement method and error-propagating evaluation software, will be delivered by the NanoDefine project (2013–2017).

3.6 CONCLUSION

It has been argued that nanomaterials should not be defined as long as we do not have a mechanistic understanding of size-related hazards, because otherwise we may fail to include some hazardous materials (Maynard 2011). Implicitly, such a definition characterizes the term nanomaterial by small size and elevated hazard, but how should one determine both elements by adopted methods if none precedes the other? The regulators have decided not to strive for a risk-based definition, and have instead established rather inclusive size-based definitions (Table 3.1). Nonetheless, size criteria (Auffan et al. 2009) together with reactivity (Zhang et al. 2012) and other triggers will be ingredients of QSARs that screen for hazardous materials among the nanomaterials (compare Chapter 16).

In the lack of technical guidance on the terms "novel," "intentional," "manufactured," "engineered," these cannot be enforced and the next iteration of guidance should replace these terms by a well-defined technical term that is at least neutral toward innovation. One can also anticipate that the development of methods and tiered guidance from now until 2017 will suggest modifications to the elements of the definitions, based on field evidence of materials, products, methods, and costs. In view of severe persisting limitations, the technical guidance must be based on a substance- and method-specific decision-tree and measurement strategy, such as developed by the NanoDefine FP7 project by 2017.

Some very first modifications to be implemented by a suitable guidance have been addressed in this chapter, including the practical but scope-limiting "best possible dispersion" approaches, the inconsistent treatment of aggregates and nanoporous materials, and the wider use of techniques other than EM, including further the general validity ranges of methods for powders and suspensions, the concepts of grouping and tiered measurement strategies with integrated validity checks.

REFERENCES

Allen T (1997a) Particle size measurement - vol. 1: Powder sampling and particle size measurement. Chapman & Hall, London.

Allen T (1997b) Particle size measurement - vol. 2: Surface area and pore size determination. Chapman & Hall, London.

Anderson W, Kozak D, Coleman VA, Jämting K, Trau M (2013) A comparative study of submicron particle sizing platforms: Accuracy, precision and resolution analysis of polydisperse particle size distributions. *J. Colloid Interfac. Sci.* 405:322–330.

Auffan M, Rose J, Bottero JY, Lowry GV, Jolivet JP, Wiesner MR (2009) Towards a definition of inorganic nanoparticles from an environmental, health and safety perspective. *Nat. Nanotech.* 4:634–641.

Baalousha M, Lead JR (2012) Rationalizing nanomaterial sizes measured by atomic force microscopy, flow field-flow fractionation, and dynamic light scattering: Sample preparation, polydispersity, and particle structure. *Environ. Sci. Technol.* 46:6134–6142.

Baalousha M, Lead JR (2013) Characterization of natural and manufactured nanoparticles by atomic force microscopy: Effect of analysis mode, environment and sample preparation. *Colloids Surf. A: Phys. Eng. Aspects* 419:238–247.

BiPRO (2013) *Study of the Scoping of a Belgian National Register for Nanomaterials and Products Containing Nanomaterials*. Federal Public Service Health, Food Chain Safety and Environment (FPS), Belgium.

Bleeker EAJ, De Jong WH, Geertsma RE, Groenewold M, Heugens EHW, Koers-Jacquemijns M, Van De Meent D, Popma JR, Rietveld AG, Wijnhoven SWP, Cassee FR, Oomen AG (2013) Considerations on the EU definition of a nanomaterial: Science to support policy making. *Regulat. Tox. Pharmacol.* 65:119–125.

Brown SC, Boyko V, Meyers G, Voetz M, Wohlleben W (2013) Towards advancing nano-object count metrology–A best practice framework. *Environ. Health Perspect* 121:1282–1291.

Calzolai L, Gilliland D, Rossi F (2012) Measuring nanoparticles size distribution in food and consumer products: A review. *Food Addit. Contam.: Part A* 29:1183–1193.

Cölfen H, Laue TM, Wohlleben W, Karabudak E, Langhorst BW, Brookes E, Dubbs B, Zollars D, Demeler B (2009) The open AUC project. *Eur. Biophys. J.* 39:347–360.

EC (2011) Commission Recommendation of 18 October 2011 on the definition of nanomaterial. http://eur-lex.europa.eu/LexUriServ/LexUriServ.do?uri = OJ:L:2011:275:0038:0040:EN:PDF (accessed January 12, 2012).

Gilliland D, Hempelmann U (2013) Inter-laboratory comparison of particle size distribution measurements applied to industrial pigments and fillers. Presented to ECHA, 02.10.2013

Hackley VA, Stefaniak AB (2013) Real-world precision, bias, and between-laboratory variation for surface area measurement of a titanium dioxide nanomaterial in powder form. *J. Nanopart. Res.* 15:1–8.

Hayashi M (2013) Tiered approach for Identifying Nanomaterials. ISO/TC 229/WG 4 N 208. 17-6-2013.

Kreyling W, Semmler-Behnke M, Chaudhry Q (2010) A complementary definition of nanomaterial. *Nanotoday* 5:154–168.

Lamberty A, Franks K, Braun A, Kestens V, Roebben G, Linsinger T (2011) Interlaboratory comparison for the measurement of particle size and zeta potential of silica nanoparticles in an aqueous suspension. *J. Nanopart. Res.* 13:7317–7329.

Lee J, Shen W, Payer K, Burg TP, Manalis SR (2010) Toward attogram mass measurements in solution with suspended nanochannel resonators. *Nano Lett.* 10:2537–2542.

Lenggoro IW, Lee HM, Okuyama K (2006) Nanoparticle assembly on patterned "plus/minus" surfaces from electrospray of colloidal dispersion. *J. Colloid Interface Sci.* 303:124–130.

Li M, Guha S, Zangmeister R, Tarlov MJ, Zachariah MR (2011) Method for determining the absolute number concentration of nanoparticles from electrospray sources. *Langmuir* 27:14732–14739.

Linsinger T, Roebben G, Gilliland D, Calzolai L, Rossi F, Gibson N, Klein C (2012) Requirements on measurements for the implementation of the European Commission definition of the term "nanomaterial." *EUR* 25404 EN.

Maynard AD (2011) Don't define nanomaterials. *Nature* 475:31.

Pease LF, Tsai DH, Hertz JL, Zangmeister RA, Zachariah MR, Tarlov MJ (2010) Packing and size determination of colloidal nanoclusters. *Langmuir* 26:11384–11390.

Planken KL, Cölfen H (2010) Analytical ultracentrifugation of colloids. *Nanoscale* 2:1849–1869.

Schäfer J, Schulze C, Marxer EEJ, Schäfer UF, Wohlleben W, Bakowsky U, Lehr CM (2012) Atomic force microscopy and analytical ultracentrifugation for probing nanomaterial protein interactions. *ACS Nano* 6:4603–4614.

Taurozzi JS, Hackley VA, Wiesner MR (2010) Ultrasonic dispersion of nanoparticles for environmental, health and safety assessment – Issues and recommendations. *Nanotoxicology* 5:711–729.

von der Kammer F, Legros S, Larsen EH, Loschner K, Hofmann T (2011) Separation and characterization of nanoparticles in complex food and environmental samples by field-flow fractionation. *Trends Anal. Chem.* 30:425–436.

Wohlleben W (2012) Validity range of centrifuges for the regulation of nanomaterials: From classification to as-tested coronas. *J. Nanopart. Res.* 14:1300.

Zhang H, Ji Z, Xia T, Meng H, Low-Kam C, Liu R, Pokhrel S, Lin S, Wang X, Liao YP, Wang M, Li L, Rallo R, Damoiseaux R, Telesca D, Mädler L, Cohen Y, Zink JI, Nel AE (2012) Use of metal oxide nanoparticle band gap to develop a predictive paradigm for oxidative stress and acute pulmonary inflammation. *ACS Nano.* 6:4349–4368.

4 Analyzing the Biological Entity of Nanomaterials

Characterization of Nanomaterial Properties in Biological Matrices

Christian A. Ruge, Marc D. Driessen, Andrea Haase, Ulrich F. Schäfer, Andreas Luch, and Claus-Michael Lehr

CONTENTS

4.1 INTRODUCTION

4.1.1 WHY IS IT IMPORTANT TO CHARACTERIZE NANOMATERIALS *IN SITU*?

The complete and careful analysis of the test item is a well-accepted paradigm, a prerequisite for any type of toxicological testing. This also applies for nanotoxicological studies. Of course, the issue of how to characterize particles appropriately is not completely new. Particle science has dealt with all aspects of particle characterization for many years, and nanotoxicology can—and should—use this expertise. As summarized by Powers et al. (2007), this starts with sampling, as the analyzed fraction of the sample needs to be a representative of the material under study. This becomes especially important when the particles have a broad size distribution. For the same reason, many particles (e.g., several thousands) need to be analyzed to obtain a reliable statistics, and the sample should be analyzed as close as possible to the relevant conditions (e.g., in biological environment), and just before the experiment (Powers 2005). However, in many of the older studies particle characterization was only conducted on a very minimalistic level. Therefore, questions were raised about the meaning of such studies in the absence of adequate material characterization (Warheit 2008). Nowadays there is a large consensus that a complete and extensive characterization of the nanoparticles is an essential starting point for any type of study. This has been intensively discussed in many publications (Jiang, Oberdörster, and Biswas 2008; Murdock et al. 2008) and summarized in several review articles (Sayes and Warheit 2009; Powers 2005; Powers et al. 2007), as well as guidance documents (OECD 2012). Most of these documents and articles give an excellent overview on which characterization data are needed and which techniques are suitable to obtain them. This refers to different physical–chemical properties, such as chemical composition, size, size distribution, shape, surface, or surface charge. But as pointed out by Warheit currently "this recommendation becomes a 'laundry list' of physicochemical characteristics and does not have adequate prioritization" (Warheit 2008).

Additionally, many studies do not pay attention to nanomaterial characterization under the conditions of use, for example, in a biological context. Instead, the nanomaterials are characterized as they have been received, in most of the cases either as a powder or as an aqueous dispersion. This is referred to as characterization of materials *as produced* or *as synthesized* (also refer to Chapter 1). However, nanomaterials may change their characteristics over their entire lifecycle. A nanomaterial that is shipped as a dry powder needs to be dispersed before any use. As soon as the nanoparticles get into contact with any type of biological matrix they will interact with biomolecules of the matrix and change their properties with respect to size (e.g., agglomeration), adsorption of biomolecules onto the surface (i.e., formation of a so-called corona), and partial dissolution (i.e., release of ionic species). This clearly means that the nanomaterial that finally reaches the cell surface will have interacted, for example, with serum-containing cell culture medium in the case of *in vitro* test conditions or with lung lining fluid, gastric juices, or other body fluids under *in vivo* conditions, and certainly the characteristics of the material will not be of the same type as it has been produced, and extensively characterized. Basically, the same holds true for any nanomaterial that is formulated into a final product, where also many changes will occur, which however will not be covered within this chapter. Most importantly, if changes in physicochemical properties of a nanomaterial *in situ* are not taken into consideration, one may actually misinterpret results from toxicity testing.

To overcome this problem, Sayes and Warheit suggest a three-phase approach for nanoparticle characterization with respect to nanotoxicological studies (Sayes and Warheit 2009). Phase 1 covers the powder state and includes the assessment of chemical composition, size, size distribution, surface area, and morphology. Phase 2 covers the characterization of the dispersions and includes tests on agglomeration or aggregation, as well as the formation of reactive species. Phase 3 then includes the nano–bio interface. Basically, it may be concluded that a nanomaterial needs to be characterized at all stages of its lifecycle starting with the synthesis, followed by particle characterization in aqueous media and then after coming into contact with biological matrices and finally even inside cells or tissues. Typically, it will be easily feasible to study nanomaterials in simulated biological fluids as we will explain later on in this chapter in detail. In contrast, characterization inside cells or *in vivo* is extremely difficult to achieve due to technical limitations. Therefore, characterization truly *in situ*, inside cells or tissues, is currently out of scope for most studies, but might become feasible in a few years.

Direct characterization of nanoparticles *in situ*, for example, directly in the test medium or simulated biological fluids has been extensively described in several publications (Sayes et al. 2008; Powers et al. 2007; Montes-Burgos et al. 2010; Maskos and Stauber 2011). This is referred to as *in situ* characterization in contrast to the characterization "*as produced*" or "*as synthesized*," although some of the techniques applied for so-called *in situ* characterization are not truly *in situ* but *ex situ* as we will explain in detail later on. To summarize, nanomaterial characterization may not only be viewed from a materials science perspective but nanomaterials in a biological environment do actually represent a biological entity, which consists of the nanomaterial with the adsorbed layers of biomolecules. This is a highly dynamic complex that will change over time.

Changes observed when introducing a nanomaterial from a pure abiotic into a biological environment are interconnected and dependent on each other. For instance, if a nanomaterial strongly interacts with proteins tightly covering its surface, this often leads to an increased particle agglomeration. On the opposite, particles that are repellent to proteins might stay more stably dispersed. All these interactions with the matrix or medium components not only change the physicochemical properties such as size or zeta-potential but also strongly influence the interactions with living systems and thus also the outcome of toxicity tests. Thus, these *in situ* properties need to be determined not only for a full characterization but also regarding a mechanistic understanding of nanomaterial behavior in interaction with living systems. The hypothesis underlying the nanoGEM work package APQ, dealing with the *in situ* characterization of nanomaterials, is that only a full characterization of nanoparticles in the biological system would enable us to derive structure–activity relationships. Before we go into details of how interactions with nanoparticles and the biological environment can be studied, and how this challenging task was dealt with in the project nanoGEM, we would like to spend a few more words on how such interactions will influence the interactions with cells.

As pointed out by the group of Kenneth Dawson, the entity that cells will finally interact with will not be the pristine nanomaterial but a nanomaterial covered by several layers of biomolecules (Lynch et al. 2007; Lynch, Salvati, and Dawson 2009; Walczyk et al. 2010). Obviously, the cells do not "see" the original nanomaterial surface; the first contact actually occurs via the biological molecules on the nanoparticle surface, and the nanoparticle–protein complex thus represents a biological entity (Lynch et al. 2007). This so-called protein corona enables completely different types of interactions compared to a pristine nanomaterial surface, for example, protein–protein interactions with different cell surface receptors, which will influence the nanoparticle uptake. Meanwhile there is clear evidence from several groups that nanomaterials uncoated with any biomolecules (for example, experiments performed *in vitro* without any serum addition to the cell culture medium) compared to nanomaterials coated with biomolecules (i.e., experiments performed *in vitro* with serum-containing cell culture medium) have a different uptake rate into cells (Petri-Fink et al. 2008; Bajaj et al. 2009; Lesniak et al. 2012). For instance, Petri-Fink and coworkers showed that for iron oxide nanoparticles (A-PVA-SPION) the presence of serum strongly inhibited cellular uptake due to formation of a protein corona (Petri-Fink et al. 2008). In contrast, Lesniak and coworkers showed that silica nanoparticles in the absence of serum adhered stronger to cell membranes and were internalized with a higher efficiency (Lesniak et al. 2012). But apart from affecting uptake levels the protein corona also influenced the intracellular nanoparticle location and the impact on cells.

Interactions of proteins with nanoparticle surfaces may lead to partial unfolding of the proteins, displaying epitopes that have previously been hidden inside the proteins on the protein surface. These so-called "cryptic epitopes" may be recognized by specific receptors, and may also lead to adverse effects (Lynch, Dawson, and Linse 2006), for example, by activation of the immune system. In addition, some limited data support the hypothesis that some nanoparticles may behave as haptens and turn antigenic when bound to a protein. Other data suggest that some nanoparticles have very efficient adjuvant properties most probably due to the fact that they create a local reservoir at the injection site leading to the protection and slow release of antigens (Zolnik et al. 2010).

Finally, the interaction with biomolecules may also influence agglomeration of nanoparticles and thus affecting sedimentation under *in vitro* or biodistribution under *in vivo* test conditions. If nanomaterials are ranked according to their toxic effects *in vitro* or *in vivo*, one needs to consider the *in situ* characterization data taking into account the dose that will finally reach a cell or a specific tissue, which is also dependent on agglomeration state, and not rank only according to the dose applied to the cell culture dish or to the animal.

To summarize, for a detailed understanding of such nano–bio interactions occurring *in vitro* and *in vivo* and therefore also for interpreting toxicity results, a complete *in situ* characterization of each nanomaterial under the specific test conditions is absolutely essential, which has been widely recognized nowadays (Warheit 2008; Sayes and Warheit 2009; Montes-Burgos et al. 2010).

In this chapter, we will highlight important issues to consider when characterizing nanomaterials *in situ*, which here refers to the conditions under which *in vitro* or *in vivo* experiments are performed. First of all, we will focus on general aspects of characterizing nanomaterials in biological fluids, and discuss parameters influencing the interactions with biomolecules as well as methods suitable to study them. Furthermore we will explain nanomaterial interactions with serum, with lung lining fluid, and with gastric juices as examples.

4.1.2 Interaction of Nanoparticles with Proteins and Influencing Factors

The fact that proteins bind to materials immediately after they have been introduced to the body is actually a common phenomenon, and this topic has been extensively investigated in the last 30–40 years, mainly with respect to implants and other medical devices (Rabe, Verdes, and Seeger 2011). Although some principal rules could be established and a few strategies such as PEGylation preventing such interactions (e.g., antifouling) are available (Walkey and Chan 2012; Morra 2000), protein adsorption to surfaces is still a complex topic and not yet fully understood (Nakanishi, Sakiyama, and Imamura 2001). Protein adsorption to nanomaterials can in principle be considered as a special case of interactions of proteins with surfaces in general such that some of the known principles may be applied here as well (Walkey and Chan 2012; Gray 2004). However, there are some unique features of nanomaterials, which may lead to important differences. For instance, nanomaterials typically have a very large specific surface area often combined with high surface curvature (leading to a higher probability of defects in crystal structure on the surface), which in combination may cause a higher surface reactivity toward biomolecules. Furthermore, nanoparticles in contrast to medical implants are mobile, that is, they can translocate to different physiological compartments and distribute within the body after entry (Walkey and Chan 2012). With this migration of nanomaterials and consequently the exposure to different physiological compartments, the composition of the protein corona will change over time. This has been referred to as "evolution" of protein corona (Dell'Orco et al. 2010; Lundqvist et al. 2011).

The pattern of adsorbed proteins depends primarily on the physicochemical properties of the nanomaterial, such as chemical composition, size, shape, roughness, effective surface charge (i.e., zeta-potential), and hydrophobicity. The sum of these properties drives and defines the interplay at the so-called "nano–bio interface." The forces governing the adsorption of protein to nanomaterials are both long range van der Waals and electrostatic interactions, as well as short range interactions such as steric, hydrophobic, or charge effects (Nel et al. 2009). The size and shape of nanomaterials—both determining their surface curvature—can largely affect the identity of adsorbed proteins (Lundqvist et al. 2008; Tenzer et al. 2011). The binding of some proteins (e.g., fibrinogen) for instance depends on the space available for a protein domain to interact with the nanomaterial (Tenzer et al. 2011). Consequently, adsorption of a protein can occur in different orientations, which can even lead to a wrapping of proteins around a nanomaterial (Roach, Farrar, and Perry 2006; Shemetov, Nabiev, and Sukhanova 2012).

Charge-driven adsorption effects are strongly dependent on the pH of the dispersion medium and can trigger adsorption effects, too. The adsorption of proteins to nanomaterials again can alter their zeta-potential, leading to a collapse of repulsive stabilizing interactions (Shemetov, Nabiev, and Sukhanova 2012). Hydrophobicity is considered as one of the most important physicochemical parameters of nanomaterial surface regarding adsorption of proteins. On the atomic level, the presence of non-polar functional groups has a large influence on protein adsorption, as hydrophobic surfaces provoke the exposure of hydrophobic protein cores toward the nanomaterial surfaces for entropic reasons (Lundqvist et al. 2008; Gessner et al. 2000). Each protein has an intrinsic stability, that is, a thermodynamically favorable state. Any additional energy added to the system will cause an alternative state, being either reversible or irreversible, whereas the latter can be considered as (partial) denaturation of the protein (Shemetov, Nabiev, and Sukhanova 2012). Hydrophobicity-driven changes in the secondary structure can expose new epitopes of a protein toward cells, which potentially cause an immunological response toward the nanomaterial–protein complex. However, hydrophilic groups can also interact with proteins via hydrogen bond formation, which in contrast to hydrophobicity-driven adsorption is less disruptive (Shemetov, Nabiev, and Sukhanova 2012). This demonstrates that the interaction of proteins with nanoparticle surfaces may also influence the conformation of proteins, which in turn can influence protein–protein interactions and then lead to fibrillation, protein aggregation, or similar processes. For instance, enhanced fibrillation has been shown to occur for β2-microglobulin when interacting with polymeric nanoparticles (Linse et al. 2007), or suppressed fibrillation can occur as shown for amyloid-β protein when interacting with the same nanoparticles (Cabaleiro-Lago et al. 2008). Nanoparticle–protein interaction may also influence enzymatic activity (Wu, Zhang, and Yan 2009; Chakraborti et al. 2010). Different enzymes may change their activities to different extents when being adsorbed to nanoparticles. When chymotrypsin was adsorbed onto multiwalled carbon nanotubes (MWCNTs), it retained only 1% of its activity whereas peroxidase retained 30% (Wu, Zhang, and Yan 2009; Karajanagi et al. 2004). In contrast, when lysozyme was adsorbed onto ZnO nanoparticles, the activity was largely retained and in addition the enzyme was partially protected against unfolding effects occurring under elevated temperature conditions or when adding chaotropic agents (Chakraborti et al. 2010). Moreover, some proteins bind in a cooperative manner, that is, secondary proteins can bind to a nanomaterial–protein complex based on protein–protein interactions (Walkey and Chan 2012). Especially in complex fluids this may significantly contribute to not only the formation of the protein corona with respect to its initial formation but also its evolution over time. For instance, it is known that in complex biological fluids (e.g., plasma), primarily the highly abundant proteins bind to a nanomaterial quickly, which is referred to as kinetic formation of the corona. Later on they can be exchanged over time by proteins with higher affinity to the nanomaterial, which is then the thermodynamically stable corona. This is also referred to as the "Vroman effect" (Vroman et al. 1980). These dynamic changes in the composition of the protein corona add another dimension to the already complex situation. As nanomaterials can move through different physiological compartments the protein composition of the corona will change and therefore it is important to study the protein corona also over time. But importantly, the corona will never be completely exchanged; traces of the "primary" corona will always be retained such that the passage of a nanomaterial may be recapitulated via identification of the adsorbed biomolecules. The protein corona thus serves as a "fingerprint" and enables us to "read" the passage route of a nanomaterial.

Studying the protein corona completely, also over time, is highly important as adsorbed proteins may shield toxic nanomaterial properties or vice versa increase toxicity (i.e., by increasing cellular uptake), may trigger cellular effects (e.g., enhanced presentation of nanomaterials to cells via adsorbed proteins), or may cause interference with cellular signaling pathways (Johnston et al. 2012). Furthermore, protein adsorption is likely to lead to altered size distribution and state of dispersion of nanomaterials (Dawson, Anguissola, and Lynch 2012; Lesniak et al. 2012). The dispersion profiles of nanomaterials in relevant biological fluids (containing protein, lipid, sugars, and ions) may differ significantly from profiles in other buffers, as biomolecules and ions can shield electrostatic interactions between nanomaterials, and thereby lead to formation of agglomerates (Nichols et al. 2002; Maskos and Stauber 2011). Especially the state of dispersion is of tremendous importance to nanotoxicological studies, as size of a nanomaterial is the key parameter that in most of the cases inversely correlates with toxicity. Agglomerated nanomaterials may behave in a completely different way in test assays compared to nanomaterials distributed in the nanometer range, as agglomerated nanomaterials will sediment and are thus most likely to be available for cells to a larger extent, which of course also influences the toxicological endpoint (Dawson, Anguissola, and Lynch 2012).

A key step toward meaningful toxicity assessment of nanomaterials is therefore sample preparation, which means dispersion of the nanomaterials into an aqueous test medium. Some additives have been identified that are capable of stabilizing nanomaterials in aqueous media, such as albumin or citric acid (Schulze et al. 2008; Ramirez-Garcia et al. 2011). However, there is still a lack of harmonization of methods for dispersing nanomaterials, and profound differences can be observed depending on the composition of the dispersion medium, as well as dispersion procedure (Dawson, Anguissola, and Lynch 2012). The major challenges in nanotoxicological studies are still the selection and application of suitable methods to obtain meaningful and conclusive data (see Section 4.1.3), as well as the choice of suitable biological matrices representing the respective physiological compartment within the human body (see Section 4.1.4).

4.1.3 Methodological Approaches for *In Situ* Characterization of Nanomaterials

Characterizing nanoparticles in a given biological environment and especially determining their biomolecule corona is an analytical challenge. One needs to decide on a case-by-case basis which technique to use as there is no one-fits-all method available.

As there are excellent review articles available (Aggarwal et al. 2009; Mahmoudi et al. 2011), we will give only a brief overview here. Basically, there are two general approaches. It is possible to study the nanomaterial directly in the test conditions (*in situ*). On the other hand, it is also possible to reisolate the nanomaterial including adsorbed biomolecules for subsequent analysis (*ex situ*) (Walkey and Chan 2012; Maskos and Stauber 2011). However, in the literature the term *in situ* characterization of nanomaterials is often used for both *in situ* and *ex situ* approaches to study nanomaterials under test conditions in contrast to nanomaterial characterization *as-synthesized*, which is typically performed in water. As each method has limitations, typically a combination of different methods (*in situ* and *ex situ*) is needed to fully describe nanomaterials in complex (i.e., biological) matrices.

Suitable *in situ* methods are limited in number. It is possible to use methods that are suitable to study size and size distribution such as dynamic light scattering (DLS) or nanoparticle tracking analysis (NTA) to check for an increase in hydrodynamic radius due to formation of a protein corona (Montes-Burgos et al. 2010). DLS uses coherent laser light and measures the angular dependent scattering occurring from particles. The fluctuations over time will depend on particle movement, which is dependent on particle size. NTA uses the same main principle but does not measure the mean scattered light instead here each particle is followed individually (referred to as "tracking" of particles). Both approaches will deliver information on the size and size distribution and eventually also on the shape of the particles or on interactions between them. Care needs to be taken as some matrices will interfere with DLS or NTA (e.g., lipid vesicles or protein aggregates may interfere with DLS or NTA). Similar approaches are analytical ultracentrifugation (AUC) or differential centrifugal sedimentation (DCS), where particles will sediment according to size during centrifugation (Wohlleben 2012; Schulze, Rothen-Rutishauser, and Kreyling 2011).

However, often the increase in hydrodynamic radius cannot simply be explained by the formation of a protein corona as it is much more pronounced. Some nanomaterials will agglomerate in biological matrices. Thus, one needs to be careful in interpreting the results from techniques that primarily determine particle size and size distribution such as DLS, NTA, AUC, or DCS.

Other useful methods are spectroscopic methods such as fluorescence spectroscopy, Raman microspectrometry, or circular dichroism (CD), which are especially useful in detecting structural changes in interacting biomolecules (e.g., structural changes in protein, partial denaturation). For instance, Raman microspectrometry may be especially useful for metal nanoparticles (in particular silver or gold), which lead to very strong Raman signals due to signal enhancement by several orders of magnitude (so called surface-enhanced Raman scattering) because of surface plasmons. This allows for studying protein corona *in situ* directly inside biological matrices, or even inside cells (Drescher et al. 2013).

Fluorescence correlation spectroscopy (FCS) has been successfully applied to study protein corona thickness and layer structure (Maffre et al. 2011; Milani et al. 2012). However this technique is limited to the study of the interactions of a single protein and a nanoparticle. In addition, the nanoparticles need to be monodisperse and with a narrow size distribution.

Other *in situ* methods are surface plasmon resonance (SPR) and isothermal titration calorimetry (ITC), which both allow for determination of thermodynamic characterization of the nanomaterial–biomolecule interaction. SPR is mainly used for studying the adsorption kinetics (i.e., association and dissociation constants, k_{on} and k_{off}), whereas ITC enables us to determine the stoichiometry of the binding (Mahmoudi et al. 2011). This technique is also limited to the study of individual proteins only.

One major disadvantage of all these approaches mentioned here is the fact that none will allow for identification of adsorbed biomolecules. Although they may deliver one piece of information such as the presence of a corona, corona thickness, or structural changes of adsorbed proteins, they will never allow mapping of adsorbed components, which is possible only with *ex situ* approaches.

The most challenging step for *ex situ* approaches is the isolation of the nanomaterial–biomolecule complex, which should be performed ideally without perturbation of the system (which is literally impossible to achieve). A common method to separate nanomaterial–biomolecule complexes from unbound material is centrifugation, followed by several washing steps (Aggarwal et al. 2009). Centrifugation, especially in combination with several washing steps, will remove weakly bound biomolecules from the nanomaterial surface. What is recovered after the subsequent washings steps is therefore often referred to as the "hard corona" (Lundqvist et al. 2008), which includes molecular species bound with high affinities to the nanomaterial. Other approaches to separate nanoparticles from nonbound biomolecules are magnetic separation, size exclusion chromatography (SEC), differential centrifugation (DC), or field-flow-fractionation (FFF) (Aggarwal et al. 2009; Walkey and Chan 2012; Maskos and Stauber 2011). Methods such as SEC or FFF may eventually allow also detection of proteins belonging to the "soft corona," which are less tightly bound and are proteins of the corona that are prone to exchange processes over time (Maskos and Stauber 2011; Cedervall et al. 2007).

Once the nanomaterial–biomolecule complex is isolated, the adsorbed biomolecules may be analyzed directly on the surface, which is possible for instance when studying lipids using secondary ion mass spectrometry (Grenha et al. 2008). Proteins may be digested directly on the surface and then analyzed in gel-free approaches or may be eluted by application of high temperature, detergents, high salt concentrations, enzymatic digestion, or extraction with organic solvents (which is relevant for lipid analysis), and then analyzed via gel electrophoretic approaches. However, there have been cases reported where complete protein elution was not possible due to strong interactions with the nanomaterial—a fact underlining the need to include meaningful controls in experiments (Schulze, Rothen-Rutishauser, and Kreyling 2011). Following isolation and elution, the most common method for protein analysis is poly(acrylamide) gel electrophoresis (PAGE), which can be performed in a one- and two-dimensional manner. Especially one-dimensional (1D) SDS-PAGE is often used, as it is relatively easy and affordable and can be combined subsequently with other highly sensitive mass spectrometric methods (Lundqvist et al. 2008; Hellstrand et al. 2009; Monopoli et al. 2011). Two-dimensional (2D) gel approaches offer the advantage of separating the complexity of bound proteins very efficiently and allow discrimination of protein species (e.g., different isoforms or different post-translational modifications), thus enabling a pattern analysis and quantification before mass spectrometric identification (Blunk et al. 1993; Gessner et al. 2000). However, these approaches are much more time and cost intensive. Bands from 1D gels or spots from 2D gels can be cut out and identified via mass spectrometry. In addition, gel-free mass spectrometric approaches have also been successfully applied to study the nanoparticle's protein corona (Tenzer et al. 2011; Tenzer et al. 2013). Proteins may also be transferred to membranes for specific immuno-detection (western blotting)—the disadvantage is that a specific antibody is needed and basically one may only confirm the presence of a specific protein but not identify all bound proteins (Pitek et al. 2012). However all these approaches have advantages and disadvantages—so ideally several of them should be combined in a complex strategy.

4.1.4 Biological Fluids as Model Systems to Study Bio–Nano Interactions

The composition of the biomolecule corona of course depends on the respective physiological compartment encountered by the nanomaterials. This means, according to the route of uptake the nanoparticle will be covered with a different set of biomolecules. Furthermore, the protein corona is a highly dynamic complex, adsorbed components will exchange over time as the nanomaterial translocates from one compartment into another depending on the presence and affinity of the respective binding biomolecules. Regarding unintentional exposure of nanomaterials, there are three main ports of entry for nanomaterials: the lungs, the gastrointestinal tract (GIT), and the skin (Figure 4.1). There is plenty of data showing that intact skin is a tight barrier, which typically prevents uptake of nanoparticles. Although uptake of nanomaterial through the skin cannot be excluded completely, the uptake via this barrier seems to be less relevant and will therefore not be discussed further in this chapter.

With respect to intentional exposure (i.e., applications in nanomedicine), intravenous (i.v.) injection is one of the main routes to introduce nanomaterials into the human body as well as oral and inhalative uptake, which have been mentioned already. Here we will now focus on three main routes of entry, namely i.v. injection leading to direct access to systemic circulation, oral uptake (GI tract barrier), and inhalation (lung barrier). For completeness it should be noted that other routes of entry also exist (e.g., through the eyes, ears, and the nose), which will not be covered here.

Depending on the uptake route the nanomaterial will first directly interact with blood plasma in the case of i.v. injection, with lung lining fluid in the case of inhalation, or with gastric fluids in the case of oral uptake. Although possible, it is technically very challenging to recover nanomaterials out of the body fluids to directly access the biomolecule corona. Therefore, another approach may be the use of simulants for nanomaterial dispersion mimicking in composition the respective body fluids such that the composition of nanomaterial biomolecule corona can be assessed in a situation resembling the *in vivo* situation. In such test media, which closely resembles the respective physiological compartment, other features such as agglomeration can also be easily analyzed. Additionally, one should mention cell

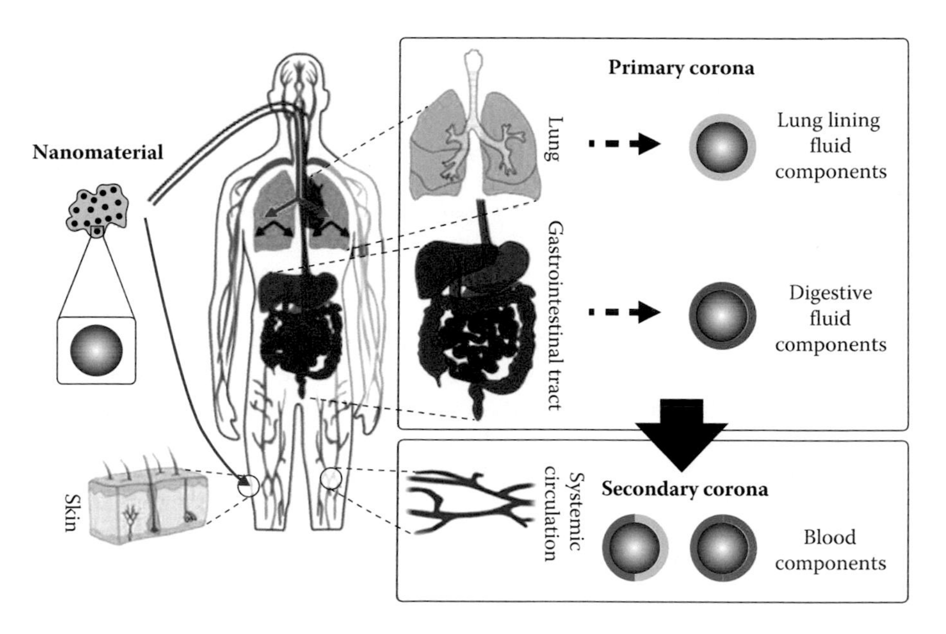

FIGURE 4.1 Ports of entry into the human body for nanomaterials and the role of bio–nano interactions. There are three main surfaces for nanomaterials to interact with the human body after unintentional exposition: the lungs, the gastrointestinal tract (GIT), and the skin. Nanomaterials contacting the skin (pink arrow) may cause local effects, but are unlikely to enter the body via this route in most cases. Inhaled and lung deposited (blue arrows) or swallowed and GIT-processed nanomaterials (orange arrows) are primarily encountered by either lung lining fluid components (e.g., surfactant lipids and proteins, pulmonary mucus) or digestive fluids from the GIT with endogenous (e.g., gastrointestinal mucus, digestive enzymes) and exogenous components (e.g., food ingredients), respectively. Biomolecules present in these primarily encountered biological compartments will interact with these foreign materials ("bio–nano interactions"), leading to the formation of a primary corona. In the case of nanomaterials injected into the systemic circulation (i.e., nanomedicines), a primary corona consisting of plasma proteins can be formed, too. This primary corona can alter physicochemical properties of the nanomaterials (e.g., size, agglomeration state, etc.), and subsequently influence their biological fate, such as uptake by epithelial cells and translocation into the systemic circulation, or clearance from the body (red arrows). After entering the systemic circulation via one of these routes, the nanomaterial–biomolecule complexes (primary corona) can be subjected to further interactions with blood components (e.g., plasma proteins) resulting in a dynamic exchange of biomolecules and eventually formation of a secondary corona consisting of various components from the different encountered compartments, which governs the final biodistribution and clearance from the body. (Reprinted from Comprehensize Biomaterials, M. Maskos and R.H. Stauber, Characterization of nanoparticles in biological environments, 329–339, Copyright 2011, with permission from Elsevier.)

culture medium containing serum as another biological test medium, which is used in *in vitro* tests.

Test media for *in situ* characterization of nanomaterial may vary in composition from simple buffers, which resemble body fluids in salt composition and sometimes may contain a few enzymes, to cell culture medium containing serum, and finally the use of fluids purified from biological systems such as the use of purified lung lining fluid. Buffers such as phosphate buffered saline (PBS) are a popular test medium for nanomaterial characterization, which are also often used in the preparation of nanomaterial for application (e.g., for i.v. injection or for instillation). In addition, some artificial fluids are defined in the literature (i.e., resembling gastric fluids or lysosomal fluid), some of which even may contain certain enzymes, and they have been shown to be a very useful tool for characterizing nanomaterials. Still, such systems do not show the same complexity seen in *in vivo* body fluids as they typically lack most of the proteins or other biomolecules.

The "gold standards" are such biological fluids that originate from the respective physiological compartment with which a nanomaterial can potentially interact. With respect to the bloodstream, the relevant biological fluid (i.e., plasma) is comparably easily available and can be used for *in vitro* studies on nanomaterial biomolecule corona formation or for analyzing dispersibility. Lung lining fluid can be purified from several sources—the most popular sources are rat or pig. However lung lining fluid is more complicated to obtain (most often via lung lavage), and is also difficult to standardize. For GIT typically only artificial fluids mentioned previously exist. Before we go into detail, now we would like to spend a few words on *in vitro* studies.

In vitro studies are typically performed with cell culture medium containing serum. Serum content can vary from 1% to 20% and may vary in origin. Most often, fetal calf serum is used but it is also possible to use horse, rat, human, or any other serum. In recent times there are discussions going on to improve *in vitro* systems in a sense that for rat cell lines, rat serum should be used; although for human cells one should use human serum, and so forth. Hitherto, most laboratories still use long established protocols using fetal calf or bovine serum. Serum may be used with or without heat inactivation for *in vitro* studies. However, one should note that the composition of the cell culture medium—especially the selection and treatment of the serum component (i.e., species, heat inactivation, concentration, etc.)—can interfere or bias an assay, for which reason various control experiments need to be performed (Lesniak et al. 2010).

In vitro studies using submersed cell culture systems may be very useful for studying interaction of nanomaterials with cells from the blood or vascular system, and for this purpose the *in vitro* system mimics the *in vivo* situation quite closely. However, for other physiological compartments, cell culture media generally lack specific components. So studies performed with submersed cell cultures using serum additives will not be able to truly mimic oral uptake or inhalation. It may be possible to improve such systems by first covering the nanomaterial for instance with lung lining fluid in the case of analyzing inhalation before their addition to cell culture systems containing lung cells. Another way would be to spike the cell culture media with additives that are specifically relevant for the respective biological fluid, for example, surfactant lipids, proteins, or mucins to mimic the lung lining fluid, or saliva components and digestive enzymes in the case of the gastrointestinal fluid (Lazzari et al. 2012). This may help to improve the *in vitro* systems in some cases, but one needs to be aware of such differences when interpreting and comparing results from different studies.

4.2 NANOMATERIAL INTERACTIONS WITH SERUM OR SERUM-CONTAINING CELL CULTURE MEDIUM

4.2.1 Physiological Considerations

Blood is most likely the route via which nanomaterials are intentionally administered, either for therapeutic purposes (e.g., imaging or cancer treatment) in humans or as part of a toxicological study in animals. In the case of unintentional exposure, nanomaterials will most likely not directly interact with the blood circuit, leaving aside of course skin injuries that would allow a direct systemic access. In unintentional exposure scenarios, nanomaterials will first need to pass body barriers such as lungs or GIT to secondarily interact with the blood. In all scenarios, blood then may facilitate distribution of nanomaterials throughout the body. When nanoparticles pass from one body compartment into another, the composition of the corona evolves (Dell'Orco et al. 2010; Lundqvist et al. 2011), meaning a significant or even major part of the corona will be exchanged as the nanomaterials encounter a different environment. However, a fingerprint of the earlier surroundings is always retained.

Analysis of nanomaterial interaction with blood is not trivial as blood is a highly complex body fluid. Blood plasma accounts for about 55% of the volume of human blood and the remaining part is composed of various cells (erythrocytes > leukocytes > thrombocytes). The blood plasma is an aqueous solution containing approximately 300 different proteins, and if taking into account the rich posttranslational modifications, it contains more than 3700 protein species as well as other biomolecules like amino acids, fatty acids, glucose, other metabolites, hormones, or vitamins (Anderson and Anderson 2002; Pieper et al. 2003; Klein 2007). Plasma proteins display a huge dynamic range, which covers approximately 10 orders of magnitude. A few highly abundant proteins such as albumin, transferrin, or immunoglobulins account for 50% of the total protein content. The most abundant protein is albumin, which has a physiological concentration of 35–50 mg/ml (35–50 × 10^9 pg/ml). In contrast, there are many low abundant proteins such as interleukin 6, which is typically present in the concentration of 0–5 pg/ml. In addition to "classical plasma proteins," plasma contains many other proteins such as different tissue proteins ("leakage markers"), foreign proteins (resulting from microorganisms or parasites occurring during infections), or aberrantly secreted proteins (e.g., tumor markers) (Anderson and Anderson 2002).

One major obstacle in serum analysis is the high dynamic range. Thus, without any additional approaches, for example, depleting highly abundant components, it will not be possible to analyze the full complexity of serum proteins and one will only detect and characterize the most abundant species interacting with nanoparticles or those that are highly enriched on the surface. Others, which may be present in tiny amounts on the surface only, may be "masked" due to high abundance of a few other proteins and thus they will not be detected although they may be functionally relevant and important as well. For instance, with a typical 2D approach it is possible to resolve only 3–4 orders of magnitude but not 10 as naturally occurring in plasma. As albumin is by far the most abundant serum protein it is not astonishing that it is often the most abundant component of nanoparticle biomolecule corona (Tenzer et al. 2011).

Taking the high complexity of plasma proteins into account, there is a huge variety of possibilities for forming a nanoparticle–protein corona. As explained in detail in the introduction, the main nanoparticle properties affecting the composition of the corona seem to be nanoparticle size and shape (affecting the curvature of the nanoparticles) and surface properties such as charge or hydrophobicity (Lundqvist et al. 2008; Tenzer et al. 2011; Gessner et al. 2000).

4.2.2 Available Models and Lessons Learned from Published Studies

The physiologically most relevant system to study would be human plasma or human serum (e.g., plasma devoid of coagulation factors), which is easily available from donations. In fact this system has been used in several studies so far (Tenzer et al. 2011; Lundqvist et al. 2008; Monopoli et al. 2011). Serum may be used either in a pure state or further diluted. For instance, Monopoli and coworkers used 3%–80% plasma for their study. PBS diluted plasma is often easier to analyze as it is less complex, allowing in addition to protein corona analysis the study of other *in situ* characteristics such as nanoparticle agglomeration, which with many techniques might not be feasible in concentrated plasma. Other studies use bovine or fetal serum diluted in cell culture medium, typically to a final concentration of 10%, which is the relevant medium for *in vitro* experiments (Maiorano et al. 2010).

Therefore the choice of which medium to use for analysis strongly depends on the questions underlying the study. For correlating results obtained from *in vitro* or *in vivo* animal testing to humans or for directly assessing nanoparticle behavior in human bodies, one certainly should aim to study nanoparticles in human plasma as well. Although for interpreting results obtained from *in vitro* studies, cell culture medium containing 10% serum is the medium of first choice for *in situ* characterization of nanoparticles. Diluted serum can be regarded as a medium complex simulant fluid, somehow mimicking pure serum. However one should keep in mind that some major differences remain. The first major issue is the difference between serum, and plasma as serum does not contain any coagulation factors that are present in plasma. The second major issue is derived from dilution, although for reducing complexity dilution may be a necessary step. The protein content in cell culture medium containing only 10% serum is around 3 mg/ml and thus is significantly lower than the protein content observed in plasma *in vivo*. Different percentages of serum may yield completely different protein coronas as it has been shown already for silica or polystyrene nanoparticles in a study by Monopoli and coworkers, with varied serum content from 3% to 80% (Monopoli et al. 2011). But being the relevant medium for *in vitro* cell culture containing all relevant serum proteins, cell culture medium containing 5%–10% serum is a very good starting point for *in situ* characterization of nanoparticles and for characterizing bio–nano interactions.

Apart from the question whether the serum should be used undiluted or diluted another aspect is even more important to consider, which is heat inactivation of serum. Typically or traditionally serum is heat inactivated for use in cell culture but actually this will certainly alter the protein conformation. Nonheat-inactivated serum resembles the *in vivo* situation more closely and most cell lines do not require the serum to be inactivated anyhow. Protein corona will drastically change when serum is heat inactivated compared to nonheat-inactivated serum, which in turn also influences the outcome of the toxicity study (Lesniak et al. 2010).

For a deeper molecular understanding of the actions taking place on the nanoparticle surface, however, one even needs to reduce the complexity of the whole system further down to a single protein–nanomaterial system. This allows us to study the interaction in detail and for instance to follow changes occurring in protein structure during interaction with nanoparticles. Therefore single nanoparticle systems prove to be very useful for certain types of analysis as well.

4.2.3 Single Protein Systems

As the surface of the nanomaterials is more or less strongly curved depending on nanomaterial size and shape most proteins will have to adapt their three-dimensional (3D) structures at least partially to fit onto that surface. Other factors determining this interaction are charge-dependent interactions as the surface charge of nanomaterials will attract oppositely charged amino acids and repel those of similar charge. Hydrophobic interaction will also contribute to partial unfolding of proteins. Eventually for some nanomaterials covalent bonds may also be formed for instance with cysteine residues. Thus taken together, often protein 3D structures will change during interaction with nanomaterials. As this has already been explained in detail earlier in this chapter we will only summarize the most important aspects as they are important to understand the studies, which will be discussed here. Interactions of proteins with nanoparticle surfaces may lead to uncovering of so-called "cryptic epitopes," which previously have been hidden inside the protein (Lynch, Dawson, and Linse 2006; Deng et al. 2010). Partial unfolding will also influence protein–protein interactions, eventually changing protein fibrillation or aggregation as it has been shown for β2-microglobulin (Lynch, Dawson, and Linse 2006) or amyloid β protein (Cabaleiro-Lago et al. 2008) interacting with polymeric nanoparticles. Of course changing protein structure will also influence protein function, which has been analyzed for instance for chymotrypsin and peroxidase interacting with MWCNTs or lysozyme interacting with ZnO (Wu, Zhang, and Yan 2009; Chakraborti et al. 2010).

Shang and coworkers could nicely show that the adsorption of BSA onto a gold nanoparticle surface changes protein secondary and tertiary structures (Shang et al. 2007). They used fluorescence spectroscopy, FT-IR, and CD-spectroscopy to prove that tryptophan residues moved into a different environment when BSA attached to the nanoparticle surface and that the overall helicity of the protein changed, depending on the pH value of the environment.

Pan and coworkers applied quantitative stopped-flow measurements to unravel the underlying kinetics of interaction of GB1 protein with latex nanoparticles to understand precisely how this interaction is established, and they furthermore analyzed how changes in pH and ionic strength would influence this system (Pan et al. 2012). They could show that upon adsorption GB1 is partially unfolded, and they could calculate the folding and unfolding rates of GB1 while in contact with the nanoparticles. Interestingly both rates are much slower than those for free GB1 in solution indicating that the free-energy barrier between folded and unfolded states is increased. In addition, in solution the folded state is more stable than the unfolded one, which is reversed upon binding to the nanoparticle surface leading to gradual denaturation of GB1 on latex beads. Further studies should unravel the protein "hot spots" directly responsible for the interaction. Exact knowledge of these aspects will not only lead to a better molecular understanding of protein binding to nanoparticles but also will ultimately enable scientists to produce "designed protein coatings" specifically aimed at lowering toxicity, altering uptake, or changing biodistribution in a defined way, for instance, for time-dependent drug release.

In other studies, FCS was successfully applied to study protein corona thickness, layer structure, or a number of adsorbed proteins (Maffre et al. 2011; Milani et al. 2012). For instance Milani and coworkers used fluorescently labeled transferrin to analyze adsorption on polystyrene beads. They could follow the formation of a first monolayer, which actually represents the "hard corona." This was followed by formation of a secondary and a tertiary layer, which represent the "soft corona" as those layers proved to be exchangeable in competing experiments when unlabeled transferrin was added and the first monolayer was not. This clearly proves that the concept of "hard" and "soft" corona holds true at least for this very simplified and ideal system and that the postulated "memory effect" (e.g., keeping traces of "primary corona" when nanoparticles move from one body compartment to another) could have a molecular explanation.

4.2.4 Interaction with Plasma or Serum

The high complexity of serum or plasma typically makes this dispersion medium less suitable to study detailed molecular mechanisms of interaction. However to determine the identity of the corona and identify which proteins do or do not interact with nanomaterial surfaces it is perfectly well suited. The study performed by Tenzer and coworkers delivers a quite comprehensive overview of protein coronas of silica nanoparticles of various sizes (8, 20, and 25 nm) in blood plasma (Tenzer et al. 2011). Although they found a significant overlap in the respective coronas indicating that a large part of the corona seems to be material specific, 37% of the corona proteins are determined only by the particle size with no apparent pattern manifesting. They showed that the occurrence of proteins in the corona does not directly reflect the amount of those proteins in plasma. Some of the corona proteins, which are close to or even below the detection limit in plasma, are highly enriched in the corona, although other proteins, which are among the most abundant plasma proteins, could still be detected in the corona but in terms of quantity they are certainly not holding the same rank as could be deduced just from their serum abundance. For example α-2-macroglobulin and complement C3 are the second and third highly abundant proteins in plasma, yet in the corona of 125-nm silica nanoparticles apolipoprotein A1 (8th place in serum) and prothrombin with a 35-fold enrichment take the second and third place. Complement C3 and α-2-macroglobulin are only the 5th and 13th most abundant proteins in the corona.

Apart from size, the surface charge is another major parameter influencing the composition of the protein corona. Influence of the surface charge of latex nanoparticles has been analyzed in several publications (Cedervall et al. 2007; Lundqvist et al. 2008; Aggarwal et al. 2009). In the study performed by Lundqvist and coworkers it was shown that the total bound protein amount for negatively charged nanoparticles (carboxyl-modified polystyrene nanoparticles) initially increased upon increasing the amount of plasma added to a certain amount of nanoparticles. But then it decreased and only increased again at much higher plasma amounts. In contrast, for positively charged nanoparticles (amine-modified polystyrene) this was not observed. Here there was a steady increase of bound protein upon increasing the amount of serum. This may be explained by the fact that most of the serum proteins are negatively charged under neutral pH conditions, and in good correlation with this the authors could furthermore show that the composition of the coronas was different for negatively and positively charged nanoparticles.

Finally, there is also correlation between the total protein adsorbed and surface hydrophobicity of nanomaterials (Gessner et al. 2002). Gessner and coworkers showed that with decreasing hydrophobicity the total amount of bound protein would decrease and in addition this also altered the corona in a qualitative manner.

Taken together all these studies nicely show that the most important nanoparticle characteristics influencing the protein adsorption and corona formation in a quantitative and qualitative way are size and shape of nanoparticles as well as surface properties such as charge and hydrophobicity.

4.2.5 Lessons Learned from NanoGEM

Although several studies have already addressed the size-dependent effects of nanomaterial interaction with serum proteins, much less is known about the other properties previously mentioned such as charge or hydrophobicity. Most of the knowledge of how charge or hydrophobicity influences the interactions with serum proteins is derived from quite ideal polymer type nanoparticles such as polystyrene or latex. When nanoGEM was started, basically no study had addressed the influence of surface functionalization on protein interaction for industrially relevant nanoparticles. This gap of knowledge was identified as one of the major research topics for the nanoGEM project. Industrially relevant nanoparticles were systematically varied on the surface using different types of functionalization. For instance, nanoscaled silicon dioxide (SiO_2) was applied in four different variants: without any specific surface functionalization (SiO_2 naked), with an overall positive charge at neutral pH (SiO_2 amino), with an overall negative charge (SiO_2 phosphate), and covered with a short PEG chain (SiO_2 PEG).

The interaction with serum proteins was studied in cell culture medium containing 10% fetal calf serum (DMEM with 10% FCS) with 1D SDS PAGE and with 2D gel electrophoresis, in combination with mass spectrometry. Proteins bound on particle surface were eluted, separated with 1D SDS PAGE, stained with Coomassie dye, and assessed by densitometry. Taking into account normalization via five different marker bands, the total amount of bound proteins could then be estimated. With that approach the time-dependent formation of protein corona had been analyzed (Figure 4.2a).

It becomes obvious that naked SiO_2 binds high amounts of proteins after 1 h, whereas PEGylation of nanoparticle surface repels proteins, at least initially during the first 2 h. However, as only 40% of the surface is covered with PEG and the PEG

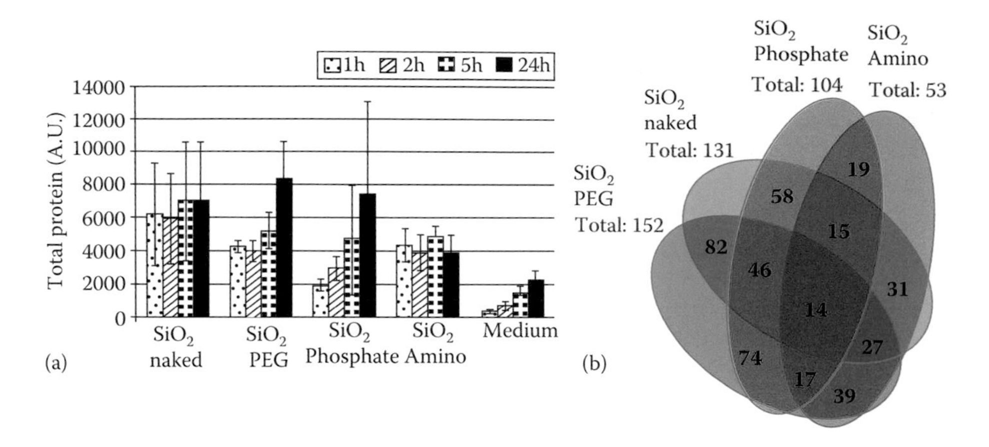

FIGURE 4.2 Analysis of protein corona for four different variants of SiO_2. (a) Four different variants of SiO_2 nanoparticles (naked, PEG, phosphate, amino) have been characterized *in situ* in DMEM cell culture medium containing 10% FCS with respect to protein corona formation. The nanoparticles have been dispersed by stirring the nanoparticles in the test medium for the indicated period of time. After that the nanoparticles were centrifuged in a table top centrifuge at 18,400 × *g*, the resulting pellet was thoroughly resuspended and washed three times in PBS. Adsorbed proteins were then eluted with Laemmli buffer and subsequently analyzed with 1D-SDS-PAGE. Protein bands were visualized by Coomassie staining and were semiquantitatively assessed with ImageLabTM Software (BioRad, Munich, Germany). Gels were normalized via five different marker bands that were present on all gels in the same amount. Total bound protein amounts were calculated by summing all band intensities of the respective lanes taking into account the normalization. (b) The protein corona of the same nanoparticle types was analyzed with 2D gel electrophoresis after eluting the corona with 2D buffer containing 7 M urea. The gels were stained with a ruthenium-based fluorescence dye and analyzed with Delta 2D™ (Decodon, Greifswald, Germany). Spots which were statistically significantly enriched in the protein corona of the particles compared to control (only DMEM with 10% FCS, no nanoparticles) were taken into account (at least 1.5 fold higher than control, $p < 0.05$). The Venn diagram shows the overlap in protein species for the different types of SiO_2 nanoparticles.

chain length is rather short (PEG 400), the protein adsorption is only time delayed but not completely avoided. Thus, after longer incubation times the differences between SiO_2 naked and SiO_2 PEG are only marginal. Furthermore 2D gel analysis was applied to separate the protein species of the protein corona (Figure 4.2b). With this approach it could be shown that a substantial amount of spots is similar for all tested nanoparticles but these spots are also present in the negative control without any nanoparticles. This background is caused by the presence of serum containing aggregates that are formed even without the presence of nanoparticles. But for each nanoparticle type we could identify unique protein species that are specifically enriched in comparison to the no nanoparticle control. Those proteins are specifically trapped on nanoparticle surfaces and are truly protein corona components. Each nanoparticle type displays a unique pattern with protein species that are unique to that particular nanoparticle type. It could be confirmed that indeed there is a large overlap in the protein corona of SiO_2 naked and SiO_2 PEG after 24 h incubation time. As explained before, this may be due to the incomplete coverage of the surface with PEG and due to the short PEG chain length. Interestingly there is only little overlap between positively charged SiO_2 (SiO_2 amino) and negatively charged SiO_2 (SiO_2 phosphate). SiO_2 amino binds more total proteins compared to SiO_2 phosphate in the first 2 h, and the total bound protein amount stays rather constant over 24 h whereas for SiO_2 phosphate it steadily increases. Eventually this may be explained by the fact that most serum proteins are negatively charged at neutral pH and thus are more likely to interact with positively charged nanoparticles. Taken together also for industrially relevant nanoparticles such as SiO_2, the surface properties strongly influence the proteins attaching to it, both in a quantitative and a qualitative way. Clearly material characterization *as-synthesized* and material characterization *in situ* need to be combined to obtain meaningful results that later on can be correlated to the outcome of toxicity testing for deriving (quantitative) structure–activity relationships.

4.2.6 Conclusions and Considerations for Future Studies

Keeping in mind how important the protein corona is for the fate of the nanoparticle, influencing cellular uptake and toxicity, one should consider presenting nanoparticles to the test system in a physiologically relevant manner being as close as possible to the *in vivo* situation. Although most cell culture protocols use fetal calf or bovine serum as this is easy to obtain and rather cheap, one should surely consider using human serum for culturing human cell lines and therefore also for *in situ* characterization of the bio–nano interface. The use of serum from a different species will introduce another variable into the system. There is a possibility that some of the adverse effects seen in *in vitro* studies may be caused by the introduction of foreign protein species into the cell. In addition, in all cases serum should not be heat inactivated as this will change the protein structure with a strong influence on the protein corona as well as on the outcome of the toxicity studies.

Batch-to-batch variations occurring for different serum batches are another aspect that deserves attention. To solve this issue, one may pool several serum batches, sometimes pooled serum is also commercially available (e.g., for cell culture purposes), or analyze several independent serum batches. One other major experimental problem is related to background, which arises from different reasons. Firstly, especially the highly abundant proteins will bind to any surface be it a nanoparticle or a test tube. Therefore, care needs to be taken when choosing plastics or glassware used for the studies. Proteins, especially highly abundant ones such as albumin, may also be trapped when pelleting nanoparticles within the nanoparticle pellet and might not be efficiently removed in washing steps in particular when nanoparticles under study are not ideal and are prone to aggregation. Of course extensive washing and centrifugation steps will certainly help to improve or overcome this issue, but finally one may only analyze the so-called "hard" corona via this approach. This is why most studies focus on "hard" corona analysis and typically include a series of washing steps until the washing solution is devoid of protein.

In addition, serum proteins are prone to aggregation, especially when the serum has been frozen. Such serum aggregates may even be visualized with DLS or

NTA—typically a cell culture medium containing 10% serum may contain 10^6 to 10^9 particles in the size range from 30 to several hundred nanometers. Of course these might be cleared before analysis starts via centrifugation or filtration. Yet, they will quickly form again typically with a time frame of a few hours, and depending on the dispersion protocol employed, it may not be possible to completely get rid of them. But as serum particle and nanoparticles, especially when being composed from metals or metal oxides, will have different densities, combination of different techniques such as chromatographic approaches like SEC/GPC with (ultra-)centrifugation could be helpful or instead one may also apply density gradient centrifugation approaches.

Summing up it should be emphasized that in an experiment the principal approach (using selected individual particles or serum, which source of serum, diluted or undiluted) as well as the techniques need to be chosen carefully on a case-by-case basis depending on the precise question. The researcher should be aware of the pitfalls as well as the strengths of each approach, and eventually some of the weaknesses of a method might not even be problematic at all in a certain type of analysis.

4.3 EXAMPLE/CASE STUDY: NANOMATERIAL INTERACTIONS WITH BIOLOGICAL FLUIDS IN THE RESPIRATORY TRACT

4.3.1 Physiological Considerations

Inhalation is one of the major routes for unintentional exposure to nanomaterials and thus deserves special attention in hazard or risk evaluation. Also there is ongoing research to use aerosol application also for certain types of nanomedicine. The lung surface is one of the largest epithelial surfaces in the human body (about 70–100 m^2) and the largest one that is in direct contact with the surrounding environment (Geiser and Kreyling 2010). Inhalation of nanomaterial aerosols may occur for consumers during use of nanoparticles in sprays and also in occupational settings. In occupational exposure scenarios, higher nanomaterial aerosol concentrations may be reached, and this can lead to significant accumulation of nanomaterials in the respiratory tract.

The major part of the pulmonary air–blood barrier (i.e., the alveoli) consists of an ultra-thin epithelial monolayer (down to 0.1 μm), providing in theory a very short distance to overcome for nanomaterials to translocate to the systemic circulation. Both local and systemic effects of inhaled nanomaterials need to be considered in hazard assessment.

Evaluations should be performed under conditions that are closely related to the physiological situation. This of course includes also a detailed characterization of the tested nanomaterial in the respective biological fluid. The epithelium of the respiratory tract is covered toward the air-side by a continuous liquid layer: the lung lining fluid. The lung lining fluid is the first biological matter encountered by inhaled nanomaterials that deposit. Therefore understanding how the nanomaterials interact with the lung lining fluid is of high relevance for interpreting results from inhalation toxicity studies. In the conducting airways, the major component of the lung lining fluid is mucus, a viscoelastic hydrogel, which is mainly composed of water and glycoproteins that form a rather thick and proper noncellular barrier on top of epithelial cells (Sanders et al. 2009; Antunes, Gudis, and Cohen 2009). Nanomaterials that deposit in this upper part of the lungs are efficiently trapped in this mucus layer, and are constantly transported in the cranial direction, where they can be swallowed and subsequently processed in the GIT (see Section 4.1). This so-called mucociliary clearance is a very efficient protective mechanism to remove particulate matter from the lungs. Thus, understanding the interactions between nanomaterial and mucus of the upper airways is only of minor relevance. Apart from the conducting airways, the lung lining fluid is mostly a very thin liquid film (0.09–0.89 μm) that spreads over the whole respiratory tract from the peripheral to the central lungs (Bastacky et al. 1995). Its major integral component is pulmonary surfactant (PS), a surface-active film located at the air–liquid interface. PS is a complex mixture of about 90% lipids and about 10% proteins. Phospholipids (PL) account for about 80% of the lipid fraction and the majority are phosphatidyl-choline (PC), -gylcerol (PG), and -inositol (PI) species (with PC accounting for the biggest PL population). The remaining 5%–10% of the lipid fraction are cholesterol, sphingolipids, glycerides, and fatty acids (Blanco and Pérez-Gil 2007). Among the protein species, which account for 5%–10% by weight, four specific surfactant proteins (SP) are known: SP-A, -B, -C, and -D (Goerke 1998).

PS fulfills two principal functions: a biophysical and an immunological function. The biophysical function—which is governed by the phospholipid fraction—is to lower the surface tension at the air–liquid interface and thereby prevent the alveoli from collapsing during exhalation (Rugonyi et al. 2008). The immunological role of PS in the host defense system of the lungs is mainly attributed to surfactant proteins, namely SP-A and -D, both belonging to the collectin protein family (Kishore et al. 2006). Their localization at the air–liquid interface makes them ideally situated to bind foreign materials (e.g., inhaled nanomaterials) to facilitate uptake by immunocompetent cells (e.g., alveolar macrophages), or to trigger and/or control inflammatory processes (Wright 2005).

Inhaled nanomaterial will be efficiently displaced into the subphase of the lung lining fluid (Schürch et al. 1990), where they can get in contact with PS components (e.g., surfactant PLs or SPs) or other constituents of the lung lining fluid (immunoglobulins, albumin, or other proteins). The bio–nano interactions occurring here might be very decisive regarding the further biological fate of the nanomaterial. It is therefore important to consider these specific components of the lung lining fluid with respect to *in situ* characterization of nanomaterials, and to study their binding to nanomaterials, as well as subsequent biological effects.

4.3.2 Available Models for Lung Lining Fluid

Unfortunately, there are no standardized models available for the lung lining fluid as test medium to study bio–nano interactions in the lungs—in contrast to, for instance, models for the systemic circulation where plasma or serum samples of standardized quality can be accessed in large quantities. The lung lining fluid is a thin liquid layer present at the air interface of the lungs, and is only accessible via broncho-alveolar lavage (BAL). This implies that access to human material is very limited, as BAL can only be performed during bronchoscopy of patients—whose lungs are often pathophysiologically altered. For this reason, animal sources are generally exploited to obtain meaningful quantities of BAL fluid from which then the native surfactant can be laboriously isolated and purified. Performing BAL means that a volume of saline buffer is instilled into the lungs and aspirated again afterwards. Consequently, the thin surface film is disrupted and diluted when the inner lung surfaces are lavaged, which not only causes dilution of the retrieved material but also leads to formation of vesicular structures, and might alter its biochemical composition, too. In addition, tissue damage may occur during the lavage leading to leakage of tissue or serum proteins that are typically not found at all or only at much lower concentrations in the lung lining fluid *in vivo*. Due to the dilution of the lung lining fluid during lavage, it is impossible to assess the original volume of the lung lining fluid and the original protein and lipid concentration taking also into account that not all components might be recovered quantitatively. Therefore, the material obtained from BAL, although derived from a biological source, should be considered as a semiartificial system.

Due to the dilution occurring during the lavage, PS—a most important biological matrix for nanomaterial–lung interactions—is only present in very low concentrations in BAL fluid, making it necessary to isolate and concentrate PS to obtain useful amounts and concentrations for studying bio–nano interaction with nanomaterials. During the last 30–40 years, extensive research has been carried out in this field to develop and improve treatment of acute respiratory distress syndrome (ARDS) (Halliday 2008). PS has been purified and analyzed in the past from various species and different isolation methodologies have been published, and even commercial clinical surfactants are available. BAL fluid can be further purified from blood contaminations (especially necessary when working with material from slaughter processes) and concentrated using ultracentrifugation and density-gradient centrifugation (Griese 1999). Native surfactant that has been purified and concentrated in such a manner from BAL is the most complete model of PS that can be isolated from the lungs, and contains physiological proportions of surfactant lipids, as well as SP-A, -B, and -C (Bernhard et al. 2000). For this reason, purified native surfactant has been regularly used as reference substance (Bernhard et al. 2000; Blanco et al. 2012). However, as native surfactant causes immunogenicity issues when applied to treat acute respiratory distress syndrome (ARDS) (mostly due to residual proteins resulting from inefficient purification and species incompatibilities), organic extractions of BAL fluid were performed to obtain semicomplex lipophilic extract surfactants, which represent the majority of available clinical surfactants on the market today. These products contain the lipophilic fraction of PS, that is, the surfactant lipids plus the hydrophobic surfactant proteins B and C but lack the hydrophilic surfactant proteins A and D. This material is biophysically active and functional on the one hand, but with defined composition on the other hand. Therefore, they are successfully used and are the products of first choice for surfactant replacement therapy in clinical treatment of ARDS. There are several such clinical surfactants on the market, which are standardized and quality controlled in terms of their production process (e.g., Alveofact®, Curosurf®, Survanta®, or Infasurf®). Due to their good accessibility, clinical surfactants have often been used in recent studies as surfactant models to investigate bio–nano interactions of nanomaterials in the lungs. However, one should be aware of the fact that these products are organic extracts, which means that they lack important components (e.g., SP-A and SP-D), and therefore do not completely reflect the physiological situation. Moreover, depending on the extraction method and origin (i.e., species), they also differ in PL composition. Hence, the most physiological model for PS is still purified native surfactant, which should be the model of choice when nanomaterials are characterized under conditions close to the *in vivo* situation. Nevertheless, there are some methodological pitfalls to consider when working with this model. The disruption of the interfacial surfactant film during isolation and purification leads to formation of vesicular structures in an aqueous environment. These large multilamellar vesicles (also referred to as "large lamellar bodies") are prone to sedimentation during centrifugation when nanomaterial–biomolecule complexes are separated. Unbound surfactant material can thereby cosediment with the nanomaterials, leading to false positive results and actually produce a very high background, which strongly interferes with most types of analysis. Furthermore, *in situ* characterization of nanomaterial sizes and size distribution in purified native surfactant is hard to accomplish, as basically for the same reason the large lamellar bodies interfere with the assays such as light scattering.

4.3.3 Lessons Learned from Published Studies

Pulmonary toxicity assessment of nanomaterials currently is focused very much on two aspects. First, biophysical investigations are performed to study the effect of nanomaterials on the functionality of PS to maintain respiratory mechanics. Second, the interaction between nanomaterials and components of the lung lining fluid is probed to study binding phenomena, and how adsorbed biomolecules trigger the biological response of the lungs toward such materials. Biophysical investigations are generally performed by mixing the nanomaterial and surfactant material in an organic phase, followed by injection of this solution onto an aqueous phase. As the solvent quickly evaporates, the remaining interfacial film (including the nanomaterial) can be compressed using a film balance (e.g., Langmuir–Blodgett film balance) to generate surfactant-like multilamellar structures. One measures the compressibility during this process, which is one parameter indicating the functionality of the surface film. After compression, such films can be transferred to mica substrates and imaged using atomic force microscopy (AFM). Several such studies have been performed in the last few years. They use either artificial surfactant models consisting of single PLs, eventually combined with purified SP-B and/or SP-C (Harishchandra, Saleem, and Galla 2010; Sachan et al. 2012), or lipophilic extracts such as Curosurf® or Survanta® (Schleh et al. 2009; Tatur and Badia 2012). Interestingly, most of the studies could demonstrate a concentration-dependent adverse effect of the respective tested nanomaterial on the biophysical function of the surfactant film. Moreover, a study by Fan and coworkers demonstrated a time-dependent adverse effect of hydroxyapatite nanoparticles on the biophysical function of PS (Infasurf) (Fan et al. 2011). However, when cellular toxicity of these nanomaterials was assessed in a human bronchial epithelial cell line, no toxic effects could be observed. Actually, insertion of nanomaterials into the surfactant layer seemed to induce structural deficiencies in close proximity to the nanomaterial, and partly even agglomeration of surfactant components was observed. Although further consequences of such effects are not yet investigated in detail, a massive disturbance of the PS system due to nanomaterial interaction might potentially lead to ARDS-like modalities, affecting the integrity of the respiratory tract. Such effects cannot be detected when *in vitro* cell cultures are the only models used, and underline the necessity to apply different, independent methods to assess the toxicity profile of nanomaterials in biological fluids. For biophysical studies of this kind, semicomplex or artificial surfactant models are sufficient, although it is not possible to exclude additional effects that might occur from other surfactant components, which for instance are not contained in lipophilic extract surfactants, but that are present *in vivo*.

Surprisingly, in most of the biophysical investigations on nanomaterial–PS interactions published so far, the aspect of biomolecule adsorption was not at all studied, although this issue is most probably the reason for the observed adverse effects on the biophysical functionality of PS. Binding of lung proteins and lipids has been so far mainly studied to investigate the effect of biomolecule adsorption to nanomaterials and the subsequent cellular effects. The characterization of nanomaterials regarding their toxic potential within the lungs is a very complex and complicated task. The epithelium of the lungs forms a large surface with an air interface. Inhaled nanomaterials are normally airborne and are wetted by the lung lining fluid after deposition. To transfer this situation to *in vitro* test systems that can be later used to characterize and study nanomaterials under physiological conditions is an immense technological challenge. In addition, controlling the applied dose and exposure time requires a close "dialog between aerosol science and biology" (Paur et al. 2011). However, after deposition and submersion of nanomaterials in the lung lining fluid, the situation can be considered as a submersed condition. Therefore, most studies published on adsorption of surfactant components to nanomaterials incubate the respective material in the biological fluid, isolating it and finally analyzing the adsorbed molecules. The latter step was performed with respect to either the adsorbed protein species (mostly by gel electrophoresis, western blotting, and mass spectrometry) or the adsorbed surfactant lipids (often performed by thin layer chromatography and LC-MS) (Salvador-Morales et al. 2007; Kendall 2007; Schleh et al. 2011; Schulze, Rothen-Rutishauser, and Kreyling 2011). In this respect, Gasser et al. were able to demonstrate that the binding pattern of surfactant PLs from Curosurf® was different for functionalized MWCNTs compared to uncoated ones, indicating that the surface chemistry of a nanomaterial dictates the identity of adsorbed biomolecule (Gasser et al. 2010). Furthermore, when the MWCNTs were precoated with PS before incubation in serum, protein adsorption profiles that differed from MWCNTs without precoating were obtained. These results indicate that nanomaterials, which enter the body via the lungs, might experience a certain "pre-conditioning" that significantly influences later secondary adsorption processes (e.g., formation of a secondary corona, Figure 4.1). In a subsequent study, PS-coated MWCNTs were found to be internalized to a higher degree in monocyte derived macrophages (MDM) compared to non-coated MWCNTs, indicating that adsorbed PS components can affect the cellular uptake of nanomaterials. Moreover, the group could show that precoating of MWCNTs with PS affected their ability to cause oxidative stress, cytokine/chemokine release, and apoptosis in MDM (Gasser et al. 2012). In another study by Kapralov et al. (2012) single-walled CNTs (SWCNTs) were instilled in mice and recovered from BAL. Analysis of adsorbed lipid components using LC-MS revealed that mainly PCs and PGs (the most abundant species) had bound to the SWCNTs and a complete coverage of the nanomaterial surface with PLs was estimated and modeled. Furthermore, SWCNTs coated with surfactant PLs were studied *in vitro* in a mouse macrophage model (RAW 264.7), both in the presence and absence of additional SP-D. Interestingly, lipid-coated SWCNTs demonstrated an increased uptake by the macrophages, whereas this was slightly enhanced when studied in the presence of SP-D.

Ruge and coworkers used magnetite nanoparticles and obtained similar results demonstrating that SP-A and SP-D can bind to such nanomaterials in material-dependent manner. Again the adsorbed surfactant proteins were able to enhance uptake of these particles by alveolar macrophages (Ruge et al. 2011; Ruge et al. 2012). The biological effects were mainly mediated via adsorbed proteins whereas protein binding was dependent on the material surface properties. This was shown by testing different types of magnetite particles, some of which were more hydrophobic and others were more hydrophilic. Interestingly, when preincubating the nanomaterials with complete native surfactant, the lipids would modulate the effects seen before, now resulting in very comparable results for the different nanomaterials. This indicates that interactions between surfactant proteins and lipids may actually counterbalance each other (Phelps 2001).

All these examples demonstrate that future studies are needed to understand the processes of nanomaterial interacting with the lung lining fluid and cells. In other words, bio–nano interactions in the lungs also are just as complex and as dynamic as in the bloodstream. In both scenarios we may observe exchange processes leading to an "evolution" of the corona but also synergistic effects may occur. PS is a lipid–protein mixture with a unique and complex 3D structure that originates from the interplay of its components (Serrano and Pérez-Gil 2006). Looking at these protein–lipid interactions, it becomes clear that nanomaterial-adsorbed molecules will also interact with nonbound molecules (which might be even part of surfactant membranes), adding another level of complexity to the situation. Other than in the bloodstream where the formation of the protein corona could be proven as an existing phenomenon, a detailed identification of a "pulmonary corona" mapping all protein and lipid compounds is still a missing milestone regarding the study of nanomaterials in the lungs, and will be the subject of future investigations.

4.3.4 Lessons Learned from NanoGEM

Regarding the *in situ* characterization of nanomaterials after contacting lung lining fluid, the availability of a suitable lung lining fluid model was one of the key issues in the nanoGEM project. To investigate interactions between nanomaterials and surfactant phospholipids, commercially available Curosurf® was chosen as a model for the lipophilic fraction of PS. However, to study the binding of surfactant associated proteins to these nanomaterials, this model is less suitable, as the majority of relevant proteins is removed from this biological matrix during the lipophilic extraction of Curosurf®. To overcome this limitation, whole native surfactant isolated from porcine lungs, containing the relevant SP-A, B, and C, while only traces of SP-D can be detected using 2D PAGE and mass spectrometry (unpublished nanoGEM data), was chosen to mimic the lung lining fluid.

Isolations were thoroughly characterized regarding their lipid as well as protein composition. Comparisons of the molar compositions of major lipid species (PC, PG, PI, and phosphatidylethanolamine) were in good correlation with values for PS described in the literature. However, isolated native surfactant forms large lamellar structures when dispersed in aqueous media, which furthermore tends to precipitate during centrifugation. Due to the physicochemical properties of the nanomaterials studied in nanoGEM, however, centrifugation was a necessary technique to separate nanomaterials from unbound biomolecules after dispersion in biological fluid. To exclude cosedimentation of the nanomaterials under investigation and unbound vesicular structures from native surfactant (possibly leading to misinterpreted adsorption results), a modified dispersion protocol was established. Particles were dispersed for 1 h in deuterated water (D_2O) at 37°C, which led to profoundly reduced amounts of sedimented surfactant material after centrifugation. After incubation in native surfactant, the binding of SP-A (as the most abundant surfactant protein) to various nanoGEM particles surfactant was studied using this improved dispersion protocol. By means of 1D SDS-PAGE with subsequent image analysis of the SP-A bands (~34 kDa), the amount of adsorbed protein could be determined. The data depicted in Figure 4.3 reveals an overall higher extent of SP-A binding for bare and amino-modified SiO_2 nanoparticles, whereas other modifications such as SiO_2 PEG or SiO_2 phosphate showed a rather moderate SP-A interaction.

On the one hand, such differences might be due to different surface properties (as for instance seen when comparing SiO_2 naked vs. SiO_2 PEG; compare also Section 5.2.3). On the other hand, due to the broad variety of tested materials and profound differences regarding their physicochemical properties (size, shape, surface area, etc.), a direct comparison is only possible within narrow limitations. Nevertheless, these data indicate that distinct variations can be observed among

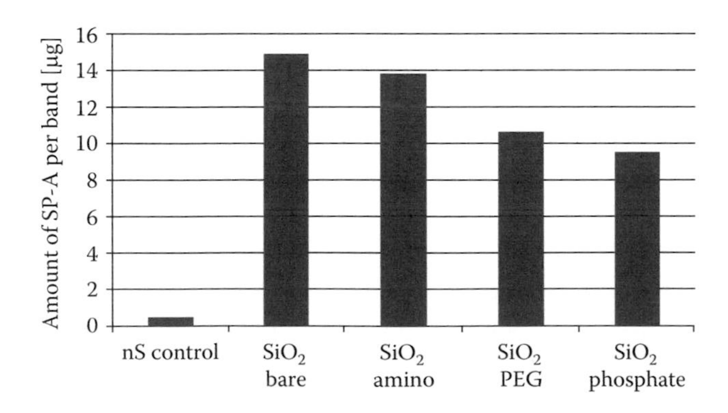

FIGURE 4.3 Adsorption of SP-A to four different variants of SiO_2. Nanoparticles were dispersed in native surfactant in deuterated water at 37°C for 1 h. After centrifugation, particles were washed once and subsequently analyzed using 1D gel electrophoresis. The SP-A band (34 kDa) was semiquantified using densitometry, and mean mass expressed in µg of adsorbed SP-A ($n = 2$).

different nanomaterials regarding their interaction with one component of the lung lining fluid, that is, SP-A, and it is interesting to note that the extremes of SiO_2 naked and SiO_2 phosphate match with *in vivo* inhalation inflammatory response (Table 8.5). However, further investigations are necessary to elucidate the patterns of adsorbed biomolecules from the complete lung lining fluid. Here, lessons learnt from the nanoGEM project regarding obstacles to overcome when *in situ* characterizing nanomaterials in the lung lining fluid (i.e., characterization of suitable biological models or methodological improvements regarding sample preparation) represent an important base of knowledge from which future investigations can be initiated.

4.3.5 Conclusions and Considerations for Future Studies

As the lung is a major uptake route with a high probability for nanomaterials to enter the body, the lung lining fluid is a highly important biological fluid to consider for *in situ* characterization of such materials. However, no model fluid samples can be obtained that are identical with the *in vivo* composition of lung lining fluid, which is in contrast to the bloodstream where serum or plasma is easily available. This makes it very difficult to choose a single good and representative biological model for the lung lining fluid. Therefore, many different models have been used so far (from single lipids, mixed single lipids up to different complex isolated surfactant preparations). The suitability of a model depends very much on the purpose of the study. The majority of studies deal with the impact of nanomaterials on the biophysical function of PS, which is an important parameter to assess. There are, however, only a few studies so far that focus on identification and mapping the binding components of PS, and on the question how biomolecule adsorption affects properties and the biological fate of nanomaterials. There are literally no studies so far on detailed analysis of the biomolecular “fingerprint” of a nanomaterial in the lung lining fluid. Successful interpretation of inhalation toxicity results and toxicity profiling will therefore very much depend on how far we will be able to understand the interactions at the bio–nano interface in the lung, which of course depends on standardized physiological models of the lung lining fluid.

4.4 EXAMPLE/CASE STUDY: NANOMATERIAL INTERACTIONS WITH BIOLOGICAL FLUIDS IN THE GASTROINTESTINAL TRACT

4.4.1 Physiological Considerations

Ingestion is considered another major route for unintentional exposure of nanomaterials. In addition, it is also a main route for intentional uptake of nanomedicine. The surface of the GIT is the largest epithelial surface in the human body (about 400 m^2) and also in terms of length scale it is one of the largest organs with the GIT being approximately 5 m long (or even up to 9 m without the muscle tonus) for a typical male human. GIT is divided into a lower and an upper part without a precise anatomical demarcation. After ingestion, the food passes from the mouth via the esophagus to the stomach and then into the duodenum, small intestine, and then into the large intestine.

Via oral ingestion humans may be exposed to several types of nanoparticles. Endogenous particles, being micro- and nano-sized, do occur naturally in food. Typically they may either be composed of inorganic salts such as calcium and phosphate or may contain food compounds such lipid vesicles or protein aggregates. Exogenous particles may be present in food as some particles are used as food additives such as TiO_2 or SiO_2. Such additives are typically microscaled but due to the large size distribution they always will contain a nanoscaled fraction. In addition one also needs to consider migration of nanoparticles for instance from food packaging materials. Similar to the formation of endogenous nanoparticles containing calcium phosphate (naturally occurring), nanoparticles may form *in vivo* also from other inorganic species such as ionic silver as shown in a recent study by van der Zande and coworkers (van der Zande et al. 2012). In this study the authors applied silver nanoparticles and also ionic silver in a 28d oral study. Dissection was performed either on day 29 or 1 or 8 weeks after last application. Interestingly at day 29 the authors could show the presence of silver containing nanoparticles in the GIT as well as in several organs (e.g., liver, spleen, and lung) not only in the nanosilver treated groups but also in the ionic silver treated group using Single Nanoparticle ICP-MS, although these particles were completely missing in control animals. This demonstrates that within the GIT, nanoparticles may be formed from various ions (endogenous or exogenous) when present in food. On the other hand, nanoparticles that may be present in food could be dissolved in gastric juices and be reformed again later on inside or outside GIT as several other studies suggest (Peters et al. 2012; Liu et al. 2012).

In recent years there is increasing evidence that human exposure to nano- or microparticles influences clinical manifestation of certain diseases. For instance Crohn's disease seems to be influenced by a combination of genetic predisposition and environmental factors. It has been hypothesized that microsized particles (100–1000 nm) might be associated with the development of this disease. Those particles might be endogenous (e.g., composed of calcium and phosphate) or exogenous (e.g., food additives like TiO_2) in origin. In a study performed by Lomer and colleagues, micro-TiO_2 and aluminosilicate accumulated in human gut associated lymphoid tissue—the tissue that is first affected in Crohn's disease (Lomer, Thompson, and Powell 2007). A diet low in endogenous and exogenous particles was then found to alleviate the disease symptoms.

4.4.2 Available Models

Of course it is not straightforward to purify original gastric juices from animal sources in the same way as this might be more or less achievable with lung fluids. But there are different types of simulants available that have already been used successfully to study nanoparticle behavior and that quite well mimic what might occur during passage of the nanoparticle through the GIT. The study performed by Peters and co-workers is a very good example, where such gastric fluid simulants were successfully used to characterize silica nanoparticle (Peters et al. 2012). In this study, the authors analyzed different types of food matrices with and without adding silica nanoparticles. They first dispersed the food matrix with and without silica nanoparticles in water to fully characterize the nanomaterials. To simulate what is going on in GIT, the food matrices were passed through an *in vitro* digestion model, which is built of three individual steps: artificial saliva (simulating the mouth), artificial gastric fluid (simulating the stomach), and artificial intestine fluid containing duodenal juice and bile juice (simulating the intestine). Basically these artificial juices are made of different salts mimicking the natural salt concentration inside these compartments and contain organic compounds that are typical for the respective compartment. The saliva has a pH of 6.8 and contains urea to simulate denaturation as well as amylase and mucins. The gastric juice has a pH of 1.3 and contains pepsin. The duodenal juice has a pH of 8.1 and contains pancreatin and lipase, whereas the bile juice has a similar pH and contains bile. Thus, these juices might very well simulate what is likely to happen *in vivo*. The food matrix is first dispersed in saliva, then passed over into the gastric fluid but still containing the saliva, and finally passed into intestinal fluid carrying on saliva and gastric fluid. The results showed that in all cases the food matrices contained 5%–40% silica nanoparticles in the size range of 5–200 nm when dispersed in saliva. However, in some of the food matrices such as coffee or instant soup those nanoparticles disappear under gastric digestion simulation at low pH and they reappear when the pH is neutralized again during simulation of intestine. Under intestinal conditions the nanoparticles are even present at a higher percentage compared to the start of the study in saliva (80% intestinal compared to 5%–40% saliva) and they also appear bigger in size indicating agglomeration. Taken together this demonstrates that nanoparticles might undergo a partial or complete dissolution and reformation when passing through the GIT. Thus it is unlikely that the particles after passing through the GIT will be the same as applied actually in all aspects, namely they may change chemical composition, size, and size distribution as well as surface.

Liu and coworkers performed a similar study using nano-silver (Liu et al. 2012). They could show that nano-silver is prone to oxidative dissolution in the GIT. Ionic silver may then interact with thiols and the resulting silver-biocomplexes may later on be photo-decomposed to inorganic Ag_2S nanoparticles. Interestingly selenide may rapidly displace sulfur leading to a selenide surface of these particles. Actually the authors then conclude the implications of these *in vitro* results for the situation *in vivo*. Nano-silver may completely dissolve upon oral uptake and may then be transported inside the human body as silver–thiol complexes. Later on, in skin, out of these silver biocomplexes nanoparticles can be reformed by photochemical reactions. In addition, those newly formed nanoparticles may be further distributed within the body. But actually the particles that are reformed will not be the same nanoparticles as before and will contain an altered chemical composition, size, and size distribution as well as surface. Both studies have important implications for *in vitro* testing of nanoparticles that should mimic oral uptake.

4.4.3 Conclusions and Considerations for Future Studies

There are several lines of evidence that changes occurring within the GIT are really fundamentally different from what can be observed in plasma or in lungs. Obviously the major issues inside GIT are not agglomeration/de-agglomeration or adsorption of biomolecules, but may be fundamental reforming of the particles. Obviously as shown for silica or silver, nanoparticles may completely dissolve inside GIT. Furthermore, the nanoparticles will not only get dissolved completely but also can be reformed later on (inside or outside the GIT), but then they would be very different in all aspects such as chemical composition, size, and size distribution as well as surface. The question that is remaining now is whether this is something general or actually does apply only for a very few selected types of nanomaterials. Most likely there should be also more biopersistent nanoparticles. When considering that *in vitro* studies are performed to predict *in vivo* outcomes and now taking into account that nanoparticles *in vivo* might be changed completely, it becomes quite evident that especially for predicting outcomes from oral studies the current design of *in vitro* studies might not be suitable, as nanoparticles are added directly into the cell culture medium just in the way they have been synthesized and not taking into account the (quite fundamental) changes that may occur during the GIT passage. A possible approach to overcome this issue would be to include something like the *in vitro* digestion model before actually adding the nanoparticle into the *in vitro* test system. Thus, one is at least mimicking the complex changes that could occur *in vivo*. Although the *in vitro* digestion protocols as such are standardized, one still needs to agree on a harmonized approach for how to use them in nano-safety testing. Only this would allow for being able to compare later on the outcome from different studies.

However, it also becomes easily obvious when analyzing the available literature that the oral route, both in terms of published toxicity studies as well as published *in situ* characterization of nanoparticles, is by far the least intensive studied route of uptake. Only a limited number of published oral studies do exist and actually even fewer deal with *in situ* characterization of the nanomaterials under conditions relevant for oral uptake. Many open questions remain and clearly much more research is needed to understand what is going on with the nanomaterials inside the GIT, also to significantly improve the quality and the relevance of *in vitro* studies for predicting outcome of oral studies.

4.5 SUMMARY AND CONCLUSION

Although careful and complete material characterization *as-synthesized* has been recognized since several years within the nanotoxicology field, the proper characterization inside the test medium or so to say the *in situ* characterization has long been ignored completely. Meanwhile consensus has grown that actually the nanomaterials need to be characterized under test conditions ("*in situ*") as well. Only when understanding the properties of the nanomaterial in the test system, be it an *in vitro* or an *in vivo* situation, one would finally be able to analyze and correlate the results from the toxicity testing and thus ultimately derive structure–activity relationships.

We have provided here several lines of evidence for different uptake routes that upon first interaction with the biological matrix, which is serum for i.v. injection, lung lining fluid for inhalation, or gastric juices for oral uptake, the nanomaterials will change significantly with respect to size (e.g., aggregation/agglomeration), adsorption of biomolecules (e.g., formation of a protein or lipid corona), or particle integrity (e.g., dissolution). This in turn has implications for nanomaterial uptake, biodistribution and also for interactions occurring inside cells and finally also for the outcome of toxicity studies. In addition, we summarized what approaches are available and useful to study nanomaterials *in situ* with respect to available model fluids and also with respect to techniques that are appropriate depending on the question underlying the study.

Studying nanoparticles in serum-containing cell culture medium is a useful and reasonable approach to understand changes likely to occur under conditions of *in vitro* testing. In addition, human plasma or serum is easily available to analyze changes in nanomaterials occurring inside the blood circulation. For the inhalation route, the picture is a little bit more diverse, as plenty of different protocols exist for obtaining more or less physiologically relevant lung lining fluids. There is a direct access to native lung lining fluid, but one needs to purify native surfactant from a diluted BAL fluid in a laborious protocol. Nevertheless, actually quite physiological model fluids are finally available for studying pulmonary interactions. However, the situation is completely different for the GIT where gastric juices are not easily accessible and need to be replaced by artificial simulant fluids.

Taken together, although several studies are available reporting on *in situ* characterization of nanomaterials in different biological environments, very few studies actually put the complete picture together, meaning only a few studies aim at full *in situ* characterization (e.g. mapping all interacting biomolecules) instead of typically focusing each study on a certain aspect. For some biomolecules such as sugars that currently may be out of technical scope as techniques to do so are not readily available but for lipids or proteins it may well be achievable. Also, only very few studies try to systematically assess nanomaterial properties *in situ* and in parallel toxicity *in vitro* and *in vivo* to correlate the results on various levels (*in vitro* versus *in vivo* toxicity; *in situ* characteristics versus *in vitro*/*in vivo* toxicity). Partially this is of course due to the fact that getting the complete picture is extremely time and cost intensive and requires technical skills that are hardly found within a single laboratory but also partly due to the fact that harmonized approaches are currently lacking. Moreover, harmonized biological model fluids are missing, which then hinders direct comparison of results that have already been published in different studies.

We provided insights into *in situ* characterization as being performed within the project nanoGEM. NanoGEM is one of the integrated efforts to obtain an as complete as possible picture for 16 different nanomaterials in terms of *in situ* characterization and also in terms of uptake into cells, toxicity *in vitro* and *in vivo* as well as toxicity mechanisms. Within nanoGEM, a first correlation of all these data will be done, however, a more detailed correlation analysis will certainly be valuable.

Finally it will be very interesting to see how hazard ranking of nanomaterials will change when in parallel to characteristics as-synthesized, the respective characteristics *in situ* are used. In addition, the *in situ* characteristics need also to be taken into consideration when developing "nanoQSAR" models. Last but not least, ultimately the characterization of nanomaterials in the respective test system in terms of agglomeration and also in terms of biomolecule interaction ("molecular fingerprint") would eventually enable prediction of biological interactions and possible toxicity of a material and thus could be a valuable independent step within a complex testing strategy.

GLOSSARY

Ex situ Analysis of nanomaterials after prior re-isolation of the nanomaterial out of a test matrix (i.e., after incubation with biomolecules, the nanomaterials are isolated to separate from non-bound materials to perform the analysis of adsorbed nanomaterials).

Ex vivo Experiments performed with whole isolated organs.

In situ Analysis of nanomaterials inside a test matrix (e.g., inside cells, tissues, or in biologically relevant test fluids) without any processing or re-isolation of the nanomaterial before analysis.

In vitro Experiments performed under simplified conditions using isolated components of the organism (cell lines or primary cells in single culture or co-culture).

In vivo Experiments performed using whole organisms (i.e., with maximal biological complexity).

REFERENCES

Aggarwal, P., Hall, J. B., McLeland, C. B. et al. (2009). Nanoparticle interaction with plasma proteins as it relates to particle biodistribution, biocompatibility and therapeutic efficacy. *Adv Drug Del Rev* 61:428–437.

Anderson, N. L., and Anderson, N. G. (2002). The human plasma proteome: history, character, and diagnostic prospects. *Mol Cell Proteomics* 2:845–867.

Antunes, M. B., Gudis, D. A., and Cohen, N. A. (2009). Epithelium, cilia, and mucus: Their importance in chronic rhinosinusitis. *Immunol Allergy Clin* 29:631–643.

Bajaj, A., Samanta, B., Yan, H. et al. (2009). Stability, toxicity and differential cellular uptake of protein passivated-Fe_3O_4 nanoparticles. *J Mater Chem* 19:6328–6331.

Bastacky, J., Lee, C. Y., Goerke, J. et al. (1995). Alveolar lining layer is thin and continuous: Low-temperature scanning electron microscopy of rat lung. *J Appl Physiol* 79:1615–1628.

Bernhard, W., Mottaghian, J., Gebert, A. et al. (2000). Commercial versus native surfactants. Surface activity, molecular components, and the effect of calcium. *Am J Respir Crit Care Med* 162:1524–1533.

Blanco, O., Cruz, A., Ospina, O. L. et al. (2012). Interfacial behavior and structural properties of a clinical lung surfactant from porcine source. *Biochim Biophys Acta* 1818:2756–2766.

Blanco, O., and Pérez-Gil, J. (2007). Biochemical and pharmacological differences between preparations of exogenous natural surfactant used to treat Respiratory Distress Syndrome: Role of the different components in an efficient pulmonary surfactant. *Eur J Pharmacol* 568:1–15.

Blunk, T., Hochstrasser, D. F., Sanchez, J. C. et al. (1993). Colloidal carriers for intravenous drug targeting: plasma protein adsorption patterns on surface-modified latex particles evaluated by two-dimensional polyacrylamide gel electrophoresis. *Electrophoresis* 14:1382–1387.

Cabaleiro-Lago, C., Quinlan-Pluck, F., Lynch, I. et al. (2008). Inhibition of amyloid beta protein fibrillation by polymeric nanoparticles. *J Am Chem Soc* 130:15437–15443.

Cedervall, T., Lynch, I., Lindman, S. et al. (2007). Understanding the nanoparticle-protein corona using methods to quantify exchange rates and affinities of proteins for nanoparticles. *Proc Natl Acad Sci USA* 104:2050–2055.

Chakraborti, S., Chatterjee, T., Joshi, P. et al. (2010). Structure and activity of lysozyme on binding to ZnO nanoparticles. *Langmuir* 26:3506–3513.

Dawson, K. A., Anguissola, S., and Lynch, I. (2012). The need for *in situ* characterisation in nanosafety assessment: funded transnational access via the QNano research infrastructure. *Nanotoxicology* 7:346–349.
Dell'Orco, D., Lundqvist, M., Oslakovic, C. et al. (2010). Modeling the time evolution of the nanoparticle-protein corona in a body fluid. *PLoS ONE* 5, e10949:1–8.
Deng, Z. J., Liang, M., Monteiro, M. et al. (2010). Nanoparticle-induced unfolding of fibrinogen promotes mac-1 receptor activation and inflammation. *Nat Nanotechnol* 6:39–44.
Drescher, D., Büchner, T., McNaughton, D. et al. (2013). SERS reveals the specific interaction of silver and gold nanoparticles with hemoglobin and red blood cell components. *Phy Chem Chem Phys* 15:5364–5373.
Fan, Q., Wang, Y. E., Zhao, X. et al. (2011). Adverse biophysical effects of hydroxyapatite nanoparticles on natural pulmonary surfactant. *ACS Nano* 5:6410–6416.
Gasser, M., Rothen-Rutishauser, B. M., Krug, H. F. et al. (2010). The adsorption of biomolecules to multi-walled carbon nanotubes is influenced by both pulmonary surfactant lipids and surface chemistry. *J Nanobiotechnol* 8(31):1–9.
Gasser, M., Wick, P., Clift, M. J. D. et al. (2012). Pulmonary surfactant coating of multi-walled carbon nanotubes (MWCNTs) influences their oxidative and pro-inflammatory potential *in vitro*. *Part Fibre Toxicol* 9(17):1–13.
Geiser, M., and Kreyling, W. G. (2010). Deposition and biokinetics of inhaled nanoparticles. *Part Fibre Toxicol* 7:1–17.
Gessner, A., Lieske, A., Paulke, B. et al. (2002). Influence of surface charge density on protein adsorption on polymeric nanoparticles: analysis by two-dimensional electrophoresis. *Eur J Pharm Biopharm* 54:165–170.
Gessner, A., Waicz, R., Lieske, A. et al. (2000). Nanoparticles with decreasing surface hydrophobicities: Influence on plasma protein adsorption. *Int J Pharm* 196:245–249.
Goerke, J. (1998). Pulmonary surfactant: Functions and molecular composition. *Biochim Biophys Acta* 1408:79–89.
Gray, J. J. (2004). The interaction of proteins with solid surfaces. *Curr Opin Struct Biol* 14:110–115.
Grenha, A., Seijo, B., Serra, C. et al. (2008). Surface characterization of lipid/chitosan nanoparticles assemblies, using X-ray photoelectron spectroscopy and time-of-flight secondary ion mass spectrometry. *J Nanosci Nanotechnol* 8:358–365.
Griese, M. (1999). Pulmonary surfactant in health and human lung diseases: state of the art. *Eur Respir J* 13:1455–1476.
Halliday, H. L. (2008). Surfactants: past, present and future. *J Perinatol* 28 Suppl 1:S47–S56.
Harishchandra, R. K., Saleem, M., and Galla, H.-J. (2010). Nanoparticle interaction with model lung surfactant monolayers. *J R Soc Interface* 7:S15–S26.
Hellstrand, E., Lynch, I., Andersson, A. et al. (2009). Complete high-density lipoproteins in nanoparticle corona. *FEBS J* 276:3372–3381.
Jiang, J., Oberdörster, G., and Biswas, P. (2008). Characterization of size, surface charge, and agglomeration state of nanoparticle dispersions for toxicological studies. *J Nanopart Res* 11:77–89.
Johnston, H., Brown, D., Kermanizadeh, A. et al. (2012). Investigating the relationship between nanomaterial hazard and physicochemical properties: Informing the exploitation of nanomaterials within therapeutic and diagnostic applications. *J Control Release* 164:307–313.
Kapralov, A. A., Feng, W. H., Amoscato, A. A. et al. (2012). Adsorption of surfactant lipids by single-walled carbon nanotubes in mouse lung upon pharyngeal aspiration. *ACS Nano* 6:4147–4156.

Karajanagi, S. S., Vertegel, A. A., Kane, R. S. et al. (2004). Structure and function of enzymes adsorbed onto single-walled carbon nanotubes. *Langmuir* 20:11594–11599.

Kendall, M. (2007). Fine airborne urban particles (PM2.5) sequester lung surfactant and amino acids from human lung lavage. *Am J Physiol Lung Cell Mol Physiol* 293:L1053–L1058.

Kishore, U., Greenhough, T. J., Waters, P. et al. (2006). Surfactant proteins SP-A and SP-D: Structure, function and receptors. *Mol Immunol* 43:1293–1315.

Klein, J. (2007). Probing the interactions of proteins and nanoparticles. *Proc Natl Acad Sci USA* 104:2029–2030.

Lazzari, S., Moscatelli, D., Codari, F. et al. (2012). Colloidal stability of polymeric nanoparticles in biological fluids. *J Nanopart Res* 14:920.

Lesniak, A., Campbell, A., Monopoli, M. P. et al. (2010). Serum heat inactivation affects protein corona composition and nanoparticle uptake. *Biomaterials* 31:9511–9518.

Lesniak, A., Fenaroli, F., Monopoli, M. P. et al. (2012). Effects of the presence or absence of a protein corona on silica nanoparticle uptake and impact on cells. *ACS Nano* 6:5845–5857.

Linse, S., Cabaleiro-Lago, C., Xue, W.-F. et al. (2007). Nucleation of protein fibrillation by nanoparticles. *Proc Natl Acad Sci USA* 104:8691–8696.

Liu, J., Wang, Z., Liu, F. D. et al. (2012). Chemical transformations of nanosilver in biological environments. *ACS Nano* 6:9887–9899.

Lomer, M. C. E., Thompson, R. P. H., and Powell, J. J. (2007). Fine and ultrafine particles of the diet: Influence on the mucosal immune response and association with Crohn's disease. *Proc Nutr Soc* 61:123–130.

Lundqvist, M., Stigler, J., Cedervall, T. et al. (2011). The evolution of the protein corona around nanoparticles: A test study. *ACS Nano* 5:7503–7509.

Lundqvist, M., Stigler, J., Elia, G. et al. (2008). Nanoparticle size and surface properties determine the protein corona with possible implications for biological impacts. *Proc Natl Acad Sci USA* 105:14265–14270.

Lynch, I., Cedervall, T., Lundqvist, M. et al. (2007). The nanoparticle-protein complex as a biological entity; a complex fluids and surface science challenge for the 21st century. *Adv Colloid Interface Sci* 134–135:167–174.

Lynch, I., Dawson, K. A., and Linse, S. (2006). Detecting cryptic epitopes created by nanoparticles. *Sci STKE* 327:14–20.

Lynch, I., Salvati, A., and Dawson, K. A. (2009). Protein-nanoparticle interactions: What does the cell see? *Nat Nanotechnol* 4:546–547.

Maffre, P., Nienhaus, K., Amin, F. et al. (2011). Characterization of protein adsorption onto FePt nanoparticles using dual-focus fluorescence correlation spectroscopy. *Beilstein J Nanotechnol* 2:374–383.

Mahmoudi, M., Lynch, I., Ejtehadi, M. R. et al. (2011). Protein-nanoparticle interactions: Opportunities and challenges. *Chem Rev* 111:5610–5637.

Maiorano, G., Sabella, S., Sorce, B. et al. (2010). Effects of cell culture media on the dynamic formation of protein-nanoparticle complexes and influence on the cellular response. *ACS Nano* 4:7481–7491.

Maskos, M., and Stauber, R. H. (2011). Characterization of nanoparticles in biological environments. In *Comprehensive Biomaterials*, 329–339. Elsevier.

Milani, S., Baldelli Bombelli, F., Pitek, A. S. et al. (2012). Reversible versus Irreversible binding of transferrin to polystyrene nanoparticles: Soft and hard corona. *ACS Nano* 6:2532–2541.

Monopoli, M. P., Walczyk, D., Campbell, A. et al. (2011). Physical–chemical aspects of protein corona: Relevance to *in vitro* and *in vivo* biological impacts of nanoparticles. *J Am Chem Soc* 133:2525–2534.

Montes-Burgos, I., Walczyk, D., Hole, P. et al. (2010). Characterisation of nanoparticle size and state prior to nanotoxicological studies. *J Nanopart Res* 12:47–53.

Morra, M. (2000). On the molecular basis of fouling resistance. *J Biomater Sci Polym Ed* 11:547–569.
Murdock, R. C., Braydich-Stolle, L., Schrand, A. M. et al. (2008). Characterization of nanomaterial dispersion in solution prior to *in vitro* exposure using dynamic light scattering technique. *Toxicol Sci* 101:239–253.
Nakanishi, K., Sakiyama, T., and Imamura, K. (2001). On the adsorption of proteins on solid surfaces, a common but very complicated phenomenon. *J Biosci Bioeng* 91:233–244.
Nel, A. E., Mädler, L., Velegol, D. et al. (2009). Understanding biophysicochemical interactions at the nano—Bio interface. *Nat Biotechnol* 8:543–557.
Nichols, G., Byard, S., Bloxham, M. J. et al. (2002). A review of the terms agglomerate and aggregate with a recommendation for nomenclature used in powder and particle characterization. *J Pharm Sci* 91:2103–2109.
OECD. (2012). Guidance on sample preparation and dosimetry for the safety testing of manufactured nanomaterials. In *Series on the Safety of Manufactured Nanomaterials Nr. 36*: OECD Environment, Health and Safety Publications:1–93.
Pan, H., Qin, M., Meng, W. et al. (2012). How do proteins unfold upon adsorption on nanoparticle surfaces? *Langmuir* 28:12779–12787.
Paur, H.-R., Cassee, F. R., Teeguarden, J. G. et al. (2011). In-vitro cell exposure studies for the assessment of nanoparticle toxicity in the lung—A dialog between aerosol science and biology. *J Aerosol Sci* 42:668–692.
Peters, R., Kramer, E., Oomen, A. G. et al. (2012). Presence of nano-sized silica during *in vitro* digestion of foods containing silica as a food additive. *ACS Nano* 6:2441–2451.
Petri-Fink, A., Steitz, B., Finka, A. et al. (2008). Effect of cell media on polymer coated superparamagnetic iron oxide nanoparticles (SPIONs): Colloidal stability, cytotoxicity, and cellular uptake studies. *Eur J Pharm Biopharm* 68:129–137.
Phelps, D. (2001). Surfactant regulation of host defense function in the lung: A question of balance. *Pediatr Pathol Mol Med* 20:269–292.
Pieper, R., Gatlin, C. L., Makusky, A. J. et al. (2003). The human serum proteome: Display of nearly 3700 chromatographically separated protein spots on two-dimensional electrophoresis gels and identification of 325 distinct proteins. *Proteomics* 3:1345–1364.
Pitek, A. S., O'Connell, D., Mahon, E. et al. (2012). Transferrin coated nanoparticles: Study of the bionano interface in human plasma. *PLoS ONE* 7, e40685:1–14.
Powers, K. W. (2005). Research strategies for safety evaluation of nanomaterials. Part VI. Characterization of nanoscale particles for toxicological evaluation. *Toxicol Sci* 90:296–303.
Powers, K. W., Palazuelos, M., Moudgil, B. M. et al. (2007). Characterization of the size, shape, and state of dispersion of nanoparticles for toxicological studies. *Nanotoxicology* 1:42–51.
Rabe, M., Verdes, D., and Seeger, S. (2011). Understanding protein adsorption phenomena at solid surfaces. *Adv Colloid Interface Sci* 162:87–106.
Ramirez-Garcia, S., Chen, L., Morris, M. A. et al. (2011). A new methodology for studying nanoparticle interactions in biological systems: Dispersing titania in biocompatible media using chemical stabilisers. *Nanoscale* 3:4617–4624.
Roach, P., Farrar, D., and Perry, C. C. (2006). Surface tailoring for controlled protein adsorption: effect of topography at the nanometer scale and chemistry. *J Am Chem Soc* 128:3939–3945.
Ruge, C. A., Kirch, J., Cañadas, O. et al. (2011). Uptake of nanoparticles by alveolar macrophages is triggered by surfactant protein A. *Nanomedicine* 7:690–693.
Ruge, C. A., Schaefer, U. F., Herrmann, J. et al. (2012). The interplay of lung surfactant proteins and lipids assimilates the macrophage clearance of nanoparticles. *PLoS ONE* 7, e40775:1–10.

Rugonyi, S., Biswas, S. C., and Hall, S. B. (2008). The biophysical function of pulmonary surfactant. *Respir Physiol Neurobiol* 163:244–255.

Sachan, A. K., Harishchandra, R. K., Bantz, C. et al. (2012). High-resolution investigation of nanoparticle interaction with a model pulmonary surfactant monolayer. *ACS Nano* 6:1677–1687.

Salvador-Morales, C., Townsend, P., Flahaut, E. et al. (2007). Binding of pulmonary surfactant proteins to carbon nanotubes; potential for damage to lung immune defense mechanisms. *Carbon* 45:607–617.

Sanders, N., Rudolph, C., Braeckmans, K. et al. (2009). Extracellular barriers in respiratory gene therapy. *Adv Drug Del Rev* 61:115–127.

Sayes, C. M., Reed, K. L., Subramoney, S. et al. (2008). Can *in vitro* assays substitute for *in vivo* studies in assessing the pulmonary hazards of fine and nanoscale materials? *J Nanopart Res* 11:421–431.

Sayes, C. M., and Warheit, D. B. (2009). Characterization of nanomaterials for toxicity assessment. *Wiley Interdiscip Rev Nanomed Nanobiotechnol* 1:660–670.

Schleh, C., Mühlfeld, C., Pulskamp, K. et al. (2009). The effect of titanium dioxide nanoparticles on pulmonary surfactant function and ultrastructure. *Respir Res* 10 (90):1–11.

Schleh, C., Rothen-Rutishauser, B., and Kreyling, W. G. (2011). The influence of pulmonary surfactant on nanoparticulate drug delivery systems. *Eur J Pharm Biopharm* 77:350–352.

Schulze, C., Schaefer, U. F., Ruge, C. A. et al. (2011). Interaction of metal oxide nanoparticles with lung surfactant protein A. *Eur J Pharm Biopharm* 77:376–383.

Schulze, C., Schulze, C., Kroll, A. et al. (2008). Not ready to use—Overcoming pitfalls when dispersing nanoparticles in physiological media. *Nanotoxicology* 2:51–61.

Schürch, S., Gehr, P., Im Hof, V. et al. (1990). Surfactant displaces particles toward the epithelium in airways and alveoli. *Respir Physiol* 80:17–32.

Serrano, A. G., and Pérez-Gil, J. (2006). Protein-lipid interactions and surface activity in the pulmonary surfactant system. *Chem Phys Lipids* 141:105–118.

Shang, L., Wang, Y., Jiang, J. et al. (2007). pH-dependent protein conformational changes in albumin:gold nanoparticle bioconjugates: a spectroscopic study. *Langmuir* 23:2714–2721.

Shemetov, A. A., Nabiev, I., and Sukhanova, A. (2012). Molecular interaction of proteins and peptides with nanoparticles. *ACS Nano* 6:4585–4602.

Tatur, S., and Badia, A. (2012). Influence of hydrophobic alkylated gold nanoparticles on the phase behavior of monolayers of dppc and clinical lung surfactant. *Langmuir* 28:628–639.

Tenzer, S., Docter, D., Kuharev, J. et al. (2013). Rapid formation of plasma protein corona critically affects nanoparticle pathophysiology. *Nat Nanotechnol* 8:772–781.

Tenzer, S., Docter, D., Rosfa, S. et al. (2011). Nanoparticle size is a critical physicochemical determinant of the human blood plasma corona: A comprehensive quantitative proteomic analysis. *ACS Nano* 5:7155–7167.

van der Zande, M., Vandebriel, R. J., Van Doren, E. et al. (2012). Distribution, elimination, and toxicity of silver nanoparticles and silver ions in rats after 28-day oral exposure. *ACS Nano* 6:7427–7442.

Vroman, L., Adams, A. L., Fischer, G. C. et al. (1980). Interaction of high molecular weight kininogen, factor XII, and fibrinogen in plasma at interfaces. *Blood* 55:156–159.

Walczyk, D., Bombelli, F. B., Monopoli, M. P. et al. (2010). What the cell "sees" in bionanoscience. *J Am Chem Soc* 132:5761–5768.

Walkey, C. D., and Chan, W. C. W. (2012). Understanding and controlling the interaction of nanomaterials with proteins in a physiological environment. *Chem Soc Rev* 41:2780–2799.

Warheit, D. B. (2008). How meaningful are the results of nanotoxicity studies in the absence of adequate material characterization? *Toxicol Sci* 101:183–185.

Wohlleben, W. (2012). Validity range of centrifuges for the regulation of nanomaterials: From classification to as-tested coronas. *J Nanopart Res* 14:1–18.

Wright, J. R. (2005). Immunoregulatory functions of surfactant proteins. *Nat Rev Immunol* 5:58–68.

Wu, Z., Zhang, B., and Yan, B. (2009). Regulation of enzyme activity through interactions with nanoparticles. *Int J Mol Sci* 10:4198–4209.

Zolnik, B. S., González-Fernández, A., Sadrieh, N. et al. (2010). Nanoparticles and the immune system. *Endocrinology* 151:458–465.

Section II

Hazard Assessment for Humans

5 Lessons Learned from Unintentional Aerosols

Joseph Brain

CONTENTS

5.1 INTRODUCTION

Much of this book focuses on nanomaterials as a new, and to some extent, unknown class of materials. Physicists, chemists, and material scientists working with entrepreneurs are designing engineered nanoparticles with novel characteristics and properties appropriate to their application including paints, coatings, structural constituents, to nanomedicines. This chapter emphasizes the reality that although many nanomaterials are novel—such as fullerenes, graphene, carbon nanotubes, and nanomedicines—others are familiar and have existed for millennia. In general, these particles, which we will call unintentional particles, are the smallest particles in a distribution of polydisperse particles produced through combustion or as a consequence of evaporation and condensation of volatized materials such as metals, organic compounds, or even sea water.

A key theme of this chapter is that polydisperse aerosols, particularly those from combustion or crushing, have always included a nanofraction that can be detected when appropriate technologies have been used to search for it. The paradigm for evaluating these incidental aerosols is common to risk characterization for both larger sizes of polydisperse distributions or of sophisticated, highly engineered nanomaterials. In all cases, we use the paradigm of hazard identification, dose-response characterization, exposure assessment, cost–benefit analysis, and ultimately, risk management. Also common is the importance of characterizing the particles in terms of chemical form, size, shape, surface area, number, density, extent of agglomeration, porosity, charge, reactivity, solubility, durability, crystalline structure, and other important and relevant physical and chemical properties.

Thus, although highly engineered nanomaterials are new and have unique and unexpected attributes, humans have always been exposed to nanoparticles as the smallest size fraction of polydisperse aerosols. This can be the result of wind, crushing, grinding, and/or combustion. Since Prometheus discovered fire, humans using fire have been exposed to nanoparticles. This is an important topic, because even though they are novel, highly engineered nanomaterials lead to exposures for relatively few individuals. Quantum dots and fullerenes, or even nanomedicines, account for relatively few human exposures. Even commodity carbon nanotubes or metal oxides rarely expose the public. In contrast, unintentional aerosols, like air pollution, indoor cooking, and natural processes, lead to huge numbers of exposures. Everyone on the planet is exposed to unintentional aerosols.

We point out that most essential questions, such as where do particles deposit, are they cleared, do particles dissolve, and do constituents of particles reach other parts of the body, and especially what biological responses are likely, are common to both nanoparticles, and larger particles. We are concerned with the same organs, such as the respiratory tract, the GI system, and skin. The laws of aerosol physics controlling particle deposition, such as gravity, diffusion, and inertial impaction, are the same. Many of the tools used in the laboratory are similar and independent of particle size. Especially, the repertoire of biologic responses is similar. Mechanisms, such as the importance of reactive oxygen species, are shared between nanoparticles and their larger and more familiar cousins.

Finally, it is also the case that many nanoparticles that meet the bright-line of size as individual particles are usually present in the form of agglomerates. They may be nano in size initially, but when they are collected, shipped, and then reused, complete dispersion of the particles is unlikely and often impossible. Thus, the public and workers exposed to nanomaterials are most likely to breathe aggregates, which are larger than the 100 nm bright-line.

It is important to understand the deposition, clearance, and translocation of unintentional nanoparticles. There is no reason to think that the physical and biological mechanisms that pertain to them are unique to unintentional aerosols. As detailed elsewhere in this book, there are considerable data on these topics in relation to nanoparticles. These principles apply equally well to unintentional aerosols as they do to highly engineered aerosols. Whether they are uniformly small or simply part of the tail of the size distribution of polydisperse aerosols, they behave similarly. Like all nanoparticles, unintentional aerosols may have an increased potential for translocation from the air to the blood. The percent of particles that cross the air–blood barrier remains small, but can be significant. Uptake by microphages generally makes it less likely that they cross epithelial and endothelial cells, which separate inspired air from circulating blood.

5.2 TYPES OF NANOPARTICLES

There are many different kinds of nanomaterials. We choose to broadly divide them into three classes. First, there are nanomaterials that have been around for many decades, such as colloidal gold and colloidal carbon (India ink). Second, many familiar materials that are small and polydisperse have particles in the nano range. For example, there is evidence that both diesel particles (Kinsey et al. 2007) and welding fume (Sowards et al. 2008) contain nanomaterials. These nanofractions appear to have an unusual ability to cross the air–blood barrier intact, where they are subsequently ingested by hepatic macrophages and other components of the mononuclear phagocyte system that have access to the circulating blood. Finally, there is the class of new, highly engineered materials such as nanofibers, fullerenes, and quantum dots, as well as abundant nanomaterials designed for oncology or other pharmaceutical purposes. These are truly novel, and their potential toxicity is largely unexplored (see Table 5.1).

Colloids are stable emulsions. For centuries, we have known of the existence of such formulations. The particles are small and electrically charged enough that, even though they may have a density greater than 1, they remain in suspension. The molecular motion of solvent molecules (Brownian motion) is sufficient to keep them suspended and their charged surfaces (repulsive forces) prevent them from aggregating when particles approach each other. India ink is a stable colloid of carbon nanoparticles. Majno and Palade used India ink, including some nanoparticles, to characterize inflammation and to study the anatomic sites responsible for increased vascular permeability (Majno and Palade 1961). Intravenously injected ^{198}Au colloids were used decades ago to treat liver cancer. Brain and Corkery used nanoparticles of ^{198}Au to quantify *in situ* lung macrophage phagocytosis rates (Brain and Corkery 1975). Unlike bare nanoparticles of gold, these nanoparticles (approximately 50 nm) were coated with collagen. This charged proteinaceous surface made them easily phagocytized by alveolar macrophages.

Further, there are many small particles, particularly those produced by combustion or condensation, which are respirable. Usually, these polydisperse aerosols contain particles that extend down into the nano range. We will emphasize these in this chapter.

TABLE 5.1
Classes of Nanomaterials

1. Stable emulsions: colloidal gold, colloidal carbon (India ink).
2. Smallest particles of familiar polydispersed materials: welding fume, smoke, and diesel particles. All have fractions less than 100 nm.
3. Highly engineered nanoparticles with unusual properties: fullerenes, quantum dots, nanomedicines, carbon nanotubes.

5.3 CHARACTERIZING UNINTENTIONAL AEROSOLS

5.3.1 Size-Selective Samplers

We need more research specifically designed to look at nanoparticles from these familiar polydisperse sources. This requires size-selective devices, which collect and characterize particles smaller than 100 nm. Second, there is a larger literature, for example, on $PM_{2.5}$ and PM_{10}, in which the epidemiologic studies for humans or inhalation toxicology papers dealing with studies for animals include nanoparticles but the major mass component is larger in size.

It is important to remember that mass is proportional to the cube of the linear dimension. Thus, the same mass of a given material has widely different numbers of particles depending on their size. If you have identical masses of 10 micron, 1 micron, 0.1 micron, and 0.01 micron particles, each order of magnitude is associated with a 1000-fold increase in the number of particles needed to produce the same weight. Thus, you need to have a million 100-nm particles to equal the mass of a single 10-micron particle. If the particles are only 10 nm in size, then a billion particles will be needed to have the same mass as a single 10-micron particle.

5.3.2 Selecting Unintentional Aerosols from Polydisperse Particles

A problem with studying the biologic responses to unintentional nanoparticles is the lack of samples of characterized reagents. By definition, we need to deal with a polydisperse aerosol and select only the ultrafine particles (UFPs) (those smaller than 0.1 μm). A key challenge is to create methods for selecting these particles and using them to expose humans and animals to "real world" ambient UFPs. A number of size selective devices exist to collect UFPs on filtered media. The challenge is then to remove them quantitatively and without altering the particles. A far better approach is to select the UFPs aerodynamically and to keep them suspended for subsequent inhalation exposures.

Demokritou et al. have developed an ultrafine particle concentrator. It has been used extensively to conduct both human and animal inhalation studies of concentrated ambient UFPs, typically urban air, which is dominated by both mobile and stationary sources (Demokritou et al. 2002a; Demokritou et al. 2002b; Gupta et al. 2004). The design first selects UFPs by their aerodynamic diameter and then grows them to a supermicron size by using them as nuclei for growth in high-humidity environments. These larger particles can then be concentrated with a series of virtual impactors. Finally, the concentrated aerosol can be returned to its initial ambient size distribution using a thermal "reshaping" process. The device operates under a wide range of ambient air temperatures and relative humidities. Particle losses in the system were about 10%.

Scientists at the United States Environmental Protection Agency (EPA), as well as at universities such as the University of Rochester and the University of Toronto, have used this system successfully. This device represents an evolution of an earlier device that characterizes ambient particles in the fine range (0.1–2.5 μm) (Demokritou et al. 2002a). This device is widely known as the Harvard Ambient Particle Concentrator. An excellent review of how to measure fine, coarse, and UFPs has been prepared by Sarnat et al. (Sarnat et al. 2003; Demokritou et al. 2003).

5.4 CLASSES OF UNINTENTIONAL AEROSOLS

5.4.1 Tobacco Smoke

Cigarette smoke is a familiar example of an aerosol with abundant nanoparticles. Many of these unintentional nanoparticles, such as those in cigarette smoke, were not known or poorly described simply because investigators lacked the technologies to identify, count, and characterize particles smaller than 100 nm. Thus, as we look at the literature for tobacco smoke or air pollution particulates, there is a gradual progression, which increasingly includes a discussion and quantification of nanomaterials. This is not because they were absent in earlier decades but that investigators failed to look for nanoparticles. They lacked adequate technologies. For example, mass sampling tells us nothing about size unless an appropriate size-selective filter is used prior to mass sampling.

Hundreds of millions of individuals are exposed to either active or passive tobacco smoke. As shown in Figure 5.1, tobacco smoke is composed of polydisperse particles, all submicronic (Ning et al. 2006). Figure 5.1 clearly shows that particle numbers depend on time for both low tar and high tar cigarettes. Particle density is so high that particle collisions and resulting agglomeration are common. As time increases by a factor of 10, the numbers of particles are substantially reduced. However, under all conditions, there are a significant number of particles smaller than 100 nm.

A recent study compared conventional cigarettes with electronic cigarettes, which generate aerosols of water, nicotine, and propylene glycol (PG) or vegetable glycerin (VG) (Zhang et al. 2013). A scanning mobility particle sizer (SMPS) was used to

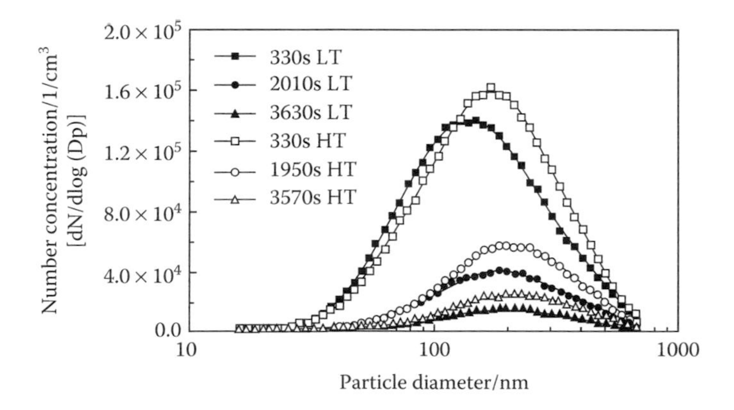

FIGURE 5.1 Nanoparticles in environmental tobacco smoke (number concentration produced by two high tar and two low tar cigarettes with increasing time). Investigators at Hong Kong Polytechnic University used a scanning mobility particle sizer (SMPS; TSI Model 3934) to characterize both size and particle number. (Reprinted from *Sci. Total Environ.*, 367, Ning, Z. et al., Experimental study of environmental tobacco smoke particles under actual indoor environment, 822–830, Copyright 2006, with permission from Elsevier.)

characterize particle number and diameter. They demonstrated considerable numbers of nanoparticles, particularly for the dominant constituent of the electronic cigarettes, PG. They conclude that *in vitro* experiments show that electronic cigarettes and conventional cigarettes produce aerosols with similar particle sizes.

5.4.2 Air Pollution

There is increasing recognition that particulate matter (PM) in ambient air both indoors and outdoors contains significant fractions of ultrafine particles (UFPs) (100 nm or less). Even though their contributions to the mass of total suspended particulates, $PM_{2.5}$ or PM_{10}, are small, they are the dominant contributors to particle number and contribute significant surface areas. A recent report by the Health Effects Institute summarizes concerns and concludes that there is significant evidence showing that the health of animals and humans can be affected by UFPs (HEI 2013).

Epidemiologic studies are more difficult because exposure to UFPs is generally not documented. However, the contribution of UFPs to overall health effects from air pollution cannot be excluded. This report and others conclude that the contribution of UFPs is not proportional to their mass. Their toxicity per unit weight is greater than that of larger particles. Another attribute of UFPs, and indeed for all nanoparticles, is that they are more persistent because they are slowly removed from ambient air by settling or by inertial impaction. They often tend to grow because of regulation mechanisms.

Figure 5.2 describes a particle size distribution in a street canyon in Shanghai, China, as a function of increasing altitude in relation to the street (Li et al. 2007). The majority of particles by number are smaller than 100 nm. The concentrations are greatest at 1.5 m from the street surface. As one moves higher and farther from the street, the concentration of mobile source emissions, resuspended road dust, and

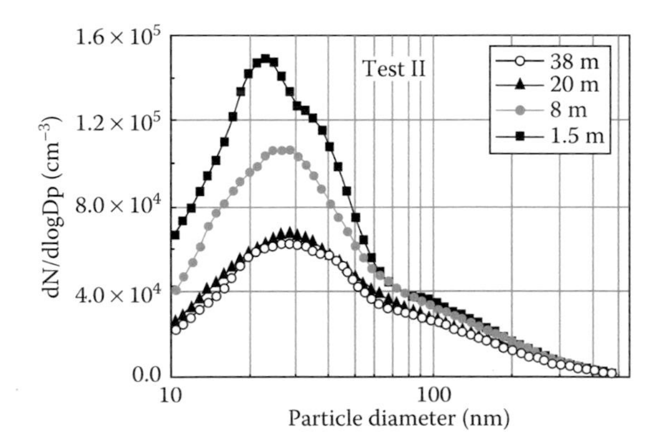

FIGURE 5.2 Nanoparticles in urban air: A street canyon in Shanghai (particle number concentration and size distribution change with increasing height). Measurements were made during daylight hours in November 2005. (Reprinted from *Sci. Total. Environ.*, 378, Li, X.L. et al., Vertical variations of particle number concentration and size distribution in a street canyon in Shanghai, China, 306–316, Copyright 2007, with permission from Elsevier.)

other particulates decreases. Particle numbers are far lower at 20 and 38 m than they are closer to the ground.

Finally, Figure 5.3 shows estimated particle emissions from idling school buses (Kinsey et al. 2007). Although emissions vary widely among buses, in all cases the majority of particles are smaller than 1 micron, and every bus produced particles in the nanometer range.

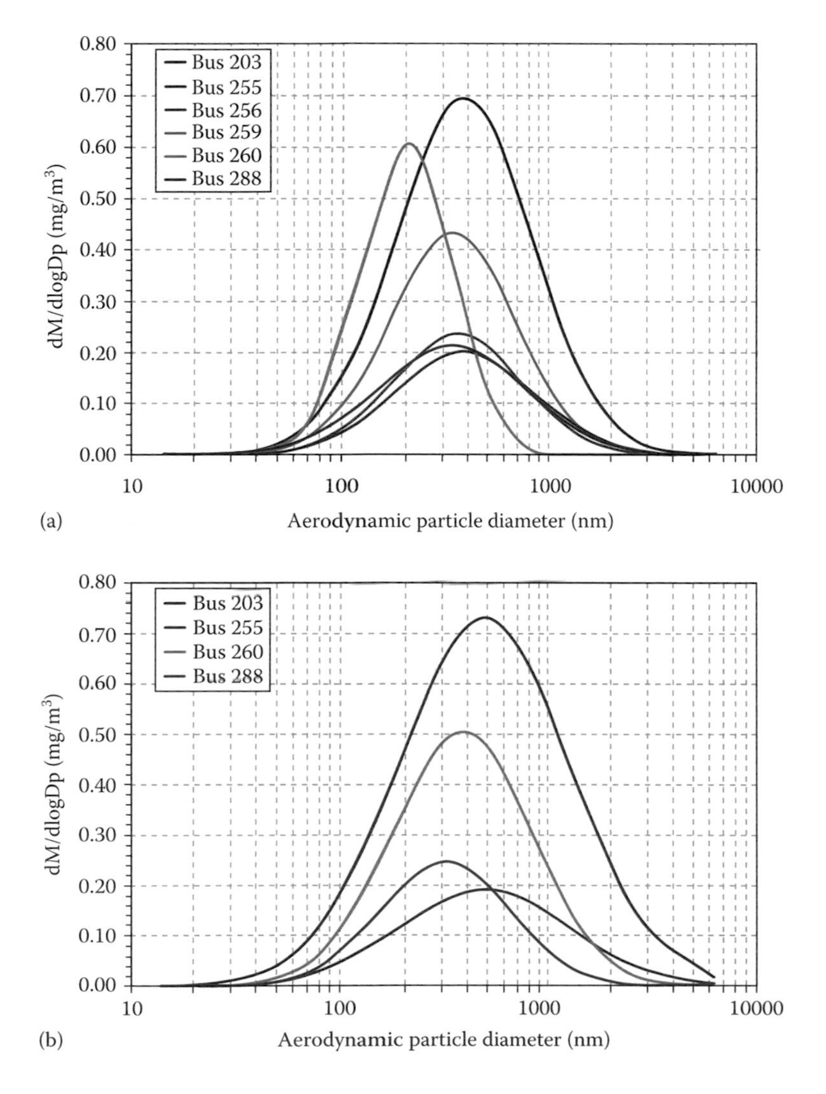

FIGURE 5.3 (a) Continuous idle and (b) Post-restart idle. Nanoparticles in diesel emissions: School bus idling. Average mass-based particle size distributions for continuous idle. The buses were manufactured between 1997 and 2004 and measured in March 2005. (Reprinted with permission from Kinsey, J.S. et al., Characterization of fine particle and gaseous emissions during school bus idling, *Environ Sci Technol* 41: 4972–4979. Copyright 2007 American Chemical Society.)

Motor vehicles are a particularly important source of nanoparticles. As detailed in a recent Health Effects Institute Report (Johnston et al. 2013), investigators developed an experimental instrument, the nano aerosol mass spectrometer (NAMS), which focused on nanoparticles smaller than 30 nm. They also analyzed the primary constituents of these particles and gathered data near a major roadway intersection.

Nanoparticles are formed during transformations of mobile and stationary sources in the atmosphere. For example, sulfur dioxide is converted into sulfurous and sulfuric acid. They typically result in many UFPs. We know that nanoparticles are abundant in diesel emissions but further research in laboratories and in human populations is needed to fully understand the extent of diesel nanoparticle emissions and especially their impact on human health.

No current regulations focus exclusively on nanoparticles (UFPs). Inevitably, regulatory standards are mass based because sampling devices measure that parameter. Neither appropriate air samplers nor data exist which permit rational standards focusing on nanoparticle emissions from gasoline or diesel engines. Another problem is that some diesel particles contain volatile material as well as solids, and most sampling techniques fail to capture and measure the volatile fraction.

5.4.3 Occupational Exposures

Smith et al. have extensive data on exposures of employees who drive trucks (Smith et al. 2006, 2012). Importantly, the most significant exposures of truck drivers occur not in their cabs while driving, but while loading and unloading materials at truck terminals. Frequently, engines continue to run and these sheltered or semisheltered areas may not have adequate air circulation to dispose off truck exhaust. Their data show that diesel particles contribute to the exposure of truck drivers and that these particles contain significant numbers of nanoparticles.

Diesel emissions have changed dramatically over the last three or four decades. Changes have been made in engine design, fuels, and after-treatment. The result has been a dramatic decrease in the mass of emissions. However, decreases in nanoparticles have been less dramatic.

Fumes, particularly metal fumes, consist of particles produced by materials that have become gaseous. As they cool, they form liquids and solids. Such nanoparticles are prominent in welding fume. These fumes consist primarily of UFPs and may lead to well-characterized responses such as metal or polymer fume fever (Drinker et al. 1927; Gordon et al. 1992).

5.4.4 Coal Dust

Underground coal mining is associated with serious lung diseases. The literature suggests that respirable particles are responsible. There is increasing evidence that regulations in the United States designed to prevent Coal Workers' pneumoconiosis (CWP) may no longer be adequate. Recent CDC data show that younger individuals, such as those working in underground mines in West Virginia and Kentucky, are getting serious lung diseases related to their occupational exposure. A key element is the increasing use of powerful machinery and the way it produces both higher concentrations of coal dust and especially smaller particles, including increasing numbers of nanoparticles.

5.4.5 Non-Anthropogenic Sources

Like their sister engineered nanomaterials, unintentional nanomaterials deposited in the deep lungs interact with surfactant film and are subject to phagocytosis by resident alveolar macrophages. Because of their small size, the receptors used may differ from those used by larger particles. Nevertheless, nanoparticles generally appear in the cell within intracellular vesicles. Peculiar to nanoparticles is their occasional appearance in the cell cytoplasm or even in the nucleus in a nonmembrane bound fashion (Geiser et al. 2005, Rothen-Rutishauser et al. 2006).

Important sources of unintentional aerosols are particles coming from volcanoes, forest fires, soil colloids (e.g., silicate clay), viruses, and sea spray. Also relevant is radon progeny. Radon gas decays into metal ions and they attach to particles already present in indoor air, especially persistent nanoparticles.

The most prominent source of unintentional nanoparticles created in the atmosphere represents condensation of gas phase molecules. This process is called nucleation and is particularly important for sulfuric acid and other compounds of low volatility which condense to small particles when they reach a critical concentration in the atmosphere. This is a process comparable to the formation of rain as water vapor condenses into cloud drops (Andreae 2013).

Polydisperse aerosols come from volcanic eruptions around the world and can travel for thousands of kilometers. For example, see Revuelta et al.'s (2012) research that measures volcanic dust in Spain, which originated 3000 km away in Iceland. Mather et al. (2004) discuss the importance of volcanic eruptions as a source of atmospheric gases and particles, and provide detailed information about concentration and particle size of two volcanoes in Chile. Typically, the technology used is not well designed to produce complete information on UFPs. They do report that the volcanic plumes are bimodal in size and that the two maxima have radii of 0.1–0.2 and 0.7–1.5 μm. Clearly, the smaller sized maximum includes particles smaller than 0.1 μm.

5.5 BIOLOGIC AND ENVIRONMENTAL RESPONSES

Pinkerton et al. deliberately produced nanoparticles of soot and showed that exposure to these particles early in neonatal development has long-term consequences on lung growth (Pinkerton et al. 2004). This is achieved by eliciting changes in cell division at critical sites in the airways during periods of rapid post-natal development. They speculate that these changes have "life-long consequences."

Nanoparticles, including unintentional ones, may be more toxic because of molecular adsorption of toxins to the particle surfaces. Given the fact that the small particles have greater surface areas per unit mass, their potential to carry toxins may be greater.

There is increasing evidence that ambient nanoparticles have consequences and may be more toxic than larger particles in the fine and coarse mode. The current terminology is that particles ranging from 2.5 to 10 μm are coarse, particles less than 2.5 μm are regarded as fine, and particles less than 0.1 μm are ultrafine. Thus, particles classified as ultrafine in the air pollution literature are nanoparticles. These nanoparticles emitted from cars, trucks, and stationary sources are of concern because they may increase the probability of heart attacks, aggravation of asthma, and resulting in increases in visits to emergency rooms and hospitals, especially by individuals with pre-existing cardiopulmonary disease (Smith et al. 2006).

In a recent editorial, Grigg emphasized how air pollution relating to traffic affects the growth of lungs *in utero*, early in childhood, and throughout adolescence (Grigg 2012). Concerns are increasing as more and more of the world's population moves from rural to urban areas.

Gong and colleagues have utilized concentrated ambient UFPs to measure health effects in humans utilizing chamber studies. Exposures were associated with significant reductions in arterial oxygen saturation as well as pulmonary function (e.g., EFV-1) (Gong et al. 2008). However, evidence of inflammation was absent, and the responses of subjects who had asthma compared to normals were similar. Comparable studies were also carried out in the EPA lab in Chapel Hill. Modest pulmonary function and cardiovascular changes were noted.

There are also profound consequences of natural aerosols on weather and other environmental problems. These natural atmospheric aerosols degrade visibility and may affect human health, as discussed earlier. Importantly, they also influence climate by reflecting and absorbing radiation from the sun. Especially, they have a key role in the formation of clouds. The size and the chemical makeup of these droplet nuclei influence weather and must be better studied in the laboratory and in environmental measurements (Zhang 2010).

In terms of biologic responses, it is interesting that as particles dissolve, they will inevitably go through a nanoparticle phase. For example, particles that are one or two microns in diameter undergoing gradual dissolution will pass through a stage when they are less than 100 nm on their way to complete dissolution. This is yet another reminder of a long history of human exposures to nanoparticles.

5.6 CONCLUSION

There is compelling evidence that although some engineered nanoparticles are novel and human exposures have not occurred before, many nanoparticles are not new. Polydisperse aerosols such as diesel particulates, silica dust, welding fume, and tobacco smoke contain abundant nanoparticles.

Typically, nanomaterials contribute the minority to exposure end-dose in terms of mass (almost always less than 10%), but they may dominate in terms of particle number and surface area. A challenge is that there is abundant literature showing health effects relating to $PM_{2.5}$, but there is little data on the extent to which these responses are attributable to the ultrafine fraction.

A challenge for the future is to learn as much as we can from these "historic particles." What properties do they have which are shared with "new" nanoparticles? Moreover, to what extent is the toxicity of familiar polydispersed materials attributable to the nanoparticle fraction? The field of unintentional nanoparticle toxicology is important and will be rewarding in terms of understanding nanotoxicology.

REFERENCES

Andreae, M. O. (2013). Atmospheric science. The aerosol nucleation puzzle. *Science* 339: 911–912.

Brain, J. D. and G. C. Corkery (1975). The effect of increased particles on the endocytosis of radiocolloids by pulmonary macrophages in vivo: competitive and toxic effects. *Inhaled Part 4 Pt 2*: 551–564.

Demokritou, P., T. Gupta, S. Ferguson and P. Koutrakis (2003). Development of a high-volume concentrated ambient particles system (CAPS) for human and animal inhalation toxicological studies. *Inhal Toxicol* 15: 111–129.

Demokritou, P., T. Gupta and P. Koutrakis (2002a). Development of a high volume Concentrated Ambient Particles Systems (CAPS) for human and animal inhalation toxicological studies. *Inhal Toxicol* 15: 101–119.

Demokritou, P., T. Gupta and P. Koutrakis (2002b). A high volume apparatus for the condensational growth of ultrafine particles for toxicological studies. *Aerosol Sci Technol* 36: 1061–1072.

Drinker, P., R. M. Thomson and F. J. L. (1927). Metal Fume Fever II: resistance acquired by inhalation of zinc oxide on two successive days. *J Ind Hyg Toxicol* 9: 98–105.

Geiser, M., B. Rothen-Rutishauser, N. Kapp, S. Schurch, W. Kreyling and H. Schulz (2005). Ultrafine particles cross cellular membranes by nonphagocytic mechanisms in lungs and in cultured cells. *Environ Health Perspect* 113: 1555–1560.

Gong, H. J., L. W. S., K. W. Clark, K. R. Anderson, C. Sioutas, N. E. Alexis, C. W. E. and R. B. Devlin (2008). Exposures of healthy and asthmatic volunteers to concentrated ambient ultrafine particles in Los Angeles. *Inhal Toxicol* 20: 533–545.

Gordon, T., L. C. Chen, J. M. Fine, R. B. Schlesinger, W. Y. Su, K. T. A. and M. O. Amdur (1992). Pulmonary effects of zinc oxide in human subjects, guinea pigs, rats and rabbits. *Am Ind Hyg Assoc J* 53: 503–509.

Grigg, J. (2012). Traffic-derived air pollution and lung function growth. *Am J Respir Crit Care Med* 186: 1208–1209.

Gupta, T., P. Demokritou, S. Ferguson and P. Koutrakis (2004). Development and performance evaluation of a high-volume ultrafine particle concentrator for inhalation toxicological studies. *Inhal Toxicol* 16: 851–862.

HEI (2013). Understanding the Health Effects of Ambient Ultrafine Particles. *HEI Perspectives 3*. Boston, Health Effects Institute.

Johnston, M. V., J. P. Klems, C. A. Zordan, M. R. Pennington and J. N. Smith (2013). *Selective Detection and Characterization of Nanoparticles from Motor Vehicles*. Boston, Health Effects Institute: 60.

Kinsey, J. S., D. C. Williams, Y. Dong and R. Logan (2007). Characterization of fine particle and gaseous emissions during school bus idling. *Environ Sci Technol* 41: 4972–4979.

Li, X. L., J. S. Wang, D. Tu, W. Liu and Z. Huang (2007). Vertical variations of particle number concentration and size distribution in a street canyon in Shanghai, China. *Sci Total Environ* 378: 306–316.

Majno, G. and G. E. Palade (1961). Studies on inflammation. 1. The effect of histamine and serotonin on vascular permeability: An electron microscopic study. *J Biophys Biochem Cytol* 11: 571–605.

Mather, T. A., V. I. Tsanev, D. M. Pyle, A. J. S. McGonigle, C. Oppenheimer and A. G. Allen (2004). Characterization and evolution of tropospheric plumes from Lascar and Villarrica volcanoes, Chile. *J Geophys Res* 109: D21303.

Ning, Z., C. S. Cheung, J. Fu, M. A. Liu and M. A. Schnell (2006). Experimental study of environmental tobacco smoke particles under actual indoor environment. *Sci Total Environ* 367: 822–830.

Pinkerton, K. E., Y. M. Zhou, S. V. Teague, J. L. Peake, R. C. Walther, I. M. Kennedy, V. J. Leppert and A. E. Aust (2004). Reduced lung cell proliferation following short-term exposure to ultrafine soot and iron particles in neonatal rats: Key to impaired lung growth? *Inhal Toxicol* 16 Suppl 1: 73–81.

Revuelta, M. A., M. Sastre, A. J. Fernandez, L. Martin, R. Garcia, F. J. Gomez-Moreno, B. Artinano, M. Pujadas and F. Molero (2012). Characterization of the Eyjafjallajokull volcanic plume over the Iberian Peninsula by lidar remote sensing and ground-level data collection. *Atmos Environ* 48: 46–55.

Rothen-Rutishauser, B. M., S. Schurch, B. Haenni, N. Kapp and P. Gehr (2006). Interaction of fine particles and nanoparticles with red blood cells visualized with advanced microscopic techniques. *Environ Sci Technol* 40: 4353–359.

Sarnat, J., P. Demokritou and P. Koutrakis (2003). Measurement of fine, coarse and ultrafine particles. *Ann 1st Super Sanita* 39: 351–355.

Smith, T. J., M. E. Davis, J. E. Hart, A. Blicharz, F. Laden and E. Garshick (2012). *Potential Air Toxics Hot Spots in Truck Terminals and Cabs*. Boston, Health Effects Institute.

Smith, T. J., M. E. Davis, P. Reaser, J. Natkin, J. E. Hart, F. Laden, A. Heff and E. Garshick (2006). Overview of particulate exposures in the US trucking industry. *J Environ Monit* 8: 711–720.

Sowards, J. W., A. J. Ramirez, J. C. Lippold and D. W. Dickinson (2008). Characterization procedure for the analysis of arc welding fume. *Welding Research* 87: 76–83.

Zhang, R. (2010). Getting to the critical nucleus of aerosol formation. *Science* 328: 1366–1367.

Zhang, Y., W. Sumner and D. Chen (2013). In vitro Particle Size Distributions in Electronic and Conventional Cigarette Aerosols Suggest Comparable Deposition Patters. *Nicotine Tobacco Res* 15: 501–508.

6 Lessons Learned from Pharmaceutical Nanomaterials

Emad Malaeksefat, Sarah Barthold, Brigitta Loretz, and Claus-Michael Lehr

CONTENTS

6.1 WHAT ARE PHARMACEUTICAL NANOMATERIALS? "NANOPARTICLES" FROM A PHARMACEUTICAL POINT OF VIEW

These days there are many controversial discussions about nanoparticles. The term nanoparticles stands for numerous objects in the nanoscale range including particle manufactured for medical use, particle generated at the industrial scale for surface coatings, electronics, energy, or consumer products, and last but not least pollution particles. Searching for "nanoparticles" via Scopus, the big citation and abstract database, presents approximately 180,000 publications of the last few decades. Only 45,000 of these publications are in medical, pharmaceutical, or biological journals. The unspecific vocabulary is not helpful in this discussion as it implies that size would be the main crucial property. The particle size is decisive for the interaction potential with biological surroundings, by determining the accessibility and the interaction surface. Other physicochemical properties of nanoparticles, however, contribute in terms of their hazard potential. Hence, being more precise in definition makes it easier to reach a realistic appreciation of both, the promises and fears connected to nanotechnology. Studies on the impact of combustion particles on environment or human health are important for risk assessment. Some of them are worst case scenarios of intoxication with nanoparticles not intended for human use. These results should not be extrapolated to particles for deliberate use in humans as is for nanomedicine. Applying nanotechnology for medical purposes means raw materials, manufacturing processes, and quality control are chosen and designed to achieve safest possible products. Nanomedicine, which is defined by the European Technology Platform on NanoMedicine as the application of nanotechnology in health care, exploits the improved and often novel physical, chemical, and biological properties of materials at the nanometric scale (NanoMedicine 2005). As nanomedicine also includes devices for medical use we term the nanoformulated drug delivery systems as "nanopharmaceuticals," which offer many advantages such as enhanced solubility, higher bioavailability, reduction of side effects, targeting specific tissues, or protection of unstable drugs. When this chapter is reporting about nanopharmaceuticals, it aims to deliver some examples of how the potential of nanotechnology can be harnessed for pharmaceutical applications. A brief description and appraisal of the current usability of nanopharmaceuticals is presented in section 6.1. Few considerations regarding the chances and risks depending on the application route are given in section 6.2. Translation into safe medicine needs careful evaluation of particle characteristics (section 6.3.1) and their biological effects (section 6.3.2), which can bring us closer to the goal of safety prediction based on physicochemical particle properties as it would be needed for quality/safety by design approaches (section 6.4). Finally, section 6.5 reviews selected nanopharmaceuticals on the market to learn from some best case examples.

6.1.1 Designs of Nanopharmaceuticals

Nanopharmaceuticals can roughly be divided into two groups: The first group includes *nanocrystals*, which are nanosized, stabilized drug particles (consisting the active ingredient itself as the main material). The second group is *drug–carrier*

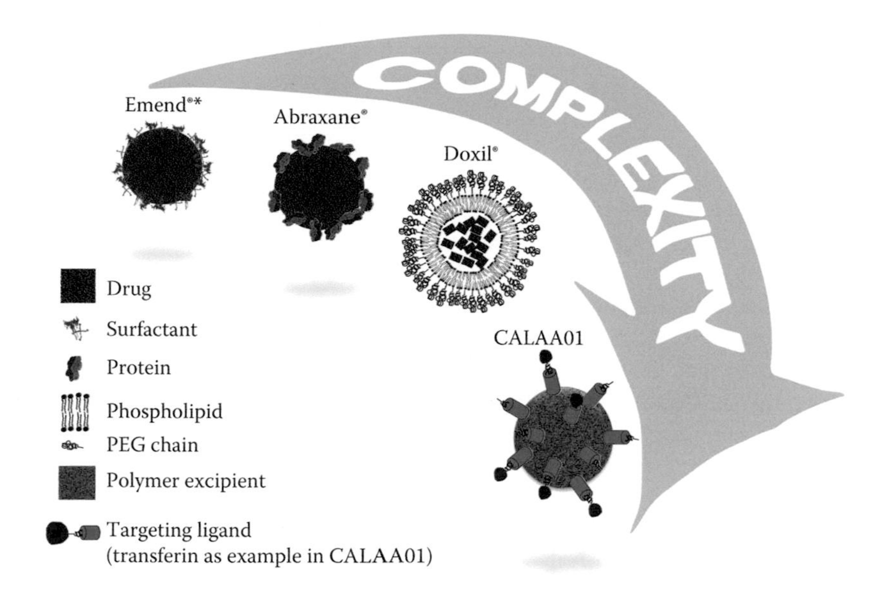

FIGURE 6.1 Schematic examples of marketed and upcoming nanopharmaceuticals ranked by increasing complexity (size dimensions not representative). *Emend: for the final medicinal product, microcrystalline cellulose is coated with the drug-surfactant nanoparticles. (Agency, European Medicines (2004) EMED formulation [Scientific Discussion], 35.)

systems, in which the drug is loaded into or onto a carrier of either one or several excipient(s), that is, typically polymers or lipids. Nowadays there is a wide variety of nanopharmaceutical formulations. Some consist of two or a few ingredients and are simple in structure. Others are complex systems consisting of many different components in often spatial arrangement (core–shell or multilayer structures). Figure 6.1 shows a few selected systems to illustrate this difference in complexity.

Reasons for the design of multicomponent systems are either the correction of a nonsatisfying drug property or the opportunity to develop combined products with multiple functions to improve the performance. The following list is only mentioning frequent examples without the aim of being complete. A polymeric particle matrix can enable a controlled release. The use of chemical cross-linkers or surfactants is a measure to enhance colloidal stability. Hydrophilic, biocompatible polymers, such as polyethylene glycol, are frequently used to coat nanoparticles for enhanced blood circulation times. Small molecular weight excipients such as sugars or amino acids are used as protection molecules for the formulation process of protein actives. Specific ligands bound to the particle surface enhance cellular uptake into specific target cells. Fluorophores or contrast agents are included to combine drug delivery with imaging for diagnostic purposes in "theranostics."

6.1.2 Into the Human Body, But Safely

For nanopharmaceuticals, it is their intended use to enter the human body. A nanotoxicological classification system (NCS) was proposed with the aim to group nanoparticles into four major safety classes (Keck and Muller 2013). The NCS may serve as a good starting point for discussion of nanopharmaceutical formulations and can be even of interest for nonpharmaceutical nanoparticles risk assessment. It is based in the first line on size and biodegradability and in the second line on biocompatibility of nanoparticles. They are categorized into four classes of increasing risk, due to size and persistency: Class 1 particles have a size above 100 nm and are biodegradable, whereas particles in class 2 are not biodegradable. Both classes show no endocytosis of particles. In classes 3 and 4, particle size is below 100 nm and particles show endocytosis. To differ between classes 3 and 4, particles in class 3 are biodegradable, whereas particles in class 4 are not. This new NCS helps to consider most relevant properties of nanoparticles related to their potential nanotoxicity. Other properties, such as solubility behavior, number and frequency of application, and medical indication, could be helpful in risk assessment. It is obvious that a single application treatment can easily tolerate low biodegradation than a repeated dosed therapy. In the development of an antibiotic carrier system with short-term dosing and a wanted immune system response, a mild activation of the immune system might even be helpful, while the same reaction might not be tolerable when developing nanopharmaceuticals for an anti-inflammatory therapy.

Besides the active pharmaceutical ingredient (API), there are also excipients in each formulation, which have to be considered often as at least partly active ingredients also since they functionally contribute to the drug effect and are not inert carriers. Therefore, materials and substances that already have a pharmaceutical acceptance are preferred. These materials should be of high purity, biocompatible, biodegradable, and nontoxic, to get a valid estimation of the safety profile of the final formulation. If not already used in some FDA (US Food and Drug Administration)-approved products, materials for drug product formulation have generally GRAS status (Generally Recognized As Safe). This status, however, depends on the application route in terms of the impact on biodistribution and biokinetics of the particles.

6.2 EFFECT OF THE ROUTE OF APPLICATION: CHANCES AND RISKS OF NANOPHARMACEUTICALS

The application route guides biodistribution, particle elimination, and thus efficacy of nanopharmaceuticals. The principal decision has to be taken if a systemic therapy or a local therapy better suits within the therapeutic concept of the target disease.

Until now the dominating route for nanopharmaceuticals is the intravenous (iv) route of application. One advantage is the certainty of the dose reaching the circulation, since the need to overcome an epithelial barrier is not given. In particular, in cancer therapy the iv administration plays an important role, as nanoparticles are able to accumulate at least to some extent in a tumor tissue, due to the enhanced permeation and retention (EPR) effect. Disadvantages of the iv route are the potential unwanted accumulation in nontarget tissues or organs. For example, particles of larger sizes or nanoparticles with some aggregation tendency can accumulate in the fine capillaries of the lungs. Particles with hydrophobic or highly charged surfaces are prone to opsonization with blood proteins and thereby enhanced uptake and elimination by macrophages (Aggarwal et al. 2009). Consequently, special surface designs have to be realized to prolong the residence time in the blood circulation. For enhanced patient compliance the oral route is most interesting for drug delivery. The pH differences, the mucus layer, the enzymatic environment, food effect, and so on make the gastrointestinal tract a real challenge for nanopharmaceuticals. The protection of nanopharmaceuticals from the hostile pH in the stomach is difficult as this means either a conversion into some solid dosage form first (often causing aggregation) or coating the nanoparticles directly with a protective layer (complete coating needed, which is difficult to achieve). The mucus barrier is easier to cross for small molecule drugs than for particulate carriers, as they face more difficulties in diffusion through the mucin mesh. The high turnover rate of the mucus layer and the uppermost epithelial cells, however, help in efficient nanoparticle elimination. Only for some nanoparticles with special properties a low level of absorption has been reported, achieving access to the circulation (Landsiedel et al. 2012). Nevertheless, some therapeutically interesting nanopharmaceutical formulations could use the pathophysiologic changes in inflammation for prolonged particle retention (Collnot, Ali, and Lehr 2012) or the uptake via M-cells for vaccination (Slutter et al. 2009).

The lungs are an interesting epithelial barrier since it is constructed as thin as possible for efficient gas exchange and is less equipped with degrading enzymes when compared to the gastrointestinal tract. In order to deliver drugs either directly to the lung tissue or without much enzymatic or metabolic loss into the circulation the pulmonary route is mostly attractive. Designing particles in the correct size range and shape for sufficient aerodynamic properties enables the avoidance of mucociliary clearance and a deposition in the alveolar region. That this concept works for protein delivery could be proven with Exubera®, the inhalable insulin marketed by Pfizer (Skyler et al. 2007). The reason for the market withdrawal of Exubera was low acceptance and high cost (Bailey and Barnett 2007). From a risk assessment point of view, however, particles deposited in the deep lungs are more difficult to remove. The clearance of particles from the alveoli occurs in first line by alveolar macrophages, while under conditions of infection neutrophiles are recruited to assist in pathogen clearance (Hasenberg, Stegemann-Koniszewski, and Gunzer 2013). The macrophage clearance is faster for microparticles than for nanoparticles (Todoroff and Vanbever 2011). Particles that cannot be completely phagocytosed by macrophages are a safety risk, as we should have learned from asbestos. Particles that are causing immune stimulation, in particular if they are more persistent, may also be considered as health risk as the immune mediators can damage the lungs or other tissues (Geiser and Kreyling 2010). As a consequence, only a few excipients are known, which are considered safe enough for drug delivery to the lungs, and so far all of them are water soluble. Particles made from water soluble materials without cross-linking simply dissolve after landing. They are not suitable to deliver drugs, which need carriers for intracellular delivery (e.g., intracellular active peptides and nucleic acid drugs). An approach to combine the higher efficacy of alveolar deposition of microparticles with the slower clearance by alveolar macrophages of nanoparticles is to form microparticles from nanoparticles (Tsapis et al. 2002).

The skin with its multilayered epidermis is a very tight barrier, which prevents unintentional permeation. There is quite a good agreement that nanoparticles do not permeate the intact *stratum corneum* to reach the blood circulation or the lymph (Labouta et al. 2011; Labouta and Schneider 2013). However, serious damage like strong sun burn might alter the penetration into the skin (Monteiro-Riviere et al. 2011). For the purpose of drug delivery, most nanopharmaceuticals are designed to stay either at the surface (in the formulation matrix) or penetrate into the epidermis (depending on their hydrophobicity and the dispersing vehicle) to act as a depot system. The toxicological risk for such systems is comparatively low and controllable. The hair follicles were reported to act as a depot reservoir enabling a prolonged residence time for particles in the suitable sub-micrometer size range (Lademann et al. 2007). Some recent work aims to make use of this as needle-free, noninvasive vaccination route (Mittal et al. 2013; Mittal, Raber, and Hansen 2013).

Nanoparticles are moreover of special interest for the delivery of active pharmaceutical ingredients to the brain. They have the ability to enhance the transport across the blood–brain barrier (formed by the endothelium of the blood vessels) for actives, which are nonpermeable in their free form. Special carrier surface properties (negative charge and polysorbate 80 coating, and specific targeting ligands) are reported to be helpful (Wohlfart, Gelperina, and Kreuter 2012). Deposition of insoluble particles in the brain is a scenario linked with the suspicion of neurodegenerative disease generation, which could be caused by reactive oxygen species or inflammation (Bondy 2011). The formulation of colloids used in the brain may therefore be critical. Perlstein evaluated dextran-coated colloids by measuring the clearance after transcranial injection or infusion. In these studies, maghemite nanoparticles had a slow clearance of 80%–90% from rat brain with some particles remaining in the brain cells (Perlstein et al. 2008). Another study evaluated dextran-stabilized superparamagnetic iron oxide particles and found a 90% clearance after 3 months (Polikarpov et al. 2013).

6.3 GET TO KNOW YOUR NANOPARTICLES

When speaking about pharmaceutical products in general, it is always required to prove not only their efficacy but also their safe use and to ensure quality. Therefore, it is a matter of course that nanopharmaceuticals are tested for their safety before entering clinical trials and eventually the market situation later on. The established methods for efficacy and safety testing were developed for small molecule actives and their formulations and are not always easy to convert to nanopharmaceuticals. The higher complexity of nanopharmaceuticals can cause variance in their spatial distribution, their chemical or physical appearance, interaction, and so on. The safety evaluation gets more demanding as the particle architecture can change according to the (biological) environment and thereby causing various types of interactions (protein binding, cellular recognition, and cellular uptake). The efficacy and safety can vary because of changed absorption, distribution, metabolism, and excretion (ADME) properties or altered drug release speed. The special problem here with nanopharmaceuticals is the difficulty to analyze small variances in the particles architecture at the nanoscale. For this reason let us have a short look at the available methods for nanoparticle characterization.

Currently, we are missing a concrete and obligatory determination if nanoparticles should be regulated separately in addition to their bulk counterpart or not. Due to the size of nanoparticles the question is arising: Are nanoparticles different from conventional drug products and do they need special testing? In spite of that, many investigate obvious characteristics of nanopharmaceuticals, such as physical/chemical parameters and their toxicological effects. Some institutions started to structure standard protocols for nanoparticle design and investigation. One pioneer, the Nanotechnology Characterization Laboratory (NCL), is trying to establish standard protocols for nanoparticles in cancer research. The NCL aims to speed up translation of nanotechnology in the field of cancer therapy. They provide the service of testing selected, promising nanopharmaceuticals in their facilities as preclinical study. Moreover, they defined an assay cascade starting with the physicochemical characterization followed by in vitro and in vivo characterization, by collection, creation, and providing standard protocols for nanoparticle characterization. In the progress of investigation, there are already some very well-defined assays and protocols for physicochemical and in vitro characterization. Nevertheless, there are several established assays missing in the in vivo area such as nanoparticle immunotoxicity, genotoxicity, and mutagenicity. Guidance documents to standardize methods and protocols (as far as the diverse nature of nanopharmaceuticals and the status of quality assurance of the assays allow this already) will simplify the development process and the comparability between formulations. The following short overview of techniques makes no claim to be complete. The aim is rather to highlight characterization possibilities, nanoscientists need to choose the best suited technique(s) from.

6.3.1 Current Status of Nanoparticle Characterization

At the nanoscale, it is not possible to directly look at and evaluate objects. To qualify prepared nanoparticles, different methods are used, and the complementary use of them describes their physical and chemical parameters. Ideally, the nanoparticle characterization consists of size, charge, shape, and chemical identity. While the following part is common for all types of nanoparticles, nanopharmaceuticals have an important additional level. The drug content, drug structure (crystal or amorphous), and localization within the carrier (surface adsorption vs. encapsulation) are important for good prediction of the drug effect. The ADME effects of the carrier have impact on the pharmacokinetic profile of the active ingredient (Zolnik and Sadrieh 2009)

6.3.1.1 Size

One of the main properties is of course the nanoparticle size. For medical/pharmaceutical applications the nanoparticle size is defined as a size range between 1 nm and 1000 nm (Wagner et al. 2008). To characterize size properties, there are different methods, which should be used complementarily. Photon correlation spectroscopy (PCS) also known as dynamic light scattering (DLS) uses scattered light of particles to determine their size distribution profile in suspensions (Berne and Pecora 1976). DLS is a method for high statistical representative size measurement. Nevertheless, it has its weaknesses in relation to size distribution of monodisperse and polydisperse samples. In order to validate a DLS size measurement, it is essential to use other size measurement methods such as Nanoparticle Tracking Analysis (NTA). NTA is based on visual detection of single nanoparticles via analyzing their size by Brownian motion related to temperature and viscosity of a suspension (Carr and Malloy 2006). NTA compared to DLS enables indirect nanoparticle visualization and detects various nanoparticle populations. In DLS measurements, small amounts of larger particles have a strong influence and shift the mean size to a higher size distribution (Filipe, Hawe, and Jiskoot 2010; James and Driskell 2013). For this reason, NTA is an important technique to complement DLS measurements. Furthermore, it is possible to choose from various other sizing techniques like size exclusion chromatography, analytical ultracentrifugation, or magnetic sedimentation. From the microscopic techniques, electron microscopies such as scanning electron microscopy (SEM) and transmission electron microscopy (TEM) provide additional information on size, mono-, and polydispersity by simultaneously visualizing the particulate morphology. In addition, atomic force microscopy (AFM) can be used (Grobelny et al. 2009).

6.3.1.2 Shape and Structure

Nanoparticles shape and their surface functionalization charge have influence, for example, on translocation into biological barriers and cell membranes. Thus, shape and charge are responsible for increasing or decreasing nanoparticle uptake (Nangia and Sureshkumar 2012). Dependent on nanoparticle composition, there are different microscopic techniques such as SEM, TEM, or AFM to clarify their three-dimensional structure. For example, a coated nanoparticle with a dense core and a polymeric shell can easily be enlightened by means of TEM, but for nanoparticle compositions of several similar raw materials, it is difficult to exhibit their structure. For this issue we need some appropriate methods for evaluating nanoparticle composition and structure.

6.3.1.3 Charge

In most cases it is important to examine the "charge" of nanoparticles. Laser doppler microelectrophoresis is a possible method to determine the electrophoretic motilities of particles in suspensions. With this analyzing method we are capable of measuring the zeta potential also named as charge of nanoparticles (Cooper 2004). Formulating new nanopharmaceuticals by using the right starting material can approximately predestinate the charge and therefore also the application. Different applications expect specific zeta potentials; for example, formation of nanoplexes by using positively charged nanoparticles combined with negatively charged nucleic acids (Nafee et al. 2012).

6.3.1.4 Chemical Identity

In comparison to clarifying the structure of nanoparticles there are more accurate techniques to determine their chemical identity. By quantifying the different spectroscopic analysis techniques such as Fourier transform infrared spectroscopy, Raman spectroscopy (Shao et al. 2007), nuclear magnetic resonance spectroscopy (Wawer and Diehl 2011), or X-ray powder diffraction (Brittain 2001), it is possible to obtain precise information on chemical components. Due to these techniques it is possible to determine impurities or reinforce the pureness of a sample. Another issue is conformation of the functional groups on the nanoparticle surface. The available groups can be identified by various staining methods, some of them used since decades. The Ellman's assay is capable of detecting sulfhydryl groups (Ellman 1959), trinitrobenzenesulfonic acid (TNBS) assay can estimate chemical free amino groups (Habeeb 1966), and surface hydrophobicity can be at least in relative terms determined via Rose Bengal adsorption (Müller 1991).

There is now a common understanding in nanomedicine that a thorough physicochemical characterization is a prerequisite for medical application and that this does include also the agglomeration/aggregation status of particles in biological fluids.

6.3.2 Particle–Cell/Tissue Interactions

There are various tests which can be used to investigate the interaction of nanoparticles with cells/tissues. These interactions are possible to be investigated at three different levels. First and most common is to test cytotoxicity, which compares living cells against dead cells and concludes cell viability. The immunostimulatory effects and the capability to induce genetic mutation received less attention, which has to be changed.

6.3.2.1 Cytotoxicity or Cell Viability

One of the first tests before using nanopharmaceuticals in clinical studies is to investigate their impact on cell viability or proliferation in cell culture. For that issue, colorimetric- and luminescence-based assays were developed to directly or indirectly measure their cytotoxicity. Assays for different cytotoxicity mechanisms are available and several should be used to verify the pharmaceutical usability of nanoparticles (Weyermann, Lochmann, and Zimmer 2005). These assays are based on association of cell death versus proliferation. One assay measures the activity of enzymes that are available to metabolize a tetrazolium salt (MTT) to its water insoluble formazan dye. Another assay enables visualization of cell membrane damage, which indicates a loss of cell membrane integrity. By cell damage a cytoplasmic enzyme, lactate dehydrogenase (LDH), is released into the extracellular supernatant. In the cell culture supernatant LDH is still active and can be used in an enzymatic assay. Just like the MTT assay a tetrazolium salt is reduced to a red formazan dye. These are two colorimetric assays. One proves cell viability (MTT assay), whereas the other proves the cell membrane damage therefore cell death. A further method exploits the consumption of the "molecular fuel" adenosine-5-triphosphate (ATP). In many cellular processes and metabolism ATP supplies the chemical energy. All viable cells can be determined by a bioluminescent method, which utilizes a catalytic enzyme (luciferase). Luciferase generates light by catalyzing ATP and the substrate luciferin. This bioluminescent assay is able to measure metabolically active cells (Crouch et al. 1993). In most cases, microtiter plates were used to have a high throughput rate using 96-well plates. This miniaturization allows us to analyze a high number of samples simultaneously.

6.3.2.2 Immunotoxicity

Complex processes such as immune response are not simple to analyze. The different types of nanopharmaceuticals and their proposed route of application can activate various immune defense mechanisms. In the past, apart from vaccine delivery, only little attention was paid to the immunostimulatory effects of nanopharmaceuticals. One opportunity to detect an immunological effect of nanopharmaceuticals is to investigate the complement system after treatment. The complement system consists of several proteins, both in serum and on cell surfaces and after activation it is capable of opsonization, chemotaxis, and cell lysis. The opportunity to measure components of the complement system is possible (Nilsson and Ekdahl 2012). Cytokine release and dendritic cell activation are suitable tests to evaluate the potential of nanopharmaceuticals to impact leukocyte recruitment and inflammation. A further strategy is to determine the antibody response. There are some assays but these are not well established yet (Jung et al. 2001). The detection of bioaccumulation and its potential also needs more attention. Although biodegradability and biocompatibility are designed from the outside, it has to be proven that the nanoparticulate formulations stay only as long as necessary and wanted in the human body.

6.3.2.3 Genotoxicity or Mutagenicity

Some materials are able to induce genetic mutations. The Organization for Economic Co-operation and Development (OECD) validated various in vitro methods for genetic toxicology testing.

Single cell gel electrophoresis also called Comet assay is a technique to analyze DNA damage in individual cells. It is sensitive to damage above 50 breaks to a maximum of 10,000 (Olive and Banath 2006). In the field of nanopharmaceuticals, the Comet assay is the most applied one until now. The Bacterial Reverse Mutation Assay also referred to as Ames test is a fast method to detect mutagenic properties of materials. Bacteria with a defect DNA repair mechanism are not able to recover DNA damage after treatment with mutagen substances, which leads to an increase in the number of revertant colonies relative to the background level. The GreenScreen HC genotoxicity assay is a yeast DNA repair reporter assay with the possibility for high throughput (Cahill et al. 2004). No single mutagenicity test so far is validated enough (in particular for nanoparticles) and provides positive controls for mutagenic nanoparticles to act as a standalone technique. The OECD recommends several other methods for determination of damages in genetic material in their "Guidelines for Testing of Chemicals."

Nanoparticle formulations could possess the capacity to act as a teratogen, which leads to the investigation of their teratogenicity. An animal model (zebra fish) was described to clarify the potential of formulations in relation to physiological maldevelopment (Brannen et al. 2010). However, it is important to mention that investigations based on animal testing should only be performed in very late stages of development and may be surrogated by alternative methods using embryonic stem cells (Scholz et al. 1999; Rohwedel et al. 2001; McNeish 2004).

6.3.2.4 Alterations in Cell Function

Alterations in cell function might also occur, for example, as a result of changes in surface proteins or shifts in gene expression patterns (as can occur via miRNA interaction), which can lead to either negative or positive results with regard to assisting or counteracting the active ingredients effect. In several cases, the particle toxicity was reported to impact not via direct contact but via mediators (Hatakeyama et al. 2011). Real-time PCR or microarrays are means to study such effects. This sort of analysis due to the complexity and difficulty to discriminate between nonrelevant and relevant changes is still at the very beginning.

In conclusion, there are more than a few methods, in particular for cytotoxicity tests of nanopharmaceuticals, but in the nanoparticle–cell interaction area we are missing established and standardized systems. One big issue is that some assay constituents interact with the nanoparticulate formulation and present false true or false negative results (Wahl et al. 2008). Already described methods have to be validated and adapted for nanopharmaceutical investigations. Validation of specific nanocharacterization assays requires more attention and maybe institutions such as the NCL or OECD are predestinated to concentrate on these topics.

6.4 SAFETY AND QUALITY BY DESIGN

The specific medical application should direct the carrier design. The physiological/pathophysiological and anatomical settings are important initial considerations. Designing a system with the minimum of needed components and fewest possible steps of robust production methods and powerful analytics for quality control are the keys for successful formulations. By starting from a known system and trying to eliminate the shortcomings by addition of further functional components, the systems are endangered to get too complicated for quality control and raise production cost, which will hamper their translation into the clinic. In an ideal world, nanopharmaceuticals would be developed following Quality by Design (QbD) strategies. These were developed for the pharmaceutical production of chemicals and are in the process of implementation in the manufacturing of biotechnological products (Rathore 2009). A prerequisite for the implementation of QbD is a sound knowledge of the dependence of critical quality attributes and the desired functional performance (e.g., medical behavior). In addition, also the impact of raw materials and the understanding of the manufacturing process steps have to be established. Following the principle of QbD the initial step is the identification of attributes with significant impact on product performance. For nanopharmaceuticals, the product performance is the combination of drug efficacy and safety. The aimed product profile is defined by its characteristics. The second step is then to identify the critical quality attributes. These are the properties, which are important to safeguard the product quality and could be of physical (size, size distribution, shape, aggregation, etc.), chemical (hydrophobicity, surface functional groups, solubility, etc.), or biological nature (toxicology, biodegradability, protein binding capacity, complement interaction, antibody generation, etc.). This includes not only the naming of the key properties but also to define the acceptable range (threshold) for each parameter. Subsequently product design space is defined. Variations in process parameters are performed using statistical methods such as Design of Experiment to gain a thorough understanding of the input variables and process parameters. Manufacturing methods with a few steps, scalability, and high reproducibility are preferable. The use of hazardous substances like solvents or chemical reagents should be minimized or replaced when possible. Purification and sterilization methods should also be taken into account. It is also important to mention the need for analytical methods, which need to be established for testing each critical quality criteria in quality and/or quantity to evaluate the fulfilling of the acceptance criteria. In the area of nanotechnology that sounds easier than it actually is. For size and size distribution there is a whole range of analytical techniques available (see also section 6.3.1) while no method is suitable as standalone analysis to deliver a full, realistic picture. The challenge here is to use a well-balanced combination of analytical methods to avoid misinterpretations. For other properties it is difficult to find suitable analytical methods at all. The measurement of hydrophobicity is such an example. The Bengal-Rose assay may serve this purpose in some formulations, but the interaction with the dye may also be influenced by other particle characteristics. The measurement of contact angle is another possibility, but not suited to high throughput analysis. Other analytical methods have to be developed together with the delivery system because they need to be tailored, for example, measurement of surface decoration with functional ligands, including the analysis of covalent binding versus adsorption. The developed control strategy is performed in the experimental pilot scale as well as in the upscaling process, and process validation is maintained as quality control finally in manufacturing. The quality of pharmaceutics is always defined by safety also, as they have no chance of passing the approval of regulatory agencies. Therefore, risk assessment is included in each QbD approach by a systematic approach of risk identification, analysis, and decision about the acceptability or need/option for reduction or elimination and the risk communication. Many nanopharmaceuticals are complex systems combining several materials. These raw materials should ideally have well defined, low variable product quality, and be analyzed in powerful standard analytics. All raw materials as well as known degradation products should be tested for biocompatibility and biodegradability. Raw materials with significant effect on the product quality should be categorized as critical raw materials, which then implicate their stronger control for each incoming batch and during the storage life-time. Whenever possible, for excipients with GRAS status and for nonbiodegradable polymers, molecular weights enabling excretion via the kidneys should be used. Polymers should ideally be well defined and be suitable for high throughput assays of quality control (batch to batch variance, purity). A difficulty in carrier design is that there are many research reports using new polymeric materials. Often the materials are chosen for smart material function (like temperature, pH, or light trigger) but do not care for toxicological considerations. Few excipients do have the GRAS status and often just for limited routes of application. With the currently available knowledge, materials, and processes, a QbD study for nanoparticulate drug carriers does not assure success. Better understanding of what quality attributes of nanoparticles have impact on drug efficacy, biodistribution, and safety implications is needed. Until a few years ago the major focus was much on the size as the main criterion. Only in recent years, researchers started to look also closer for further properties like hydrophobicity, shape, protein-binding profile, and so on. An appraisement of carrier properties as critical properties and less critical properties with impact on various host reactions and their assignment to the respective host reaction is needed. The range of critical particle property needs to be evaluated for their tunability. Currently, we manage to tune size and/or shape by the use of special methods. As a goal our particle manufacturing methods need to be able to control all crucial properties at the same time.

Following as far as possible and reasonable QbD attempts will lead to the creation of high quality results in research. The sharing of these results in publications and databases with open access will help to organize and sort the gained knowledge. Academic partners alone cannot do this—large cooperation projects best with industrial partners or public funded institution would be needed for such work. Statistical analysis, modeling, and prediction (e.g., NanoQSAR [Puzyn et al. 2011]) would be of high benefit in such interdisciplinary and wide scientific area. The development of high throughput analytical methods to manage the large number of needed tests for process validation is a further need in nanomaterial science.

From the plurality of developed drug carriers in the controlled environment of in vitro experiments many smart drug delivery systems perform well. Maintenance of the performance in the environment of complex biological systems (i.e., in vivo) is another category of challenge. The systems which already managed the translation into the clinics are often systems which not only take account of the physiology but also sometimes even draw advantages of it as, for example, Abraxane® gaining from the binding of albumin.

6.5 NANOPHARMACEUTICALS STATE OF THE ART

Contemplating today's pharmaceutical industry, nanoparticulate systems showed importance on the market over the last few decades and promise rising potential for the future. Birrenbach and Speiser (1977) started to synthesize "nanoparts" for pharmaceutical application as early as in 1976. They distinguished between nanocapsules, such as shell-like polymer constructions, and nanopellets also known as compact polymer particles. Materials and techniques used at that time proved to be problematic in terms of toxicity. As a consequence, preparation methods and materials had to be substituted, further optimized, or newly developed. From nineties onwards, various nanoparticulate formulations for pharmaceutical applications have been approved by the FDA and are currently on the market.

6.5.1 Application Fields: Why to Use Nanopharmaceuticals

One reason to design nanoparticulates for pharmaceutical applications is their potential to successfully reformulate traditional formulations to overcome articulation problems, due to poor water solubility of the drug (Bawa 2008). An example for optimizing the performance of APIs is the Nanocrystal® technology: Reducing the size of particles to the nano range via wet-milling and subsequent stabilization of the drug particles results in a stable product with substantially increased surface area, improved water solubility, and hence bioavailability (Merisko-Liversidge, Liversidge, and Cooper 2003; Liversidge 1992; Liversidge and Cundy 1995). Sirolimus (Rapamune®), approved by the FDA in 1999, is the first marketed drug, developed with this technology. Fenofibrate (TriCor®) and Aprepitant (Emend®) are other examples of reformulated FDA-approved therapeutics, which employ this technology. Reformulation also leads to less variability in the fasted and fed states of patients, thus food effect decreased compared to conventional formulations (Bawa 2009; Moschwitzer 2013).

Another aspect of nanopharmaceuticals is the improved safety profile compared to traditional formulations. Size and surface properties largely dictate the in vivo behavior of nanopharmaceuticals. Bawa showed that tuning size and surface of nanocarrier systems alters the pharmacokinetic profile compared to their parent counterpart (Bawa 2009), so that adverse effects occurring with traditional formulations are reduced, according to their adjustable release kinetics, as can be seen for AmBisome® (Adler-moore and Proffitt 1993; Bekersky et al. 2002; Boswell et al. 1998).

The importance of carrier systems for pharmaceutical applications is also based on their targeting potential and potential to overcome biological barriers. Although the structure of the blood brain barrier (BBB) makes most of the small-molecule drugs and practically all of macromolecules not suitable for traversing it, some nanoparticulate therapeutics, however, are able to cross this barrier without the need for opening it beforehand (Pardridge 2003; Tosi et al. 2008). Tosi et al. showed various strategies for crossing the BBB using modified nanoparticles, such as magnetic nanoparticles, the use of surfactants for coating of the nanoparticles, or the covalent conjugation of the particles with specific ligands. They offer a more specific and selective drug delivery to the central nervous system. If ligands are chosen, they have to show appropriate characteristics for taking advantage of receptor-mediated transcytosis or endocytosis. Molecules such as transferrin, insulin, or thiamin and also small peptides have been tested successfully (Tosi et al. 2008). Maeda and Matsumura described a passive tumor-targeting mechanism, which is known as the EPR effect (Matsumura et al. 1986): tumor capillaries show higher endothelial fenestrations and architectural anarchy compared to healthy tissue. Yuan et al. found liposomes of up to 400 nm in diameter to be able to permeate tumor vessels, suggesting the cutoff size of pores to be between 400 and 600 nm in diameter (Yuan et al. 1995). Consequently, injected agents with the "right size" are able to accumulate at tumor sites, if not previously cleared by the immune system. Further, surface characteristics are important to escape mononuclear phagocyte system (MPS) capture, which represents one part of immune response. Nanoparticles with hydrophilic surface (e.g., via PEGylation) have a higher chance of escaping macrophage clearance (Moghimi and Szebeni 2003). Once a particle fulfills size and surface characteristics requirements, it will circulate for a longer period of time in the bloodstream reaching and accumulating in tumor tissue. An example of such a marketed formulation is Doxil®/Caelyx®. A comparable effect, which could be applicable to passive targeting via nanoparticulate formulations, was observed for rheumatoid arthritis: Results of various studies suggest that retention of prodrugs at inflamed joints occurs due to similar effects, shown for the EPR effect. A prolonged half-life of prodrugs, in combination with leaky vasculature, leads to accumulation at the site of action (Quan et al. 2010; Wang et al. 2004; Wunder et al. 2003). Hence, enhanced endocytic capacity is exhibited in synoviocytes and explains the sustained suppression of inflammation (Wunder et al. 2003; Wang and Goldring 2011; Quan et al. 2010). This effect was recently termed ELVIS "Extravasation through Leaky Vasculature and the subsequent Inflammatory cell-mediated Sequestration" by Yuan et al. They suspect that the ELVIS effect can be transformed to other inflammatory tissue as well, when they share similar pathophysiological features, such as vascular leakage and activated inflammatory cells (Yuan et al. 2012). Another opportunity to make use of passive targeting via pathophysiological changes is the particle accumulation in inflamed areas of the intestinal epithelium (Schmidt et al. 2013; Lamprecht, Schafer, and Lehr 2001).

To summarize, unique size and surface properties of nanoparticles give them the chance to accumulate, for example, in tumor tissue, due to endothelial defectiveness. Models like the EPR and also the ELVIS effect give nanoparticulate carrier systems the possibility of passively targeting the site of action.

In contrast to passive targeting, active targeting, known as the modification of carriers' surface with appropriate ligands to specifically target required pathological sites in the body, is a more advanced approach. In addition, nanopharmaceuticals offer unique formulation possibilities for the delivery of biopharmaceuticals, which are often limited by chemical instability during formulation and storage, enzymatic degradation *in vivo*, thus poor bioavailability. Besides stability issues of biopharmaceutics, bioavailability is also diminished due to the low permeability of biopharmaceutics across cell barriers (Cleland, Daugherty, and Mrsny 2001). Due to the progress in biotechnology, biopharmaceutics, such as proteins, peptides, and nucleic acids, are currently representing a large fraction of compounds in drug development pipelines (Wong 2009). Although parenteral application is the typical route to deliver biopharmaceuticals and is so far the least expensive and quickest strategy for commercialization, noninvasive routes of administration would still be favored. Therefore, nanotechnology can play an important role in the delivery of such compounds. They can not only serve as delivery agents but also protect the drug from degradation and can furthermore control release and enhance permeation via cell membranes (Almeida and Souto 2007).

Although there are only a few formulations on the market yet (a selection can be found in Table 6.1) the impact of reformulated or novel nanoparticle-based formulations on medicine and health care is more than promising. Figure 6.2 provides a scheme to summarize the achievable advantages via nanoformulations.

6.5.2 Nanopharmaceuticals on the Market

In the following section, examples of regulatory approved and thus marketed nanomedicine products will be presented to illustrate their use for pharmaceutical applications. Examples of nanocrystals will be shown first, followed by two examples of liposomal formulations (carrier systems). In particular, their formulation as well as their advantage over conventional (i.e., non-nano) formulations will be explained. In addition, an overview of various marketed formulations is shown in Table 6.1.

6.5.2.1 Rapamune® (Sirolimus)

Although sirolimus failed in its original purpose as an antibiotic, it is today the most potent immunosuppressive agent for the prevention of graft rejection in organ transplantation. Its highly hydrophobic structure makes it practically insoluble in water. Hence, bioavailability is greatly reduced. An oral lipid-based solution of

TABLE 6.1
Approved Examples of Nanomedicine Products

Brand Name	Nanoparticle Drug Component	Delivery Route	Company	FDA Approved Indication(s)	FDA Approval Date
Abraxane®	Paclitaxel bound albumin NP	IV	Abraxis BioScience, AstraZeneca	Various cancers	Jan 2005
AmBiSome®	Amphotericin B liposomes	IV	Gilead Sciences	Fungal infections, leishmaniasis	Aug 1997
Doxil/Caelyx®	PEGylated doxorubicin*HCl liposomes	IV	OrthoBiotech Schering-Plough	Metastatic ovarian cancer and AIDS related Kaposi's sarcoma	Nov 1995
DaunoXome®	Encapsulated daunorubicin citrate liposomes	IV	Gilead Sciences	Advanced HIV related Kaposi's sarcoma	Apr 1996
Emend®	Nanocrystal aprepitant	Oral	Merck, Elan	Nausea in chemotherapy patients	Mar 2003
Rapamune®	Nanocrystal sirolimus	Oral	Wyeth, Elan	Immunosuppressant for kidney transplantation	Sep 1999
TriCor®	Nanocrystal fenofibrate	Oral	Abbott, Elan	Primary hypercholesterolemia, mixed lipidemia, and hypertriglyceridemia	Nov 2004
Estrasorb®	Estradiol hemihydrate micellar NP (emulsion)	Trans-dermal	Novavax	Reduction of vasomotor symptoms, such as hot flushes and night sweats in menopausal women	Oct 2003
Elestrin®	Estradiol gel (0.06%) incorpor. calcium phosphate NP	Trans-dermal	BioSanté	Treatment of moderate to severe hot flushes in menopausal women	Dec 2006

Source: Bawa, R., *Touch Briefings,* 2009, 6:122–127; and Möschwitzer, J.P., *Int J Pharm,* 2013, 453(1):142–156.
Note: Drug products shown in bold are explained in the text in more detail.

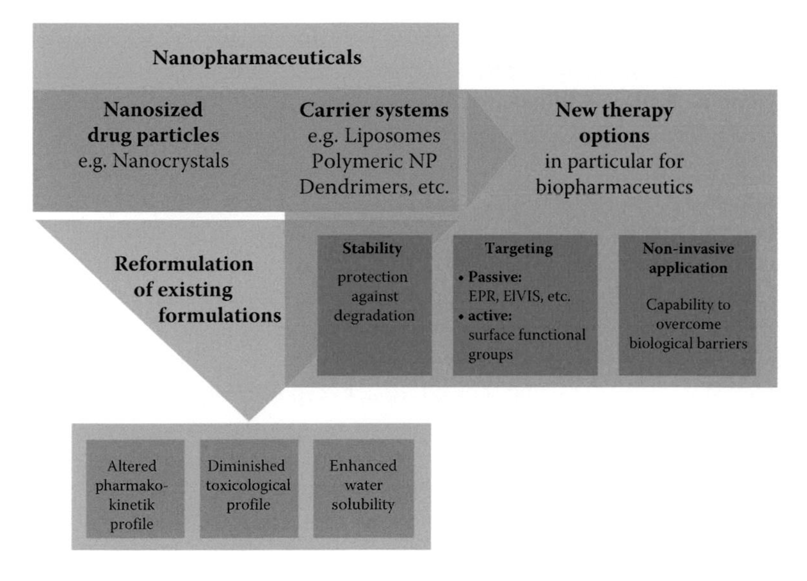

FIGURE 6.2 Chances for nanopharmaceuticals—an overview.

sirolimus was developed to overcome this issue. Nevertheless, bioavailability was still only 14% and a strong influence of high-fat food on drug uptake could be observed (Lampen et al. 1998; Zimmerman et al. 1999). Besides that, the solution had an unpleasant taste and was unstable at room temperature; thus had to be stored refrigerated. Wyeth successfully developed a tablet formulation, which is comparable to the oral solution in terms of therapeutic issues. Moreover, this new formulation provides stability at room temperature and improves palatability (Mathew et al. 2006). The new formulation is based on a wet media-based milling technique (Nanocrystal Technology) (Liversidge and Cundy 1995; Liversidge 1992).

6.5.2.2 Abraxane® (Paclitaxel)

With the disruption of microtubule function, taxanes represent an important class of antitumor agents (Schiff and Horwitz 1980; Rowinsky, Cazenave, and Donehower 1990; Ringel and Horwitz 1991). The clinical advances are limited by the highly hydrophobic character of the molecules, causing poor water solubility and hence poor bioavailability. For example, the conventional solvent-based paclitaxel formulation (Taxol®) contains high amounts of Cremophor EL® as surfactant to improve the solubility of paclitaxel. This formulation requires a long infusion period and is associated with severe hypersensitivity reactions. Premedication with corticosteroids and antihistamines is therefore required (Subramaniam et al. 2003). Additionally, due to the formation of micelles in the plasma compartment that entrap paclitaxel, formulations with Cremophor EL show nonlinear pharmacokinetics, meaning the given dose of paclitaxel is not linear to the plasma drug concentration (Sparreboom et al. 1999). Abraxane, a new Cremor EL®-free formulation of paclitaxel, is based on the nanoparticle albumin-bound (nab™) technology to improve solubility of paclitaxel and decrease side effects of older formulations (Hawkins, Soon-Shiong, and

Desai 2008). Albumin is a natural carrier protein of endogenous hydrophobic molecules. It binds them in a reversible and noncovalent fashion (Hawkins, Soon-Shiong, and Desai 2008; Paal, Muller, and Hegedus 2001; Purcell, Neault, and Tajmir-Riahi 2000). The formulation takes advantage of the following mechanism: it consists of paclitaxel nanoparticles stabilized with human serum albumin through hydrophobic interactions. Its size of about 130 nm allows it to form a colloidal suspension in saline solution (0.9% NaCl) for iv injection. Desai et al. showed a higher penetration of Abraxane into tumor cells compared to an equal dose of standard paclitaxel, probably due to the fact that Cremophor EL inhibits transcytosis of paclitaxel (Desai et al. 2006). Clinical studies of the formulation showed lower toxicity, shorter infusion times, and less hypersensitivity reactions, meaning no premedication is required. Moreover, Abraxane showed higher response rates and increased survival (Gradishar et al. 2005; Ibrahim et al. 2002).

6.5.2.3 AmBisome® (Amphotericin B)

The use of amphotericin B, which offers a broad fungicidal activity, is limited by its toxicity. Although topical administration is well tolerated, acute toxic reactions and chronic nephrotoxicity can occur when the drug is administered via infusion (Adler-Moore and Proffitt 1993).

AmBisome® is a formulation of amphotericin B incorporated into unilamellar liposomes with a diameter of less than 100 nm. The antifungal drug is integrated tightly within the liposomal membrane by forming a noncovalent charge-mediated complex with the membrane material, which consists of various lipids, phospholipids, and cholesterol. The special properties of this liposomal formulation, such as small particle size and formation of a noncovalent charge complex between drug and carrier, result in a prolonged circulation of the intact liposome in the bloodstream. Therefore, its clinical pharmacokinetic profile is different from other amphotericin B formulations: a prolonged half-life, combined with a lower toxicological profile, shows that AmBisome has potential therapeutic advantage compared to other amphotericin B formulations (Adler-Moore and Proffitt 1993; Bekersky et al. 2002; Boswell et al. 1998).

6.5.2.4 Doxil/Caelyx® (Doxorubicin)

Doxorubicin shows antitumor activity by intercalating DNA strands. Doxil (named Caelyx outside the US) is a unilamellar liposomal formulation, with a size of less than 100 nm, which is sterically stabilized with polyethylene glycol (PEG) (Lasic 1996). PEGylation can improve the therapeutic value of drugs (Veronese and Pasut 2005). Coating with inert polymers, such as PEG, can effectively reduce nonspecific interactions of liposomes with their milieu, hence prolonging their blood circulation times. These STEALTH® liposomes can target tumor tissue via the EPR effect more efficiently, as they are not recognized by the MPS as fast as without PEGylation (Vaage et al. 1994; Lasic 1995). Studies on Doxil®/Caelyx® in rabbits and dogs have shown decreased cardiotoxicity compared to free doxorubicin (Working et al. 1999).

6.5.2.5 First siRNA Carrier in Clinical Trials (CALAA-01)

The delivery technology is based on a cyclodextrin containing cationic polymer, which builds a charge-based complex with nucleic acid actives. The addition of a further hydrophilic polymer (PEG) with an adamantane end and an optional targeting ligand (transferrin in the case of CALAA-01) enables iv administration and tumor targeting (Davis 2009). Calando Pharmaceuticals Inc. as part of Arrowhead Research Corporation took this nanopharmaceutical in a clinical phase 1 study as therapy for solid tumors. This study not only demonstrated knock-down efficacy of the target but also observed adverse effects (Davis et al. 2010). The Phase 1b study evaluated the possibilities to overcome the adverse effects with a pretreatment. Recently, Arrowhead decided not to take CALAA-01 into phase II studies (genomeweb.com 2013).

6.5.3 Requirements for Nanopharmaceuticals

The key for the success of the aforementioned formulations and pharmaceutical nanoparticulate formulations in general lies within their benefits for therapeutic outcome (Figure 6.1) combined with their safe use and cost-effective production.

International regulations on medicinal and pharmaceutical products are rather strict, putting high demands on proving their *quality*, *efficacy*, and *safety,* before getting marketing approval or even entering clinical trials. As nanopharmaceuticals are understood as a subgroup of pharmaceutical products, the same regulations not only hold for "normal" drug products but also for nanopharmaceuticals in particular. Safety is typically tested first, usually in vitro and in vivo before entering clinical trials, while demonstrating efficacy is typically done afterwards. A basis for safety is of course the use of appropriate materials and the appropriate testing of final formulation. Quality is another factor that has to be taken into account while developing a nanopharmaceutical formulation for market purposes. Generally, consistent quality is achieved in the manufacturing process and the characterization of the formulation. With respect to the possibility of designing and tuning properties of nanoparticulate formulations various drug delivery systems are available: platforms such as liposomes and albumin-based particles are already approved, various polymer–drug composites are now in clinical trials, and dendrimers are in *pre*clinical trials (Davis, Chen, and Shin 2008). Taking this into consideration, it is obvious that the more complicated a nanoparticulate formulation is (nanocrystals of one material vs. multicomponent nature) the more complicated is its manufacturing and characterization. Manufacturing from an industry point of view includes the ability to scale-up processes, developed at small scale in the lab, as well as cost management. Although proof-of-concept studies are widely published, there is a difference in producing multifunctional nanoparticles in milligram quantities or in gram to kilogram quantities for clinical trials with batch to batch consistency (McNeil 2009). On the one hand, processes such as wet-ball-milling are well established, easy to scale up, and used for nanoparticle production on the market today (e.g., Nanocrystal Technology). On the other hand, processes such as precipitation are known for encapsulation of biopharmaceuticals and applied at the lab scale. Industry production is challenging, as batch to batch variation can be quite high. Specific good-manufacturing procedures (GMP), as they are known for common medicinal products production, have not yet been established for nanoparticle production and the current GMP guidelines cannot readily be applied to multifunctional nanoparticle production (McNeil 2009).

Characterization of nanopharmaceuticals is an important point in nanopharmaceuticals production. As characterization of a simple formulation is easier compared to a more difficult constructed one, the complexity (structure and composition) of marketed formulations so far has been generally simple. Although there are a couple of characterization methods, nanoparticulate formulations can be challenging to characterize (Oberdörster, Oberdörster, and Oberdörster 2005; Oberdörster et al. 2005; Powers et al. 2006). Besides formal characterization techniques for size, size distribution, and so on, there is a special need for characterizing the toxicological hazard, which is difficult due to the wide range of different engineered nanomaterials. A team of the International Life Sciences Institute Research Foundation/Risk Science Institute developed a screening strategy for the hazard identification of nanomaterials for early stage risk assessment.

6.6 CONCLUSION

The first nanopharmaceuticals, which have already reached the market, demonstrate that there is a chance of success for such products. But these relatively few positive cases do not reflect the full potential of this technology yet. There are still many unmet medical needs, where nanopharmaceuticals hold promises to either improve efficiency and/or reduce toxicity, or even open pathways to entirely new therapeutic modalities. With the rise of macromolecular biopharmaceuticals (i.e., peptide or nucleotide based drugs), there is also a need for new delivery agents that not only deliver the active pharmaceutical ingredient to its target but also preserve its stability during preparation, storage, and circulation in the body. Sound knowledge of the materials, their toxicity, biocompatibility, and biodegradability is a basis for successful nanoparticulate formulations (Cleland, Daugherty, and Mrsny 2001; Miele et al. 2009). Statements that simpler carrier systems are much easier in translation to clinics do by no way implicate that more complicated systems may not have their advantages. Scientists working on combinatorial systems, however, should also work on methods for high precision fabrication and analytical methods, which are suitable not only for elucidation of the exact structure but also for enabling a quality control.

There is no clear regulatory and safety guideline explicitly for nanopharmaceuticals. At the moment, they are treated like every other medicinal product to get approval for the market. Most likely, however, nanopharmaceuticals could get faster in their translation, if we manage to identify rules helping in prediction of nanoparticles risks based on the input parameters (materials, manufacturing methods, purification, etc.) and their resulting particle properties (size, zeta potential, surface functionalities, etc.). In this scenario, we hopefully know the impact of the carrier and thus have to characterize only for biokinetics (comparable to what we perform now with tablets or capsules).

The multidisciplinary nature of the field cannot be reduced. Thus, it is vital for scientists working in the field of nanopharmaceuticals to have a basic understanding of the physicochemical behavior, nanotechnological methods as well as biological targets, physiological environment, and toxicological possibilities—at least in such an extent to manage communication to experts of the other subdisciplines. In addition, we may need communication/computer scientists to manage the organization of the huge data sets. In a best case scenario all the subdisciplines coevolve and assist each other. The way we perform current studies, even when not intended for use in humans, should follow current best methods and protocols in particle characterization and toxicity testing. Only then the data we create are of use in data collections, which can be the starting point of the risk prediction. Larger cooperation projects and academia–industry cooperation are important to take the lead in organization and management of such data collections.

There is no need to exaggerate—neither potentially achievable advantages nor possible risk. A realistic appraisal of both sides is the order of the day. Fatal errors with too fast marketing of nanotechnology-based products have to be avoided as they can complicate or even stop the development process and marketing chances of many other products by nonrational reasons. For nanopharmaceuticals, however, we have very stringent regulatory systems in place evaluating not only before regulatory approval but also monitoring in clinical use.

REFERENCES

Adler-Moore, J. P. and R. T. Proffitt (1993) Development, characterization, efficacy and mode of action of ambisome, a unilamellar liposomal formulation of Amphotericin B. *J Liposome Res* 3 (3):429–450.

Agency, European Medicines (2004) EMED formulation (Scientific Discussion). 35.

Aggarwal, P., J. B. Hall, C. B. McLeland, M. A. Dobrovolskaia, and S. E. McNeil (2009) Nanoparticle interaction with plasma proteins as it relates to particle biodistribution, biocompatibility and therapeutic efficacy. *Adv Drug Deliv Rev* 61 (6):428–437.

Almeida, A. J. and E. Souto (2007) Solid lipid nanoparticles as a drug delivery system for peptides and proteins. *Adv Drug Deliv Rev* 59 (6):478–90.

Bailey, C. J. and A. H. Barnett (2007) Why is Exubera being withdrawn? *BMJ* 335 (7630):1156–1156.

Bawa, R. (2008) Nanoparticle-based therapeutics in humans: A survey. *Nanotechnol Law Business* 5 (2):135–155.

Bawa, R. (2009) Nanopharmaceuticals for drug delivery—A review. *Touch Briefings* 6:122–127.

Bekersky, I., R. M. Fielding, D. E. Dressler, J. W. Lee, D. N. Buell, and T. J. Walsh (2002) Pharmacokinetics, excretion, and mass balance of liposomal amphotericin B (AmBisome) and amphotericin B deoxycholate in humans. *Antimicrob Agents Chemother* 46 (3):828–833.

Berne, B. and R. Pecora (1976) *Dynamic light scattering with application to chemistry, biology and physics*. J. Wiley & Sons: New York.

Birrenbach, G., P. Speiser (1977) Microcapsules in the nanometric range and a method for their production. US-Patent 4,021,364.

Bondy, S. C (2011) Nanoparticles and colloids as contributing factors in neurodegenerative disease. *Int J Environ Res Public Health* 8 (6):2200–2211.

Boswell, G. W., I. Bekersky, D. Buell, R. Hiles, and T. J. Walsh (1998) Toxicological profile and pharmacokinetics of a unilamellar liposomal vesicle formulation of amphotericin B in rats. *Antimicrob Agents Chemother* 42 (2):263–268.

Brannen, K. C., J. M. Panzica-Kelly, T. L. Danberry, and K. A. Augustine-Rauch (2010) Development of a zebrafish embryo teratogenicity assay and quantitative prediction model. *Birth Defects Res B Dev Reprod Toxicol* 89 (1):66–77.

Brittain, H. G. (2001) X-ray diffraction III: Pharmaceutical applications of x-ray powder diffraction. *Spectroscopy* 16 (7):14–18.

Cahill, P. A., A. W. Knight, N. Billinton, M. G. Barker, L. Walsh, P. O. Keenan, C. V. Williams, D. J. Tweats, and R. M. Walmsley (2004) The GreenScreen genotoxicity assay: A screening validation programme. *Mutagenesis* 19 (2):105–119.

Carr, B. and A. Malloy (2006) NanoParticle Tracking Analysis–The Nanosight System:1–73.

Cleland, J. L., A. Daugherty, and R. Mrsny (2001) Emerging protein delivery methods. *Curr Opin Biotechnol* 12 (2):212–219.

Collnot, E. M., H. Ali, and C. M. Lehr (2012) Nano- and microparticulate drug carriers for targeting of the inflamed intestinal mucosa. *J Control Release* 161 (2):235–246.

Cooper, A. (2004) Measuring Zeta Potential using Laser Doppler Electrophoresis. *Malvern Technical Note*:1–2.

Crouch, S., R. Kozlowski, K. Slater, and J. Fletcher. (1993) The use of ATP bioluminescence as a measure of cell proliferation and cytotoxicity. *J Immunol Methods* 160 (1):81–88.

Davis, M. E (2009) The first targeted delivery of siRNA in humans via a self-assembling, cyclodextrin polymer-based nanoparticle: From concept to clinic. *Mol Pharm* 6 (3):659–68.

Davis, M. E., Z. G. Chen, and D. M. Shin (2008) Nanoparticle therapeutics: An emerging treatment modality for cancer. *Nat Rev Drug Discov* 7 (9):771–782.

Davis, M. E., J. E. Zuckerman, C. H. Choi, D. Seligson, A. Tolcher, C. A. Alabi, Y. Yen, J. D. Heidel, and A. Ribas (2010) Evidence of RNAi in humans from systemically administered siRNA via targeted nanoparticles. *Nature* 464 (7291):1067–1070.

Desai, N., V. Trieu, Z. Yao, L. Louie, S. Ci, A. Yang, C. Tao, T. De, B. Beals, D. Dykes, P. Noker, R. Yao, E. Labao, M. Hawkins, and P. Soon-Shiong (2006) Increased antitumor activity, intratumor paclitaxel concentrations, and endothelial cell transport of cremophor-free, albumin-bound paclitaxel, ABI-007, compared with cremophor-based paclitaxel. *Clin Cancer Res* 12 (4):1317–1324.

Ellman, G. L. (1959) Tissue sulfhydryl groups. *Arch Biochem Biophys* 82 (1):70–77.

Filipe, V., A. Hawe, and W. Jiskoot (2010) Critical evaluation of Nanoparticle Tracking Analysis (NTA) by NanoSight for the measurement of nanoparticles and protein aggregates. *Pharm Res* 27 (5):796–810.

Geiser, M. and W. G. Kreyling (2010) Deposition and biokinetics of inhaled nanoparticles. *Part Fibre Toxicol* 7 (2):1–17.

Gene Silencing News | RNAi | GenomeWeb (2013) Available from http://www.genomeweb.com/rnai/roche-delivery- tech-advancing-arrowhead-drops-first-generation-cancer-drug (accessed last on 10.07.14).

Gradishar, W. J., S. Tjulandin, N. Davidson, H. Shaw, N. Desai, P. Bhar, M. Hawkins, and J. O'Shaughnessy (2005) Phase III trial of nanoparticle albumin-bound paclitaxel compared with polyethylated castor oil-based paclitaxel in women with breast cancer. *J Clin Oncol* 23 (31):7794–7803.

Grobelny, J., F. Del Rio, N. Pradeep, and D. I. Kim (2009) Size measurement of nanoparticles using atomic force microscopy. *NIST–NCL Method PCC-6.*

Habeeb, A. F. (1966) Determination of free amino groups in proteins by trinitrobenzenesulfonic acid. *Anal Biochem* 14 (3):328–336.

Hasenberg, M., S. Stegemann-Koniszewski, and M. Gunzer (2013) Cellular immune reactions in the lung. *Immunol Rev* 251 (1):189–214.

Hatakeyama, H., E. Ito, M. Yamamoto, H. Akita, Y. Hayashi, K. Kajimoto, N. Kaji, Y. Baba, and H. Harashima (2011) A DNA microarray-based analysis of the host response to a nonviral gene carrier: A strategy for improving the immune response. *Mol Ther* 19 (8):1487–198.

Hawkins, M. J., P. Soon-Shiong, and N. Desai (2008) Protein nanoparticles as drug carriers in clinical medicine. *Adv Drug Deliv Rev* 60 (8):876–885.

Ibrahim, N. K., N. Desai, S. Legha, P. Soon-Shiong, R. L. Theriault, E. Rivera, B. Esmaeli, S. E. Ring, A. Bedikian, G. N. Hortobagyi, and J. A. Ellerhorst (2002) Phase I and pharmacokinetic study of ABI-007, a Cremophor-free, protein-stabilized, nanoparticle formulation of paclitaxel. *Clin Cancer Res* 8 (5):1038–1044.

James, A. E. and J. D. Driskell (2013) Monitoring gold nanoparticle conjugation and analysis of biomolecular binding with nanoparticle tracking analysis (NTA) and dynamic light scattering (DLS). *Analyst* 138 (4):1212–1218.

Jung, T., W. Kamm, A. Breitenbach, K. D. Hungerer, E. Hundt, and T. Kissel (2001) Tetanus toxoid loaded nanoparticles from sulfobutylated poly(vinyl alcohol)-graft-poly(lactide-co-glycolide): Evaluation of antibody response after oral and nasal application in mice. *Pharm Res* 18 (3):352–360.

Keck, C. M. and R. H. Muller (2013) Nanotoxicological classification system (NCS)—a guide for the risk-benefit assessment of nanoparticulate drug delivery systems. *Eur J Pharm Biopharm* 84 (3):445–448.

Labouta, H. I., L. K. el-Khordagui, T. Kraus, and M. Schneider (2011) Mechanism and determinants of nanoparticle penetration through human skin. *Nanoscale* 3 (12):4989–4999.

Labouta, H. I. and M. Schneider (2013) Interaction of inorganic nanoparticles with the skin barrier: Current status and critical review. *Nanomedicine* 9 (1):39–54.

Lademann, J., H. Richter, A. Teichmann, N. Otberg, U. Blume-Peytavi, J. Luengo, B. Weiss, U. F. Schaefer, C. M. Lehr, R. Wepf, and W. Sterry (2007) Nanoparticles—An efficient carrier for drug delivery into the hair follicles. *Eur J Pharm Biopharm* 66 (2):159–164.

Lampen, A., Y. Zhang, I. Hackbarth, L. Z. Benet, K. F. Sewing, and U. Christians (1998) Metabolism and transport of the macrolide immunosuppressant sirolimus in the small intestine. *J Pharmacol Exp Ther* 285 (3):1104–1112.

Lamprecht, A., U. Schafer, and C. M. Lehr (2001) Size-dependent bioadhesion of micro- and nanoparticulate carriers to the inflamed colonic mucosa. *Pharmaceut Res* 18 (6):788–793.

Landsiedel, R., E. Fabian, L. Ma-Hock, B. van Ravenzwaay, W. Wohlleben, K. Wiench, and F. Oesch (2012) Toxico-/biokinetics of nanomaterials. *Arch Toxicol* 86 (7):1021–1060.

Lasic, D. (1995) The "Stealth" Liposome: A Prototypical Biomaterial. *Chem Rev* 95 (8):2601–2628.

Lasic, D. D. (1996) Doxorubicin in sterically stabilized liposomes. *Nature* 380 (6574):561–562.

Liversidge, G. G. and K. C. Cundy (1995) Particle size reduction for improvement of oral bioavailability of hydrophobic drugs: I. Absolute oral bioavailability of nanocrystalline danazol in beagle dogs. *Int J Pharm* 125 (1):91–97.

Liversidge, G. G., K. C. Cundy, J. F. Bishop, and D. A. Czekai (1992) Surface modified drug nanoparticles. US-Patent 5145684, Issued on September 8, 1992.

Mathew, T. H., C. Van Buren, B. D. Kahan, K. Butt, S. Hariharan, and J. J. Zimmerman (2006) A comparative study of sirolimus tablet versus oral solution for prophylaxis of acute renal allograft rejection. *J Clin Pharmacol* 46 (1):76–87.

Matsumura, Y. and H. Maeda (1986) A new concept for macromolecular therapeutics in cancer chemotherapy: Mechanism of tumoritropic accumulation of proteins and the antitumor agent smancs. *Cancer Res* 46:6387–6392.

McNeil, S. E (2009) Nanoparticle therapeutics: A personal perspective. *Wiley Interdiscip Rev Nanomed Nanobiotechnol* 1 (3):264–271.

McNeish, J (2004) Embryonic stem cells in drug discovery. *Nat Rev Drug Discov* 3 (1):70–80.

Merisko-Liversidge, E., G. G. Liversidge, and E. R. Cooper (2003) Nanosizing: A formulation approach for poorly-water-soluble compounds. *Eur J Pharm Sci* 18 (2):113–120.

Miele, E., G. P. Spinelli, E. Miele, F. Tomao, and S. Tomao (2009) Albumin-bound formulation of paclitaxel (Abraxane® ABI-007) in the treatment of breast cancer. *Int J Nanomedicine* 4 (4):99–105.

Mittal, A., A. S. Raber, and S. Hansen (2013) Particle based vaccine formulations for transcutaneous immunization. *Hum Vaccin Immunother* 9 (9):1950–1955.

Mittal, A., A. S. Raber, U. F. Schaefer, S. Weissmann, T. Ebensen, K. Schulze, C. A. Guzman, C. M. Lehr, and S. Hansen (2013) Non-invasive delivery of nanoparticles to hair follicles: A perspective for transcutaneous immunization. *Vaccine* 31 (34):3442–3451.

Moghimi, S. M. and J. Szebeni (2003) Stealth liposomes and long circulating nanoparticles: Critical issues in pharmacokinetics, opsonization and protein-binding properties. *Prog Lipid Res* 42 (6):463–478.

Monteiro-Riviere, N. A., K. Wiench, R. Landsiedel, S. Schulte, A. O. Inman, and J. E. Riviere (2011) Safety evaluation of sunscreen formulations containing titanium dioxide and zinc oxide nanoparticles in UVB sunburned skin: An in vitro and in vivo study. *Toxicol Sci* 123 (1):264–280.

Moschwitzer, J. P (2013) Drug nanocrystals in the commercial pharmaceutical development process. *Int J Pharm* 453 (1):142–156.

Müller, R. H. 1991. *Colloidal carriers for controlled drug delivery and targeting: Modification, characterization and in vivo distribution*. Taylor & Francis: USA

Nafee, N., M. Schneider, K. Friebel, M. Dong, U. F. Schaefer, T. E. Murdter, and C. M. Lehr (2012) Treatment of lung cancer via telomerase inhibition: Self-assembled nanoplexes versus polymeric nanoparticles as vectors for 2'-O-Methyl-RNA. *Eur J Pharm Biopharm* 80 (3):478–489.

Nangia, S. and R. Sureshkumar (2012) Effects of nanoparticle charge and shape anisotropy on translocation through cell membranes. *Langmuir* 28 (51):17666–17671.

NanoMedicine ETP (2005) Vision paper and basis for a strategic research agenda for nanomedicine. *European Technology Platform on NanoMedicine*:1–40.

Nilsson, B. and K. N. Ekdahl (2012) Complement diagnostics: Concepts, indications, and practical guidelines. *Clin Dev Immunol* 2012:1–11.

Oberdörster, G., A. Maynard, K. Donaldson, V. Castranova, J. Fitzpatrick, K. Ausman, J. Carter, B. Karn, W. Kreyling, D. Lai, S. Olin, N. Monteiro-Riviere, D. Warheit, and H. Yang (2005) Principles for characterizing the potential human health effects from exposure to nanomaterials: Elements of a screening strategy. *Part Fibre Toxicol* 2 (1):1–35.

Oberdörster, G., E. Oberdörster, and J. Oberdörster (2005) Nanotoxicology: An emerging discipline evolving from studies of ultrafine particles. *Environ Health Perspect* 113 (7):823–839.

Olive, P. L. and J. P. Banath (2006) The comet assay: A method to measure DNA damage in individual cells. *Nat Protoc* 1 (1):23–29.

Paal, K., J. Muller, and L. Hegedus (2001) High affinity binding of paclitaxel to human serum albumin. *Eur J Biochem* 268 (7):2187–2191.

Pardridge, W. M (2003) Blood-brain barrier drug targeting: The future of brain drug development. *Mol Interv* 3 (2):90–105.

Perlstein, B., Z. Ram, D. Daniels, A. Ocherashvilli, Y. Roth, S. Margel, and Y. Mardor (2008) Convection-enhanced delivery of maghemite nanoparticles: Increased efficacy and MRI monitoring. *Neuro Oncol* 10 (2):153–161.

Polikarpov, D. M., V. M. Cherepanov, R. R. Gabbasov, M. A. Chuev, I. N. Mischenko, V. A. Korshunov, and V. Y. Panchenko (2013) Efficiency analysis of clearance of two types of exogenous iron from the rat brain by Mössbauer spectroscopy. *Hyperfine Interactions* 218 (1–3):83–88.

Powers, K. W., S. C. Brown, V. B. Krishna, S. C. Wasdo, B. M. Moudgil, and S. M. Roberts (2006) Research strategies for safety evaluation of nanomaterials. Part VI. Characterization of nanoscale particles for toxicological evaluation. *Toxicol Sci* 90 (2):296–303.

Purcell, M., J. F. Neault, and H. A. Tajmir-Riahi (2000) Interaction of taxol with human serum albumin. *Biochim Biophys Acta* 1478 (1):61–68.

Puzyn, T., B. Rasulev, A. Gajewicz, X. K. Hu, T. P. Dasari, A. Michalkova, H. M. Hwang, A. Toropov, D. Leszczynska, and J. Leszczynski (2011) Using nano-QSAR to predict the cytotoxicity of metal oxide nanoparticles. *Nat Nanotechnol* 6 (3):175–178.

Quan, L. D., P. E. Purdue, X. M. Liu, M. D. Boska, S. M. Lele, G. M. Thiele, T. R. Mikuls, H. Dou, S. R. Goldring, and D. Wang (2010) Development of a macromolecular prodrug for the treatment of inflammatory arthritis: Mechanisms involved in arthrotropism and sustained therapeutic efficacy. *Arthritis Res Ther* 12 (5):1–10.

Rathore, A. S (2009) Roadmap for implementation of quality by design (QbD) for biotechnology products. *Trends Biotechnol* 27 (9):546–553.

Ringel, I. and S. B. Horwitz (1991) Studies with RP 56976 (taxotere): A semisynthetic analogue of taxol. *J Natl Cancer Inst* 83 (4):288–291.

Rohwedel, J., K. Guan, C. Hegert, and A. M. Wobus (2001) Embryonic stem cells as an in vitro model for mutagenicity, cytotoxicity and embryotoxicity studies: Present state and future prospects. *Toxicol* in vitro 15 (6):741–753.

Rowinsky, E. K., L. A. Cazenave, and R. C. Donehower (1990) Taxol: A novel investigational antimicrotubule agent. *J Natl Cancer Inst* 82 (15):1247–1259.

Schiff, P. B. and S. B. Horwitz (1980) Taxol stabilizes microtubules in mouse fibroblast cells. *Proc Natl Acad Sci U S A* 77 (3):1561–1565.

Schmidt, C., C. Lautenschlaeger, E. M. Collnot, M. Schumann, C. Bojarski, J. D. Schulzke, C. M. Lehr, and A. Stallmach (2013) Nano- and microscaled particles for drug targeting to inflamed intestinal mucosa: A first in vivo study in human patients. *J Control Release* 165 (2):139–145.

Scholz, G., I. Pohl, E. Genschow, M. Klemm, and H. Spielmann (1999) Embryotoxicity screening using embryonic stem cells in vitro: Correlation to in vivo teratogenicity. *Cells Tissues Organs* 165 (3–4):203–211.

Shao, L., M. J. Pollard, P. R. Griffiths, D. T. Westermann, and D. L. Bjorneberg (2007) Rejection criteria for open-path Fourier transform infrared spectrometry during continuous atmospheric monitoring. *Vib Spectrosc* 43 (1):78–85.

Skyler, J. S., L. Jovanovic, S. Klioze, J. Reis, and W. Duggan (2007) Two-year safety and efficacy of inhaled human insulin (Exubera) in adult patients with type 1 diabetes. *Diabetes Care* 30 (3):579–585.

Slutter, B., L. Plapied, V. Fievez, M. A. Sande, A. des Rieux, Y. J. Schneider, E. Van Riet, W. Jiskoot, and V. Preat (2009) Mechanistic study of the adjuvant effect of biodegradable nanoparticles in mucosal vaccination. *J Control Release* 138 (2):113–121.

Sparreboom, A., L. Van Zuylen, E. Brouwer, W. J. Loos, P. De Bruijn, H. Gelderblom, M. Pillay, K. Nooter, G. Stoter, and J. Verweij (1999) Cremophor EL-mediated alteration of paclitaxel distribution in human blood: Clinical pharmacokinetic implications. *Cancer Res* 59 (7):1454–1457.

Subramaniam, V., H. Li, M. Wong, R. Kitching, L. Attisano, J. Wrana, J. Zubovits, A. M. Burger, and A. Seth (2003) The RING-H2 protein RNF11 is overexpressed in breast cancer and is a target of Smurf2 E3 ligase. *Br J Cancer* 89 (8):1538–1544.

Todoroff, J. and R. Vanbever (2011) Fate of nanomedicines in the lungs. *Curr Opin Colloid In* 16 (3):246–254.

TAXOL (paclitaxel) Injection (Patient Information Included) (2011) Available from packageinserts.bms.com/pi/pi_taxol.pdf (accessed last on 10.07.14).

Tosi, G., L. Costantino, B. Ruozi, F. Forni, and M. A. Vandelli (2008) Polymeric nanoparticles for the drug delivery to the central nervous system. *Expert Opin Drug Deliv* 5 (2):155–174.

Tsapis, N., D. Bennett, B. Jackson, D. A. Weitz, and D. A. Edwards (2002) Trojan particles: Large porous carriers of nanoparticles for drug delivery. *Proc Natl Acad Sci U S A* 99 (19):12001–12005.

Vaage, J., E. Barbera-Guillem, R. Abra, A. Huang, and P. Working (1994) Tissue distribution and therapeutic effect of intravenous free or encapsulated liposomal doxorubicin on human prostate carcinoma xenografts. *Cancer* 73 (5):1478–1484.

Veronese, F. M. and G. Pasut (2005) PEGylation, successful approach to drug delivery. *Drug Discov. Today* 10 (21):1451–1458.

Wagner, V., B. Hüsing, S. Gaisser, and A. Bock (2008) Nanomedicine: Drivers for development and possible impacts. *JRC-IPTS*. Luxembourg: Spain.

Wahl, B., N. Daum, H. L. Ohrem, and C. M. Lehr (2008) Novel luminescence assay offers new possibilities for the risk assessment of silica nanoparticles. *Nanotoxicology* 2 (4):243–251.

Wang, D. and S. R. Goldring (2011) The bone, the joints and the Balm of Gilead. *Mol Pharm* 8 (4):991–993.

Wang, D., S. C. Miller, M. Sima, D. Parker, H. Buswell, K. C. Goodrich, P. Kopeckova, and J. Kopecek (2004) The arthrotropism of macromolecules in adjuvant-induced arthritis rat model: A preliminary study. *Pharm Res* 21 (10):1741–1749.

Wawer, I. and B. Diehl (2011) NMR spectroscopy in pharmaceutical analysis. Elsevier: Amsterdam.

Weyermann, J., D. Lochmann, and A. Zimmer (2005) A practical note on the use of cytotoxicity assays. *Int J Pharm* 288 (2):369–376.

Wohlfart, S., S. Gelperina, and J. Kreuter (2012) Transport of drugs across the blood-brain barrier by nanoparticles. *J Control Release* 161 (2):264–273.

Wong, G. (2009) Biotech scientists bank on big pharma's biologics push. *Nat Biotechnol* 27:293–296.

Working, P. K., M. S. Newman, T. Sullivan, and J. Yarrington (1999) Reduction of the cardiotoxicity of doxorubicin in rabbits and dogs by encapsulation in long-circulating, pegylated liposomes. *J Pharmacol Exp Ther* 289 (2):1128–233.

Wunder, A., U. Muller-Ladner, E. H. Stelzer, J. Funk, E. Neumann, G. Stehle, T. Pap, H. Sinn, S. Gay, and C. Fiehn (2003) Albumin-based drug delivery as novel therapeutic approach for rheumatoid arthritis. *J Immunol* 170 (9):4793–4801.

Yuan, F., M. Dellian, D. Fukumura, M. Leunig, D. A. Berk, V. P. Torchilin, and R. K. Jain (1995) Vascular permeability in a human tumor xenograft: Molecular size dependence and cutoff size advances in brief vascular permeability in a human tumor xenograft: Molecular size dependence and cutoff size. *Cancer Res* 55 (17):3752–3756.

Yuan, F., L.-D. Quan, L. Cui, S. R. Goldring, and D. Wang (2012) Development of macromolecular prodrug for rheumatoid arthritis. *Adv Drug Deliv Rev 64* (12):1205–1219.

Zimmerman, J. J., G. M. Ferron, H-K Lim, and V. Parker (1999) The effect of a high-fat meal on the oral bioavailability of the immunosuppressant sirolimus (rapamycin). *J Clin Pharmacol* 39 (11):1155–1161.

Zolnik, B. S. and N. Sadrieh (2009) Regulatory perspective on the importance of ADME assessment of nanoscale material containing drugs. *Adv Drug Deliv Rev* 61 (6):422–427.

7 Measurement of Nanoparticle Uptake by Alveolar Macrophages

A New Approach Based on Quantitative Image Analysis

Darius Schippritt, Hans-Gerd Lipinski, and Martin Wiemann

CONTENTS

7.1 INTRODUCTION

The lung is the major entrance route for air borne nanoparticles (NP) into the body, some of which have been shown to have considerable biologic effects (Oberdörster et al. 2005a; Xia et al. 2009). Consequently, inhalation experiments appear best suited to the toxicological testing of NP, since particles are distributed to all parts of the lung in a life-like manner (Oberdörster et al. 2005b). However, the distribution of nanoparticles in the lungs does not remain homogeneous. Several studies have shown—and were also confirmed in the course of the NanoGEM project—that the vast majority of the nanomaterial reaching the alveolar region appears in alveolar macrophages (AM) (Gosens et al. 2009). The underlying mechanisms may involve agglomeration of NP due to an interaction with components of the lung surfactant or the lung lining fluid or pathways through the epithelium and/or interstitium (Rothen-Rütishauser et al. 2007; Semmler-Behnke et al. 2007). In any case, particle laden AM may reach the mucociliary escalator or travel toward mediastinal lymph nodes, if their motility is not disturbed by phagocytized particles (Oberdörster et al. 1992). However, AM will ingest particles also under cell culture conditions and it has been shown that well designed in vitro experiments with AM are of some predictive value with respect to the lung toxicity of micron-sized as well as nanosized particles (Lanone et al. 2009; Rehn et al. 1999; Wiemann and Bruch 2009). The cellular particle burden of AM may be adapted to the situation in vivo where a mean dose of 60–90 pg/AM resulted from a 4 week inhalation exposure to 30 mg/m^3 (Pauluhn 2009). In this chapter, a light microscopic method is described with the help of which the exposure of macrophages to (nano)particles as well as their uptake can be quantified in vitro. The topic of in vitro particokinetics becomes increasingly important for correctly describing the cellular dose of cells exposed to nanoparticles (Teeguarden et al. 2007). Particles with a hydrodynamic diameter larger than ca. 300 nm are subject to instantaneous gravitational settling, whereas smaller nonagglomerated particles stay suspended. Based on the mathematical description of diffusion, agglomeration, and sedimentation the ISDD model was proposed which allowed calculating as to whether a particle might sediment, which eases its contribution to the cellular dose (Hinderliter et al. 2010). Data requirements for the model were not only temperature, viscosity, particle size, and density but, in the case of agglomerate formation, also the fractal dimension and packing factors. Importantly, it was shown that the "difference between equivalent nominal media concentration exposures on a μg/mL basis and target cell doses on a particle surface area or number basis can be as high as three to six orders of magnitude." Consequently, researchers have to be aware that there will be a strong albeit indirect impact of particle properties on dose rate, that is, the increment of cellular particle burden over time.

In this chapter, we will introduce an alternative approach to describe gravitational settling of nanoparticles during in vitro testing. The method primarily takes advantage of phase contrast micrographs serially collected with a time-lapse imaging device operated under cell culture conditions. Agglomerated particles can be viewed and counted in the lowermost optical plane in the absence or presence of cells. Due to optical resolution the method is confined to agglomerates large enough to be viewed with phase contrast optics (>0.3 μm) and has to be supplemented by analyzing the supernatant for remaining nanoparticles, for example, by tracking analysis. The combination of both methods delivers a fairly complete picture of particokinetics in cell culture experiments and allows us to calculate the mass of settled or even ingested particles in ideal cases, that is, if the end point of sedimentation is known. Then the measure "particle mass per mL" can be converted into "particles per area." The development of the computer program was based not only on particle and macrophage experiments but also on numerous simulations to define a sedimentation model to correct for particle overlap (Schippritt et al. 2010). The major features of the software, its development as well as some results will be described in this chapter.

7.2 MATERIALS AND METHODS

7.2.1 Cells

Experiments were carried out with two macrophage cell lines namely the murine cell line RAW 264.7 and the rat AM-derived cell line NR8383. RAW 264.7 macrophages are fast growing cells, which are strongly dependent on serum and virtually show no locomotion. They were cultivated in DMEM supplemented with 10% fetal calf serum (FCS), 2 mM glutamine, penicillin (100 U/mL), streptomycin (100 μg/mL), and passaged twice a week. In contrast, NR8383 cells are highly motile cells, which actively collect particles while moving around. They were cultured in F12-K supplemented with 15% FCS, 2 mM glutamine, penicillin (100 U/mL), and streptomycin (100 μg/mL). Cultures were passaged once a week.

7.2.2 Setting up Observation Experiments

Observation experiments were carried out with a fully automated, inverted microscope (Nikon, BioStation IM) operated under cell culture conditions (37°C, 5% CO_2, 100% humidity). To achieve highest optical resolution, glass bottom dishes (Ibidi, Munich, Germany) were used. The filling height of the Petri dish was a critical parameter and was set to 6 mm (3.88 mL total volume). This equaled the filling height obtained with 200 μL pipetted into a well of 96-well plate, which is routinely used for cell culture experiments. Since inhaled particles inside alveoli are not primarily covered with proteins, experiments with macrophages were conducted in the absence of serum. Particle concentrations were 11.25–180 μg/mL as indicated in the following and reflected the conditions hitherto used for biological testing. For a typical experiment, cells had been cultivated in a Petri dish at low density. The medium was then replaced by serum-free F12-K and the dish was transferred into the observation chamber. The filling height was adapted to in vitro testing conditions (6 mm). All media and suspensions were used at 37°C, which was found indispensable to keep the focal plane stable. Time-lapse imaging was started using a z-stack option to take micrographs at 1 μm intervals. Images were captured at—three to four sites every 30 s with a 20-fold objective. This allowed observation of—three to six cells per image and optical recording of settling particles in cell-free regions. Particles were added by replacing half of the fluid with the prewarmed, double concentrated particle suspension. This procedure maintained the correct focal plane and led to a more even distribution of particles in the chamber. Addition of particles was achieved within 1 min. All steps were equally carried out in cell-free experiments to study particle sedimentation.

7.2.3 Particle Suspensions

In this study, we analyzed the sedimentation of AlOOH and CeO_2 nanoparticle agglomerates under cell culture conditions. The primary particle size was 40 nm and 70 nm for AlOOH and CeO_2, respectively. Further characterization data were taken from Kroll et al. (2011) and are shown in Table 7.1. To suspend AlOOH or CeO_2, powders were mixed into a serum-free medium at a concentration of 180 µg/mL, briefly vortexed, and ultrasonicated for 10 s using a 3-mm probe (Vibra Cell, Sonics and Materials, Danbury, CT) operated at 50 W. Suspensions were further diluted in the F12-K medium to obtain concentrations as indicated in the text.

7.2.4 Scanning Electron Microscopy

To view single nanoparticles, AlOOH and CeO_2 powders were suspended in double distilled H_2O by ultrasonication (45 µg/mL), and concentrated on 0.2-µm Teflon® filters (Millipore). Agglomerates formed in the cell culture medium were enriched on 0.45-µm carbonate filters and briefly washed with dH_2O. All samples were air dried, sputtered with gold (Sputter Coater S150B, Edwards, North Walsham, UK), and examined with a Gemini DSM 982 (ZEISS, Oberkochen, Germany) operated at 15 kV.

TABLE 7.1
Particle Properties

	AlOOH Nanoparticles	CeO_2 Nanoparticles
Primary particle size (nm)	40	70
Particle size, (d50) (nm)	40	40
BET (m^2/g)	47	33
Density (g/cm^3)	2.85	—
Crystallinity	Orthorhombic, Böhmit-like	Cerianite, cubic
Particle morphology	Irregular spherical	Irregular spherical
Purity (%)	82.7	99
Surface modification	None	None
Surface chemistry (%)	O (61.6), Al (31.7), C (6.7)	O (53), Ce (26), C (20), Cl (0.6)
ζ-Potential (pH 7) (mV)	34.0 (in 1 mmol KCl)	23.7 (in H_2O)
ζ-Potential DMEM, 10% FBS (mV)	−7.6 ± 0.8	−7.3 ± 1.1
Solubility DMEM, 10% FBS (mg/kg)	Al: 0.185 ± 0.012	Ce: <1

Note: AlOOH and CeO_2 values are from Pauluhn (2009) and Kroll et al. (2011), respectively.

7.2.5 Particle Tracking Analysis

Supernatants from sedimentation experiments were subjected to particle tracking analysis to look for nonagglomerated nanoparticles. A volume of 200 μL was retrieved from particle-containing media at the end of the observation period and pipetted onto the stage of a NanoSight Instrument (NanoSight LM10) equipped with a green laser device (530 nm) and an intensified CCD camera (Andor-DL-658M-OEM). Particle sizes were analyzed with NTA software 2.1.

7.2.6 General Requirements for the Software Development

To analyze particle sedimentation and phagocytosis of particles by macrophages we developed a semiautomated software tool for image analysis. The application was developed with Visual Studio 2010 based on the .NET-Framework 2.0 and C# as a programming language. Special emphasis was put on automatic object detection: Although nanoparticle agglomerates could be detected as dark objects in phase contrast images in the optical plane, cell bodies of macrophages appeared encircled by a bright halo with fine pseudopodia, which were only slightly darker than the background. To identify the complete outer contour of a macrophage the image software had to consider both types of contrasts, such that special algorithms had to be programmed. In general, the program comprises sequential processing steps, such as harmonization of luminance throughout the whole image stack. Function parameters such as gray levels or filter settings can be adjusted either manually or automatically by an integrated fuzzy logic. The binarization of particles and macrophages is followed by an object detection step. Finally, the number of objects is counted and the particle-covered area is calculated in user-defined regions. Using well tested settings, the program successfully processed complete image stacks and automatically counted particles ingested by macrophages.

7.3 RESULTS

7.3.1 Sedimentation of Agglomerated Nanoparticles in Cell Culture Medium

Nanoparticles composed of AlOOH or CeO_2, when exposed to physiological fluids or cell culture medium, tend to readily agglomerate. This is mainly, though not exclusively, due to the breakdown of the zeta-potential upon ion binding to charged particle surfaces, such that van der Waals forces become dominant. As shown in Figure 7.1, these agglomerates are up to several micrometers large and, therefore, will undergo gravitational settling (cf. Teeguarden et al. 2007). The next chapters will describe how the sedimentation and uptake of these agglomerates can be quantified.

7.3.1.1 Quantity of Sedimented Agglomerates

In a first step we simulated more than 100 sedimentation scenarios using different shapes, sizes, and agglomeration rates of particles. From these results we defined a sedimentation model, which described the time-dependent change in the single

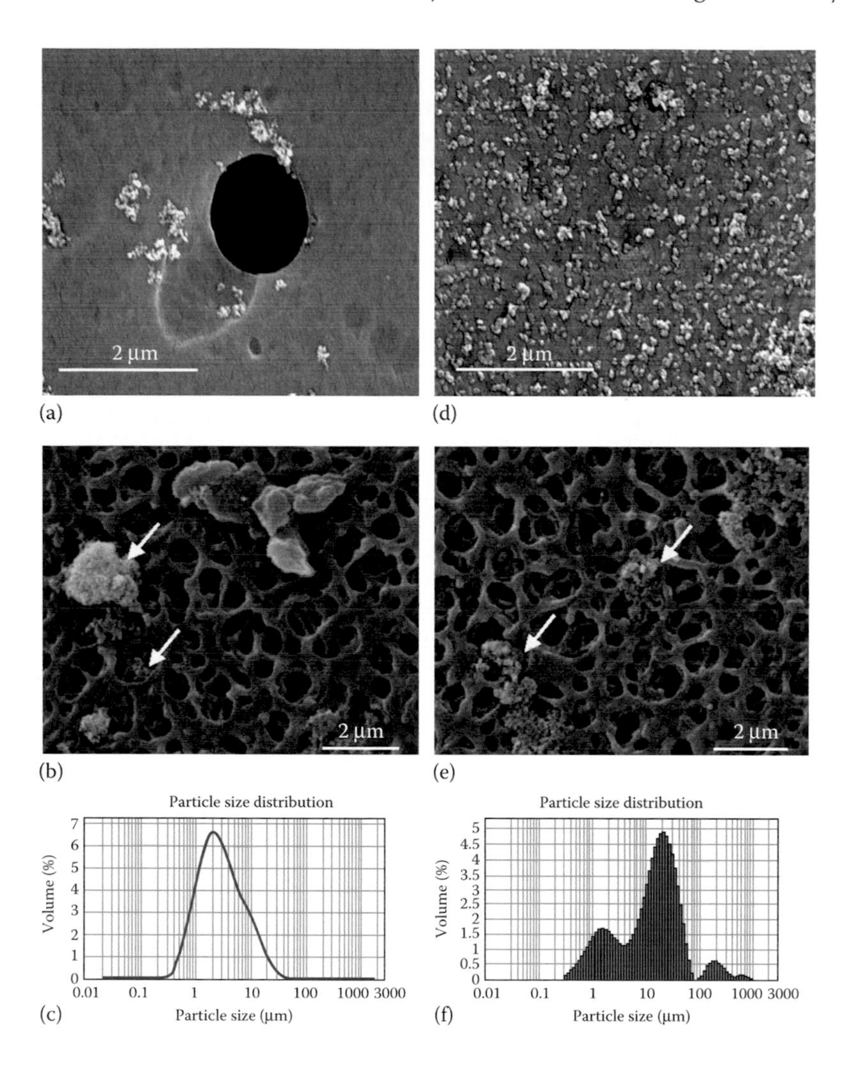

FIGURE 7.1 Structure of nanoparticles used for sedimentation and cell culture experiments. (a–c) AlOOH nanoparticles, (d–f) CeO_2 nanoparticles. Nanoparticles were dispersed by ultrasonication in H_2O and spread on Teflon® filters with a pore size of 1 μm (a: AlOOH, d: CeO_2). Typical agglomerates (arrows) as formed in cell culture medium were enriched on a polycarbonate filter with a pore size of 0.22 μm (b: AlOOH, e: CeO_2). Size distribution histograms of particle size in cell culture medium (Malvern Mastersizer) reveal that smallest particles are in the range of 0.3 μm for both AlOOH (c) and CeO_2 (f).

particle count $N(t)$ at the bottom of the culture vessel. These simulations showed that $N(t)$ increased rapidly in the beginning until the rising number of particles led to a partial overlap. Consequently, the increase in $N(t)$ was attenuated or, later on, even decreased, whereas both the apparent mean size of particles and the total area covered by particles increased continuously. The kinetic sedimentation is described by the following equation:

$$N(t)=\begin{cases} N_\infty\cdot(1-e^{-q\cdot t})\cdot e^{-\beta\cdot t} & if \quad 0\le t<t_0 \\ N(t_0)=N_0 & if \quad t_0\le t \\ 0 & if \quad t<0 \end{cases} \tag{7.1}$$

Although $N(t)$ is the total count of singular particles on the bottom, q represents the increase in particles per unit time, and β stands for the overlap of particles. N_∞ describes the fictive, calculable number of all sedimented particles for the case that no overlapping occurs ($\beta = 0$), whereas N_0 gives the countable number of particles at the end of the sedimentation process at time t_0. The kinetic sedimentation model (based on Equation 7.1) is illustrated in Figure 7.2 which shows two virtual border case components $N_q(t)$ and $N_\beta(t)$: Although $N_q(t)$ represents the sedimentation of particles without any particle overlap, $N_\beta(t)$ comprises a certain degree of overlap leading to optically merged particles and to an apparently lower number of particles (see Figure 7.3 for definition of $N\infty$, N_0, and N^* and associated t-values). A realistic scenario certainly depends on particle concentration and is expected to lie between both curves. Figure 7.3 shows the settling of agglomerated AlOOH nanoparticles. The measured number of sedimented particles can also be described by a modeled curve (dotted line) from which a deviation of less than 2% was obtained. The accuracy of the sedimentation model was evaluated by fitting it to the measured values of several sedimentation experiments. We discovered that the sedimentation parameters N_0, N_∞, q, and β were normally distributed; the maximum deviation from the particular parameter average value was approximately 5%.

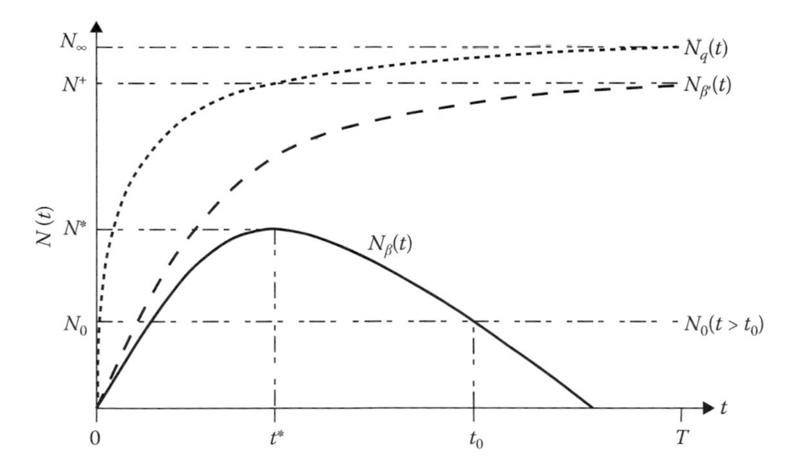

FIGURE 7.2 Limit cases of the kinetic sedimentation model. The quantity of virtually sedimented particles N at the bottom is shown as a function of time t ($0 \le t \le T$). The upper curve $N_q(t)$ shows the virtual situation without overlapping particles ($\beta = 0$); particle numbers converge against the maximum value N_∞. The lower curves $N_\beta(t)$ and $N_{\beta'}(t)$ show what happens to particle numbers in case that particles overlap ($0 < \beta' < \beta$). At low overlap (β') a pseudo-plateau value of N^+ occurs; at a marked overlap, a maximum number N^* is reached at t^*, whereas the final value N_0 (when the sedimentation process is finished) is lower and is reached at time t_0.

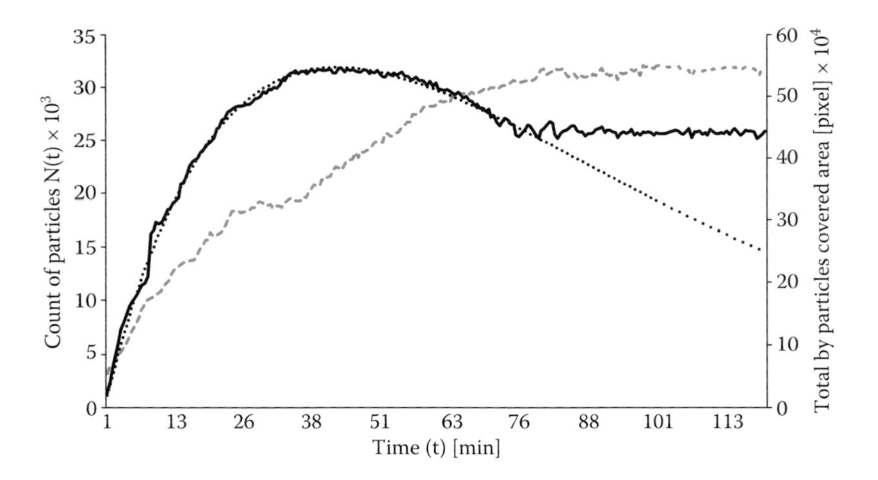

FIGURE 7.3 Sedimentation of agglomerated AlOOH nanoparticles and approximation by modeled data. Gray curve (▬▬) shows measured values of *N*, dotted curve (- - -) shows data from the model, and dashed curve (■ ■ ■) gives the total pixel area covered by particles, covered area pixel) in dependency of time *t*. The model parameters N_0, N_∞, q, and β were estimated by least-square approximation with $q = 0.00495$/min; $\beta = 0.0117$/min; $N_0 = 25673$; $N_\infty = 75267$; and $t_0 = 77$ min. The AlOOH suspension contained 180 µg/mL; the smallest recognized particle size was 1 pixel.

7.3.1.2 Conversion of Measured Particle Size into Particle Mass

The precision of the sedimentation model allowed us to predict the number of sedimented particles per area at each point in time during an ongoing sedimentation process. Provided that the initial concentration of particles is known and the complete particle mass underwent gravitational settling, the mass per particle agglomerate could be calculated. At this point, however, the overlap of particle agglomerates becomes an issue and has to be corrected for: In Figure 7.4 (upper curve) the increase in particle-covered area is shown for the virtual case that an overlap does not occur (this can be implemented in the model). Then a conventional simulation with overlapping particles was carried out, which better reflects a real situation. Particle-covered areas were determined for both cases and the degree of particle overlap (μ_0) was calculated. Figure 7.4 shows that the value for μ_0 was ca. 28% (with a standard deviation of 0.6%), if the particle-covered area amounts to 40%. In other words, the true area which would have been covered by all sedimented particles is underestimated by 28% due to particle overlap. We further varied size (and shape) of the simulated particles from 0.1 to 3-fold the size of real AlOOH or CeO_2 nanoparticle agglomerates (assuming a particle diameter of 1.2 µm) and found no significant effect on μ_0, which remained at 28% (at 40% particle-covered area). However, a massive increase in particle size to 400 µm increased μ_0 up to 55% (at 10% particle-covered area). These examples show that the particle overlap μ_0 can be retrieved from modeled data, and that μ_0 needs to be considered (and corrected for) if the measured particle-covered area is converted into real values or particle mass, respectively.

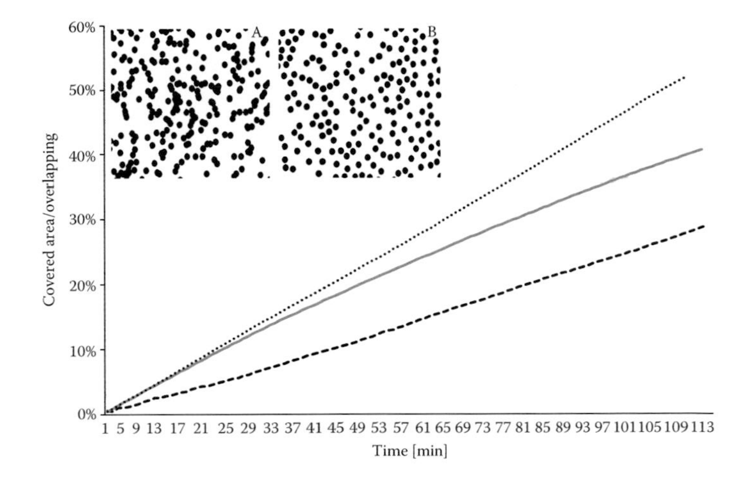

FIGURE 7.4 Degree of particle overlap modeled for a particle size of >5 px (>0.5 μm): The upper curve shows the particle-covered area in the absence of overlap. The middle curve shows the particle-covered area including overlap (▬▬), and the lower curve displays the calculated rate of overlap μ_0 (▬ ▬ ▬). Insets show particles arranged with (left) and without (right) overlap.

7.3.1.3 Gravitational Settling of ZrO_2 Nanoparticles

Within the NanoGEM project, in vitro and in vivo experiments were carried out with differentially functionalized ZrO_2 nanoparticles, namely a pegylated (ZrO_2-PGA), an acrylated (ZrO_2-Acryl), an aminated (ZrO_2-APTS), and an acid stabilized form (ZrO_2-TODS) (compare surface characteristics in Table 1.2). When these ZrO_2 NP were immersed in serum-free cell culture medium, as was used for experiments with AM (see Chapter 8), agglomeration followed by complete gravitational settling occurred, as was revealed by particle tracking analysis of the supernatant (data not shown). However, to compare the dose rates in vitro, the time courses of sedimentation, that is, the appearance of particles at the bottom of the culture dish, were analyzed with the program "AM tracking" (see the following). Slight differences were found between the four types of ZrO_2 NP as is shown in Figure 7.5. Such data have to be taken into account when toxicity data of agglomerating nanoparticles are evaluated (cf. Teeguarden et al. 2007; Hinderliter et al. 2010). It should be stressed that other particles used in NanoGEM such as silver and SiO_2 particles did not agglomerate or only partially agglomerated within the observation time. An alternative method to directly observe the uptake of single particles by light microscopy is therefore described in the last paragraph.

7.3.2 Phagocytosis of Agglomerated Nanoparticles

To analyze the phagocytosis of agglomerated AlOOH and CeO_2 nanoparticles by macrophages we performed numerous sedimentation experiments in the presence of NR8383 macrophages. Time-lapse image series were collected from up to four sites

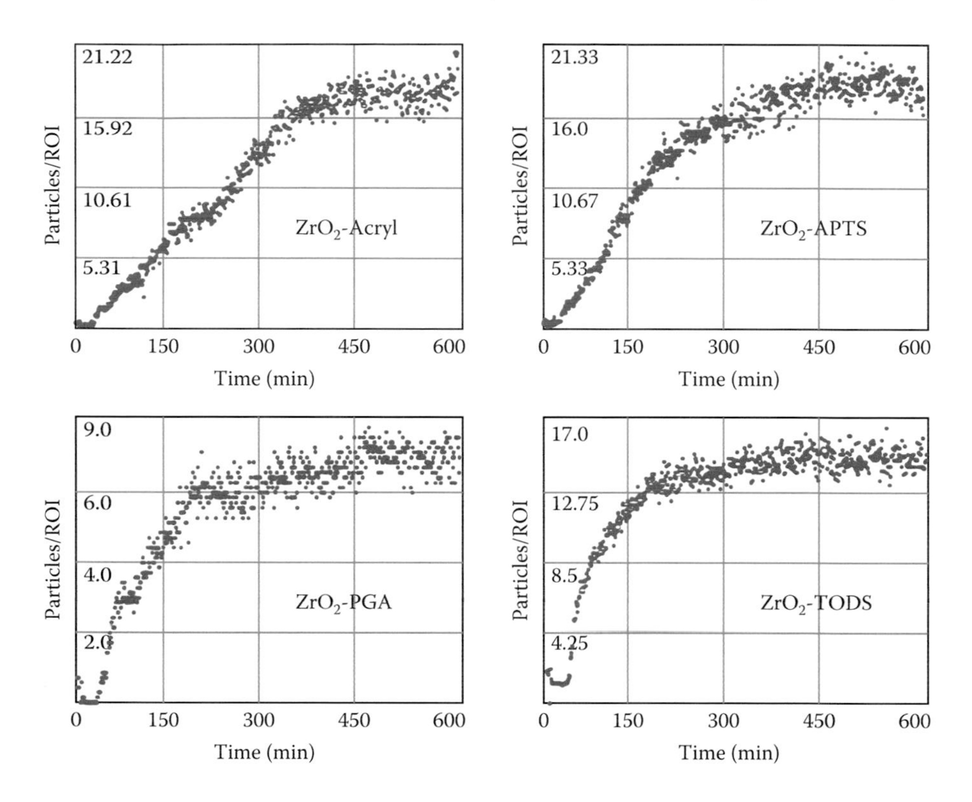

FIGURE 7.5 Comparison of the gravitational settling of four differentially functionalized ZrO_2 nanoparticles. Experiments were carried out in the absence of cells (5.5 µg particulate mass per mL F12K medium, 37°C, 5% CO_2). Steepness of the curves indicates the degree of gravitational settling.

during a single experiment. The data were used to develop and refine the program settings, and to collect first data on particle uptake.

7.3.2.1 Detection and Tracking of Macrophages

Most macrophages were actively moving single cells, which were automatically identified and tracked by the program (see the following). The initial identification of each cell to be studied had to be made by the user, namely by placing a region of interest (ROI) over the cell (see Figure 7.6). Also at least one cell-free region had to be defined to monitor particle sedimentation. Regions of interest with and without macrophages will be referred to as "activity ROI" and "control ROI," respectively. The diameter of an active ROI was set to span approximately 3-fold the size of the cell's diameter. This allowed detection of all particles within the reach of a macrophage and minimized the risk of interference between active ROIs.

The next step was to define the size ranges needed for the detection of particles and macrophages. The most versatile solution was to define a minimum and maximum size (in pixels) for both types of objects. As default values the program uses a size range from 5 to 399 and 400 up to 100,000 pixels for particle and macrophages, respectively.

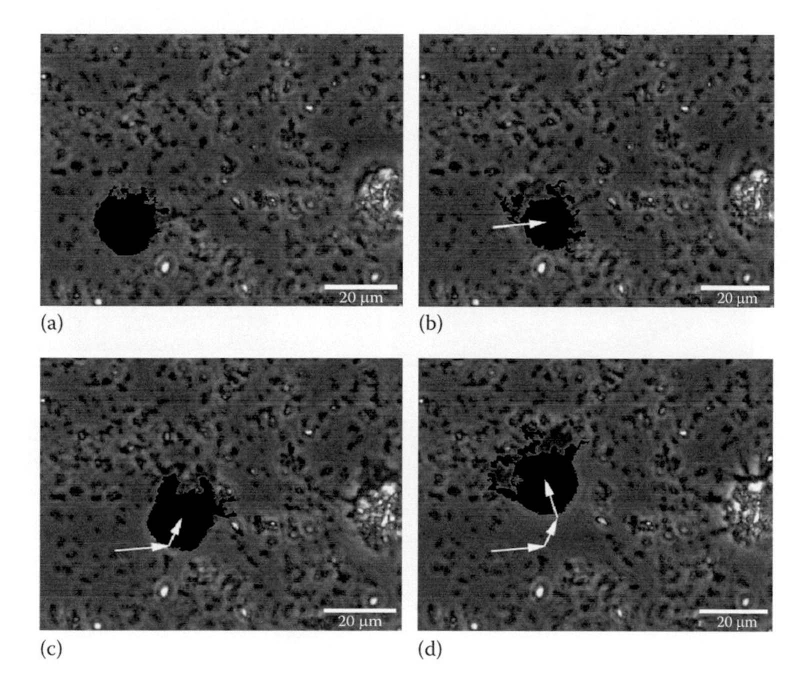

FIGURE 7.6 A moving macrophage ingesting settled particles. Four different positions (a–d) are shown at time intervals of 2 min each. Arrows track the center of the macrophage, whose contour is filled and overlaid in black.

A major issue during program development was to properly identify the objects prior to their binarization. We found that the well-known procedures for object edge detection such as Laplace or Marr-Hildreth operators delivered only poor results, as they were unable to draw coherent outer contours due to the varying contrast around macrophages. Therefore, a new algorithm needed to be developed, which expanded the possibilities and functions of digital image processing. To identify the contour of a macrophage (an object with more than 400 pixels) we developed a new algorithm. This algorithm uses the center point of the macrophage area from where it automatically constructs a set of radii. The intersection of each radius with a so far identified object edge defines a set of reference points (Figure 7.7a). The edge of the object will then be completed between the reference points. Therefore, a new search algorithm was developed: starting simultaneously at two neighboring reference points, the algorithm wanders alongside the edge of the object and completes the outer contour upon collision. This algorithm considers the grayscale difference between the actual pixel and its neighbors as well as the grayscale difference between both reference points. This mode of operation allows us to draw complex contours and is visualized in Figure 7.7c and d. It can be seen that the "clipping" of image data and the detection of island-like structures by the Laplace method are overcome by the algorithm. Eventually, results from the algorithm based on reference points and Laplace filtering were combined to compute the final contour, which was then used to calculate the perimeter from single pixels (Figure 7.7b).

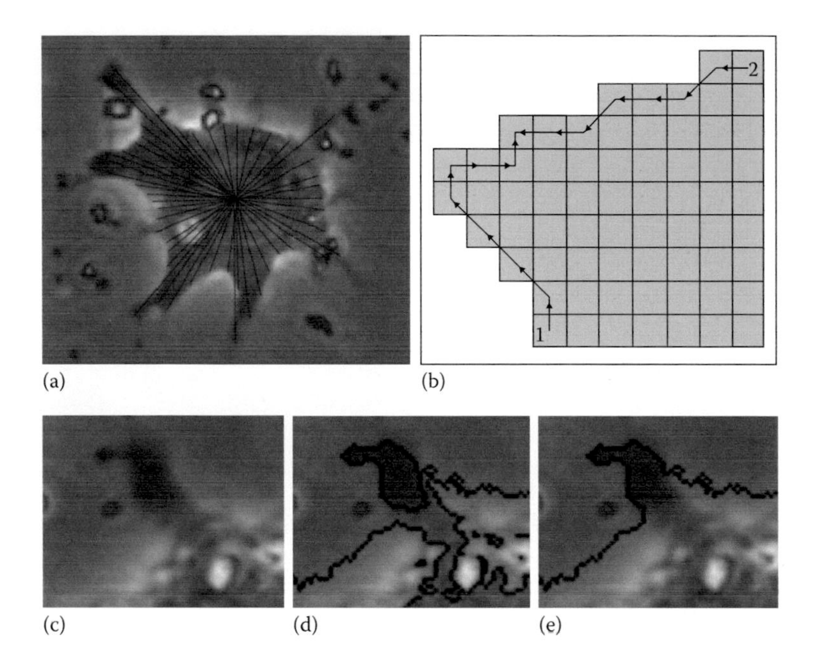

FIGURE 7.7 Method developed for an improved edge detection of macrophages. Cell is viewed with phase contrast optics. (a) Starting from an automatically found center point numerous radii are constructed toward the edge of the macrophage. End of each radius is defined by the maximum alteration of the gradient (derived from the gray values along the radius) and defines a set of "reference points." (b) An algorithm starts from neighboring reference points, travels along the contrasted edge, and demarcates the border of the cell. The perimeter of the cell is calculated from pixel size; a diagonal step is multiplied by $\sqrt{2}$. (c) and (d) Details of a macrophage with a pseudopodium; (d) edge detection by standard Laplace filtering, (e) edge detection with a Laplace filter plus detection point method. Magnification: (a) 400×, (c–e) 1200×.

After differentiation of background, particles, and macrophages, and the correct recognition of the macrophage's edge, the next step was to track moving macrophages such as NR8383 macrophages. Therefore, the active ROI needed to be adjusted in every image and around each single macrophage. Based on the correctly detected edges, the active ROI was centered around the balance point of the macrophage. This allowed us to track the highly motile and variously shaped macrophages and to calculate the migration distance in a properly defined manner (Figure 7.6).

7.3.2.2 Measuring the Uptake of Agglomerated Particles

In principle, the quantification of particle uptake by macrophages would be a simple task if highly contrasted particles had settled down *before* macrophages were added. Under these conditions, each cell would clear a defined region from particles and the mass of engulfed particles could be deduced from the number of missing particles. However, the sequence of events is vice versa in cell culture testing as it is in vivo. Figure 7.6 shows what is to be expected and illustrates the strategy, which has been chosen to quantify particle uptake under in vitro conditions: Starting with a single macrophage and a few particles spread over the scene (Figure 7.8a) the number of sedimented particles will increase over time in the control ROI (black circle), but will be diminished in the active ROI (red circle) as long as the macrophage is able to phagocytize. The area through which the macrophage has driven its comigrating active ROI is shown as a hatched field (Figure 7.8b–d).

The number of particles ingested by a single macrophage, therefore, can be described as the difference between two cumulative curves obtained from the control and the active ROI (Figure 7.9), provided that: (a) both regions are equally large; (b) particles are equally distributed; and (c) the area occupied by the macrophage is regarded as cleared. The last point cannot be controlled but is a well justified assumption because particles contacting the macrophage surface are readily ingested or roll off the surface and appear in the vicinity. Although the strategy to subtract the

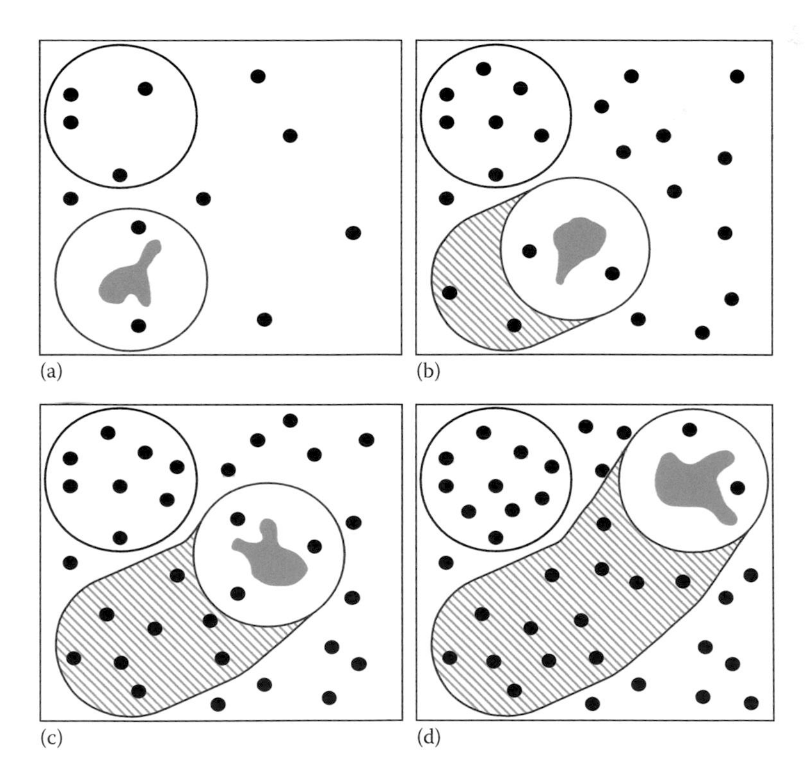

FIGURE 7.8 Procedure to analyze particle uptake under dynamic conditions, as used in the program "AM Tracking." Four consecutive virtual scenarios are shown (a–d). Settling particles are shown as black dots, a motile macrophage is represented by the gray area; control and macrophage regions are shown as black and red circles, respectively. The hatched area represents the field the macrophage had partially cleaned from particles. Particles in the control region increased from 4 (a) to 10 (d). Assuming an even distribution of particles, there should have been 37 particles within the hatched plus the macrophage area. As only 15 remain, 22 particles had been ingested. Note that this number could not be retrieved if only the final scenario (d) was considered.

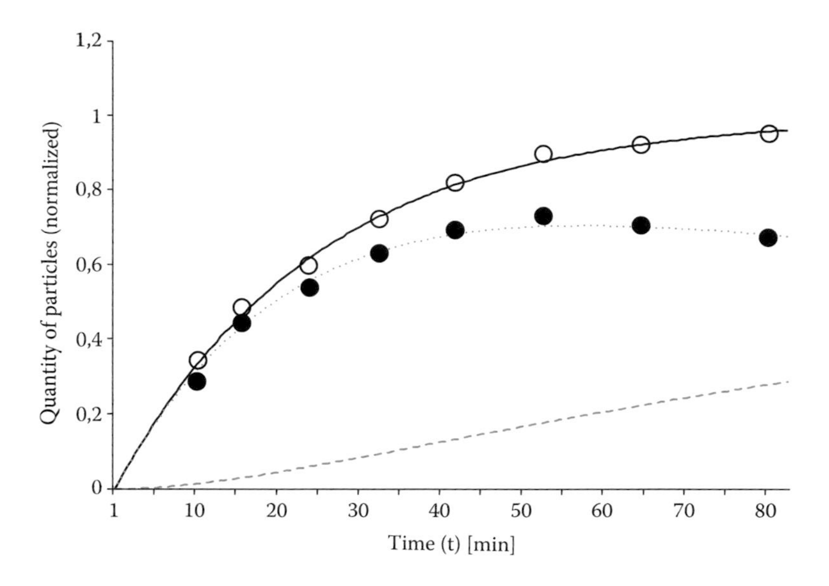

FIGURE 7.9 Sedimentation and uptake of AlOOH (22.5 µg/mL) by a NR8383 macrophage. Development of particle numbers was normalized to the maximum and smoothed by modeling. The parameters $q = 0.04$/min and $\tau = 234$ min were derived from the best fit of modeled and measured AlOOH data. Control ROI, N_C (▬▬); active ROI, N_P (▪▪▪); •, measured values, number of phagocytized particles (▬ ▬ ▬).

counting results of an active ROI from those of a control ROI is feasible in principle, there may be a considerable variation in counting results. This can be due to small particle numbers or due to a movement of particles caused by advection or mixing (Hinderliter et al. 2010). Therefore, it appeared advisable to include into calculation: (a) the variance of particle quantity inside the control ROI and (b) the description of the sedimentation process according to the sedimentation model. As outlined earlier, this model allows us to calculate the termination of the sedimentation process (t_0) for a defined particle mass per volume and also to predict the number of particles per area for every point in time. Thus, the particle counts within the control ROIs served for adjusting the modeled values to the real measurement result. To automatically analyze this process on a single cell basis, all steps previously outlined were integrated into a computer program termed "AM Tracking," which standardized the image analysis to quantify particle phagocytosis of macrophages from time-lapse images.

7.3.2.3 Model-Based Description of Phagocytosis

The increase in particle quantity over time in a control ROI (N_C) and the particle quantity in an active ROI (N_P) can be described through a kinetic model, where N_P is the ratio of time-dependent particle number $N(t)$ and maximum number N_0 of observed particles with $0 \leq N_C \leq 1$. The kinetic model is based on a simple relation between the rate dN_P/dt and the actual number of particles $N_C(t)$. We empirically found that the rate dN_C/dt (approximately calculated from the change in particle number ΔN_C within a constant time interval Δt) linearly decreased with increasing N_C:

$$\frac{d}{dt}N_C(t) = q \cdot (1 - N_C(t)) \approx \frac{\Delta N_C}{\Delta t}. \tag{7.2}$$

From this equation, which describes the sedimentation rate of the particles in the control region (N_C) and is primarily based on observation, a simple kinetic model can be derived:

$$N_C(t) = 1 - \exp(-q \cdot t) \tag{7.3}$$

This equation describes the increase in particles number in the control ROI and is valid for the case that no overlap occurs between sedimented particles (see Equation 7.1 case $\beta \approx 0$). The value of q represents the so-called sedimentation parameter. It describes how fast the maximum number of particles is reached at the bottom of the observation chamber. The q-value derives either from Formula 7.3 or from the initial increase in N_C according to Formula 7.2:

$$q = \frac{\text{In}(1 - N_C(\text{t}))}{t} \equiv \frac{dN_C(t=0)}{dt} \approx \frac{\Delta N_C(t=0)}{\Delta t} \tag{7.4}$$

Active macrophages will take up particles and this will reduce the particle number N_C within the macrophage area with increasing time. We supposed a monoexponential decrease resulting in the actual particle number N_P:

$$N_P(t) = N_C(t) \cdot \exp(-\frac{t}{\tau}) = (1 - \exp(-q \cdot t)) \cdot \exp(-\frac{t}{\tau}) \tag{7.5}$$

The time constant τ is a value for the phagocytosis rate of the macrophage and describes its "phagocytosis ability." Its value derives from the maximum of the kinetic model 7.4:

$$\tau = \frac{\exp(q \cdot t^*) - 1}{q} \tag{7.6}$$

Figure 7.8 shows an experiment with AlOOH nanoparticles (22.5 µg/mL): The development of particle numbers (normalized to the maximum) in a control ROI and in an active ROI was calculated according to our model (7.6). It can be seen that small fluctuations based on particle movement are eliminated using an arithmetic average function. The quantity of particles inside the control ROI rises to a maximum after 80 min. This point in time marks the end of the sedimentation process. The parameters q = 0.04/min and τ = 234 min were derived from the best fit of model and measured AlOOH data. The second graph of Figure 7.9 represents the quantity of particles inside an active ROI, which underscores the control ROI throughout the experiment due to ongoing particle uptake. It can also be seen that after 80 min the macrophage had phagocytized nearly 28% of the particles of the control region.

7.3.2.4 Experimental Data of Particle Uptake

In this chapter, we describe in vitro experiments carried out with AlOOH and CeO_2 particles applied to macrophage cultures, using the software and counting algorithms previously developed with simulated data. As is shown in the upper two curves of Figure 7.10, the program gives the counting results for control and active ROIs. Furthermore, the total particle number of an active ROI can be displayed, if particles concealed by the macrophage were calculated: Assuming normally distributed particles, the program uses the mean particle counts of the control region but adapts that value to the size of the macrophage (Figure 7.10, lower curve). Changes in the mean size of sedimented particles is also indicated (Figure 7.10, inset). This option not only allows us to demonstrate some inhomogeneity of a particle population in the course of an experiment but furthermore shows that larger particles appear early in the optical plane and, therefore, may dominate biologic effects. This is of pivotal importance if mixed populations of nanoscaled and microscaled particles are being investigated.

Based on the number of phagocytized particles and on their 2-dimensional dimension (projected size) the program calculated mass, volume, and BET surface of particles ingested by a macrophage. This, however, required the manual input into the program of: (a) initial particle concentration, (b) particle density, (c) particles BET value, (d) pixel size, and (e) the filling height of the vessel. A typical result is shown in Figure 7.11.

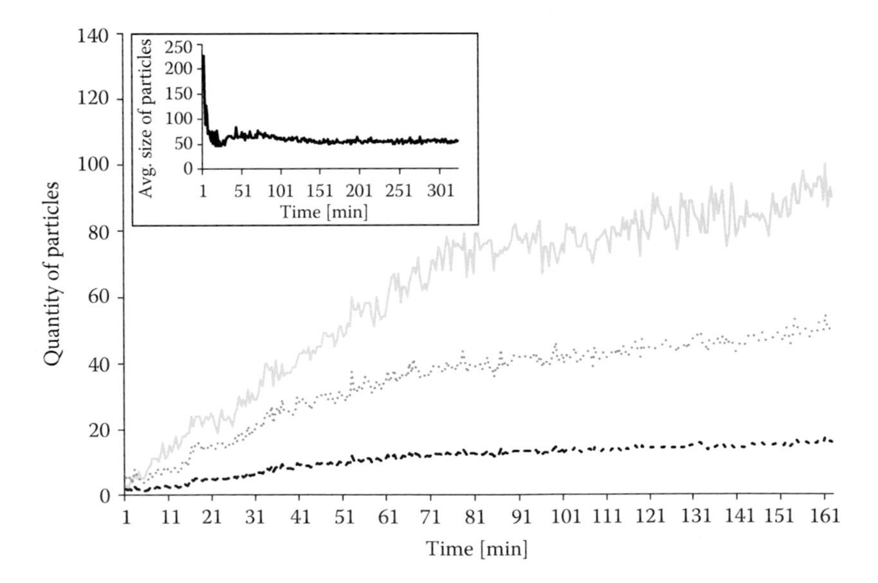

FIGURE 7.10 Measurement of particle uptake by a NR8383 macrophage treated with AlOOH nanoparticles (45 μg/mL). The ordinate gives the value for the number of visible particle agglomerates in the control region (▬▬), the calculated density of particles within the macrophage region (▬ ▬; $\times 10^{-2}$), and the total count of pixels representing the particles in the control region (▪▪▪, $\times 10^{-2}$). The inset shows the decreasing average size of particles in the control region over time (▬▬, in pixels).

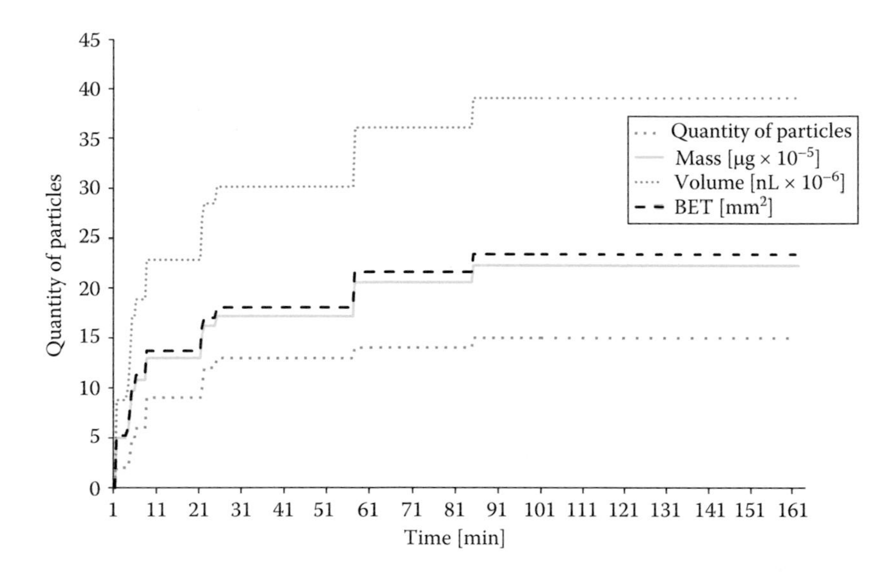

FIGURE 7.11 Uptake of AlOOH (45 μg/mL) particles by a NR8383 macrophage and calculated conversion into mass, volume, and BET surface. Results are based on particle numbers counted by the program AM Tracking. Density and BET surface of AlOOH nanoparticles were 2.85 g/cm³ and 105 m²/g, respectively (Pauluhn 2009). Particle count (■ ■); mass (——); volume (▪ ▪ ▪); and BET surface (— —).

TABLE 7.2
Uptake of AlOOH and CeO_2 NP Agglomerates by NR8383 Alveolar Cells

Particle	Conc. μg/mL	n	N_C (max)	N_P (max)	N_C (final)	N_P (final)	$\overline{q}$	$SD_{\overline{q}}$	$\overline{\tau}$	$SD_{\overline{\tau}}$
AlOOH	11.25	7	11	8	11	8	0.0045	0.000117	51	1.328
AlOOH	22.50	10	23	16	23	15	0.0230	0.000667	176	5.087
AlOOH	45.00	8	62	48	62	48	0.0350	0.000735	309	6.521
AlOOH	90.00	6	133	114	133	114	0.0500	0.002400	542	26.001
CeO_2	22.50	7	33	23	31	23	0.0300	0.001050	255	8.960
CeO_2	45.00	5	125	113	93	95	0.1400	0.003640	833	21.843
CeO_2	90.00	6	123	113	76	103	0.1350	0.005130	987	37.711

Note: Maximum (max) and final number of particles after 83 min.
Abbreviations: n, number of experiments; N_C, control ROIs; N_P, active ROIs; $\overline{q}$ and $\overline{\tau}$, mean values of sedimentation parameters; SD, standard deviation.

The automated method for the analysis and interpretation of phagocytosis was then applied to a concentration series of AlOOH and CeO_2 NP. Table 7.2 shows average values from 5 to 10 experiments. The observation period was limited to 83 min for technical reasons, such that the sedimentation had not finished in some cases. As all ROIs had the same size counting results were directly comparable. It becomes clear from particle number (N) that for a concentration of 11.25 μg/mL and 22.5 μg/mL the quantity of AlOOH and CeO_2 particles inside the active ROI (macrophage area) amounted to ca. 70% the control ROI. Consequently, 30% of both types of particles had been ingested. Importantly, no difference between AlOOH and CeO_2 particles could be discerned with respect to phagocytosis behavior at low concentration. At 45 and 90 μg/mL the sedimentation rate (q) of AlOOH increased whereas the relative amount of phagocytized particles decreased to 22% (45 μg/mL) and 15% (90 μg/mL), indicating that the uptake of particle could no longer keep pace with particle sedimentation. Interestingly, the sedimentation rate observed for high concentrations of CeO_2 (45 and 90 μg/mL) was markedly higher compared to low concentrations, most likely reflecting a pronounced agglomeration. Consequently, the sedimentation process was finished after 80–90 min. Another consequence of pronounced agglomeration and particle overlap was that the quantity of small single particles decreased. It was an intriguing finding that the particle number in active ROIs increased. However, as evidenced directly from the micrographs, macrophages continued to engulf preferentially large or overlapping particles. This effect counteracted the agglomeration in active ROIs and led to the seemingly paradox finding of increased particle numbers. This example shows that the careful evaluation of phagocytosis data should consider effects of agglomeration and needs to be confirmed by a direct interpretation of micrographs.

7.4 DISCUSSION AND PERSPECTIVES

In this study, particle phagocytosis by macrophages was quantified from phase contrast images, taken from the lowermost optical plane, within which agglomerated particles assemble by gravitational settling. The appearance of particle agglomerates between macrophages is a prerequisite for their uptake and determines the dose rate of an in vitro experiment. Unlike alternative approaches, which, for example, try to describe this sedimentation process indirectly on the basis of particle characteristics (Hinderliter et al. 2010), we propose an image based and, therefore, more direct method to quantify particle sedimentation. To this aim, we developed a special program to meet the most necessary requirements. The program development was based on modeled or simulated data as well as on the experiences obtained with real measurements. The program termed "AM Tracking" is able to automatically detect the border of a macrophage in a complex surrounding, to track the cells through the series of time-lapse images, and finally to count the number of ingested particles. This allows us to directly describe the dose rate for single cells and to study the impact of particle uptake on macrophage motility, which is of major importance for lung clearance processes. Furthermore, the sedimentation of particles can be described by a sedimentation model, which is part of the program and which is especially useful if dose rates of particles applied to cell cultures shall be investigated.

The accuracy of the measurements is, of course, limited by many factors and has not yet been confirmed by other methods such as the quantification of particles in single cells, for example, by mass spectroscopy methods. However, first results showed that the phagocytized mass for AlOOH and CeO_2 amounted to 150 pg/cell when the cells had a nonlimited access to particles. This is a plausible result as it is in the range formerly known from in vitro experiments in which AM engulfed up to 240 pg/cell (Rehn et al. 1999) and even in the range of in vivo results, in which AM were loaded with up to 90 pg (Pauluhn 2009). Thus, the proposed concept of measuring the uptake of agglomerated particles via digital image processing appears sound. Nevertheless, results have to be discussed with respect to possible sources of error. The first is the assumption that >99% of the suspended particle mass agglomerates and sediments onto the bottom of the measurement chamber (Hinderliter et al. 2010). At least for AlOOH and CeO_2 NP this could be confirmed as no NP were detected in the supernatant by NanoSight tracking analysis. Also an inhomogeneous distribution of particles at the bottom may strongly influence the outcome, but this obstacle has been made unlikely by the mode of particle application and was further on controlled by using different sites to measure. A small fraction of particles may also bind to the vessel wall. However, given the dimension of the Petri dish also this mistake appears negligible.

Another source of error emanates from data processing itself. We estimate that up to 1% of the mass equivalent particles may have been removed from evaluation by smoothing digital images. However, the correct setting of parameters for the automatic analysis or compensating the influence of particle overlap is certainly more important. The correct identification of particles, macrophages, and cell contours, therefore, was a special issue since it strongly influences all calculations. For all these parameters, a constant accuracy of 95%–99% was reached if the outcome was compared to the detection result of an experienced user. As outlined earlier, a review of the optical information by the user is mandatory for a sound interpretation of single results. However, despite some uncertainties with respect to absolute values, the method allows a rapid comparison of the uptake kinetics for many different particles or cell types and, by this, enables a better interpretation of biologic effects.

Statistic evaluation was needed at several steps of evaluation and helped to better analyze the data. The developed sedimentation model enables us to characterize the sedimentation process and to precompute the quantity of particles at every point in time inside the ROI as it considered the collected sedimentation parameters from the image data (Schippritt et al. 2010). The sedimentation model not only eliminated indispensable statistic variance as it eliminated noise from advection- or diffusion-based dislocation of particles but also was helpful to better calculate the in vitro dose for macrophage experiments before or after an experiment. For the determination of phagocytized particles the difference between control ROIs (without macrophage) and active ROIs (with macrophage) was used. In this context, it was helpful that the characteristics of particle quantity for the control ROI (formula 7.3) and the active ROI (cf. formula 7.5) could be described by kinematic models.

What are the future perspectives of light microscopy and image analysis methods for describing particle uptake by cells in vitro? As previously outlined, the approach described here is largely limited to agglomerated NP, that is, to study their sedimentation and uptake kinetics. However, single NP close to or inside cells may be viewed by light microscopy as well. Recently, a method based on dark field

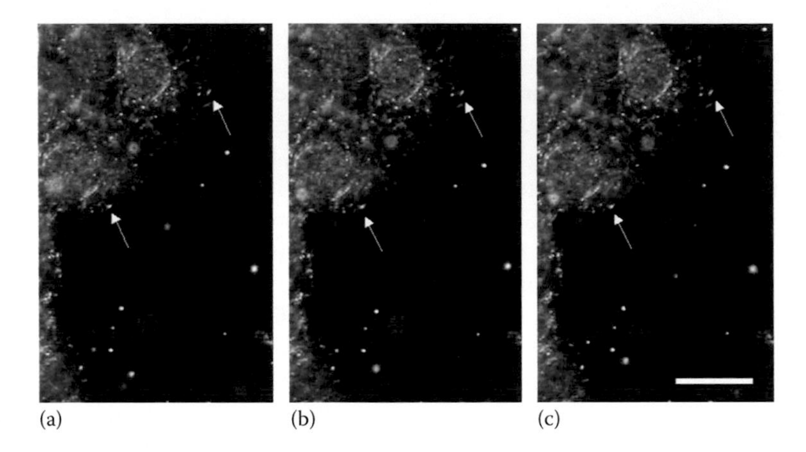

FIGURE 7.12 Silver nanoparticles (Ag50.PVP) diffusing close to a group of cultured V79 cells, which were immersed in physiologic saline (KRPG-buffer) and viewed with dark field microscopy. Particles appear blue, green, and yellow due to their different plasmon resonance. In parts (a), (b), and (c) (taken at intervals of ca. 0.3 s) particles outside cells appear at different positions due to Brownian motion. Arrows point to particles attached to cell surface. Bar: 20 μm.

microscopy was introduced, which allows us to view single 60 nm gold NP on color images of cultured cells (Wang et al. 2010). Provided that cells can be maintained under sufficient culture conditions, it may be used for directly observing particles together with cells in special experimental chambers over many hours. Within the NanoGEM project we adapted and improved this method so that V79 cells could be observed together with diffusing silver nanoparticles (Ag50.PVP, see Figure 7.12). By this, the diffusion coefficient of single particles close to cells could be measured. Future experiments will now give insight into preferential sites of particle adherence and/or particle uptake. However, compared to the approach described earlier, information retrieved from 3-dimensional dark field microscopy is far more complex because time-lapse series of z-stacks need to be evaluated. Furthermore, color information appears important at least if gold or silver NP are investigated, whose plasmon resonance is subject to change upon particle uptake in vesicles (Wang et al. 2010). Nevertheless, future image analysis methods should allow us to quantify the uptake of NP into single cells. In a next step this methodology could then be used to attribute biologic effects to the particle load at the single cell level.

ACKNOWLEDGMENTS

The work was supported by grants of the German Ministry of Education and Science (BMBF, 03X0105) and the Northrhine Westphalian Ministry of Innovation, Science, Research and Technology (FH-GP 2008-9/Lipinski).

REFERENCES

Gosens, I., Post, J. A., de la Fonteyne, L. J. J. et al., 2010. Impact of agglomeration state of nano- and sub-micron sized gold particles on pulmonary inflammation. Part Fibre Toxicol 7:3–7.

Hinderliter, P. M., Minard, K. R., Orr, G. et al., 2010. ISDD: A computational model of particle sedimentation, diffusion and target cell dosimetry for in vitro toxicity studies. Part Fibre Toxicol 7:36–55.

Kroll, A., Dierker, C., Rommel, C. et al., 2011. Cytotoxicity screening of 23 engineered nanomaterials using a test matrix of ten cell lines and three different assays. Part Fibre Toxicol 8:9.

Lanone, S., Rogerieux, F., Geys, J. et al., 2009. Comparative toxicity of 24 manufactured nanoparticles in human alveolar epithelial and macrophage cell lines. Part Fibre Toxicol 6:14–25.

Oberdörster, G., Ferin, J., Gelein, R. et al., 1992. Role of the alveolar macrophage in lung injury: Studies with ultrafine particles. Environ Health Perspect 97:193–199.

Oberdörster, G., Maynard, A., Donaldson, K. et al., 2005a. Principles for characterizing the potential human health effects from exposure to nanomaterials: Elements of a screening strategy. Part Fibre Toxicol 2:8–42.

Oberdörster, G., Oberdörster, E., Oberdörster, J., 2005b. Nanotoxicology: An emerging discipline evolving from studies of ultrafine particles. Environ Health Perspect 113:823–839.

Pauluhn, J., 2009. Pulmonary toxicity and fate of agglomerated 10 and 40 nm aluminum oxyhydroxides following 4-week inhalation exposure of rats: Toxic effects are determined by agglomerated, not primary particle size. Toxicol Sci 109:152–167.

Rehn, B., Rehn, S., Bruch, J., 1999. Ein neues In-vitro-Prüfkonzept (Vektorenmodell) zum biologischen Screening und Monitoring der Lungentoxizität von Stäuben. Gefahrstoffe-Reinhaltung der Luft 59:181–189.

Rothen-Rütishauer, B., Mühlfeld, C., Blank, F. et al., 2007. Translocation of particles and inflammatory responses after exposure to fine particles and nanoparticles in an epithelial airway model. Part Fibre Toxicol 4:9–17.

Schippritt, D, Wiemann, M., Lipinski, H. G., 2010. Sedimentation of agglomerated nanoparticles under cell culture conditions studied by image based analysis. Proc SPIE 7715: 77153A–77153A.

Semmler-Behnke, M., Takenaka, S., Fertsch, S. et al., 2007. Efficient elimination of inhaled nanoparticles from the alveolar region: Evidence for interstitial uptake and subsequent reentrainment onto airways epithelia. Environ Health Perspect 115:728–733.

Teeguarden, J. G., Hinderliter, P. M., Orr, G. et al., 2007. Particokinetics in vitro: Dosimetry considerations for in vivo nanoparticle toxicity assessments. Toxicol Sci 95:300–312.

Wang, S. -H, Lee, C. -W., Chiou, A. et al., 2010. Size-dependent endocytosis of gold nanoparticles studied by three-dimensional mapping of placmonic scattering images. J Nanobiotechnol 8:33–45.

Wiemann, M., Bruch, J., 2009. Comparison of in vitro and in vivo findings. In *NanoCare. Health Related Aspects of Nanomaterials*. Final Scientific Report, ed. T.A.J. Kuhlbusch, H.F. Krug, and K. Nau. Dechema e.V., Frankfurt, Germany.

Xia, T., Li, N., Nel, A., 2009. Potential health impact of nanoparticles. Public Health 30:137–150.

8 Toxicological Effects of Metal Oxide Nanomaterials

Daniela Hahn, Martin Wiemann, Andrea Haase, Rainer Ossig, Francesca Alessandrini, Lan Ma-Hock, Robert Landsiedel, Marlies Nern, Antje Vennemann, Marc D. Driessen, Andreas Luch, Elke Dopp, and Jürgen Schnekenburger

CONTENTS

8.1 INTRODUCTION

Metal oxide nanoparticles exhibit unique properties that largely differ from those displayed by their larger counterparts. These properties are highly desirable, rendering these nanomaterials very useful for many different types of applications such as consumer products like cosmetics or textiles and also paints or fuel additives. An overview of nanomaterial containing consumer products in the US declared by the manufacturer is available at the Wilson Center Project on Emerging Nanotechnologies Consumer Products Inventory (http://www.nanotechproject.org/cpi).

On the other side, nanomaterials may have unpredictable effects on human health and the environment. Hazard information generated for bulk metal oxides is usually not suitable for characterizing the toxic potential of nanoparticles. Titanium dioxide (TiO_2) nanoparticles, for instance, have been found to induce a much greater pulmonary inflammatory response in rats when compared to fine-sized TiO_2 materials (Oberdorster, Oberdorster, and Oberdorster 2005). In general, nanomaterials with a primary particle size below 20 nm seem to be particularly critical because this can lead to enhancement of interfacial reactivity with impact on biological effects (Nel et al. 2009). Apart from primary particle size, other properties such as shape, surface charge, agglomeration state, crystalline phase, or the ability to trigger the generation of reactive oxygen species (ROS) may influence the toxic behavior of metal oxide nanoparticles. Moreover, some metal oxide nanoparticles can release ions that, in turn, may exert detrimental effects on cells. The contribution of ions to nanomaterial toxicity has been studied in detail for ZnO (Cho et al. 2012; Xia et al. 2011; Kao et al. 2012). Since nanoscaled metal oxides have a larger surface area and a greater tendency to release ions than their larger counterparts, toxic effects emanating from released ions will be enhanced and smaller particles may therefore display greater toxic potential.

In addition, many metal oxide nanoparticles exhibit high surface energy and may interact differently with biomolecules such as proteins and DNA. In biological fluids, they adsorb proteins leading to the formation of a dynamic protein corona, which may influence particle effects in multiple ways. Intrinsic physicochemical properties (size, surface, surface charge, structure, and shape) and aggregation of nanoscaled particles will change when they become coated with proteins (Cedervall et al. 2007; Maiorano et al. 2010; Wells et al. 2012). Conversely, nanoparticles may alter the structure of biological macromolecules with important impact on normal cell functions, for example (Deng et al. 2011; Xu, Wang, and Gao 2010), see Chapter 5 for further details.

Correlating physicochemical properties with toxic effects of metal oxide nanoparticles is a major challenge for nanotoxicology. Currently, each nanomaterial type has to be tested individually and carefully characterized with respect to its physicochemical properties. Even if the size and chemical composition of nanoparticles are the same, other attributes of nanoparticles produced by different manufacturers may differ such that even different batches of the same or only slightly different nanoparticles need to be tested individually.

Available *in vivo* data are much less numerous compared to published *in vitro* results and are mainly limited to acute or subchronic effects, hardly any chronic effects have been addressed. In contrast, numerous *in vitro* studies have been published; however, the relevancy of the results has to be carefully analyzed since some approaches may not reflect realistic nanoparticle exposure concentrations and/or routes of entry to cells and tissues. In addition, the individual *in vitro* assay concepts and conditions should be validated for the use with a specific nanomaterial. Especially for cell culture-based studies, it has been demonstrated that metal oxide nanoparticles at concentrations higher than 50 μg/mL tend to interfere with assay components or the assay readout procedure thereby introducing artifacts (Monteiro-Riviere, Inman, and Zhang 2009; Kroll et al. 2012).

The main exposure routes for humans are dermal, oral uptake, and inhalation. When assessing nanoparticle toxicity *in vitro*, cell systems of the skin, gut, and lung should thus be applied. Recent evidence suggests that particles may translocate from their site of exposure and reach or even accumulate in remote tissues including lymph nodes, liver, spleen, and kidney. Thus, *in vitro* studies focusing on cell cultures carefully selected to represent these organs can provide useful toxicity data as well. Standardized cell lines provide high reproducibility but are often dedifferentiated and transformed tumor cells with low responsiveness to toxic agents. Furthermore, single cell types can hardly reflect the complex response of tissues leading, for example, to an inflammatory process or to a slowly proceeding neoplastic alteration such as lung fibrosis.

This chapter deals with metal oxide nanoparticles of widespread use in consumer, medical, and industrial applications such as ZnO, TiO_2, ZrO_2, SiO_2, or CeO_2. Many metal oxide nanoparticle types have been found to exhibit no or only very low toxicity in numerous *in vitro* and *in vivo* studies. However, it is not possible to categorically define metal oxide nanoparticles as toxic or nontoxic merely by considering their chemical composition. TiO_2 nanoparticles, for instance, cannot generally be regarded as biologically inactive *per se* since a few TiO_2 nanoparticle types with definite crystalline structure (anatase) can cause adverse effects in contrast to the majority of the TiO_2 nanoparticle types tested so far that displayed no toxicity. This chapter will provide an overview on the effects of metal oxide nanomaterials and review data published in the literature as well as data obtained in the German projects nanoGEM and NanoCare. Initially, *in vitro* results will be summarized focusing on biological effects of metal oxide nanoparticles on cultured cell lines with respect to different toxicological end points followed by a section on more elaborated *in vitro* models using primary cells such as primary alveolar macrophages. Finally, this chapter reviews results from *in vivo* studies with a special emphasis on pulmonary toxicity revealed via different experimental models.

Bearing in mind that most metal oxide nanoparticles do not cause toxic effects when applied at reasonable concentrations, this chapter focuses on the physicochemical characteristics that seem to contribute to nanoparticle toxicity under experimental conditions and highlight those metal oxide particle types that have been shown to exert toxic effects in cell or animal models. The nanoparticle properties that seem to contribute to the biological effects reported here are marked in italics.

8.2 EFFECTS OF METAL OXIDE NANOPARTICLES ON CULTURED CELLS

In comparison to animal models, *in vitro* studies allow for a simpler, faster, and more cost-efficient determination of defined toxicity end points. In addition, *in vitro* studies help to elucidate the mechanisms of toxicity and affected pathways. On the other hand, one of the major disadvantages of *in vitro* assays is that a single specific cell type cannot represent the numerous characteristics and features of a complex organ.

Cultured cells may respond differentially compared to whole organs or organisms. For instance, it appears not justified to assume that an *in vitro* effect of one cell type fully reflects a pathophysiological outcome or toxic response *in vivo*. Effects observed *in vitro* may for instance not reflect adaptation or clearance processes, thus leading to an overestimation of adverse effects. On the other side, *in vitro* models consisting of one cell type may lack important metabolic properties and/or the interplay between different cells typically occurring in organized tissue such as the liver. This in turn may lead to false negative results. In general, it is therefore recommended to use a number of different cell lines and various assay readouts to collect more relevant data (Kroll et al. 2011). Many of the established *in vitro* test systems were initially designed for testing soluble chemical substances and are often not suitable to assess the toxic potential of nanomaterials since these can interfere with assay reagents or detection systems thereby influencing the toxicological outcome (Kroll et al. 2012).

In a previous German project (NanoCare), a set of metal oxide nanoparticles were used to identify those particles that are most influential in driving nanomaterial toxicity. During the project, an *in vitro* test matrix initially comprising 10 standardized cell lines of diverse tissue origin and a set of standardized assays for different end points was used (Kroll et al. 2011). Due to the combination of several cell lines from different tissues and assays reporting on the impairment of various cell functions this approach largely increases the reliability and predictability of an *in vitro* test and overcomes the problem that nanomaterials are heterogeneous in cell type and end point activation. The combination of highly sensitive cytotoxicity test systems (see Kroll et al. 2009) may therefore be used to prescreen nanomaterials for their potential toxicity and to identify candidates that require more detailed testing and thereby reducing the number of animal studies. The test matrix was shown to function as an effective prescreening for the selection of nanomaterial candidates for an *in vivo* assessment and may allow an extrapolation of results to the *in vivo* situation to inform about potential adverse health effects in humans.

In the following German project nanoGEM, the *in vitro* test matrix was continued using a selection of the most responsive cell lines. Here, the test matrix includes human A549 and rat RLE-6TN cells, which display features of lung alveolar epithelial cells as *in vitro* models for inhalation. HaCaT keratinocytes were derived from normal human skin and have retained the full epidermal differentiation capacity. As cellular models to assess toxic effects of metal oxide nanoparticles that have been distributed via the blood system to secondary target organs, kidney cell lines (MDCK, NRK-52E) as well as fibroblasts (NIH-3T3) and macrophages (RAW264.7) were used (for further details see Table 8.1).

In both projects, NanoCare and nanoGEM, the physicochemical properties of the nanoscaled metal oxide dispersions were thoroughly characterized, and standardized protocols as well as reference materials were used to improve the comparability between *in vivo* and *in vitro*. The nanoGEM project focused on the impact of different surface coatings and included four types of SiO_2 nanoparticles (SiO_2.naked, SiO_2.PEG, SiO_2.amino, or SiO_2.phosphate), and four types of ZrO_2 nanoparticles (ZrO_2.TODS, ZrO_2.PEG, ZrO_2.amino, or ZrO_2.acryl) (see Chapter 1, Table 1.2 for particle details). This chapter will provide an overview of the physicochemical

TABLE 8.1
Origin of the Cell Lines Used within the nanoGEM Project

Cell Line	Species	Origin	Provider	Catalogue Number
A549	*Homo Sapiens*	Lung adenocarcinoma, alveolar epithelium-like	ATCC	CCL-185
HaCaT	*H. sapiens*	Skin, keratinocytes	CLS	–
MDCK (NBL-2)	*Canis familiaris*	Kidney, epithelium-like	ATCC	CCL-34
NIH-3T3	*Mus musculus*	Embryo, fibroblasts	DSMZ	ACC 59
NRK-52E	*Rattus norvegicus*	Kidney, epithelium-like	DSMZ	ACC 199
RAW 264.7	*R. norvegicus*	Macrophages, AML virus transformed	ATCC	TIB-71
RLE-6TN	*R. norvegicus*	Lung, alveolar epithelium-like	ATCC	CRL-2300

properties that drive the metal oxide nanoparticles' cytotoxicity with emphasis on the results obtained by these two projects as well as from other studies that have been conducted using well-characterized nanomaterials and multiple test systems. The effects of metal oxide nanoparticles and the physicochemical properties identified to be responsible for these effects are summarized in Table 8.2.

8.2.1 Oxidative Stress

Recent evidence suggests that oxidative stress initiated by the formation of ROS is a key route, thereby inducing cell damage. The ROS generating capacity of metal oxide nanoparticles seems to correlate with their ability to induce cellular inflammation and DNA damage (Xu et al. 2009; Sayes et al. 2006; Sayes, Reed, and Warheit 2007).

The exposure of BEAS-2B cells to ZnO nanoparticles, for example, led to an increase in intracellular ROS levels and decreased cell viability over time and this cytotoxic effect was reduced by *N*-acetyl cysteine treatment (Huang et al. 2010). This suggests that oxidative stress is closely related to the cytotoxicity of nanoparticles. Thus, measuring the oxidative stress potential of metal oxide nanoparticles becomes mandatory in the nanoparticle toxicity assessment.

To determine the impact of the *chemical composition/identity* and its catalytic activity on ROS formation, Limbach et al. (2007) exposed human lung epithelial cells (A549) to thoroughly characterized particles of similar morphology, comparable size, shape, and degree of agglomeration. Their studies suggest that the chemical composition of nanoparticles is a decisive factor influencing ROS formation in lung epithelial cells (Limbach et al. 2007). This could be further confirmed by a study comparing mixed metal oxide nanoparticles of titanium dioxide and zirconium dioxide (TiO_2-ZrO_2). The nanoparticle mixtures containing the largest amount of TiO_2 (90%) induced ROS formation in seven of ten cell lines tested, whereas Ti–Zr mixed nanoparticles with lower amounts of TiO_2 (10%, 50%) did not lead to an increase in ROS production in any of the cell lines tested (Kroll et al. 2011). The induction of oxidative stress was also evaluated in primary mouse

TABLE 8.2
Summary of Effects of Metal Oxide Nanoparticles on Cultured or Primary Cells

Nanoparticle	Effect	Cell Lines Tested	Property Involved	Reference
TiO_2, Fe_2O_3, CO_3O_4, Mn_3O_4	Oxidative stress	A549	Chemical composition	Limbach et al. 2007
TiO_2–ZrO_2	Oxidative stress	See Table 8.1, $CaCO_2$, CaLu3, MDCK II	Chemical composition	Kroll et al. 2011
ZnO, SiO_2	Oxidative stress	Primary mouse fibroblasts	Chemical composition	Yang et al. 2009
ZnO	Oxidative stress	BEAS-2B, RAW 264.7	Solubility	Xia et al. 2008
ZnO	Oxidative stress	BEAS-2B		Huang et al. 2010
Rutile TiO_2, anatase TiO_2	Oxidative stress	A549, human dermal fibroblasts (HDF)	Crystalline structure	Sayes et al. 2006
Anatase TiO_2	Oxidative stress	See Table 8.1, $CaCO_2$, CaLu3, MDCK II	Surface modification	Kroll et al. 2011
TiO_2	Oxidative stress	Primary dermal fibroblasts	Surface modification	Pan et al. 2009
CeO_2	Oxidative stress	See Table 8.1, $CaCO_2$, CaLu3, MDCK II	Surface chemistry	Kroll et al. 2011
SiO_2.naked	Induction of protein carbonyls	NRK-52E	Surface modification	nanoGEM, final report
ZnO, TiO_2	Induction of protein carbonyls	NRK-52E		nanoGEM, final report
TiO_2, Al_2O_3	Cell viability/ metabolic activity	A549	Chemical composition	Simon-Deckers et al. 2008
SiO_2	Cell viability/ metabolic activity	EAHY926, HUVEC, human blood cells		Jaganathan and Godin 2012
SiO_2	Cell viability/ metabolic activity	A549, EAHY926, J774	Size/surface area	Lin et al. 2006
ZnO, CuO, Cu_2O, Zn_2TiO_4	Cell viability/ metabolic activity	A549, THP-1	Solubility	Lanone et al. 2009
ZnO	Cell viability/ metabolic activity	L2 lung epithelial cells	Solubility	Sayes, Reed, and Warheit 2007
ZnO	Cell viability/ metabolic activity	Neuro-2A	Solubility	Jeng and Swanson 2006
ZnO	Cell viability/ metabolic activity	HK-2 human renal cells	Solubility	Kermanizadeh et al. 2013
Rutile TiO_2, anatase TiO_2	Cell viability/ metabolic activity	A549	Crystalline structure	Simon-Deckers et al. 2008
ZnO	Cell viability/ metabolic activity	See Table 8.1		nanoGEM, final report
SiO_2.naked	Cell viability/ metabolic activity	RAW 264.7	Surface modification	nanoGEM, final report

(*Continued*)

TABLE 8.2 (*Continued*)
Summary of Effects of Metal Oxide Nanoparticles on Cultured or Primary Cells

Nanoparticle	Effect	Cell Lines Tested	Property Involved	Reference
ZrO_2.amino	Cell viability/ metabolic activity	RLE-6TN	Surface modification	nanoGEM, final report
TiO_2	Genotoxicity	MEF mouse primary embryo fibroblasts		Xu et al. 2009
ZnO	Genotoxicity	HepG2		Sharma, Anderson, and Dhawan 2012
TiO_2	Genotoxicity	BEAS-2B	Size/surface area	Gurr et al. 2005
ZnO	Genotoxicity	Primary nasal mucosa cells	Size/surface area	Hackenberg et al. 2011b
Rutile TiO_2, anatase TiO_2	Genotoxicity	HepG2	Crystalline structure	Petkovic et al. 2011
SiO_2.naked, SiO_2.phosphate	Genotoxicity	3D bronchial models EpiAirway™, MatTEK	Surface modification	nanoGEM, final report
TiO_2	Effects on gene expression	BEAS-2B	Chemical composition	Park et al. 2008b
TiO_2 (P25)	Inflammatory response	A549	Size/surface area	Singh et al. 2007
Rutile TiO_2, anatase TiO_2	Inflammatory response	A549, HDF	Crystalline structure	Sayes et al. 2006

embryo fibroblasts (MEF) for ZnO and SiO_2 nanoparticles (Yang et al. 2009). Both nanoparticle types induced significant ROS generation in a dose-dependent manner, however, ZnO displayed much stronger effects than SiO_2. Since the SiO_2 and ZnO nanoparticles used in this study had a similar crystal shape and particle size this further suggests that the chemical composition has an impact on ROS generation (Yang et al. 2009). ZnO nanoparticles also induced oxidative stress in RAW 264.7 macrophages and BEAS-2B bronchial cells (Xia et al. 2008). However, it is not yet clear whether ZnO nanoparticles directly induced the formation of ROS or if ROS was formed as a consequence of apoptotic cell death as it was observed recently in Jurkat cells exposed to ZnO nanoparticles (Buerki-Thurnherr et al. 2013). It is well documented that ZnO nanoparticles display *solubility* in aqueous solutions and recent evidence suggests that detrimental effects of ZnO nanoparticles can be mainly attributed to dissolved zinc ions (Gilbert et al. 2012; Buerki-Thurnherr et al. 2013; Kao et al. 2012; Xia et al. 2008; Cho et al. 2011).

A correlation between the *crystalline structure* and the ROS generating capacity has been found for TiO_2 nanoparticles (Sayes et al. 2006). In dermal fibroblasts, rutile TiO_2 particles produced two orders of magnitude less ROS than anatase TiO_2 particles of similar size (Sayes et al. 2006). However, anatase TiO_2 nanoparticles with an organic surface modification did not trigger ROS formation in 10 different cell lines tested, whereas nonmodified TiO_2 nanoparticles of similar size and surface area and an even lower anatase content induced a concentration-dependent increase in ROS in all cell lines (Kroll et al. 2011).

These findings suggest that *particle coatings* can have a strong impact on the biological effects of nanoparticles. In line with that notion, Pan et al. reported that intracellular ROS formation was reduced to control levels when rutile TiO_2 particles were coated with a dense polymer brush (Pan et al. 2009).

In a comprehensive study conducted to identify physicochemical properties that modulate nanoparticle toxicity, four nearly identical CeO_2 nanoparticles (A–D) from the same production process with only slight variations were generated and assessed for their ROS generating potential (Kroll et al. 2011). Whereas three of the four CeO_2 nanoparticle dispersions (A–C) significantly induced ROS in the majority of the cell lines tested, one CeO_2 nanoparticle (D) failed to evoke a significant increase in ROS in any of the cell lines tested. CeO_2-D nanoparticles were almost identical to CeO_2-A nanoparticles regarding size and surface area but displayed slight differences in the surface chemistry due to some variation in gassing with carbon dioxide during the production process. These differences in *surface chemistry* may lead to direct or indirect effects. Among the latter changes in agglomeration and gravitational settling of particles should be considered (see Chapter 7, Schippritt et al.). In addition, a different adsorption of serum proteins from the cell culture medium may be involved. It has been speculated that the composition of the nanoparticle protein corona influences the cytotoxicity of nanoparticles (Lundqvist et al. 2008). Indeed, it was demonstrated that the absorption of CeO_2-A and CeO_2-D nanoparticles to bovine serum albumin differed significantly (Schaefer et al. 2012). Thus, a disparity of the protein composition adsorbed by CeO_2-A and CeO_2-D nanoparticles may be responsible for their different effects on cells.

In the nanoGEM project, cellular ROS formation in response to metal oxide nanomaterial exposure was tested based on the *in vitro* test matrix using the DCF assay. An increase in intracellular ROS levels could not be observed directly after exposure to any of the naked or surface-modified SiO_2 or ZrO_2 nanomaterials in any of the cell lines tested (see Table 8.1). In contrast, P25 TiO_2, which was used as a reference nanomaterial at a concentration of 32 μg/mL, was shown to increase intracellular ROS levels in most cell lines of the test matrix with the exception of HaCaT (nanoGEM, unpublished results). In nanoGEM, we futhermore established an indirect measure of oxidative stress through the detection of oxidative protein modifications (i.e., protein carbonylation). This assay proved to be highly sensitive and was well suited to analyze the potential of nanoparticles for triggering oxidative stress. Protein carbonyls are detected via coupling with dinitrophenyl (DNP) hydrazine. The resulting hydrazone can be detected in a 1D or 2D western blot using an anti-DNP antibody (Robinson et al. 1999). In some studies such an approach was already successfully used (Haase et al. 2011; Haase et al. 2012; Sun et al. 2013). Here, we applied this assay for assessing the oxidative stress potential of all nanoGEM nanomaterials in NRK-52E cells. The results showed that 8 out of 16 nanoGEM nanoparticles induced protein carbonylation (Driessen et al., unpublished).

8.2.2 Cell Viability

A number of studies have been conducted to assess the effects of metal oxide nanoparticles on cell viability. In addition to this chapter, several excellent review articles summarize the current knowledge on TiO_2 (e.g., Johnston et al. 2009; Iavicoli et al. 2011; Shi et al. 2013) or on SiO_2 (e.g., Jaganathan and Godin 2012). In comparison to carbon nanotubes (CNTs), metal oxide nanoparticles (rutile or anatase TiO_2 and Al_2O_3) were found to be less toxic in A549 human lung epithelium cells (Simon-Deckers et al. 2008). Although all nanoparticles were efficiently internalized in A549 cells, TiO_2-mediated cytotoxicity was generally low with a maximum cell death rate of 25%. Since TiO_2 and Al_2O_3 particles were of similar size and shape but differed in their cytotoxic behavior (maximum cell death rate of 3% for Al_2O_3 compared to 25% for TiO_2), this study again confirmed that nanoparticles' toxicity can be associated with their *chemical composition.* A time- and dose-dependent effect of 20 nm CeO_2 particles on the viability of A549 cells was reported by Lin et al. In their studies, cell viability decreased to 53.9% when 23.3μg/mL CeO_2 was applied (Lin et al. 2006).

SiO_2 nanoparticles also were found to reduce cell viability in a number of different studies (for review, see Jaganathan and Godin 2012). For instance, the cytotoxicity of 15 nm and 46 nm SiO_2 nanoparticles was investigated in human bronchoalveolar carcinoma-derived cells by using crystalline SiO_2 as a positive control. Both SiO_2 nanoparticles were more cytotoxic than the bulk material (Lin et al. 2006). Reduced cell viability elicited by SiO_2 nanoparticles in A549, endothelial EAHY926 cells, and J774 monocytes/macrophages was found to be dependent on the total *mass, number, and surface area* as well as on the concentration (Lin et al. 2006).

A comparative toxicity study of 24 manufactured nanoparticles including CuO, Cu_2O, ZnO, TiO_2, Zn_2TiO_4, anatase TiO_2, and ZrO_2 revealed that copper- and zinc-based nanoparticles exerted the most toxic effects on the viability of lung epithelial cells (A549 and THP-1, as determined via MTT, Neutral Red assay), presumably due to their high *solubility* (Lanone et al. 2009). Similarly, using LDH assays nanosized and fine-sized ZnO particles were found to be more cytotoxic in L2 lung epithelial cells than SiO_2 particles (Sayes, Reed, and Warheit 2007). In addition, ZnO nanoparticles have been shown to decrease redox activity in murine neuroblastoma cells, whereas other metal nanoparticles such as Fe_3O_4, TiO_2, Al_2O_3, and CrO_3 revealed no measurable effect (Jeng and Swanson 2006). In a very recent study, Kermanizadeh et al. assessed the effects of different nanomaterials including two variants of ZnO and five variants of TiO_2 on human renal proximal tubule epithelial cells (HK-2). They found no cytotoxic effects for TiO_2 nanoparticles, but both ZnO particle types displayed significant cytotoxicity (Kermanizadeh et al. 2013). TiO_2 nanoparticles also lacked any effect on A549 cells in different cell viability assays (LDH and MTT) (Duffin et al. 2007).

However, in line with the data mentioned earlier (Sayes et al. 2006) it was found that anatase TiO_2 was slightly more toxic than rutile TiO_2 demonstrating an impact of the *crystalline structure* on cytotoxicity (Simon-Deckers et al. 2008), which may be explained by the ROS generating capacity of anatase TiO_2 as mentioned previously.

In the nanoGEM project, exposure of cultured cells to ZnO nanoparticles was shown to adversely affect cell viability to a substantial extent. Cytotoxicity caused by 32 μg/mL ZnO is in the range of 25% to 95%, depending on the specific cell lines tested. None of the SiO_2 and ZrO_2 nanoparticles from the nanoGEM project (applied at concentrations between 0.32 μg/mL and 32 μg/mL) induced an increase in LDH release in any of the cell lines tested, at least in the presence of serum proteins. By contrast, ZnO nanoparticles have been shown to almost completely reduce viability of all cell lines tested in WST-8 assays. Only A549 cells displayed a slightly weaker reactivity toward ZnO nanoparticles with an inhibition of proliferation of 70% to 80%. SiO_2.naked nanoparticles were found to affect proliferation of RAW264.7 macrophage-like cells. Conversely, none of the surface-modified SiO_2 particles (SiO_2.PEG, SiO_2.amino, or SiO_2.phosphate) were found to affect proliferation of this cell type. When applied at 32 μg/mL, SiO_2.naked nanoparticles significantly reduced the total metabolic activity of the macrophage cell culture, and microscopic inspection confirmed a reduction in cell numbers. Surface modification of the SiO_2 nanoparticles (as shown for the listed nanoGEM materials SiO_2.PEG, SiO_2.amino, or SiO_2.phosphate) could be demonstrated to clearly influence their cytotoxicity. The finding that fluorescently labeled SiO_2 nanoparticles with unmodified surface (SiO_2.FITC) also affected the proliferation of RAW264.7 cells (measured via WST8 assay) further supported these results. Similarly, when using the WST-8 assay, a cell type specific effect for ZrO_2.amino nanoparticles could be observed. After addition of 32 μg/mL of these nanoparticles to a growing culture of RLE-6TN lung epithelium-like cells, reduced cell proliferation became obvious. Yet, all other tested cell lines (see Table 8.1) were not affected. Moreover, at concentrations of up to 32 μg/mL no other types of ZrO_2 nanoparticles (ZrO_2.acryl, ZrO_2.PEG, and ZrO_2.TODS) could impair the proliferation of the cell lines tested (Ossig and Schnekenburger, unpublished results from nanoGEM).

These data verify that specific surface modifications of nanomaterials can modulate their cytotoxic activity and that selected cell types may be uniquely sensitive for such adverse effects.

8.2.3 Genotoxicity

For any comprehensive evaluation of genotoxicity and the mutagenic potential of a substance, information is required on its capability to induce gene mutations, structural (clastogenicity) and numerical (aneugenicity) chromosome aberrations. Recent studies demonstrate that nanomaterials do not reveal a new quality of genotoxicity, but rather display effects well known from other genotoxins, for example, gene mutations, clastogenic or aneugenic effects, either through direct or indirect mechanisms (Pfuhler et al. 2013; Oesch and Landsiedel 2012). An overview on genotoxicity studies with engineered nanoparticles is provided by Magdolenova et al. (2013).

Test methods preferred for the application in Regulatory Toxicology are officially adopted through OECD testing guidelines (TG). The Ames test (OECD TG471 "Bacterial Reverse Mutation Test" Adopted 21 July, 1997) is commonly used to detect substances, which cause two classes of gene mutations, base pair substitution and small frame shifts. However, it has been questioned several times whether the

Ames test can be used for nanomaterial testing or not. In some cases, the Ames test failed to detect genotoxicity whereas other genotoxicity tests were positive with basically the same nanomaterial. In particular, Al_2O_3 nanoparticles did not display mutagenic effects on cells when assessed by the Ames test, but were found to be genotoxic when using micronucleus (MN) and comet assays *in vitro* as well as the MN assay *in vivo* (Balasubramanyam et al. 2010; Di Virgilio et al. 2010; Balasubramanyam et al. 2009). A possible explanation might be that larger nanoparticles are unable to cross the bacterial cell wall, or, if they are able to cross the cell wall, they might interfere with histidine synthesis thereby inducing false-negative or false-positive results (Magdolenova et al. 2013). The *in vitro* MN assay (OECD TG487 "*In vitro* mammalian cell micronucleus test," Adopted 22 July, 2010) detects both clastogenic as well as aneugenic substances and is a sensitive cellular test system capable of detecting DNA damage in mammalian cells. However, a number of studies resulted in conflicting data, and the response depended on the cell type used, the particle characteristics such as surface coating (Landsiedel et al. 2009; Warheit and Donner 2010). Therefore, further validation work is needed to clarify the predictive value of the MN test for determining the genotoxicity of nanoparticles. As the time period during which cells are exposed to nanoparticles is mostly limited to several hours, sedimentation and uptake of particles may be a major issue.

Detecting mammalian DNA damage *in vitro* via the comet assay is the most commonly used approach to assess chemical-mediated genotoxicity. However, the mechanisms of nanoparticle-mediated genotoxicity are not yet clear. Since several metal oxide nanoparticles have been found to enter the nucleus (Ahlinder et al. 2013), genotoxic effects may be triggered by direct particle–DNA interactions. However, increasing evidence also suggests that genotoxicity may result from indirect damage via oxidative stress induced by metal oxide nanoparticles. These indirect genotoxic effects have been further elucidated by Xu et al. (2009). They found anatase TiO_2 particles (5 nm and 40 nm) to increase the mutation rate in primary MEF with proper dose dependence. Both particle sizes led to the formation of peroxynitrite anions and induced deletions in the kbp range that could be protected by antioxidants. The level of DNA damage could be reduced via suppression of cyclooxygenase-2 (COX-2). COX-2 plays an important role in cellular inflammation and genomic instability, thus particle-induced oxidative stress may therefore trigger COX-2 signaling pathways (Xu et al. 2009). Since the ROS generating potential of metal oxide nanoparticles depends on their *chemical composition* (Limbach et al. 2007; Kroll et al. 2011) at least indirect genotoxicity of nanoparticles may thus be attributed to their chemistry.

The importance of *size* for the induction of DNA damage has been demonstrated in the case of TiO_2 (Gurr et al. 2005) and ZnO (Hackenberg et al. 2011a). In comparison to particles of >200 nm, 10 and 20 nm TiO_2 induced a greater level of oxidative DNA damage in human bronchial epithelial cells (BEAS-2B) (Gurr et al. 2005). While bulk ZnO did not induce genotoxicity (comet assay) in nasal mucosa cells, a significant level of DNA damage was observed for ZnO nanoparticles at a concentration of 10 µg/mL (Hackenberg et al. 2011b).

At present, it is unclear whether *solubility* plays an important role in nanoparticle-induced genotoxicity. Sharma, Anderson, and Dhawan (2012) demonstrated that ZnO nanoparticles were internalized by human epidermal keratinocytes and then

TABLE 8.3
Genotoxicity Testing of Naked and Coated Metal Oxide Nanoparticles

Particles	Test	Cytotoxic/Toxic to Bacteria	Result
AlO(OH)	Micronucleus (MN) assay *in vitro*	>50 μg/mL	Not genotoxic
	Ames test	No	Not mutagenic
	Comet assay	No	Not genotoxic
SiO_2	MN assay *in vitro*	>250 μg/mL	Not genotoxic
	Ames test	No	Not mutagenic
	Comet assay	No	Genotoxic
SiO_2.phosphate	MN assay *in vitro*	>100 μg/mL	Not genotoxic
	Ames test	No	Not mutagenic
	Comet assay	No	Genotoxic
SiO_2.PEG	MN assay *in vitro*	No	Not genotoxic
	Ames test	No	Not mutagenic
	Comet assay	No	Not genotoxic

induced DNA damage (detected via the comet assay). They suggested that intrinsic particle properties rather than soluble ions were responsible for the genotoxic effects (Sharma, Anderson, and Dhawan 2012). This was also discussed by other groups (Gojova et al. 2007; Lin et al. 2009).

Differences in the genotoxic responses of HepG2 cells after exposure to 25 nm anatase and 100 nm rutile TiO_2 were suggested to be dependent not only on their different particle size but also on their different *crystalline structure* (Petkovic et al. 2011). However, other studies could not clearly demonstrate an impact of nanoparticles' crystal structure on genotoxicity. Exposure of human peripheral blood lymphocytes to 15–30 mm anatase TiO_2, for instance, did not lead to genotoxic effects (comet assay) albeit the nanoparticles were detected in the nucleus (Hackenberg et al. 2011b). Similarly, other authors reported that neither rutile TiO_2 NP nor P25 TiO_2 particles induce chromosomal aberrations or mutagenicity (Warheit et al. 2007; Bhattacharya et al. 2009).

In the nanoGEM project, the metal oxide nanomaterials AlO(OH) (37 nm), SiO_2.naked (15 nm), SiO_2.PEG (15 nm), and SiO_2.phosphate (15 nm) (compare properties in Table 1.2), some of which are of commercial relevance, were investigated for their ability to induce chromosomal breaks, aneuploidy, or point mutations in the absence or presence of an extrinsic metabolizing system using the MN and the Ames test, respectively. This was one of the first studies in which both assays were performed according to OECD guidelines (Ames: OECD TG471, MN: OECD TG487) upon adaptation/modification of the testing procedure to meet special requirements of the manufactured nanomaterials. In addition, the comet assay was performed using 3D bronchial models (EpiAirway™, MatTek, Ashland, USA).

The results show that none of the investigated nanoparticles, independent of their size, chemical composition, and coating, were able to induce mutagenic effects in either the MN assay or the Ames test. However, two of the SiO_2 nanoparticles tested (SiO_2.naked and SiO_2.phosphate) displayed genotoxic effects when using the comet assay (Table 8.3). The Comet assay has previously been performed with SiO_2 nanoparticles of different sizes (30 nm, 80 nm, and 400 nm), but failed to reveal genotoxic effects in 3T3-L1 fibroblasts (Barnes et al. 2008). Using another SiO_2 nanomaterial, Wang et al. reported opposing results. Their study identified genotoxicity of SiO_2 nanomaterials in the MN but not the Comet assay (Wang, Wang, and Sanderson 2007). Recent results from Downs et al. highlighted the mechanisms of SiO_2-induced genotoxicity. They found no effect of these particles in *in vitro* MN assays, but demonstrated in parallel *in vivo* studies a secondary, inflammatory response-dependent genotoxicity of SiO_2 (Downs et al. 2012).

8.2.4 Gene Expression Profiles

DNA microarray analyses revealed the gene expression profiles of human keratinocyte HaCaT cells that were exposed in the dark to anatase TiO_2 nanoparticles of different (7 nm, 20 nm, and 200 nm) average sizes (Fujita et al. 2009). The data suggest that TiO_2 nanoparticles, in the absence of illumination, have no significant impact on ROS-associated oxidative damage. However, these particles seem to affect the cell–matrix adhesion and thus extracellular matrix remodeling of keratinocytes. In addition, mouse cDNA microarrays revealed that TiO_2 nanoparticles induced differential gene expression of hundreds of genes including activation of pathways involved in cell cycle control, apoptosis, chemokine, and complement cascades. *In vivo* it was found that TiO_2 nanoparticles can induce severe lung emphysema, which may be triggered via activation of placental growth factor and related inflammatory pathways (Chen et al. 2006). Studies assessing the toxic effects of P25 TiO_2 and CeO_2 nanoparticles of different sizes (15, 25, 30, and 45 nm) in bronchial epithelial cells (BEAS-2B) revealed that both particle types induced oxidative stress-related gene expression whereas inflammation-related genes were only induced by TiO_2 nanoparticles (Park et al. 2008a; Park et al. 2008b). This again suggests that the *chemical composition* of the particles decisively influences their inherent toxicity.

8.2.5 Inflammatory Response

For evaluating immunotoxic effects induced by nanomaterials, the production of inflammatory markers such as the chemokines, interleukin8 (IL-8), TNF-α, or IL-6 are usually measured in cell culture supernatants using the enzyme-linked immunosorbent assay (ELISA). TiO_2 nanoparticles (P25) but not fine TiO_2 particles were found to trigger Il-8 release in A549 cells indicating a *size*-dependent effect of immunotoxicity. Since the P25 TiO_2 nanoparticles remained highly aggregated in cell culture as well as inside the cells, inflammatory properties of TiO_2 particles appear to be driven by their *specific surface area* (Singh et al. 2007). In a comprehensive study aimed to determine the importance of surface area and surface reactivity of particles to induce inflammatory responses, Duffin et al. used a variety of manufactured particles, such as TiO_2, CB, and metal nanoparticles (Ni and Co), both for instillation and for treatment of A549 cells. They observed a correlation between particle surface area dose, specific surface activity, and the proinflammatory effects *in vivo* and *in vitro*. Their study also demonstrated the utility of *in vitro* assays for predicting the ability of nanoparticles to cause inflammation *in vivo* on the basis of their surface area and reactivity (Duffin et al. 2007). In addition, a concordance between surface area dose of low-solubility, low-toxicity particles (LSLTP) that produce inflammation *in vivo* and the surface area dose of LSLTP that produce an IL-8 release in A549 cells *in vitro* was observed (Donaldson et al. 2008). These studies suggest *in vitro* studies as a valuable complement to animal studies.

In line with the results from assessing ROS formation and decrease in cell viability, TiO_2 nanoparticles displayed an overall greater immunotoxicity in A549 cells when compared to rutile TiO_2 nanoparticles demonstrating the importance of *crystalline structure* in nanoparticle-induced inflammatory responses. In particular, the release of IL-8 observed in human dermal fibroblasts and A549 cells after exposure to anatase TiO_2 nanoparticles was found to be significantly lower when the cells were exposed to rutile TiO_2 nanoparticles (Sayes et al. 2006).

8.3 EFFECTS OF METAL OXIDE NANOPARTICLES ON PRIMARY ALVEOLAR MACROPHAGES

Many types of nanoparticles form agglomerates inside the lung alveoli (Gosens et al. 2010; Morfeld et al. 2012), where they are ingested by alveolar macrophages. Measuring the *in vitro* cytotoxicity with freshly isolated primary alveolar macrophages (AM) therefore gives first insight into whether a particle is cytotoxic to this particular cell type, which is chiefly responsible for clearing the lung from particle loads. Furthermore, a particle-induced release of ROS or of proinflammatory cytokines (such as TNFα) from AM can be analyzed. This yields plausible information to what extent cells of the lung epithelium may be indirectly compromised by particle-loaded AM. However, to produce meaningful *in vitro* results, which may be even predictive for *in vivo* findings, several requirements have to be met, two of which appear to be of major importance: first, the mean particle load of AM *in vitro* should be adapted to what is obtained during an inhalation experiment. As AM in the lung actively collects particles, surprisingly high mass per cell values are reached and up to 90 pg/cell have been found secondary to inhalation exposure to AlOOH nanoparticles (Pauluhn 2009). Therefore, *in vitro* tests are carried out with fixed numbers of AM per culture well in order to achieve a defined loading with particles of up to 120 pg/cell after 16 h (Rehn, Rehn, and Bruch 1999; Bruch et al. 2004). Second, *in vitro* tests with AM should be carried out in the absence of proteins, as inhaled particles are primarily not covered by proteins. Furthermore, serum-free conditions allow us to study the primary effects of particles on AM.

In the nanoGEM project, the biological effects of different *surface coatings* were investigated for ZrO_2 nanoparticles, which are widely used as a ceramic filler material. ZrO_2 nanoparticles are insoluble, and exhibit no measurable ROS generation at their surface (in contrast to, e.g., TiO_2 nanoparticles), which predestines them for investigations on the biological effects of surface modifications. In particular, 10 nm ZrO_2 nanoparticles were modified on their surface either with a caustic polyacrylate (ZrO_2.acryl), a polyoxocarbonic acid (ZrO_2.TODS), an amino silane (ZrO_2.amino), or with a polyethylene glycol (PEG) (ZrO_2.PEG) (compare properties in Table 1.2). All types of ZrO_2 nanoparticles agglomerated under serum-free cell culture conditions. This led to a complete gravitational settling of fine grains, which were visible by phase contrast microscopy (compare Chapter 7). Particles were quantitatively ingested by AM, and nearly no nanoparticle remained dispersed in the supernatant. Consequently, a mean dose in picograms per macrophage could be calculated. The cytotoxic effects of all ZrO_2 nanoparticles, measured as release of LDH, were largely similar, and started upon a dose of approximately 30 pg/cell (corresponding to 45 μg/ml), comparable to the lowest observed adverse effect level of TiO_2 NM105, but nearly an order of magnitude above ZnO NM110. Minor differences were observed among modifications, with ZrO_2.acryl being the most bioactive particle: ZrO_2.acryl led to a release of TNFα and glucuronidase and extracellular H_2O_2 formation (Wiemann and Vennemann, unpublished results from the nanoGEM study).

8.4 EFFECTS OF METAL OXIDE NANOPARTICLES IN ANIMALS

Compared to the *in vitro* assessment of nanoparticle toxicity, *in vivo* investigations are highly expensive and most often require the sacrifice of animals. *In vivo* studies are the gold standard of toxicity testing, however, there is only a limited number of studies available that evaluate the effects of metal oxide nanoparticles *in vivo*. *In vivo* studies generally focus on a single route of exposure and by far the greatest number of available studies concentrate on the pulmonary exposure of nanomaterials. Consequently, this chapter will focus on intratracheal instillation and rat lung inhalation studies using metal oxide nanoparticles. End points of toxicity for pulmonary exposure range from oxidative stress, cell proliferation to markers of inflammation and histopathology of the lung.

8.4.1 Biological Activity of Metal Oxide Nanoparticles in Rat Lungs after Intratracheal Instillation

Based on intratracheal instillation studies and literature review, it was concluded that, in general, "ultrafine particles" are more inflammogenic in the rat lung than fine, respirable particles made from the same material, which is driven by their *surface area* (Donaldson et al. 2008). Recently, Sun et al. found intratracheally (i.t.) instilled TiO_2 nanoparticles of 5–6 nm to elicit pulmonary inflammation via induction of inflammatory cytokines (Sun et al. 2012). Likewise, intratracheal instillation of anatase TiO_2 nanoparticles for 24 h led to pulmonary inflammation (Saber et al. 2012). Since DNA damage (detected by the comet assay) could not be observed in this study, the authors suggested that inflammation is not linked to DNA damage when using TiO_2 nanoparticles. Similar results were observed in a recent instillation study with 5 nm anatase TiO_2 nanoparticles, which were found not to be genotoxic as detected by the comet assay (Naya et al. 2012). However, an early report demonstrated *in vivo* genotoxicity of TiO_2 nanoparticles. After intratracheal instillation of 18 nm anatase TiO_2 nanoparticles in rats, hypoxanthine phosphoribosyltransferase (HPRT) mutation frequency was increased in alveolar type II cells (Driscoll et al. 1997).

For nanoGEM *in vivo* experiments based on intratracheal instillation, ZrO_2 nanoparticles had to be suspended in such a way that (a) a coarse agglomeration is avoided and (b) the suspension does not interfere with the physiologic requirements of the lung. In the case of acid- or alkaline-stabilized ZrO_2 nanoparticle suspensions, a protein precoating with rat serum (Bihari et al. 2008) followed by a transfer to bicarbonate buffer was used to prevent agglomeration. This type of coating and buffering allowed two investigated surface modifications of ZrO_2 nanoparticles. The bioactivity of ZrO_2.acryl was tested *in vivo* and compared to ZrO_2.TODS, which was the least active modification *in vitro*. However, to circumvent agglomeration prior to intratracheal instillation, ZrO_2.acryl and ZrO_2.TODS had to be coated with rat serum. Unbound proteins were removed and particle fractions were applied to rat lungs in three doses (0.6, 1.2, and 2.4 mg/lung). Effects were studied in bronchoalveolar fluid (BALF) and also histologically after 3 and 21 days. The distribution of fluorescence labeled ZrO_2.Acryl and ZrO_2.TODS was carried out in parallel and revealed a progressive uptake into AM (nanoGEM, not shown). Effects on BALF parameters (cell count, increase in PMN) commenced at 0.6–1.2 mg/lung; this dose range properly reflects the LOAEL observed *in vitro* if the LOAEL (30 pg/cell) is multiplied by the macrophage number per lung (2×10^7). However, slight differences seen between ZrO_2.acryl and ZrO_2.TODS *in vitro* were largely absent *in vivo* (Wiemann and Venneman, unpublished results from nanoGEM). Although further research, for example, on oxidative damage is needed, it may be speculated that this may be due to the precoating with proteins. In addition, nanoparticle surfaces could be further changed upon contact with the lung lining fluid. Based on these results obtained for modified ZrO_2-NP it can be stated that the influence of surface modifications is obvious in highly loaded AM *in vitro*, but may be blunted under *in vivo* conditions.

8.4.2 Biological Activity of Metal Oxide Nanoparticles in Rat Lungs after Inhalation

In rat lung inhalation studies, ZnO, TiO_2, CeO_2, ZrO_2, and SiO_2 nanoparticles of similar size and shape differed largely in their toxicity (Landsiedel et al. 2010). ZnO and CeO_2 nanoparticles were found to be most toxic whereas SiO_2 and ZrO_2 did not show any adverse effect at the highest tested aerosol concentration of 10 mg/m^3, which was higher than the general thresholds for fine dusts (Landsiedel et al. 2010). Based on the level of no adverse effects in inhalation, decreasing effects were found in the following order: CeO_2 > ZnO > TiO_2 > SiO_2, ZrO_2 (Landsiedel et al. 2010). Similar to the results from the *in vitro* studies mentioned earlier, the *chemical composition* influenced the toxicity and not—or not exclusively—the size or shape of the material. Moreover, a contribution of *solubility* to nanomaterial toxicity that has been found *in vitro* for ZnO (Buerki-Thurnherr et al. 2013; Xia et al. 2011; Kao et al. 2012) has also been suggested in studies based on intratracheal instillation (Cho et al. 2011) and in rat inhalation studies (Landsiedel et al. 2010). In rat lungs, ZnO induced a concentration-related inflammation reaction in addition to necrosis, which was detected in the lung and the nose. At least the observed necrosis could be attributed to soluble zinc ions released from the ZnO nanoparticles (Landsiedel et al. 2010).

ZnO and TiO_2 particles were tested as nanosized and as fine-sized particles in different inhalation studies. Most of the nanomaterials agglomerated to particles similar in size to the fine-sized particles, resulting in similar particle sizes for both materials. However, the agglomerates of TiO_2 and ZnO nanosized particles exhibited stronger effects than the solid fine-sized particles of TiO_2 and ZnO, respectively. Studies with TiO_2 did not reveal any deagglomeration in the body and hence the differences in toxicity between solid particles and agglomerates are most likely due to the fact that the agglomerates are built from nanoscaled primary particles with a different inner structure and higher specific *surface area* of the agglomerates compared to the solid particles (for review, see Landsiedel et al. 2010).

In line with the results from *in vitro* studies mentioned earlier, rutile and surface-treated TiO_2 showed only transient and fast reversible pulmonary inflammatory responses, whereas mixed anatase/rutile TiO_2 nanoparticles induced inflammatory and adverse lung effects (Landsiedel et al. 2010). This again demonstrates the impact of *crystalline structure* on nanoparticle toxicity.

Within the framework of the nanoGEM project several short-term inhalation studies were performed to examine the influence of surface properties on pulmonary toxicity of amorphous silica and zirconium oxide particles (compare surface properties in Table 1.2). During the studies, test atmospheres were generated by spraying the particle suspension by an atomizer using pressured air. The atmospheres were characterized thoroughly by off-line gravimetrical measurements, online scattered-light photometers, and several devices for particle size measurements (see Ma-Hock et al. 2007 for further details).

In particular, male Wistar rats were head–nose exposed for 6 h/day for five consecutive days. As a rule, a concurrent clean air control group was included in each study. Details of the studies and examination time points are described in Table 8.4.

Several biochemical and cytological parameters were examined in lavage fluid (protein, enzyme activity of selected indicator enzymes, selected cytokines, total cell count, macrophage, and neutrophils). They indicate an early inflammation process in the lung. Moreover, the respiratory tract of exposed and control animals was examined histologically to detect morphological changes that manifested during exposure and recovery period. To evaluate the potential progression or recovery of the effects, all examinations were performed in satellite animals after 3 weeks of

TABLE 8.4
Design of Rat Short-Term Inhalation Studies Conducted within the nanoGEM Project

Study Day	0	1	2	3	4	5	6	7	8–24	25	26	27	28
In-life phase	X	X	X	X	X	R	R	R	R	R	R	R	R
Examinations					E			L		E			L

X, exposure; R, recovery; E, histopathology (5 rats/group/time point) and examination of the organ burden (3 rats/group/time point); L, lung lavage (5 rats/group/time point).

recovery period. Lung, lung-associated lymph nodes, and spleen were analyzed for the test substance. The target concentrations were 2, 10, and 50 mg/m^3.

The comparison of pulmonary toxicity of particles with different surface modification showed particle type-dependent effects of the surface modification. The two surface-modified ZrO_2 nanoparticles did not induce any adverse effects up to an atmospheric concentration of 50 mg/m^3. The switch of surface charge (see Table 1.2) from strongly negative (acrylic acid) to neutral/positive (TODS) did not elicit significant changes in the inhalation toxicity of ZrO_2-based particles (Table 8.5), except for a transient increase in neutrophils directly after exposure.

TABLE 8.5
Effects of 50 mg/m^3 Metal Oxide Nanoparticles Observed in Short-Term Inhalation Studies (nanoGEM) Directly after Exposure or after an Additional 3-Week Recovery Period

Property or Effect	Time Point	Units	SiO_2. Naked	SiO_2. PEG	SiO_2. Amino	SiO_2. Phosph.	ZrO_2. Acryl	ZrO_2. TODS
Lung burden	After exposure	μg/lung	343	836	743	500	169	694
	After recovery	μg/lung	209	371	475	304	190	521
Bronchoalveolar fluid (BALF) cell counts	After exposure	Fold of control	1.2	0.8	0.8	1.3	1	0.8
	After recovery	Fold of control	1	1.1	0.9	1	0.9	0.6
BALF neutrophils	After exposure	Fold of control	11.2**	2.3	4.2	2	1.3	8*
	After recovery	Fold of control	21.8**	1	1.8	0.3	2.2	0.8
BALF lymphocytes	After exposure	Fold of control	5.6**	1.8	1	3	3.8	1.4
	After recovery	Fold of control	4.8**	0.7	2.4	0.6	0.7	0.5
Total protein	After exposure	Fold of control	1	1.1	0.7	1.1	1	0.8
	After recovery	Fold of control	1.2	0.83	0.9	1.2	1.2	0.8
MCP-1 in BALF	After exposure	Fold of control	0.9	1	1	1	1.1	1.7
	After recovery	Fold of control	1.8	0.9	1.1	1	1	1.1
CINC-1/IL-8 in BALF	After exposure	Fold of control	1.2	1.2	0.8	1.1	0.7	1.7
	After recovery	Fold of control	1.5	1	0.9	1.2	0.9	1.4

(Continued)

TABLE 8.5 (*Continued*)
Effects of 50 mg/m³ Metal Oxide Nanoparticles Observed in Short-Term Inhalation Studies (nanoGEM) Directly after Exposure or after an Additional 3-Week Recovery Period

Property or Effect	Time Point	Units	SiO_2. Naked	SiO_2. PEG	SiO_2. Amino	SiO_2. Phosph.	ZrO_2. Acryl	ZrO_2. TODS
(Multi)focal inflammation	After exposure	–	0	(1)	0	0	0	0
	After recovery	–	2‡	0	0	0	0	0
Macrophage aggregation with eosinophilic content	After exposure	–	2‡	(1)	0	0	0	0
	After recovery	–	2‡	0	0	0	0	0
Comet assay	5d inhal., 21d rec., 50 mg/m³	% tail intensity	n.e.	n.e.	2.0%–3.9%	n.e.	3.2%–5.2%	n.e.
	Concurrent control	% tail intensity	n.e.	n.e.	1.3%–3.3%	n.e.	2.7%–3.2%	n.e.

Abbreviations: n.e., not examined; **, statistically significant $p < 0.01$; *, statistically significant $p < 0.05$; ‡, adverse effect.

With regard to SiO_2-based particles, adverse effects were only observed in animals exposed to SiO_2 Levasil®200 nanoparticles. SiO_2.PEG induced a few adaptive effects in lung and larynx without any changes in lavage parameters. SiO_2.amino induced a weak and transient influx of neutrophils directly after exposure, confirming the effect of neutral/positive charge observed on ZrO_2.TODS. SiO_2.phosphate did not cause any effects on pathology and clinical chemistry (Table 8.5). These data showed that the *surface modification* did reduce the inhalation toxicity of SiO_2 particles. An influence of surface modification on clearance and translocation behavior could not be detected within this project. The lung clearance of all SiO_2variants was faster than expected for granular poorly soluble particles (t½ approximately 20 versus 60 days) (Morrow 1988). While SiO_2 may partially dissolve in specific cellular compartments, ZrO_2 is almost insoluble under all relevant conditions, including pH value inside phagolysosomes, and was rapidly cleared from the lung with about the same half-time. It is interesting to note that the affinity to surfactant proteins from the lung lining fluid (Figure 4.3) provides the same identification of SiO_2.naked and SiO_2.phosphate as extreme cases.

Based on the examination of hematological and clinical chemical parameters in blood, there was no indication of potential systemic toxicity for any of the tested substances. The MN assay in bone marrow and the comet assay in lung showed that ZrO_2.acryl and SiO_2.amino were not mutagenic in bone marrow and did not cause

DNA damage in the lung (Table 8.5). Similar results have been observed for 50 nm pure rutile TiO_2 and 30–200 nm ZnO nanoparticles. Landsiedel et al. reported that these particles did not display genotoxicity neither *in vitro* (Ames' Salmonella gene mutation 22 test and V79 MN chromosome mutation test) nor *in vivo* (mouse bone marrow MN test 23 and comet DNA damage assay in lung cells from rats exposed by inhalation) (Landsiedel et al. 2010). Also for smaller TiO_2 nanoparticles with higher anatase content (21 nm, 74% anatase, 26% rutile), no DNA damage in lung epithelial cells or genotoxic effects on micronuclei in bone marrow polychromatic erythrocytes could be observed after inhalation exposure for 5 days (0.8, 7.2, 28.5 mg/cm^{-3}, 4 h/day) (Lindberg et al. 2012). Even after chronic inhalation for 2 years to 10.4 mg/m^3 of 15–20 nm TiO_2 nanoparticles, no DNA adduct formation could be detected in the peripheral lung tissue (Gallagher et al. 1994). However, for P25 nanoparticles and for 33 nm anatase TiO_2 nanoparticles genotoxic effects detected by the comet assay and the MN test have been found after oral gavage to mice for 5 days (500 mg P25/kg body weight) or 7 days (40, 200, 1000 mg 33 nm TiO_2/kg body weight), respectively (Trouiller et al. 2009; Sycheva et al. 2011), for review see Magdolenova et al. (2013).

To summarize on surface properties, none of the surface-coated SiO_2 or ZrO_2 materials exhibited significant adverse effects in inhalation. In contrast, the functionalization of SiO_2.naked nanoparticles decreased inflammatory effects. Other studies with different TiO_2 materials also demonstrated that particle surface property rather than particle form or size determines the inhalation toxicity of particles in the lung (Warheit et al. 2006, Warheit et al. 2007, Madl and Pinkerton 2009). Weak and transient effects were observed exclusively for SiO_2.amino and ZrO_2.TODS, which are the only particles with neutral/positive surface charge. On the other extreme, $BaSO_4$ and ZrO_2.acryl carried the highest negative charge and had the lowest toxic potency in the lung. The analysis of the interaction of the particles with proteins, including the formation of lipid/protein corona (Chapter 4) and association with cell membranes and internalization as well as intracellular trafficking (Chapter 7), does not necessarily simplify the grouping of materials, if as-produced parameters (Chapter 1) such as charge were considered sufficient to prioritize materials for testing (Chapter 16).

8.4.3 Effects of Metal Oxide Nanoparticles on Animals with Allergic Airway Inflammation

Active AM and an effective epithelial layer can usually prevent adverse effects of nanomaterials in intact lungs. These protective mechanisms are impaired when nanomaterial exposure occurs in a lung damaged by a mild or severe disease. This situation was studied in a model of allergic airway inflammation.

Allergic airway inflammation is characterized by inflammatory cell infiltration, mucus cell hyperplasia, and airway hyperresponsiveness. These effects are closely linked and interdependent. Lymphocytes and eosinophils are considered to be the key effector cells: soluble factors derived principally from T cells play an essential role in recruiting inflammatory cells in the lungs (Gavett et al. 1994). Eosinophil accumulation and activation with the subsequent release of cationic proteins in the airways have been implicated in many pathological processes and in the subsequent modification of airway function (Humbles et al. 2004). Previous studies have shown that exposure to ultrafine particles exerts strong adjuvant effect on the manifestation of allergic airway inflammation (Alessandrini et al. 2006). A mouse model of allergic airway inflammation was used to assess the potential health hazard of SiO_2 NPs exposure in allergic airway disease.

This model was designed to induce a mild inflammatory response for evaluating potential enhancing effects of nanoparticle inhalation on the allergen-induced inflammation of the lung. Allergen sensitization and challenge occurred as previously described (Alessandrini et al. 2006). Shortly, mice were sensitized by several repetitive intraperitoneal injections of 1 µg ovalbumin/alum over 52 days and sera were checked for specific immunoglobulins. Mice were then i.t. instilled with SiO_2 naked, SiO_2.PEG, SiO_2.amino, or SiO_2.phosphate at a concentration of 50 µg/mouse, followed by ovalbumin aerosol challenge (20 min; 10 mg/ml). Five days later, allergic inflammation was evaluated by cell differentiation of BALF, histology, lung function analysis, and determination of cytokines/chemokines mRNA level in lung tissue. BALF cell profiles from allergic phenotypes are characterized by an early peak of neutrophils 24 h after allergen challenge and by a late increase in eosinophils, lymphocytes, and AM starting 48 h after allergen challenge.

In sensitized mice, intratracheal instillation of 50 µg SiO_2.naked and SiO_2.PEG induced an increase in both lymphocytes, and to a higher extent of eosinophils in the BALF. Lymphocytes were increased 2.8 fold and 3.4 fold by SiO_2.naked and SiO_2.PEG, respectively, compared to respective controls (sensitized mice exposed to nanoparticles supernatant). Eosinophils were increased 4.4 fold and 8.4 fold by SiO_2.naked and SiO_2.PEG, respectively, compared to respective controls (sensitized mice exposed to nanoparticles supernatant). The intratracheal instillation of the same amount of SiO_2.amino or SiO_2.phosphate had no significant effect, either on lymphocyte or on eosinophil recruitment in the BALF. In nonsensitized mice, some effects on BALF cell recruitment following intratracheal instillation of all nanoparticles tested were significant, but they were still significantly lower compared to the corresponding effects on sensitized mice. Histological evaluation of lung tissue showed that epithelial mucus cell hyperplasia was increased in lungs of sensitized mice instilled with SiO_2.naked and SiO_2.PEG prior to allergen challenge compared to sensitized mice instilled with respective supernatant prior to allergen challenge.

Airway hyperresponsiveness was also compromised in mice exposed to SiO_2.naked and SiO_2.PEG: Following a standard protocol of methacholine provocation, airway resistance was significantly increased in mice instilled with SiO_2.PEG and dynamic compliance was significantly reduced in mice instilled with SiO_2.naked. No significant effects on lung function were evaluated in sensitized mice exposed to SiO_2.amino or to SiO_2.phosphate prior to allergen challenge.

Evaluation of both Th1 and Th2 cytokines and eosinophil activation markers in the lungs of sensitized mice showed that SiO_2.naked and SiO_2.PEG NPs induced a significant increase in the expression of IL-13, a cytokine which plays an important role in airway hyperreactivity and eosinophilia (Wills-Karp and Chiaramonte 2003), but no alteration of Th1 cytokines. In addition, both nanoparticles induced a significant expression of Ear11, an important eosinophil activation marker in asthma

(Di Valentin et al. 2009) whereas SiO_2.amino and SiO_2.phosphate had no effect on the expression of Th1 and Th2 cytokines and of eosinophil activation markers (Marzaioli and Alessandrini, unpublished results).

This study demonstrates that SiO_2 nanoparticles exert an adjuvant effect in an *in vivo* model of allergic airway inflammation and that the functionalization of these particles through PEG does not modulate their adjuvant activity. In contrast, *surface coating* of SiO_2 nanoparticles with amino or phosphate groups effectively mitigates the SiO_2 nanoparticles-induced adjuvant effects in allergic airway inflammation.

8.5 CASE STUDY ON CORRELATIONS OF *IN VIVO* AND *IN VITRO* DATA

The following section intends to show that it is possible to correlate the gradual responses from *in vitro* and *in vivo* experimentation under certain conditions. Therefore, experiments with all four surface modifications of identical SiO_2 core NP (Table 1.2) were carried out on AM (primary cells as well as NR8383 cell line) under serum-free conditions, as introduced in section 8.3. Furthermore, particles were used for intratracheal instillation into the rat lung. Interestingly, gradual responses observed with both approaches were highly similar with respect to the rank order of inflammatory response or cytotoxicity, provided that the doses used were at or beyond the threshold for biological effects.

In vitro testing revealed a clear cytotoxic effect of SiO_2.naked on primary and cultured AM (release of LDH and glucuronidase) starting at 11.25 µg/ml and being fully developed upon 45 µg/ml under serum-free conditions. Within the same concentration range the effect of SiO_2.amino on release of LDH or glucuronidase was less pronounced, followed by SiO_2.PEG and SiO_2.phosphate, which had nearly no effect up to 45 µg/ml. Similarly, we obtained a severe induction of TNFα upon SiO_2.naked followed by SiO_2.amino and SiO_2.PEG, which was equally noneffective as was SiO_2.phosphate. All effects were largely diminished, although not abolished, when tests were run in the presence of 10% (v/v) fetal calf serum, albumin, or in the presence of native surfactant preparation (compare section 4.3.4). Under these conditions a precipitation of particles was observed. Together these results suggested that the cytotoxicity of SiO_2 NP is strongly influenced by their surface chemistry and, at least in part, seems to be competitively inhibited by the binding of protein to the particles' surface, as also suggested by others (Panas et al. 2013).

Previous *in vivo* experiments with amorphous silica carried out by us and others revealed high inflammatory potential of SiO_2 NP which, however, is transient in nature (Arts et al. 2007). We, therefore, restricted our *in vivo* experiments with SiO_2 NP to a short incubation period (3 days after intratracheal instillation) in order to observe early, albeit fully developed signs of inflammation. The instilled dose was set 0.36 mg per rat lung as this corresponded to the calculated and also measured lung burden obtained upon inhalation exposure after a five day exposure to 50 mg/m^3 (see Table 8.5).

We found that, after 3 days, SiO_2.FITC was nearly exclusively incorporated within CD68 positive alveolar macrophages. No enrichment of fluorescence such as fluorescent patches or punctuate labeling was found within alveolar septae. Considering the fact that also SiO_2.FITC particles were fairly well suspended and truly nanosized during application, this finding highlights the role of AM for clearing the lung from nanoparticulate matter. The degree of inflammation caused by the SiO_2 particles displayed a clear rank order of biologic activity (SiO_2.naked > SiO_2.amino > SiO_2.PEG > SiO_2.phosphate), which was reflected by the concentration of total protein in the BALF, the total number of lavageable cells, and by the percentage of neutrophilic granulocytes (Wiemann and Vennemann, unpublished results from the nanoGEM study). The obvious similarity between these *in vivo* findings and the *in vitro* experiments with AM led us to calculate various correlations between both types of experiments. In fact, we found positive correlation between the amount of bioactive TNFα released *in vitro* by AM (NR8383) and the occurrence of PMN or protein in BALF (Figure 8.1). Similarly, there was a correlation between the *in vitro* release of LDH or glucuronidase and the *in vivo* parameters describing inflammation.

These results are not in contrast to other findings obtained in nanoGEM for the same types of particles, although most types of SiO_2 NP showed no overt effect on various cell lines. This is because experiments described earlier were conducted at lower concentration and in the presence of serum, which lowers toxicity. With respect to the discrepancies between the results from the instillation and inhalation studies on SiO_2 NP (see Table 8.5) one should consider that the dose rate during inhalation is far lower compared to that of the bolus application during instillation. Binding of SiO_2 NP to lung proteins and/or surfactant components, *in situ* agglomeration, and ongoing clearance by AM most likely counteract the effects of accumulating particles, such that an inhaled lung burden of SiO_2 particles ranging between 343 μg/lung and 836 μg/lung may cause no, or, in the case of SiO_2.naked and SiO_2.amino, minor effects (see Table 8.5).

Considering the previously outlined results, one major conclusion from the nanoGEM experiments on SiO2 NP is that the comparison between *in vitro* macrophage responses and the *in vivo* response upon intratracheal instillation is a versatile tool

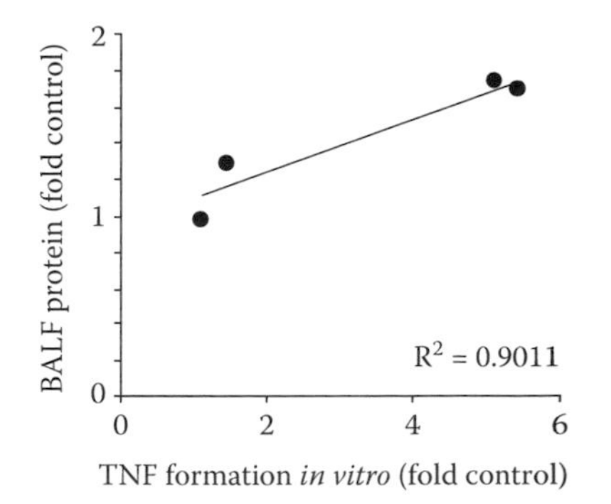

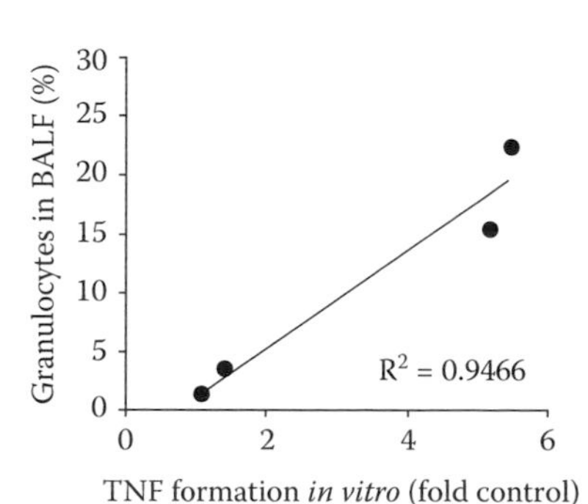

FIGURE 8.1 *In vitro–in vivo* correlation of the effects of SiO_2.naked, SiO_2.amino, SiO_2.PEG, and SiO_2.phosphate. *In vitro* results were obtained with AM (NR8383) subjected to nanoparticle suspended in cell culture medium (45 μg/ml) for 16 h. TNFα was measured indirectly (loss of L-929 cells upon treatment with the macrophage supernatant) and expressed as fold control values. *In vivo* results (bronchoalveolar fluid protein, Lowry method) were from rats treated with 360 μg/lung for 3 days.

and well suited to correlate *in vitro* and *in vivo* data, if experiments were carried out with supra-threshold concentrations. As such this approach appears to be able to predict from *in vitro* assays whether or not a functionalization of the surface of a nanoparticle may alter the acute biological activity or the toxicity of a given type of particle *in vivo*.

8.6 SUMMARY AND CONCLUSION

In this chapter, we have described the effects of metal oxide nanoparticles on human cells and rodent lungs. Albeit the majority of metal oxide nanoparticle types tested so far display no or only moderate toxicity, we have selected reports from those particle varieties that display adverse effects in an effort to illustrate what attributes are most influential in driving nanoparticle toxicity. Since for many different nanoparticle types adverse effects could not be observed *in vitro* and *in vivo* up to a certain dose, it is reasonable to assume that the nanodimension *per se* does not account for toxicity. In fact, a larger specific surface area associated with the nanosize leads to a greater surface activity and affects particle toxicity indirectly, for example, by enhancing potentially harmful particle effects such as metal ion release, ROS generation, or protein adsorption.

The toxicity of certain metal oxide nanoparticles has been demonstrated to have inflammogenic, oxidative, and genotoxic consequences. While the studies presented here cannot be considered to be representative for all metal oxide nanoparticles, it becomes apparent that the biological effects of metal oxide nanoparticles are not solely determined by their chemical nature but rather can be mediated by several physicochemical properties. Apart from chemical composition, crystalline structure, solubility, and surface chemistry influence nanoparticle toxicity both *in vitro* and *in vivo*. The contribution of crystalline structure to toxic behavior has been demonstrated for TiO_2 nanoparticles with anatase TiO_2 inducing more adverse effects than rutile TiO_2 *in vitro* and *in vivo*. Solubility of metal oxide nanoparticles (ZnO) seems to be responsible for inducing cytotoxic and inflammatory responses both *in vitro* and *in vivo*. From the *in vitro* studies in correlation with the inhalation studies reported here, one can conclude that nanoparticles with a high oxidative potential generally display a greater toxic potential confirming the influence of chemical composition on nanoparticle toxicity. Surface chemistry (CeO_2) and organic surface modification (TiO_2, SiO_2) were found to be associated with biological activity in the studies conducted within the NanoCare and nanoGEM projects. Previously, surface coatings have directly been associated with a relative decrease in toxicity for SiO_2 nanoparticles (Pan et al. 2009). Considering the overall *in vitro* and *in vivo* data obtained for SiO_2 nanoparticles with different surface modifications within the nanoGEM project, the influence of surface modifications on toxic behavior could be clearly confirmed.

It is of vital importance that the physicochemical properties responsible for the toxicity of metal oxide nanoparticles are used to generate structure–activity models, which can provide guidance for a safe design of nanomaterials. Taking into account the nanoGEM nanomaterials *in vitro* and *in vivo* toxicity results, we were able to separate nanoparticles with considerable toxic potential from those displaying no or only very low toxicity based on a principle component analysis (Figure 8.2).

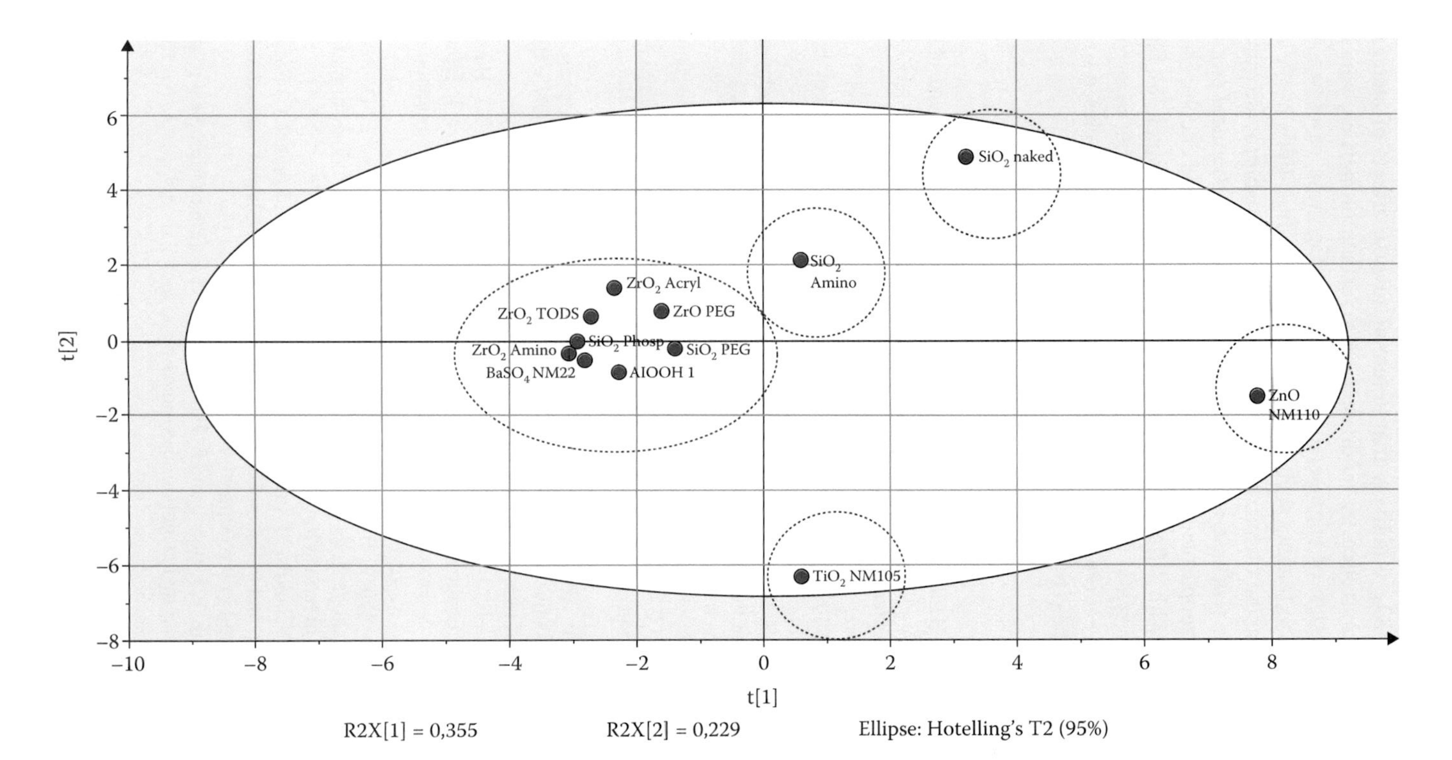

FIGURE 8.2 Principal component analysis of total *in vitro* and *in vivo* data from nanoGEM using the SIMCA-Software (Umetrics AB, Umea, Sweden). The divergence of the particle attributes is reflected by the spatial distance of the data points obtained. Most of the nanoparticle types tested displayed low or no toxicity and are arranged in a cluster, whereas nanoparticles displaying distinct biological effects are localized separately and distant from the cluster.

A clear correlation of specific physicochemical properties with the resulting biological effects required for QSAR modeling is not available. However, based on the data obtained for SiO_2, it is likely that potential adverse effects of metal oxide nanoparticles can be controlled by modifying their surface properties, which can be helpful for safer design of engineered nanoparticles for those cases where the product performance can tolerate such relatively drastic modifications of the nanomaterial properties. The data from nanoGEM and NanoCare also provide guidance for a future nanoparticle risk assessment. Since different physicochemical properties seem to contribute to nanoparticle toxicity, a testing strategy should encompass multidisciplinary working with different characterization tools (see Chapter 16). Finally, the correlations of *in vivo* and *in vitro* data presented here suggest that *in vitro* tests may allow an initial screening to preselect potentially toxic nanoparticles, a strategy that eventually should help to minimize the use of animals.

REFERENCES

Ahlinder, L., B. Ekstrand-Hammarström, P. Geladi, L. Osterlund. 2013. Large uptake of titania and iron oxide nanoparticles in the nucleus of lung epithelial cells as measured by raman imaging and multivariant classification. *Biophysical* J 105 (2):310–319.

Alessandrini, F., H. Schulz, S. Takenaka, B. Lentner, E. Karg, H. Behrendt, and T. Jakob. 2006. Effects of ultrafine carbon particle inhalation on allergic inflammation of the lung. *J Allergy Clin Immunol* 117 (4):824–830.

Arts, J.H.E., H. Muijser, E. Duistermaat, K. Junker, and C. F. Kuper. 2007. Five-day inhalation toxicity study of three types of synthetic amorphous silicas in Wistar rats and post-exposure evaluations for up to 3 months. *Food Chem Toxicol* 45 (10):1856–1867.

Balasubramanyam, A., N. Sailaja, M. Mahboob, M. F. Rahman, S. M. Hussain, and P. Grover. 2009. In vivo genotoxicity assessment of aluminium oxide nanomaterials in rat peripheral blood cells using the comet assay and micronucleus test. *Mutagenesis* 24 (3):245–251.

Balasubramanyam, M., A. Adaikalakoteswari, Z. Sameermahmood, and V. Mohan. 2010. Biomarkers of oxidative stress: Methods and measures of oxidative DNA damage (COMET assay) and telomere shortening. *Methods Mol Biol* 610:245–261.

Barnes, C. A., A. Elsaesser, J. Arkusz, A. Smok, J. Palus, A. Lesniak, A. Salvati, J. P. Hanrahan, W. H. Jong, E. Dziubaltowska, M. Stepnik, K. Rydzynski, G. McKerr, I. Lynch, K. A. Dawson, and C. V. Howard. 2008. Reproducible comet assay of amorphous silica nanoparticles detects no genotoxicity. *Nano Lett* 8 (9):3069–3074.

Bhattacharya, K., M. Davoren, J. Boertz, R. P. F. Schins, E. Hoffmann and E. Dopp. 2009. Titanium dioxide nanoparticles induce oxidative stress and DNA-adduct formation but not DNA-breakage in human lung cells. *Part Fibre Toxicol* 6 (1):17.

Bihari, P., M. Vippola, S. Schultes, M. Praetner, A. G. Khandoga, C. A. Reichel, C. Coester, T. Tuomi, M. Rehberg, and F. Krombach. 2008. Optimized dispersion of nanoparticles for biological in vitro and in vivo studies. *Part Fibre Toxicol* 5 (1):14.

Bruch, J., S. Rehn, B. Rehn, P. J. Borm, and B. Fubini. 2004. Variation of biological responses to different respirable quartz flours determined by a vector model. *Int J Hyg Environ Health* 207 (3):203–216.

Buerki-Thurnherr, T., L. Xiao, L. Diener, O. Arslan, C. Hirsch, X. Maeder-Althaus, K. Grieder, B. Wampfler, S. Mathur, P. Wick, and H. F. Krug. 2013. In vitro mechanistic study towards a better understanding of ZnO nanoparticle toxicity. *Nanotoxicology* 7 (4):402–416.

Cedervall, T., I. Lynch, S. Lindman, T. Berggard, E. Thulin, H. Nilsson, K. A. Dawson, and S. Linse. 2007. Understanding the nanoparticle-protein corona using methods to quantify exchange rates and affinities of proteins for nanoparticles. *Proc Natl Acad Sci U.S.A.* 104 (7):2050–2055.

Chen, H. W., S. F. Su, C. T. Chien, W. H. Lin, S. L. Yu, C. C. Chou, J. J. Chen, and P. C. Yang. 2006. Titanium dioxide nanoparticles induce emphysema-like lung injury in mice. *FASEB J* 20 (13):2393–2395.

Cho, W. S., R. Duffin, S. E. Howie, C. J. Scotton, W. A. Wallace, W. Macnee, M. Bradley, I. L. Megson, and K. Donaldson. 2011. Progressive severe lung injury by zinc oxide nanoparticles; the role of Zn2+ dissolution inside lysosomes. *Part Fibre Toxicol* 8 (1):27.

Cho, W. S., R. Duffin, C. A. Poland, A. Duschl, G. J. Oostingh, W. Macnee, M. Bradley, I. L. Megson, and K. Donaldson. 2012. Differential pro-inflammatory effects of metal oxide nanoparticles and their soluble ions in vitro and in vivo; zinc and copper nanoparticles, but not their ions, recruit eosinophils to the lungs. *Nanotoxicology* 6 (1):22–35.

Deng, Z. J., M. Liang, M. Monteiro, I. Toth, and R. F. Minchin. 2011. Nanoparticle-induced unfolding of fibrinogen promotes Mac-1 receptor activation and inflammation. *Nat Nanotechnol* 6 (1):39–44.

Di Valentin, E., C. Crahay, N. Garbacki, B. Hennuy, M. Gueders, A. Noel, J. M. Foidart, J. Grooten, A. Colige, J. Piette, and D. Cataldo. 2009. New asthma biomarkers: Lessons from murine models of acute and chronic asthma. *Am J Physiol Lung Cell Mol Physiol* 296 (2):L185–L197.

Di Virgilio, A. L., M. Reigosa, P. M. Arnal, and M. Fernandez Lorenzo de Mele. 2010. Comparative study of the cytotoxic and genotoxic effects of titanium oxide and aluminium oxide nanoparticles in Chinese hamster ovary (CHO-K1) cells. *J Hazard Mater* 177 (1–3):711–718.

Donaldson, K., P. J. Borm, G. Oberdorster, K. E. Pinkerton, V. Stone, and C. L. Tran. 2008. Concordance between in vitro and in vivo dosimetry in the proinflammatory effects of low-toxicity, low-solubility particles: The key role of the proximal alveolar region. *Inhal Toxicol* 20 (1):53–62.

Downs, T. R., M. E. Crosby, T. Hu, S. Kumar, A. Sullivan, K. Sarlo, B. Reeder, M. Lynch, M. Wagner, T. Mills and S. Pfuhler. 2012. Silica nanoparticles administered at the maximum tolerated dose induce genotoxic effects through an inflammatory reaction while gold nanoparticles do not. *Mutat Res Genetic Toxicol Environ Mutagen* 745 (1–2):38–50.

Driscoll, K. E., L. C. Deyo, J. M. Carter, B. W. Howard, D. G. Hassenbein, and T. A. Bertram. 1997. Effects of particle exposure and particle-elicited inflammatory cells on mutation in rat alveolar epithelial cells. *Carcinogenesis* 18 (2):423–430.

Duffin, R., L. Tran, D. Brown, V. Stone, and K. Donaldson. 2007. Proinflammogenic effects of low-toxicity and metal nanoparticles in vivo and in vitro: Highlighting the role of particle surface area and surface reactivity. *Inhal Toxicol* 19 (10):849–856.

Fujita, K., M. Horie, H. Kato, S. Endoh, M. Suzuki, A. Nakamura, A. Miyauchi, K. Yamamoto, S. Kinugasa, K. Nishio, Y. Yoshida, H. Iwahashi, and J. Nakanishi. 2009. Effects of ultrafine TiO_2 particles on gene expression profile in human keratinocytes without illumination: Involvement of extracellular matrix and cell adhesion. *Toxicol Lett* 191 (2–3):109–117.

Gallagher, J., U. Heinrich, M. George, L. Hendee, D. H. Phillips, and J. Lewtas. 1994. Formation of DNA adducts in rat lung following chronic inhalation of diesel emissions, carbon black and titanium dioxide particles. *Carcinogen* 15 (7):1291–1299.

Gavett, S. H., X. Chen, F. Finkelman, and M. Wills-Karp. 1994. Depletion of murine CD4+ T lymphocytes prevents antigen-induced airway hyperreactivity and pulmonary eosinophilia. *Am J Respir Cell Mol Biol* 10 (6):587–593.

Gilbert, B., S. C. Fakra, T. Xia, S. Pokhrel, L. Madler, and A. E. Nel. 2012. The fate of ZnO nanoparticles administered to human bronchial epithelial cells. *ACS Nano* 6 (6):4921–4930.

Gojova, A., B. Guo, R. S. Kota, J. C. Rutledge, I. M. Kennedy, and A. I. Barakat. 2007. Induction of inflammation in vascular endothelial cells by metal oxide nanoparticles: Effect of particle composition. *Environ Health Perspect* 115 (3):403–409.

Gosens, I., J. A. Post, L. J. de la Fonteyne, E. H. Jansen, J. W. Geus, F. R. Cassee, and W. H. de Jong. 2010. Impact of agglomeration state of nano- and submicron sized gold particles on pulmonary inflammation. *Part Fibre Toxicol* 7 (1):3–7.

Gurr, J. R., A. S. Wang, C. H. Chen, and K. Y. Jan. 2005. Ultrafine titanium dioxide particles in the absence of photoactivation can induce oxidative damage to human bronchial epithelial cells. *Toxicology* 213 (1–2):66–73.

Haase, A., H. F. Arlinghaus, J. Tentschert, H. Jungnickel, P. Graf, A. Mantion, F. Draude, S. Galla, J. Plendl, M. E. Goetz, A. Masic, W. Meier, A. F. Thunemann, A. Taubert, and A. Luch. 2011. Application of laser postionization secondary neutral mass spectrometry/time-of-flight secondary ion mass spectrometry in nanotoxicology: Visualization of nanosilver in human macrophages and cellular responses. *ACS Nano* 5 (4):3059–3068.

Haase, A., S. Rott, A. Mantion, P. Graf, J. Plendl, A. F. Thunemann, W. P. Meier, A. Taubert, A. Luch, and G. Reiser. 2012. Effects of silver nanoparticles on primary mixed neural cell cultures: Uptake, oxidative stress and acute calcium responses. *Toxicol Sci* 126 (2):457–468.

Hackenberg, S., G. Friehs, M. Kessler, K. Froelich, C. Ginzkey, C. Koehler, A. Scherzed, M. Burghartz, and N. Kleinsasser. 2011a. Nanosized titanium dioxide particles do not induce DNA damage in human peripheral blood lymphocytes. *Environ Mol Mutagen* 52 (4):264–268.

Hackenberg, S., F. Z. Zimmermann, A. Scherzed, G. Friehs, K. Froelich, C. Ginzkey, C. Koehler, M. Burghartz, R. Hagen, and N. Kleinsasser. 2011b. Repetitive exposure to zinc oxide nanoparticles induces DNA damage in human nasal mucosa mini organ cultures. *Environ Mol Mutagen* 52 (7):582–589.

Huang, C. C., R. S. Aronstam, D. R. Chen, and Y. W. Huang. 2010. Oxidative stress, calcium homeostasis, and altered gene expression in human lung epithelial cells exposed to ZnO nanoparticles. *Toxicol* In Vitro 24 (1):45–55.

Humbles, A. A., C. M. Lloyd, S. J. McMillan, D. S. Friend, G. Xanthou, E. E. McKenna, S. Ghiran, N. P. Gerard, C. Yu, S. H. Orkin, and C. Gerard. 2004. A critical role for eosinophils in allergic airways remodeling. *Science* 305 (5691):1776–1779.

Iavicoli, I., V. Leso, L. Fontana, and A. Bergamaschi. 2011. Toxicological effects of titanium dioxide nanoparticles: A review of in vitro mammalian studies. *Eur Rev Med Pharmacol Sci* 15 (5):481–508.

Jaganathan, H. and B. Godin. 2012. Biocompatibility assessment of Si-based nano- and microparticles. *Adv Drug Deliv Rev* 64 (15):1800–1819.

Jeng, H. A. and J. Swanson. 2006. Toxicity of metal oxide nanoparticles in mammalian cells. *J Environ Sci Health A Tox Hazard Subst Environ Eng* 41 (12):2699–2711.

Johnston, H. J., G. R. Hutchison, F. M. Christensen, S. Peters, S. Hankin, and V. Stone. 2009. Identification of the mechanisms that drive the toxicity of $TiO_{(2)}$particulates: The contribution of physicochemical characteristics. *Part Fibre Toxicol* 6 (1):33.

Kao, Y. Y., Y. C. Chen, T. J. Cheng, Y. M. Chiung, and P. S. Liu. 2012. Zinc oxide nanoparticles interfere with zinc ion homeostasis to cause cytotoxicity. *Toxicol Sci* 125 (2):462–472.

Kermanizadeh, A., G. Pojana, B. K. Gaiser, R. Birkedal, D. Bilanicova, H. Wallin, K. A. Jensen, B. Sellergren, G. R. Hutchison, A. Marcomini, and V. Stone. 2013. In vitro assessment of engineered nanomaterials using a hepatocyte cell line: Cytotoxicity, proinflammatory cytokines and functional markers. *Nanotoxicology* 7 (3):301–313.

Kroll, A., C. Dierker, C. Rommel, D. Hahn, W. Wohlleben, C. Schulze-Isfort, C. Gobbert, M. Voetz, F. Hardinghaus, and J. Schnekenburger. 2011. Cytotoxicity screening of 23 engineered nanomaterials using a test matrix of ten cell lines and three different assays. *Part Fibre Toxicol* 8 (1):9.

Kroll, A., M. H. Pillukat, D. Hahn, and J. Schnekenburger. 2009. Current in vitro methods in nanoparticle risk assessment: Limitations and challenges. *Eur J Pharm Biopharm* 72 (2):370–377.

Kroll, A., M. H. Pillukat, D. Hahn, and J. Schnekenburger. 2012. Interference of engineered nanoparticles with in vitro toxicity assays. *Arch Toxicol* 86 (7):1123–1136.

Landsiedel, R., M. D. Kapp, M. Schulz, K. Wiench and F. Oesch. 2009. Genotoxicity investigations on nanomaterials: Methods, preparation and characterization of test material, potential artifacts and limitations-Many questions, some answers. *Mutat Res Rev Mutat Res* 681 (2–3):241–258.

Landsiedel, R., L. Ma-Hock, A. Kroll, D. Hahn, J. Schnekenburger, K. Wiench, and W. Wohlleben. 2010. Testing metal-oxide nanomaterials for human safety. *Adv Mater* 22 (24):2601–2627.

Lanone, S., F. Rogerieux, J. Geys, A. Dupont, E. Maillot-Marechal, J. Boczkowski, G. Lacroix, and P. Hoet. 2009. Comparative toxicity of 24 manufactured nanoparticles in human alveolar epithelial and macrophage cell lines. *Part Fibre Toxicol* 6 (14):1–12.

Limbach, L. K., P. Wick, P. Manser, R. N. Grass, A. Bruinink, and W. J. Stark. 2007. Exposure of engineered nanoparticles to human lung epithelial cells: Influence of chemical composition and catalytic activity on oxidative stress. *Environ Sci Technol* 41 (11):4158–4163.

Lin, W., Y. W. Huang, X. D. Zhou, and Y. Ma. 2006. Toxicity of cerium oxide nanoparticles in human lung cancer cells. *Int J Toxicol* 25 (6):451–457.

Lin, W. S., Y. Xu, C. C. Huang, Y. F. Ma, K. B. Shannon, D. R. Chen, and Y. W. Huang. 2009. Toxicity of nano- and micro-sized ZnO particles in human lung epithelial cells. *J Nanopart Res* 11 (1):25–39.

Lindberg, H. K., G. C. Falck, J. Catalan, A. J. Koivisto, S. Suhonen, H. Jarventaus, E. M. Rossi, H. Nykasenoja, Y. Peltonen, C. Moreno, H. Alenius, T. Tuomi, K. M. Savolainen, and H. Norppa. 2012. Genotoxicity of inhaled nanosized $TiO_{(2)}$ in mice. *Mutat Res* 745 (1–2):58–64.

Lundqvist, M., J. Stigler, G. Elia, I. Lynch, T. Cedervall and K. A. Dawson. 2008. Nanoparticle size and surface properties determine the protein corona with possible implications for biological impacts. *Proc Natl Acad Sci U S A* 105 (38):14265–14270.

Madl, A. K. and K. E. Pinkerton. 2009. Health effects of inhaled engineered and incidental nanoparticles. *Crit Rev Toxicol* 39 (8):629–658.

Magdolenova, Z., A. Collins, A. Kumar, A. Dhawan, V. Stone, and M. Dusinska. 2013. Mechanisms of genotoxicity. A review of in vitro and in vivo studies with engineered nanoparticles. *Nanotoxicology* 8 (3):233–278.

Ma-Hock, L., A. O. Gamer, R. Landsiedel, E. Leibold, T. Frechen, B. Sens, M. Linsenbuehler, and B. van Ravenzwaay. 2007. Generation and characterization of test atmospheres with nanomaterials. *Inhal Toxicol* 19 (10):833–848.

Maiorano, G., S. Sabella, B. Sorce, V. Brunetti, M. A. Malvindi, R. Cingolani, and P. P. Pompa. 2010. Effects of cell culture media on the dynamic formation of protein-nanoparticle complexes and influence on the cellular response. *ACS Nano* 4 (12):7481–7491.

Monteiro-Riviere, N. A., A. O. Inman and L. W. Zhang. 2009. Limitations and relative utility of screening assays to assess engineered nanoparticle toxicity in a human cell line. *Toxicol Appl Pharmacol* 234 (2):222–235.

Morfeld, P., S. Treumann, L. Ma-Hock, J. Bruch, and R. Landsiedel. 2012. Deposition behavior of inhaled nanostructured TiO_2 in rats: Fractions of particle diameter below 100 nm (nanoscale) and the slicing bias of transmission electron microscopy. *Inhal Toxicol* 24 (14):939–951.

Morrow, P. E. 1988. Possible mechanisms to explain dust overloading of the lungs. *Fundam Appl Toxicol* 10 (3):369–384.

Naya, M., N. Kobayashi, M. Ema, S. Kasamoto, M. Fukumuro, S. Takami, M. Nakajima, M. Hayashi, and J. Nakanishi. 2012. In vivo genotoxicity study of titanium dioxide nanoparticles using comet assay following intratracheal instillation in rats. *Regul Toxicol Pharmacol* 62 (1):1–6.

Nel, A. E., L. Madler, D. Velegol, T. Xia, E. M. Hoek, P. Somasundaran, F. Klaessig, V. Castranova, and M. Thompson. 2009. Understanding biophysicochemical interactions at the nano-bio interface. *Nat Mater* 8 (7):543–557.

Oberdorster, G., E. Oberdorster, and J. Oberdorster. 2005. Nanotoxicology: An emerging discipline evolving from studies of ultrafine particles. *Environ Health Perspect* 113 (7):823–839.

OECD TG471. Bacterial reverse mutation test, Adopted 21 July, 1997.

OECD TG487. In vitro mammalian cell micronucleus test, Adopted 22 July, 2010.

Oesch, F. and R. Landsiedel. 2012. Genotoxicity investigations on nanomaterials. *Arch Toxicol* 86 (7):985–994.

Pan, Z., W. Lee, L. Slutsky, R. A. Clark, N. Pernodet, and M. H. Rafailovich. 2009. Adverse effects of titanium dioxide nanoparticles on human dermal fibroblasts and how to protect cells. *Small* 5 (4):511–520.

Panas, A., C. Marquardt, O. Nalcaci, H. Bockhorm, W. Baumann, H. R. Paur, S. Mülhopt, S. Diabate, C. Weiss. 2013. Screening of different metal oxide nanoparticles reveals selective toxicity and inflammatory potential of silica nanoparticles in lung epithelial cells and macrophages. *Nanotoxicology* 7 (3):259–273

Park, E. J., J. Choi, Y. K. Park, and K. Park. 2008a. Oxidative stress induced by cerium oxide nanoparticles in cultured BEAS-2B cells. *Toxicology* 245 (1–2):90–100.

Park, E. J., J. Yi, K. H. Chung, D. Y. Ryu, J. Choi, and K. Park. 2008b. Oxidative stress and apoptosis induced by titanium dioxide nanoparticles in cultured BEAS-2B cells. *Toxicol Lett* 180 (3):222–229.

Pauluhn, J. 2009. Retrospective analysis of 4-week inhalation studies in rats with focus on fate and pulmonary toxicity of two nanosized aluminum oxyhydroxides (boehmite) and pigment-grade iron oxide (magnetite): The key metric of dose is particle mass and not particle surface area. *Toxicology* 259 (3):140–148.

Petkovic, J., T. Kuzma, K. Rade, S. Novak, and M. Filipic. 2011. Pre-irradiation of anatase TiO_2 particles with UV enhances their cytotoxic and genotoxic potential in human hepatoma HepG2 cells. *J Hazard Mater* 196:145–152.

Pfuhler, S., R. Elespuru, M. J. Aardema, S. H. Doak, E. Maria Donner, M. Honma, M. Kirsch-Volders, R. Landsiedel, M. Manjanatha, T. Singer, and J. H. Kim. 2013. Genotoxicity of nanomaterials: Refining strategies and tests for hazard identification. *Environ Mol Mutagen* 54 (4):229–239.

Rehn, B., S. Rehn, and J. Bruch. 1999. Ein neues In-vitro-Prüfkonzept (Vektormodell) zum biologischen Screening und Monitoring der Lungentoxizität von Stäuben. *Gefahrstoffe - Reinhaltung der Luft* 59:181–188.

Robinson, C. E., A. Keshavarzian, D. S. Pasco, T. O. Frommel, D. H. Winship, and E. W. Holmes. 1999. Determination of protein carbonyl groups by immunoblotting. *Anal Biochem* 266 (1):48–57.

Saber, A. T., N. R. Jacobsen, A. Mortensen, J. Szarek, P. Jackson, A. M. Madsen, K. A. Jensen, I. K. Koponen, G. Brunborg, K. B. Gutzkow, U. Vogel, and H. Wallin. 2012. Nanotitanium dioxide toxicity in mouse lung is reduced in sanding dust from paint. *Part Fibre Toxicol* 9 (4):8977–8979.

Sayes, C. M., K. L. Reed, and D. B. Warheit. 2007. Assessing toxicity of fine and nanoparticles: Comparing in vitro measurements to in vivo pulmonary toxicity profiles. *Toxicol Sci* 97 (1):163–180.

Sayes, C. M., R. Wahi, P. A. Kurian, Y. Liu, J. L. West, K. D. Ausman, D. B. Warheit, and V. L. Colvin. 2006. Correlating nanoscale titania structure with toxicity: A cytotoxicity and inflammatory response study with human dermal fibroblasts and human lung epithelial cells. *Toxicol Sci* 92 (1):174–185.

Schaefer, J., C. Schulze, E. E. Marxer, U. F. Schaefer, W. Wohlleben, U. Bakowsky, and C. M. Lehr. 2012. Atomic force microscopy and analytical ultracentrifugation for probing nanomaterial protein interactions. *ACS Nano* 6 (6):4603–4614.

Sharma, V., D. Anderson, and A. Dhawan. 2012. Zinc oxide nanoparticles induce oxidative DNA damage and ROS-triggered mitochondria mediated apoptosis in human liver cells (HepG2). *Apoptosis* 17 (8):852–870.

Shi, H., R. Magaye, V. Castranova, and J. Zhao. 2013. Titanium dioxide nanoparticles: A review of current toxicological data. *Part Fibre Toxicol* 10 (1):15.

Simon-Deckers, A., B. Gouget, M. Mayne-L'hermite, N. Herlin-Boime, C. Reynaud, and M. Carriere. 2008. In vitro investigation of oxide nanoparticle and carbon nanotube toxicity and intracellular accumulation in A549 human pneumocytes. *Toxicology* 253 (1–3):137–146.

Singh, S., T. Shi, R. Duffin, C. Albrecht, D. van Berlo, D. Hohr, B. Fubini, G. Martra, I. Fenoglio, P. J. Borm, and R. P. Schins. 2007. Endocytosis, oxidative stress and IL-8 expression in human lung epithelial cells upon treatment with fine and ultrafine TiO_2: Role of the specific surface area and of surface methylation of the particles. *Toxicol Appl Pharmacol* 222 (2):141–151.

Sun, Q., D. Tan, Y. Ze, X. Sang, X. Liu, S. Gui, Z. Cheng, J. Cheng, R. Hu, G. Gao, G. Liu, M. Zhu, X. Zhao, L. Sheng, L. Wang, M. Tang, and F. Hong. 2012. Pulmotoxicological effects caused by long-term titanium dioxide nanoparticles exposure in mice. *J Hazard Mater* 235–236:47–53.

Sun, W., A. Luna-Velasco, R. Sierra-Alvarez, and J. A. Field. 2013. Assessing protein oxidation by inorganic nanoparticles with enzyme-linked immunosorbent assay (ELISA). *Biotechnol Bioeng* 110 (3):694–701.

Sycheva, L. P., V. S. Zhurkov, V. V. Iurchenko, N. O. Daugel-Dauge, M. A. Kovalenko, E. K. Krivtsova, and A. D. Durnev. 2011. Investigation of genotoxic and cytotoxic effects of micro- and nanosized titanium dioxide in six organs of mice in vivo. *Mutat Res* 726 (1):8–14.

Trouiller, B., R. Reliene, A. Westbrook, P. Solaimani, and R. H. Schiestl. 2009. Titanium dioxide nanoparticles induce DNA damage and genetic instability in vivo in mice. *Cancer Res* 69 (22):8784–8789.

Wang, J. J., H. Wang, and B. J. S. Sanderson. 2007. Ultrafine quartz-induced damage in human lymphoblastoid cells in vitro using three genetic damage end-points. *Toxicol Mechanisms Meth* 17 (4):223–232.

Warheit, D. B. and E. M. Donner. 2010. Rationale of genotoxicity testing of nanomaterials: Regulatory requirements and appropriateness of available OECD test guidelines. *Nanotoxicology* 4 (4):409–413.

Warheit, D. B., T. R. Webb, K. L. Reed, S. Frerichs, and C. M. Sayes. 2007. Pulmonary toxicity study in rats with three forms of ultrafine-TiO_2 particles: Differential responses related to surface properties. *Toxicology* 230 (1):90–104.

Warheit, D. B., T. R. Webb, C. M. Sayes, V. L. Colvin, and K. L. Reed. 2006. Pulmonary instillation studies with nanoscale TiO_2 rods and dots in rats: Toxicity is not dependent upon particle size and surface area. *Toxicol Sci* 91 (1):227–236.

Wells, M. A., A. Abid, I. M. Kennedy, and A. I. Barakat. 2012. Serum proteins prevent aggregation of Fe2O3 and ZnO nanoparticles. *Nanotoxicology* 6 (8):837–846.

Wills-Karp, M. and M. Chiaramonte. 2003. Interleukin-13 in asthma. *Curr Opin Pulm Med* 9 (1):21–27.

Xia, T., M. Kovochich, M. Liong, L. Madler, B. Gilbert, H. Shi, J. I. Yeh, J. I. Zink, and A. E. Nel. 2008. Comparison of the mechanism of toxicity of zinc oxide and cerium oxide nanoparticles based on dissolution and oxidative stress properties. *ACS Nano* 2 (10):2121–2134.

Xia, T., Y. Zhao, T. Sager, S. George, S. Pokhrel, N. Li, D. Schoenfeld, H. Meng, S. Lin, X. Wang, M. Wang, Z. Ji, J. I. Zink, L. Madler, V. Castranova, and A. E. Nel. 2011. Decreased dissolution of ZnO by iron doping yields nanoparticles with reduced toxicity in the rodent lung and zebrafish embryos. *ACS Nano* 5 (2):1223–1235.

Xu, A., Y. Chai, T. Nohmi, and T. K. Hei. 2009. Genotoxic responses to titanium dioxide nanoparticles and fullerene in gpt delta transgenic MEF cells. *Part Fibre Toxicol* 6 (3).

Xu, Z., S. L. Wang, and H. W. Gao. 2010. Effects of nano-sized silicon dioxide on the structures and activities of three functional proteins. *J Hazard Mater* 180 (1–3):375–383.

Yang, H., C. Liu, D. Yang, H. Zhang, and Z. Xi. 2009. Comparative study of cytotoxicity, oxidative stress and genotoxicity induced by four typical nanomaterials: The role of particle size, shape and composition. *J Appl Toxicol* 29 (1):69–78.

9 Toxicological Effects of Metal Nanomaterials

Rainer Ossig, Daniela Hahn, Martin Wiemann, Marc D. Driessen, Andrea Haase, Andreas Luch, Antje Vennemann, Elke Dopp, Marlies Nern, and Jürgen Schnekenburger

CONTENTS

9.1 INTRODUCTION

Metallic nanoparticles are used in a variety of industrial applications and consumer products. Although metallic nanoparticles represent only a small group among the plethora of engineered nanomaterials, they display by far the greatest number of applied nanomaterials in consumer products which have been introduced into the market, primarily because of their antimicrobial, electrical, optical, and/or magnetic properties. This chapter will focus on engineered metal nanomaterials with widespread application in many industrial fields, in particular, nanosized silver, gold, copper, and platinum. Biological effects induced by these nanomaterials *in vitro* and *in vivo* will be discussed with special reference to their unique physicochemical properties. In addition, we will describe important findings from the German project nanoGEM, in which different variants of well-characterized silver nanoparticles were analyzed for their biological effects using multiple test systems.

Among the metals, the most frequently employed nanomaterial is nanosilver (see Woodrow Wilson database of nanotechnology-based products http://www.nanotechproject.org/cpi/products/). Silver acts as an effective antibacterial agent and constitutes an integral part of many consumer products, ranging from cosmetics, cleaning disinfectants to sport textiles. Moreover, nanosilver has found widespread use in medical products, such as adhesive tape, wound dressings as well as medical instruments. The antimicrobial behavior of nanosilver is derived from its interaction with bacterial proteins resulting in enzyme inactivation or protein denaturation. However, this effect does not seem to be restricted to microorganisms because silver nanoparticles can also bind to free thiol groups of cysteines and sulfhydryl groups of proteins in mammalian cells (Chen et al. 2008).

Gold nanomaterials can be easily synthesized in multifaceted shapes, with variant sizes and different surface modifications. Apart from colloidal gold, gold nanoshells, nanorods, and nanowires are the most popular. Engineered gold nanoparticles display excellent absorbance and scattering of light making them very useful for biomedical applications like medical imaging, biological diagnostics, and biosensorics. In recent years, gold nanoparticles have also been tested in *in vivo* models for drug therapy as well as in rheumatoid arthritis and hyperthermia treatment for tumor destruction (Khlebtsov and Dykman 2010).

In contrast to silver and gold nanoparticles, copper nanoparticles are predominantly present in industrial applications; for example, as additives for lubricant oil, metallic coatings and inks, and lithium ion batteries, as well as polymers and plastics. Nanosized copper particles were also listed as skin product components in the US Woodrow Wilson database of nanotechnology-based products (http://www.nanotechproject.org/cpi/products). This might raise concerns because copper has strong cytotoxic effects mediated at least in part by binding to sulfur-based polypeptides thereby leading to protein denaturation (Jeon, Jeong, and Jue 2000; Cecconi et al. 2002).

Due to their high conductance and reactivity which is strongly enhanced in nanosized materials, platinum nanoparticles are employed in many industrial products and processes, for example, as catalysts for fuel cells. In addition, platinum nanoparticles are also announced as supplements in some cosmetics and personal care products in the United States, advertising their antioxidative potential and protective effect against reactive oxygen species (http://www.nanotechproject.org/cpi/products).

The wide range of applications and the intended use of metal-based nanoparticles in personal care products pose the risk of human exposure in environmental, occupational, and consumer settings. Consequently, the possible hazards associated with exposure to these nanomaterials have to be carefully evaluated on the basis of toxicological investigations. With regard to metal nanoparticles, the largest number of toxicological studies have been conducted using silver and gold nanoparticles, whereas relatively few studies were devoted to copper or platinum nanoparticles. Recent reviews summarize the toxicity of nanosilver or nanogold (Johnston et al. 2010; Khlebtsov and Dykman 2011; Pan, Bartneck, and Jahnen-Dechent 2012; Reidy et al. 2013; Schäfer et al. 2013). As already pointed out in Chapter 8, the toxic potential of nanoparticles cannot always be extrapolated from the toxicity of the corresponding bulk material. The unique properties that drive the exploitation of metal nanoparticles in industry might also entail unpredictable effects of these materials on cells and organs. This emphasizes the need for a thorough understanding of the correlative relationship between the nanoparticles' physicochemical properties and their biological effects on the cellular and systemic levels. From a toxicological view, it is therefore decisive to address the question of how specific physicochemical properties affect the toxic potential of nanoparticles, for example, by modifying molecular interaction with biomolecules or by influencing cellular behavior and uptake of the nanomaterials.

Especially for gold nanoparticles many studies report toxic effects and the migration through biological barriers by particles smaller than 10 nm (Semmler-Behnke et al. 2008). Most of these materials were only intended for research use. The small size and the simplicity of labeling—for example, with proteins—enable studies on the uptake and biodistribution of gold particles. Here, we focus on metal nanomaterials manufactured for industrial use which range from 3 to 90 nm (see Woodrow Wilson database of nanotechnology-based products http://www.nanotechproject.org/cpi/products/). The biological effects of metal nanomaterials will be outlined in three main sections. In the first section, we will summarize studies on cultured cells. Subsequently, the impact of metal nanomaterials on alveolar macrophages (AM) is described. The third main section focuses on *in vivo* studies in nonmammalians, rodents, and humans with respect to the different testing strategies used.

9.2 EFFECTS OF METAL NANOMATERIALS ON CULTURED HUMAN CELLS

Inflammation, oxidative stress, cytotoxicity, and genotoxicity have been detected after exposure to silver nanoparticles in different cell lines and all these effects seem to be inherently linked to one another (Johnston et al. 2010). However, data are not complete and not consistent. Nanosilver exists in manifold sizes and shapes and is further functionalized with a multitude of surface coatings such as citrate, organic polymers, peptides, or sugars, all of which might influence the biological effects (Johnston et al. 2010).

9.2.1 Oxidative Stress/Cell Viability/Inflammation

Various publications report on cytotoxicity and oxidative stress upon uptake of silver nanoparticles. For instance, Arora et al. used spherical silver (7–20 nm) in primary fibroblast and liver cells, where both cytotoxicity and oxidative stress responses could be detected (Arora et al. 2009). A size- and coating-dependent effect for cytotoxicity and oxidative stress (ROS formation) was detected with different sized silver nanoparticles (10, 15, 25–30, and 80 nm) and two differently coated nanosilver particles in spermatogonal stem cells (Braydich-Stolle et al. 2010). Trickler et al. have included three sizes (28, 48, and 102 nm) and could detect a size-dependent cytotoxicity in primary rat brain microvessel endothelial cells as well as a size-dependent effect on inflammation (IL-1β and TNF-α). In CaCo-2 cells, the effects of 20, 34, 61, and 113 nm nanosilver were assessed and again a size-dependent effect on cytotoxicity was observed (Trickler et al. 2010). In addition, other authors analyzed transcriptomic responses using an *in vitro* model for the human intestinal epithelium. They found upregulation of genes involved in several stress responses (oxidative stress, endoplasmic stress, and apoptosis). These effects were at least partially triggered by the released silver ions, demonstrating the contribution of solubility to nanoparticle toxicity (Bouwmeester et al. 2011). A correlation of ROS induction and apoptosis was found by Foldbjerg et al. using two sizes of nanosilver (60–70 and 149 nm) and A549 cells (Foldbjerg, Dang, and Autrup 2011). Haase et al. assessed four different silver nanoparticle variants, a 40 nm sized silver nanoparticle as well as three variants of 20 nm sized nanosilver (uncoated, citrate-coated, and peptide-coated) in THP-1-derived macrophages. The 20 nm nanosilver displayed stronger cytotoxic effects and induced a stronger oxidative stress response compared to the 40 nm sized silver nanoparticles. Similar to the results of Braydich-Stolle et al., the observed effects depended on the coating, in which the influence of citrate coating was less pronounced (Braydich-Stolle et al. 2010; Haase et al. 2011). In summary, nanosilver has the potential to induce cytotoxic effects *in vitro*, which seem to be ion and size dependent and which often involve oxidative stress and inflammation. However, none of these studies attributed effects to the intracellular concentration of particles or dissolved silver species which would be an important issue in the understanding of these effects.

The increasing use of gold nanomaterials in biomedical applications has been paralleled by an increase in research activity in the biodistribution and toxicity of gold nanoparticles. Albeit most of the *in vivo* studies conducted so far has focused on the biodistribution of gold nanomaterials, a few cytotoxicity studies have comprehensively assessed the impact of size (Pan et al. 2007), surface modification (Uboldi et al. 2009; Connor et al. 2005; Sun, Liu, and Wang 2008), and aggregation (Albanese and Chan 2011).

Different groups have adjusted the perception that gold nanoparticles can be considered as inert like their bulk counterparts. Li et al. demonstrated that gold nanoparticles of 20 nm diameter induce oxidative stress and inhibit cell proliferation in lung fibroblasts (MRC-5 cells; Li et al. 2008). Likewise, Pan et al. have shown size- and concentration-dependent cytotoxicity of gold nanoparticles (ranging from 0.8 to 15 nm). Gold clusters of 1.4 nm were found to induce cell death in four different cell lines (HeLa cells, J774A1 mouse macrophages, L929 mouse fibroblasts, and SK-Mel-28 melanoma cells), with IC_{50} values ranging from 30 μM to 56 μM, whereas IC_{50} values of 0.8, 1.2, and 1.8 nm gold clusters were considerably higher (250, 140, and 230 μM, respectively) and 15 nm sized colloidal gold particles (Au15MS) were found to be completely nontoxic up to 6,300 μM (Pan et al. 2007). By studying the major cell death pathways, Pan et al. found indicators of oxidative stress and proposed that the toxicity of the 1.4 nm gold nanoparticles depended on their ability to trigger the intracellular formation of ROS from dioxygen (Pan et al. 2009). In another study, different aggregates of transferrin-coated gold nanoparticles of 16 nm size were generated and found to display different uptake patterns in HeLa cells, A549 lung epithelial cells, and melanoma cells (MDA-MB-435; Albanese and Chan 2011). These results demonstrate that cell type and mechanism of interactions may contribute to nanoparticle uptake and toxicity. Connor et al. studied the metabolic activity of a human leukemic cell line (K562) after three days of continuous exposure to gold nanoparticles of average diameters of 4, 12, or 18 nm with different surface modifications (Connor et al. 2005). They found citrate- and biotin-modified particles at concentrations of up to 250 μg/ml as well as preparations with glucose or cysteine surface modifiers, or with a reduced gold surface (up to 25 μM) to be nontoxic. Gold nanomaterials are often synthesized using seed-mediated methods comprising the cationic surfactant cetyltrimethylammonium bromide (CTAB) in growth solution, leaving an original coating of assembled CTAB molecules on the particle surface (Takahashi et al. 2006; Hauck, Ghazani, and Chan 2008; Wang et al. 2013a). Several studies reported that the modified surface of gold nanomaterials by capping with the surfactant CTAB has a negative impact on cellular functions. However, after a washing procedure, gold nanoparticles with surface-bound CTAB did not cause a change in cellular metabolic activity, whereas the surfactant CTAB in solution was found to be highly cytotoxic. The removal of excess soluble CTAB by washing completely abolished the cytotoxicity that was demonstrated for the initial preparation of CTAB-containing gold nanoparticles (Connor et al. 2005). Similar results of further studies illustrated that human colon carcinoma cells show significantly reduced viability when exposed to 0.4 nM of CTAB-capped gold nanorods (Alkilany et al. 2009). The authors concluded that these effects were mainly caused by excess CTAB that desorbs from the surface of the capped rods or by residual free contaminations, rather than by the CTAB-capped nanorods themselves. Overcoating of the CTAB-capped gold nanorods with two different polyelectrolytes, either with the net negatively charged polyacrylic acid (PAA), thereby converting the surface charge, or a further layering with the positively charged poly(allylamine) hydrochloride, was found to be relatively nontoxic, implicating that the initially observed cytotoxicity is not correlated to the particle surface charge. Moreover, supernatants of the CTAB-capped gold nanorod preparations were shown to induce nearly the same cytotoxicity as the original particle-containing dispersions (Alkilany et al. 2009).

A recently published study reported that CTAB-capped 30 nm gold nanorods exposed in serum-free culture-medium generate defects in the cell membranes and induce cell death (Wang et al. 2013). The authors attributed these effects mainly to the bilayer structure of CTAB on the surface of the rods rather than to their surface charge. Other positively charged surface modifications using polyelectrolytes exhibited minor adverse effects to the cells (Wang et al. 2013b). An extraction of the CTAB surface capping and replacement with phospholipids like phosphatidylcholine was able to reduce the cytotoxicity for HeLa cells (Takahashi et al. 2006). Coating with polyethylene glycol (PEG) has also been described as a useful method to confer protective effects against cytotoxicity to gold nanorods. PEGylated gold nanorods have been found to reveal superior biocompatibility, compared to polystyrene sulfonate-coated nanorods, which facilitated cellular uptake and substantially reduced cell viability, demonstrated in four cell lines of different origin (Rayavarapu et al. 2010). These and other data again demonstrate that surface modifications, coatings, and surface charge, indeed play substantial roles for cytotoxic properties of gold nanomaterials, as well as for their cellular uptake (Hauck, Ghazani, and Chan 2008). Furthermore, in a physiological environment the nanomaterials will change their physiochemical properties by surface interaction and coatings with biomolecules. For example, the binding of serum proteins has been shown to alter the surface charge of gold nanorods, and a serum albumin protein corona on CTAB-coated gold nanorods was found to confer improved biocompatibility to the nanomaterials by reducing their potentially destructive effects for the cell membrane. A serum protein corona may also facilitate the receptor-mediated endocytosis and the subsequent intracellular transport of gold nanorods to endosomes and lysosomes (Alkilany et al. 2009; Wang et al. 2013a).

Only a few studies have been published assessing the toxic potential of gold nanoshells. Most of these studies reported a noncytotoxic behavior of these nanomaterials in different cell lines, such as human hepatocellular carcinoma cells (Liu et al. 2010) or human prostate cancer cells (Stern et al. 2007). Likewise, PEG-coated silica-gold nanoshells did not display cytotoxic effects in human breast adenocarcinoma cells (SK-BR-3) (Hirsch et al. 2003) nor did immunoglobulin-coated nanoshells (Loo et al. 2005).

Taken together, the cytotoxicity studies of gold nanomaterials clearly demonstrate that in contrast to the inertness of gold in bulk form, nanosized gold particles can induce cytotoxic effects dependent on their size and surface coatings. This emphasizes the need to investigate the toxic potential of engineered gold nanoparticles on a case-by-case basis.

In an interdisciplinary research effort, Midander et al. investigated cytotoxicity of micro- and nanosized copper particles in relation to the particle properties (Midander et al. 2009). The nanosized copper particles (100 nm, 80 µg/mL, 18 h) were found to induce a significantly higher percentage of cell death in lung epithelial cells (A549) in comparison to their microsized counterparts. Moreover, the nanoparticles released greater amounts of copper per quantity of particles compared to the microsized copper particles. The effects, however, could be correlated only to a small extent to the fraction of released copper, indicating that the cytotoxic effects were caused mainly by the particles themselves (Midander et al. 2009). A correlation between size and toxicity for copper nanoparticles was also found by Prabhu et al. using 40, 60, and 80 nm copper particles at concentrations of 10–100 µM to expose dorsal root ganglion (DRG) cells. Although all copper nanoparticles displayed cytotoxicity, the effects of 40 nm sized copper particles on cell death (LDH assay) and cellular metabolic activity (MTS-assay) were much more pronounced (Prabhu et al. 2010).

Regarding toxicological studies of platinum nanoparticles, current literature data are limited. Nanoscaled platinum particles of different shapes in the size range of 11–35 nm and their effects on A549 cells and human umbilical vein endothelial cells (HUVEC) have been investigated by Elder et al. (2007). Colloidal platinum nanoflowers, multipods, and spheres were taken up by the cells but did not induce significant cytotoxic effects, such as oxidative stress (DCF assay), cell death (LDH assay), or inflammatory markers (Il-6-ELISA) at concentrations up to 250 µg/ml. Similar results were found for 5–10 nm platinum nanoparticles of 100% purity when these were exposed to A549 or HaCaT cells in a more recent study (Horie et al. 2011). Cell viability (MTT assay), cell proliferation (clonogenic assay), apoptosis induction (caspase-3 activity), intracellular ROS level (DCFH assay), and lipid peroxidation level (DPPP assay) were not influenced by these pristine platinum nanoparticles although the nanoparticles were internalized by the cells (Horie et al. 2011). Likewise, nanoscaled platinum particles of different sizes (12 and 66 nm) did not induce oxidative stress in human colon carcinoma cells (HT29; Pelka et al. 2009). Albeit caution is advised to draw general conclusions from the still incomplete data, the studies presented here suggest that the biological effects of platinum nanoparticles are generally low compared to other nano-sized metals such as silver and copper.

In the nanoGEM project, a number of different silver nanoparticles were investigated regarding their cytotoxic potential. Ag50.EO and Ag50.Citrate nanoparticles were selected to address the influence of particle surface modifications whereas Ag50.PVP and Ag200.PVP nanoparticles should illustrate the role of particle size (see Table 1.2). As the sensitivity toward nanomaterial exposure not only is cell type-specific but also depends on the particle type as well as on the addressed end point and method, to determine cytotoxicity we applied a multiplex *in vitro* testing strategy. This approach is based on a test matrix that includes both a series of suitable cell lines and a set of standardized cytotoxicity assays to obtain a more comprehensive view of the biological impact of nanomaterials. The well characterized standard cell lines representing various routes of particle exposure (see Table 8.1) and three *in vitro* toxicity assays were selected and adapted specifically for the application of nanomaterials as already described in more detail earlier (see Chapter 8).

The LDH assay was applied to detect cellular necrosis and displayed a cytotoxic effect of Ag50.EO nanoparticles for the lung epithelial cell line RLE-6TN at a concentration of 32 μg/mL. These cells were also affected during the exposure of surface-modified nanomaterial Ag50.citrate; however, the detected necrosis was less pronounced compared to the effect of Ag50.EO. The silver nanoparticles Ag50.PVP and Ag200.PVP exhibit a size distribution around 50 nm and 200 nm, respectively, and were not found to induce LDH release by cell death when applied in concentrations up to 32 μg/mL. Intracellular reactive oxygen species (ROS) were not significantly enhanced during the exposure of any tested silver nanomaterials, as determined by the DCF *in vitro* assays.

When using the tetrazolium salt WST-8 to determine the metabolic activity of proliferating tissue culture cells, clear adverse effects on cellular functions could be demonstrated for the application of each of the silver nanoparticles. However, not all of the used cell lines were sensitive to these metal nanomaterials. Alveolar epithelium-like RLE-6TN, and kidney epithelium-like NRK-52E, as well as mouse embryo fibroblasts NIH-3T3, cells were more severely affected by Ag50.EO at 0.32 μg/mL, compared to Ag50.Citrate at concentrations of 3.2 μg/mL. The test cell lines HaCaT (human skin keratinocytes) and MDCK-NBL2 (kidney epithelium-like cells) were affected by Ag50.EO at 32 μg/mL, but no measurable adverse effects could be determined for Ag50.citrate with these cells. In a similar manner, the silver ENMs Ag50.PVP and Ag200.PVP displayed differing cytotoxic potential with these cell lines. At the concentration of 3.2 μg/mL or 32 μg/mL nanosilver Ag50.PVP was shown to impair proliferation of NRK-52E, MKDCK-NBL2, and NIH-3T3 or RLE-6TN and HaCaT cell lines, respectively. Instead, 160 μg/mL of Ag200.PVP was needed to affect the RLE-52E cell line, and other cells lines were less severely affected by Ag200.PVP compared to Ag50.PVP at the same concentration.

In parallel, we could establish an indirect assessment of oxidative stress via the detection of protein modification (i.e., protein carbonylation), which proved to be very suitable to analyze the oxidative stress potential of nanomaterials (see Chapter 8). Here, we applied this assay to assess the oxidative stress potential of nanoGEM nanomaterials in NRK-52E cells and figured out that 8 out of 16 nano-GEM nanoparticles did induce protein carbonyls. The strongest effects within the silver group were detected for Ag50.PVP. Basically, both 50 nm variants of silver nanoparticles were positive (Ag50.PVP and Ag50.Citrate) but Ag50.Citrate showed much less pronounced effects compared to Ag50.PVP. Ag200.PVP was not positive and did not induce protein carbonyls.

In summary, a more articulated cytotoxic effect was observed for Ag50.EO compared to the surface modified Ag50.Citrate, and severe adverse effects of Ag50.PVP, which were less pronounced or not measurable during the exposure of the larger Ag200.PVP.

To conclude, the biological activity of the nanoGEM Ag materials was size and surface dependent.

9.2.2 Genotoxicity

Genotoxicity of silver nanoparticles has been analyzed *in vitro* (Kim et al. 2011a; AshaRani et al. 2009; Ahamed et al. 2008) and *in vivo* (Kim et al. 2008; Kim et al. 2011b; Tiwari, Jin, and Behari 2011). Kim and coworkers analyzed cyto- and genotoxicity of silver nanoparticles in L5178Y and BEAS-2B cells using comet assay and a gene mutation assay (thymidine kinase) (Kim et al. 2011) and found a positive result in comet but a negative result in the gene mutation assay. AshaRani and coworkers used silver nanoparticles (6–20 nm, starch coated) and assessed the effects in lung fibroblasts (IMR-90) and human glioblastoma cells (U251) for up to 72 h and found a dose-dependent genotoxicity for nanosilver in both the comet and the micronucleus assay (AshaRani et al. 2009). In another study, 25 nm polysaccharide capped and uncapped silver were analyzed and an increased protein phosphorylation of p53, Rad51, and H2AX, all of which are related to DNA damage or repair of DNA double strand breaks, was observed, with stronger effects being caused by the uncapped silver (Ahamed et al. 2008). These results indicate a positive genotoxic effect *in vitro*. In two *in vivo* rat studies genotoxicity of nanosilver was assessed after 28 days of oral administration (Kim et al. 2008) or 90 days of inhalation (Kim et al. 2011) and no genotoxic effect could be detected with *in vivo* micronucleus assay in the bone marrow according to the OECD guideline 474. Other tissues (e.g., lungs) have not been assessed. Another *in vivo* study was performed by Tiwari et al. with repeated intravenous injection in rats (Tiwari, Jin, and Behari 2011). Here, genotoxic effects of silver nanoparticles (15–40 nm) via an alkaline comet assay were detectable. Thus, for nanosilver the genotoxicity *in vitro* and *in vivo* seems to be dependent on many aspects such as specific particle characteristics (size, coating), the assays that have been used, or the addressed tissues or target cells which were analyzed.

In contrast to nanosilver, most of the gold nanomaterials tested so far do not seem to induce oxidative stress in human cells and therefore only a few reports on the potential genotoxicity of these particles can be found in the literature. Li et al. demonstrated that 20 nm colloidal gold nanoparticles could induce oxidative DNA damage in human fetal lung fibroblasts (MRC-5). When the fibroblasts were exposed to 1 nM gold nanoparticles, a significant increase in 8 hydroxydeoxyguanosine (8OHdG), a marker of oxidative stress, was found. In addition, they detected a down-regulation of DNA repair genes suggesting that gold nanoparticles at a concentration of 1 nM may interact directly or indirectly with regulators of genomic integrity (Li et al. 2008). A comparative proteomic analysis by two-dimensional gel electrophoresis further demonstrated the upregulation of a range of oxidative stress-related proteins. The results obtained from the comet assay confirmed DNA damage in MR-5 fibroblasts after treatment with 1 nM gold nanoparticles (Li et al. 2011).

In a study characterizing size-dependency of copper nanoparticles, Midander et al. reported that nanosized copper particles (100 μm, 80 μg/mL) caused a higher degree of DNA damage (comet assay) in human lung cells (A549) than their micro-sized counterparts (Midander et al. 2009). The effect seemed to be predominately mediated by particles because no DNA damage was induced when using the fraction containing only released copper. These findings suggest that the genotoxicity of copper nanoparticles is indeed particle related rather than being caused by released metal ions, and also that genotoxicity of the nanomaterials is dependent on the particle size.

As mentioned earlier, platinum nanoparticles were found to be of low or no cytotoxicity. However, in some reports, genotoxic behavior of these metal nanoparticles has been described. For instance, platinum nanoparticles of 5–8 nm in size were found to induce p53 activation and DNA damage in human cells (Asharani et al. 2010). Comparing platinum nanoparticles of different sizes, Pelka et al. could correlate an impairment of DNA integrity with particle size in an inverse manner. The smaller particles (<20 nm and <100 nm) displayed stronger DNA damage than their micro-sized counterparts (<100 nm). Because ROS formation could not be observed, a contribution of oxidative stress to the DNA-damaging properties of the nanoscaled platinum particles could be virtually excluded (Pelka et al. 2009). This demonstrates the need for investigating different toxicological end points to obtain reliable information on nanoparticle toxicity.

In the nanoGEM project, we analyzed Ag50.PVP and Ag200.PVP for their ability to induce chromosome breakage, aneuploidy, or point mutagenicity in the absence or presence of an extrinsic metabolizing system using the micronucleus assay and the Ames test, respectively. This was the first study in which the micronucleus assay on V79 cells and the Ames test were performed according to OECD guidelines (AMES: OECD 471, MN: OECD 487) considering the adaptation of the testing guidelines to manufactured nanomaterials in particular elements. In addition, the comet assay was performed using 3D bronchial models (EpiAirway™, MatTek Corp., Ashland, USA).

All Ames tests performed with silver nanoparticles have been negative so far. The results obtained in the nanoGEM project with the micronucleus test revealed increased genotoxic effects in V79 cells when silver nanoparticles were used at concentrations >11 μg/mL (Ag50.PVP) and >27.5 μg/mL (Ag200.PVP), respectively (Table 9.1). Testing of supernatants of suspended Ag200.PVP and Ag50.PVP nanomaterials after high-speed centrifugation also showed increased genotoxic effects in V79 cells. This suggests that free PVP or silver ions, released from the nanomaterial, contribute to the increased toxicity of silver nanoparticles.

TABLE 9.1
Genotoxicity of Silver Nanoparticles Tested within the NanoGEM Project

Particles	Test	Cytotoxic/Toxic to Bacteria	Result
Ag200.PVP	Micronucleus assay *in vitro*	>55 μg/mL	Genotoxic > 27.5 μg/mL
	Ames test	>110 μg per plate	Not mutagenic
	Comet assay	Not detected up to 50 μg/cm^2	Not genotoxic
Ag50.PVP	Micronucleus assay *in vitro*	>11 μg/ml	Genotoxic >11 μg/mL
	Ames test	>160 μg per plate	Not mutagenic
	Comet assay	Not detected up to 50 μg/cm^2	Not genotoxic

9.3 EFFECTS OF METAL NANOMATERIALS ON ALVEOLAR MACROPHAGES

The model of primary alveolar macrophages (AM) was introduced in Chapter 8. In the nanoGEM project, 50 nm Ag nanoparticles coated with either a polyether (Ag50.EO, dispersed in Disperbyk 190), citrate (Ag50.citrate), or polyvinylpyrrolidone (Ag50.PVP, dispersed in Luvitek K90) were assessed for toxic effects using AM. The latter variant was also used as a 200 nm silver particle (Ag200.PVP). Under serum-free cell culture conditions, Ag nanoparticles tended to incompletely agglomerate such that dark silver grains were visible by light microscopy outside and inside AM, a phenomenon also found to occur *in vivo* secondary to intratracheal instillation of silver nanoparticle suspension into the lungs (Figure 9.1).

A concentration range of up to 45 μg/mL was tested. However, no "mean mass per cell" values could be calculated because a considerable fraction of nanoparticles remained dispersed in the supernatant. Based on the release of lactate dehydrogenase

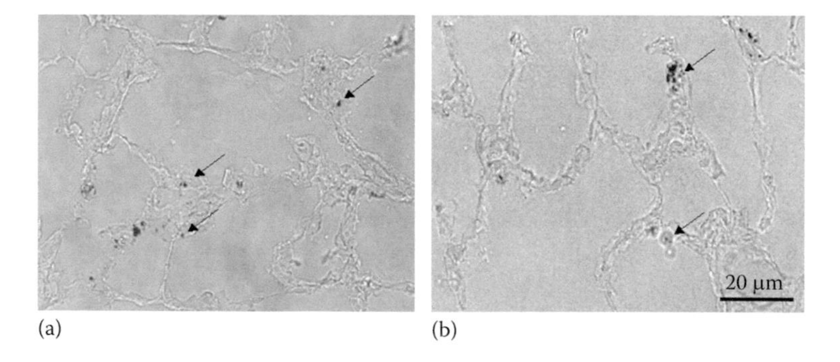

FIGURE 9.1 Bright field microscopy of unstained cryo-sections of snap-frozen rat lungs intratracheally instilled with silver nanoparticles. The lungs were removed 30 min (a) and 3 days (b) post instillation of 0.6 mg Ag50 per lung. There is a progressive concentration of agglomerated silver grains (arrows) in macrophage-like structures after 3 days.

TABLE 9.2
Comparison of *In Vitro* and *In Vivo* (Instillation) Effects of Various Types of Silver Nanoparticles, Compared to Benchmark Metal Oxide Nanoparticles (See Also Chapter 8)

		Ag50. EO	Ag5. PVP	Ag200. PVP	Ag50. citrate	ZrO_2 acryl	ZnO NM110	TiO_2 NM105
In vivo instillation	mg per rat lung	0.6	0.6	0.6	0.6	0.6		
BALF	AM (3 days)		++	+/–	+++	+		
	AM (21 days)		+++	++	++	+/–		
	PMN (3 days)		++++	++++	++++	+		
	PMN (21 days)		++++	++	++++	+/–		
	NP in AM	y	y	y	y	n.o.		
	Free agglomerates	n	n	n	y	n		
Histology	Local necrosis	y	y	y	y	n		
	Particles in AM	y	y	y	y	n.o.		
	NP in tissue AM	y	y	y	y	y		
In vitro macrophages	pg/cell	1.9–45	1.9–45	1.9–45	1.9–45	15–120		
	LDH	++++	+++	++	+	+		
	LOAEL (pg/cell)	1.9–3.8	7.5–15	>30	15–30	30	3.8–7.5	30–60
	LOAEL (µg/mL)	2.8–5.6	11.3–22.5	>45	22.5–45	45	5.7–11.3	45–90

Abbreviations: AM, alveolar macrophages; PMN, granulocytes; LDH, lactate dehydrogenase; LOAEL, low observed effect level; +/–, +, ++, +++, ++++, no, small, moderate, large, or excessive effect; n, no; y, yes; n.o., likely but not directly observable due to weak contrast.

(LDH) and glucuronidase the following ranking of cytotoxicity was obtained: Ag50.EO > Ag50.PVP > Ag50.citrate > Ag200.PVP. In general, all effects could be attributed to the presence of particles because particle-free supernatants evoked no apparent cytotoxicity. Significant cytotoxic effects of silver NPs started in the low micro-molar range, but differed with respect to surface coating (Table 9.2). Interestingly, Ag50.citrate at nontoxic concentrations induced more TNFα than all other tested Ag NPs, suggesting a "particle-to-cell-surface" mediated effect (M. Wiemann and A. Venneman, unpublished observation).

9.4 *IN VIVO* TOXICITY OF METAL NANOMATERIALS

The *in vitro* studies can act as a prescreening for nanoparticles with biological effects but they cannot simulate the complexity of a human organ. A reliable hazard assessment therefore requires *in vivo* studies. Most of the animal studies published so far have focused on biodistribution of metal nanoparticles, whereas current data regarding their toxicity *in vivo* are limited. In this section we will summarize reports obtained from nonmammalian (zebrafish), rodent, and human exposure.

9.4.1 *In Vivo* Studies in Non-Mammalian Animal Models

Comparative toxicity studies of silver, gold, and platinum nanoparticles in zebrafish models have found silver nanoparticles to exert the strongest adverse effects. Exposure of zebrafish embryos to nanosilver (10–100 μg/mL, 5–35 nm in size) resulted in a concentration-dependent mortality rate and hatching delay, whereas platinum nanoparticles (10–100 μg/mL, 3–10 nm) only induced hatching delays and gold nanoparticles (10–100 μg/mL, 5–35 nm) did not show any toxic behavior (Asharani et al. 2011). Similarly, other groups reported that gold nanoparticles of different sizes (11, 3, 5, 50, and 100 nm) did not show any adverse effects on embryo development in zebrafish (Browning et al. 2009; Bar-Ilan et al. 2009), albeit all sizes of silver nanoparticles (3, 5, 50, and 100 nm) tested generated a variety of embryonic morphological malformations and mortality (Bar-Ilan et al. 2009). These results demonstrate the stronger hazard potential of nanosilver in comparison to other metal nanoparticles, which is in good agreement with the toxicological findings obtained by *in vitro* studies.

Exposure of zebrafish to copper nanoparticles (80 nm) revealed gill injury and acute lethality with a 48 h LC_{50} concentration of 1.5 mg/L (Griffitt et al. 2007). Although the mechanisms of toxicity are still unclear, the authors demonstrated that the effects were not solely due to dissolution, because copper nanoparticles produced different morphological effects and global gene expression patterns than soluble copper (Griffitt et al. 2007). Similar findings were recently obtained for silver nanoparticles and other aquatic model organisms—for example, *Daphnia magna* (Georgantzopoulou et al. 2013). Georgantzopoulou et al. found silver nanoparticles of different sizes (20, 23, 27, and 200 nm) to exert toxic effects on aquatic organisms, with 23 nm silver particles being the most potent. These effects could not be explained solely by soluble silver suggesting that the particulate material itself contributes to toxicity (Georgantzopoulou et al. 2013). Such effects may, however, also be indirect as they could be linked to a size- or surface-dependent biodistribution.

9.4.1.1 *In Vivo* Studies in Mammals

9.4.1.1.1 Oral Toxicity

For studies on oral toxicity, nanosilver particles of four different sizes (22, 42, 71, and 323 nm) were used to feed mice at 1 mg/kg body weight/day for 14 days with no significant toxic effects. However, when mice were fed for 28 days with the 42 nm silver particles, an increase in inflammatory cytokines, signs of hepatotoxicity (i.e., increased alkaline phosphatase [ALP], aspartate aminotransferase [AST], alanine aminotransferase [ALT]), and histopathological changes in the kidney were reported with a NOAEL of 0.50 mg/kg body weight per day (Park et al. 2010). Likewise, mice exposed to 60 nm silver particles (with 300 or 1,000 mg/kg body weight) in a 28-day repeat oral gavage study showed significantly enhanced values of ALP and cholesterol in the blood, suggesting liver damage (Kim et al. 2008). This may indicate the impact of exposure time on the toxicological outcome.

A study by Zhang et al. investigated the toxic effects of gold nanoparticles of 13.5 nm diameter in mice following oral administration (Zhang et al. 2010). Mice were fed over 14 to 28 days with different nanoparticle concentrations ranging from 137.5 to 2,200 μg/kg body weight, and animal survival rates, animal mass, and effects on hematology and morphology were investigated. At low concentrations, the gold nanoparticles did not display toxic behavior, whereas at higher concentrations (1,100 μg/kg) a significant decrease in body weight, thymus and spleen index, and red blood cells was observed. The authors also compared other administration routes (intraperitoneal and tail vein injection) and found orally and intraperitoneally administered gold nanoparticles to yield the highest toxicity (Zhang et al. 2010).

Chen et al. reported an effect of size on acute toxicity of copper particles in mice (Chen et al. 2006). Oral exposure of mice to 23.5 nm copper nanoparticles led to injuries in the kidney, liver, and spleen with LD_{50} values of 413 mg/kg body weight (i.e., moderately toxic), whereas microsized copper particles (17 μm) were virtually nontoxic with LD_{50} values of >5,000 mg/kg body weight (Chen et al. 2006).

9.4.1.1.2 Pulmonary Toxicity

The *in vivo* studies addressing the inhalation route of exposure generally investigate pulmonary toxicity after instillation of nanoparticles into the animal lungs or after animal exposure via inhalation. Silver nanoparticles have been shown to induce pulmonary inflammation in different inhalation studies. Sung et al., for instance, conducted a 90-day inhalation exposure of rats to 20 nm silver nanoparticles (6 h/day) and found inflammatory responses and alterations in the lung function at low, medium, and high particle doses (48.94, 133.19, or 514.78 μg/m^3; Sun, Liu, and Wang 2008; Sung et al. 2009). They determined a NOAEC of 100 μg/m^3. Similar signs of toxicity and a similar NOEAC (117 μg/m^3) were recently obtained in a 12-week inhalation study conducted in rats with 14–15 nm silver nanoparticles (Song et al. 2013). In contrast, in a subacute murine inhalation model, a 10-day exposure (4 h/day) to small silver nanoparticles (5 ± 2 nm primary size) at a dose of 3.3 mg/m^3 demonstrated only minimal pulmonary toxicity (Stebounova et al. 2011), and no significant toxic effects were reported in rats exposed to nanosilver (15 nm) at concentrations ranging from 0.48 to 61.24 μg/m^3 for 14 days (6 h/day) (Ji et al. 2007). As a whole these results indicate that—depending on the duration of exposure—the accumulated lung burden, the biodistribution of silver nanoparticles, and, most likely, the subsequent release of ions from the secondary sites have an influence on the response to inhaled nanosilver.

For other metal nanoparticles, information on pulmonary toxicity is very limited. Intratracheal instillation of platinum nanoparticles in mice has been reported to induce different inflammatory markers (Park et al. 2010). Intratracheal instillation of a single dose (1 mg/kg body weight) of nanosized platinum (21 nm) led to a significant increase in inflammatory cytokines (IL-1, TNFα, IL-4, IL-5, IL-6, and TGF-β) in the BAL fluid for up to 28 days after instillation. Moreover, histopathological investigations revealed infiltration of macrophages and neutrophils in the lungs. Based on these findings, the authors suggest that platinum nanoparticles may induce inflammatory responses in mice (Park et al. 2010). In a comparative study of iron and copper nanoparticles, Pettibone et al. used 25 nm copper nanoparticles for both subacute and acute inhalation studies. For acute exposure, the animals were exposed to copper for 4 h/day and sacrificed within an hour after exposure. For the subacute exposure, mice were exposed for 4 h/day for 2 weeks (5 days/week) and necropsied within 1 h or 3 weeks post exposure. No significant signs of toxicity were found following acute exposure. However, immediately after the subacute exposure to copper or iron nanoparticles, an increase in inflammatory cytokines was observed, with copper being significantly more effective than iron nanoparticles. The authors discussed a contribution of copper ions to the observed effects, because copper was found to partially dissolve in the solutions used, whereas dissolved iron was not detected (Pettibone et al. 2008).

In the nanoGEM project, silver nanoparticles of different size and coatings (see the aforementioned) were exposed to rats via intratracheal instillation. Among these, Ag50.PVP, Ag200.PVP, and Ag50.citrate (0.6 mg/lung) elicited very strong signs of inflammation: Differential cell counts of the broncho-alveolar lavage fluid (BALF) revealed a significant increase in macrophage and granulocyte (PMN) counts after 3 days. After 21 days the effects on PMN were even more pronounced for Ag50.PVP and Ag50.citrate (Table 9.2) but partially reversed for Ag200.PVP. The BALF from Ag50.PVP- and Ag200.PVP-treated animals contained numerous large macrophages laden with dark silver grains, whereas Ag50.citrate-treated animals showed agglomerates of silver particles predominantly outside the cells. The rank order of *in vivo* toxicity was Ag50.PVP > Ag50.citrate > Ag200.PVP. In a second study, we investigated the concentration dependence effects upon application of Ag50.PVP. Significant changes in cell counts and protein concentration in BALF appeared for 0.075 mg per rat lung. This value was below what might have been expected from the *in vitro* studies with AM, which had suggested effects to begin in the range of 0.3 mg/lung.* However, as the uptake of Ag nanoparticles *in vitro* was incomplete (Hinderliter et al. 2010), cytotoxicity measurements underestimated the effect *in vivo*, where the complete dose is delivered to the organ. This example underlines the necessity to determine the cellular dose when AM or other cells are used to predict effects on the whole lung.

9.4.1.1.3 Intraperitoneal/Intravenous Injections

Due to the exploitation of *gold nanomaterials* in therapeutic and diagnostic fields of nanomedicine, their biodistribution following intravenous or intraperitoneal injections of rodents has been intensively studied. A few of these studies, however, have assessed the *in vivo* toxic effects of these nanomaterials. Using intravenous injection, Cho et al. administered PEG-coated gold nanoparticles (13 nm) at single doses of 0.17, 0.85, and 4.26 mg/kg body weight to mice (Cho et al. 2009). For up to 7 days post injection, they observed acute inflammation and apoptosis within the liver, which was the primary site of accumulation. They found increased cytokine levels, infiltration of neutrophils,

* The mean dose per cell found at the LOAEL *in vitro* may be multiplied by 2×10^7, which is the mean number of alveolar macrophages per rat lung (Rehn et al. 1992), to obtain a first estimation of a critical dose for an instillation experiment.

and an increased cell adhesion as signs of inflammation (Cho et al. 2009). In another study, the group investigated PEG-coated gold nanoparticles of 4 and 100 nm size. These were intravenously administered by single injection at a dose of 4.26 mg/kg body weight. After 30 min, mice were sacrificed and histological and genetic effects were analyzed. Whereas no pathological changes were found in the liver tissues, expression of genes involved in inflammation, metabolism, and apoptosis was significantly changed. The changes seemed to be independent of the particle size, because both the 4 and 100 nm gold particles displayed similar effects and altered the expression of the same gene categories (Cho et al. 2009). The impact of size on toxicity was studied in detail by Chen et al. using 3, 5, 8, 12, 17, 37, 50, and 100 nm gold nanoparticles for intraperitoneal injections of 8 mg/kg body weight/week to Balb/C mice. No evidence of toxicity was found for 3, 5, 50, and 100 nm particles. In contrast, all other gold nanoparticles used (8, 12, 17, and 37 nm) induced severe toxic effects in the liver, spleen, and the lungs accompanied by fatigue, weight loss, and death within 21 days (Chen et al. 2009). These findings suggest that the toxicity of gold nanomaterials is not solely driven by their size but also by other particle-specific properties. This seems to become more apparent by the data obtained by Lasagna-Reeves et al., who found no signs of toxicity for gold nanoparticles of 13 nm size after intraperitoneal administration to mice. It is, however, difficult to compare these findings to the data obtained for 12 nm gold nanoparticles because the doses used here (0.04, 0.2, and 0.4 mg/kg body weight for 8 days) were 3–30 times lower compared to those used by Chen et al. (Lasagna-Reeves et al. 2010; Chen et al. 2009). In summary, from the few studies on *in vivo* toxicity of gold nanoparticles conducted so far it is impossible to correlate their physicochemical properties to their toxic effects.

9.4.2 Effects of Metal Nanoparticles on Humans

A health surveillance study conducted by Lee et al. determined silver deposition in blood and urine of two workers from a facility manufacturing *silver nanoparticles* of 20–30 nm. In one of the two workers (who worked there for 7 years), silver was detected in the blood at a concentration of 0.034 and 0.0135 μg/dL and in the urine at a concentration of 0.043 μg/dL, whereas in the other worker, silver was found neither in the blood nor in the urine. Adverse health effects could not be detected and blood chemistry and hematology data were within the normal range (Lee et al. 2012).

Moreover, silver is known to display wound healing effects which are exploited in wound dressings, and following exposure to damaged skin, silver has been reported to become systemically available (Johnston et al. 2010). In a clinical study, Vlachou et al. treated 30 burn patients with nanosilver (15 nm) containing wound dressings for 28 days or less and determined the systemic availability and potential toxic effects. They found serum levels of silver being increased during exposure to the dressings and up to 6 months after cessation of the treatment indicating that silver ions or nanoparticles become systemically available. However, no signs of toxicity on the biochemical or hematological level could be found (Vlachou et al. 2007). Silver deposition after treatment with wound dressings has also been reported in another case study in which a burn patient displayed argyria (Trop et al. 2006) a gray discoloration of the skin presumably caused by the formation of metallic silver by UV light

(Chang, Khosravi, and Egbert 2006). Apart from skin discoloration, Trop et al. also found increased levels of liver enzymes indicating hepatotoxicity as a consequence of the treatment with wound dressings for 6 days (Trop et al. 2006). These data do not yet allow for a risk–benefit assessment but safe utilization of nanosilver-based wound healing products requires further studies.

9.5 SUMMARY AND CONCLUSION

The summary of the presented metal nanoparticle data shows a simpler picture compared to metal oxide nanomaterials. Metal nanoparticles tested so far are often much smaller and have well-defined sizes, which allow for a better correlation of size and biological effects. Moreover, the tested metal NM exposure routes differ from those of metal oxides. Inhalation studies on metal particles are less frequent and do not yet cover all aspects necessary for a sound evaluation of potential health risks.

A comparison between *in vitro* and *in vitro* studies, as outlined in Chapter 8 for metal oxide nanomaterials, has been carried out for at least some types of silver NPs in the nanoGEM project. It appears that a ranking of acute toxic effects can be similarly made from *in vitro* and *in vivo* studies. Nonetheless, evidence is accumulating that particle size and biodistribution seem to contribute to nanoparticle toxicity with smaller particles generally displaying greater effects. This has also been repeatedly reported for metal oxide nanoparticles (see Chapter 8). To compare nanomaterials of different sizes, it is, however, of relevance to apply particles at an equal mass-dose, which has not always been considered in the studies published so far. Generally, doses are calculated on the mass of particles applied per ml particle suspension (cell culture) or per kg body weight (animals). For a reliable assessment of the impact of size on the toxicity of metal nanoparticles, the data obtained also have to be compared on an equal particle number or on the basis of equal surface area. This requires comprehensive data on particle size, number, and surface area and emphasizes the need for interdisciplinary studies. Importantly, such data would also be necessary after application to the biological system, that is, when NPs are located in or close to cells in a possibly agglomerated form. Moreover, because some of the metal nanoparticles dissolve in solution, it will be of relevance to determine the dissolution rates *in situ*. This appears especially important for nanosilver and copper nanoparticles which partially dissolve in solution. Currently, knowledge is still lacking regarding the toxic effects contributed by locally released metal species, a phenomenon often referred to as the "trojan horse effect." Albeit soluble particle fractions are known to be involved in the antimicrobial behavior of silver and copper, some groups reported that the toxicity of these nanomaterials is also reliant on the particle as a whole (Midander et al. 2009; Bouwmeester et al. 2011). It is, however, still unclear to what extent the toxicity of particles can be explained by the soluble or released metal fraction or by the specific particle. The most important nanoparticle property or a combination of these that governs toxicity has not yet been sufficiently explored. But from the literature presented here, it is likely that both size and solubility act in concert to mediate metal oxide toxicity. In addition, studies with surface-modified particles clearly demonstrate the impact of surface-coatings, -charge, and -impurities on the toxic effects observed for metal nanoparticles (Wang et al.

2013a; Alkilany et al. 2009). The coating chemicals may act as toxic agents themselves, diminish the release of ions, or alter the adsorption of biomolecules such as proteins or lipids, which then could influence the particle–cell interaction.

Taken together, the picture emanating from the findings described earlier is that the toxicity of metal nanoparticles is driven by physicochemical properties of nanoparticles. Thus, size, solubility, and surface coating may act in concert on biodistribution and secondary biological responses, which share many components with metal ion toxicity.

REFERENCES

Ahamed, M., M. Karns, M. Goodson, J. Rowe, S. M. Hussain, J. J. Schlager, and Y. Hong. 2008. DNA damage response to different surface chemistry of silver nanoparticles in mammalian cells. *Toxicol Appl Pharmacol* 233 (3):404–10.

Albanese, A., and W. C. Chan. 2011. Effect of gold nanoparticle aggregation on cell uptake and toxicity. *ACS Nano* 5 (7):5478–89.

Alkilany, A. M., P. K. Nagaria, C. R. Hexel, T. J. Shaw, C. J. Murphy, and M. D. Wyatt. 2009. Cellular uptake and cytotoxicity of gold nanorods: Molecular origin of cytotoxicity and surface effects. *Small* 5 (6):701–8.

Arora, S., J. Jain, J. M. Rajwade, and K. M. Paknikar. 2009. Interactions of silver nanoparticles with primary mouse fibroblasts and liver cells. *Toxicol Appl Pharmacol* 236 (3):310–8.

Asharani, P. V., Y. Lianwu, Z. Gong, and S. Valiyaveettil. 2011. Comparison of the toxicity of silver, gold and Platinum nanoparticles in developing zebrafish embryos. *Nanotoxicology* 5 (1):43–54.

Asharani, P. V., G. Low Kah Mun, M. P. Hande, and S. Valiyaveettil. 2009. Cytotoxicity and genotoxicity of silver nanoparticles in human cells. *ACS Nano* 3 (2):279–90.

Asharani, P. V., N. Xinyi, M. P. Hande, and S. Valiyaveettil. 2010. DNA damage and p53-mediated growth arrest in human cells treated with Platinum nanoparticles. *Nanomedicine (Lond)* 5 (1):51–64.

Bar-Ilan, O., R. M. Albrecht, V. E. Fako, and D. Y. Furgeson. 2009. Toxicity assessments of multisized gold and silver nanoparticles in zebrafish embryos. *Small* 5 (16):1897–910.

Bouwmeester, H., J. Poortman, R. J. Peters, E. Wijma, E. Kramer, S. Makama, K. Puspitaninganindita, H. J. Marvin, A. A. Peijnenburg, and P. J. Hendriksen. 2011. Characterization of translocation of silver nanoparticles and effects on whole-genome gene expression using an in vitro intestinal epithelium coculture model. *ACS Nano* 5 (5):4091–103.

Braydich-Stolle, L. K., B. Lucas, A. Schrand, R. C. Murdock, T. Lee, J. J. Schlager, S. M. Hussain, and M. C. Hofmann. 2010. Silver nanoparticles disrupt GDNF/Fyn kinase signaling in spermatogonial stem cells. *Toxicol Sci* 116 (2):577–89.

Browning, L. M., K. J. Lee, T. Huang, P. D. Nallathamby, J. E. Lowman, and X. H. Xu. 2009. Random walk of single gold nanoparticles in zebrafish embryos leading to stochastic toxic effects on embryonic developments. *Nanoscale* 1 (1):138–52.

Cecconi, I., A. Scaloni, G. Rastelli, M. Moroni, P. G. Vilardo, L. Costantino, M. Cappiello, D. Garland, D. Carper, J. M. Petrash, A. Del Corso, and U. Mura. 2002. Oxidative modification of aldose reductase induced by copper ion. Definition of the metal-protein interaction mechanism. *J Biol Chem* 277 (44):42017–27.

Chang, A. L., V. Khosravi, and B. Egbert. 2006. A case of argyria after colloidal silver ingestion. *J Cutan Pathol* 33 (12):809–11.

Chen, Y. S., Y. C. Hung, I. Liau, and G. S. Huang. 2009. Assessment of the in vivo Toxicity of Gold Nanoparticles. *Nanoscale Res Lett* 4 (8):858–64.

Chen, Z., H. Meng, G. Xing, C. Chen, Y. Zhao, G. Jia, T. Wang, H. Yuan, C. Ye, F. Zhao, Z. Chai, C. Zhu, X. Fang, B. Ma, and L. Wan. 2006. Acute toxicological effects of copper nanoparticles in vivo. *Toxicol Lett* 163 (2):109–20.

Chen, Z., S. Xie, L. Shen, Y. Du, S. He, Q. Li, Z. Liang, X. Meng, B. Li, X. Xu, H. Ma, Y. Huang, and Y. Shao. 2008. Investigation of the interactions between silver nanoparticles and Hela cells by scanning electrochemical microscopy. *Analyst* 133 (9):1221–8.

Cho, W. S., M. Cho, J. Jeong, M. Choi, H. Y. Cho, B. S. Han, S. H. Kim, H. O. Kim, Y. T. Lim, and B. H. Chung. 2009. Acute toxicity and pharmacokinetics of 13 nm-sized PEG-coated gold nanoparticles. *Toxicol Appl Pharmacol* 236 (1):16–24.

Cho, W. S., S. Kim, B. S. Han, W. C. Son, and J. Jeong. 2009. Comparison of gene expression profiles in mice liver following intravenous injection of 4 and 100 nm-sized PEG-coated gold nanoparticles. *Toxicol Lett* 191 (1):96–102.

Connor, E. E., J. Mwamuka, A. Gole, C. J. Murphy, and M. D. Wyatt. 2005. Gold nanoparticles are taken up by human cells but do not cause acute cytotoxicity. *Small* 1 (3):325–7.

Elder, A., H. Yang, R. Gwiazda, X. Teng, S. Thurston, H. He, and G. Oberdorster. 2007. Testing nanomaterials of unknown toxicity: An example based on Platinum nanoparticles of different shapes. *Adv Mater* 19 (20):3124.

Foldbjerg, R., D. A. Dang, and H. Autrup. 2011. Cytotoxicity and genotoxicity of silver nanoparticles in the human lung cancer cell line, A549. *Arch Toxicol* 85 (7):743–50.

Georgantzopoulou, A., Y. L. Balachandran, P. Rosenkranz, M. Dusinska, A. Lankoff, M. Wojewodzka, M. Kruszewski, C. Guignard, J. N. Audinot, S. Girija, L. Hoffmann, and A. C. Gutleb. 2013. Ag nanoparticles: size- and surface-dependent effects on model aquatic organisms and uptake evaluation with NanoSIMS. *Nanotoxicology* 7:1168–78.

Griffitt, R. J., R. Weil, K. A. Hyndman, N. D. Denslow, K. Powers, D. Taylor, and D. S. Barber. 2007. Exposure to copper nanoparticles causes gill injury and acute lethality in zebrafish (Danio rerio). *Environ Sci Technol* 41 (23):8178–86.

Haase, A., H. F. Arlinghaus, J. Tentschert, H. Jungnickel, P. Graf, A. Mantion, F. Draude, S. Galla, J. Plendl, M. E. Goetz, A. Masic, W. Meier, A. F. Thunemann, A. Taubert, and A. Luch. 2011. Application of laser postionization secondary neutral mass spectrometry/time-of-flight secondary ion mass spectrometry in nanotoxicology: Visualization of nanosilver in human macrophages and cellular responses. *ACS Nano* 5 (4):3059–68.

Hauck, T. S., A. A. Ghazani, and W. C. Chan. 2008. Assessing the effect of surface chemistry on gold nanorod uptake, toxicity, and gene expression in mammalian cells. *Small* 4 (1):153–9.

Hinderliter, P. M., K. R. Minard, G. Orr, W. B. Chrisler, B. D. Thrall, J. G. Pounds, and J. G. Teeguarden. 2010. ISDD: A computational model of particle sedimentation, diffusion and target cell dosimetry for in vitro toxicity studies. *Part Fibre Toxicol* 7 (1):36.

Hirsch, L. R., J. B. Jackson, A. Lee, N. J. Halas, and J. L. West. 2003. A whole blood immunoassay using gold nanoshells. *Anal Chem* 75 (10):2377–81.

Horie, M., H. Kato, S. Endoh, K. Fujita, K. Nishio, L. K. Komaba, H. Fukui, A. Nakamura, A. Miyauchi, T. Nakazato, S. Kinugasa, Y. Yoshida, Y. Hagihara, Y. Morimoto, and H. Iwahashi. 2011. Evaluation of cellular influences of Platinum nanoparticles by stable medium dispersion. *Metallomics* 3 (11):1244–52.

Jeon, K. I., J. Y. Jeong, and D. M. Jue. 2000. Thiol-reactive metal compounds inhibit NF-kappa B activation by blocking I kappa B kinase. *J Immunol* 164 (11):5981–9.

Ji, J. H., J. H. Jung, S. S. Kim, J. U. Yoon, J. D. Park, B. S. Choi, Y. H. Chung, I. H. Kwon, J. Jeong, B. S. Han, J. H. Shin, J. H. Sung, K. S. Song, and I. J. Yu. 2007. Twenty-eight-day inhalation toxicity study of silver nanoparticles in Sprague-Dawley rats. *Inhal Toxicol* 19 (10):857–71.

Johnston, H. J., G. Hutchison, F. M. Christensen, S. Peters, S. Hankin, and V. Stone. 2010. A review of the in vivo and in vitro toxicity of silver and gold particulates: Particle attributes and biological mechanisms responsible for the observed toxicity. *Crit Rev Toxicol* 40 (4):328–46.

Khlebtsov, N., and L. Dykman. 2011. Biodistribution and toxicity of engineered gold nanoparticles: A review of in vitro and in vivo studies. *Chem Soc Rev* 40 (3):1647–71.

Kim, H. R., M. J. Kim, S. Y. Lee, S. M. Oh, and K. H. Chung. 2011a. Genotoxic effects of silver nanoparticles stimulated by oxidative stress in human normal bronchial epithelial (BEAS-2B) cells. *Mutat Res* 726 (2):129–35.

Kim, J. S., J. H. Sung, J. H. Ji, K. S. Song, J. H. Lee, C. S. Kang, and I. J. Yu. 2011b. in vivo Genotoxicity of Silver Nanoparticles after 90-day Silver Nanoparticle Inhalation Exposure. *Saf Health Work* 2 (1):34–8.

Kim, Y. S., J. S. Kim, H. S. Cho, D. S. Rha, J. M. Kim, J. D. Park, B. S. Choi, R. Lim, H. K. Chang, Y. H. Chung, I. H. Kwon, J. Jeong, B. S. Han, and I. J. Yu. 2008. Twenty-eight-day oral toxicity, genotoxicity, and gender-related tissue distribution of silver nanoparticles in Sprague-Dawley rats. *Inhal Toxicol* 20 (6):575–83.

Lasagna-Reeves, C., D. Gonzalez-Romero, M. A. Barria, I. Olmedo, A. Clos, V. M. Sadagopa Ramanujam, A. Urayama, L. Vergara, M. J. Kogan, and C. Soto. 2010. Bioaccumulation and toxicity of gold nanoparticles after repeated administration in mice. *Biochem Biophys Res Commun* 393 (4):649–55.

Lee, J. H., J. Mun, J. D. Park, and I. J. Yu. 2012. A health surveillance case study on workers who manufacture silver nanomaterials. *Nanotoxicology* 6 (6):667–9.

Li, J. J., S. L. Lo, C. T. Ng, R. L. Gurung, D. Hartono, M. P. Hande, C. N. Ong, B. H. Bay, and L. Y. Yung. 2011. Genomic instability of gold nanoparticle treated human lung fibroblast cells. *Biomaterials* 32 (23):5515–23.

Li, J. J., L. Zou, D. Hartono, C. N. Ong, B. H. Bay, and L. Y. L. Yung. 2008. Gold nanoparticles induce oxidative damage in lung fibroblasts in vitro. *Advanced Materials* 20 (1):138.

Liu, S. Y., Z. S. Liang, F. Gao, S. F. Luo, and G. Q. Lu. 2010. in vitro photothermal study of gold nanoshells functionalized with small targeting peptides to liver cancer cells. *J Mater Sci Mater Med* 21 (2):665–74.

Loo, C., L. Hirsch, M. H. Lee, E. Chang, J. West, N. Halas, and R. Drezek. 2005. Gold nanoshell bioconjugates for molecular imaging in living cells. *Opt Lett* 30 (9):1012–4.

Midander, K., P. Cronholm, H. L. Karlsson, K. Elihn, L. Moller, C. Leygraf, and I. O. Wallinder. 2009. Surface characteristics, copper release, and toxicity of nano- and micrometer-sized copper and copper (II) oxide particles: A cross-disciplinary study. *Small* 5 (3):389–99.

Pan, Y., M. Bartneck, and W. Jahnen-Dechent. 2012. Cytotoxicity of gold nanoparticles. *Methods Enzymol* 509:225–42.

Pan, Y., A. Leifert, D. Ruau, S. Neuss, J. Bornemann, G. Schmid, W. Brandau, U. Simon, and W. Jahnen-Dechent. 2009. Gold nanoparticles of diameter 1.4 nm trigger necrosis by oxidative stress and mitochondrial damage. *Small* 5 (18):2067–76.

Pan, Y., S. Neuss, A. Leifert, M. Fischler, F. Wen, U. Simon, G. Schmid, W. Brandau, and W. Jahnen-Dechent. 2007. Size-dependent cytotoxicity of gold nanoparticles. *Small* 3 (11):1941–9.

Park, E. J., E. Bae, J. Yi, Y. Kim, K. Choi, S. H. Lee, J. Yoon, B. C. Lee, and K. Park. 2010. Repeated-dose toxicity and inflammatory responses in mice by oral administration of silver nanoparticles. *Environ Toxicol Pharmacol* 30 (2):162–8.

Park, E. J., H. Kim, Y. Kim, and K. Park. 2010. Intratracheal instillation of Platinum nanoparticles may induce inflammatory responses in mice. *Arch Pharm Res* 33 (5):727–35.

Pelka, J., H. Gehrke, M. Esselen, M. Turk, M. Crone, S. Brase, T. Muller, H. Blank, W. Send, V. Zibat, P. Brenner, R. Schneider, D. Gerthsen, and D. Marko. 2009. Cellular uptake of Platinum nanoparticles in human colon carcinoma cells and their impact on cellular redox systems and DNA integrity. *Chem Res Toxicol* 22 (4):649–59.

Pettibone, J. M., A. Adamcakova-Dodd, P. S. Thorne, P. T. O'Shaughnessy, J. A. Weydert, and V. H. Grassian. 2008. Inflammatory response of mice following inhalation exposure to iron and copper nanoparticles. *Nanotoxicology* 2 (4):189–204.

Prabhu, B. M., S. F. Ali, R. C. Murdock, S. M. Hussain, and M. Srivatsan. 2010. Copper nanoparticles exert size and concentration dependent toxicity on somatosensory neurons of rat. *Nanotoxicology* 4 (2):150–160.

Rayavarapu, R. G., W. Petersen, L. Hartsuiker, P. Chin, H. Janssen, F. W. van Leeuwen, C. Otto, S. Manohar, and T. G. van Leeuwen. 2010. in vitro toxicity studies of polymer-coated gold nanorods. *Nanotechnol* 21 (14):145101.

Rehn, B., J. Bruch, T. Zou, and G. Hobusch. 1992. Recovery of rat alveolar macrophages by bronchoalveolar lavage under normal and activated conditions. *Environ Health Perspect* 97:11–6.

Reidy, B., A. Haase, A. Luch, K. A. Dawson, and I. Lynch. 2013. "Mechanisms of Silver Nanoparticle Release, Transformation and Toxicity: A Critical Review of Current Knowledge and Recommendations for Future Studies and Applications." *Materials* 6 (6):2295–350.

Schäfer, B., J. V. Brocke, A. Epp, M. Gotz, F. Herzberg, C. Kneuer, Y. Sommer, J. Tentschert, M. Noll, I. Gunther, U. Banasiak, G. F. Bol, A. Lampen, A. Luch and A. Hensel. 2013. "State of the art in human risk assessment of silver compounds in consumer products: A conference report on silver and nanosilver held at the BfR in 2012." *Arch Toxicol* 87 (12):2249–62.

Semmler-Behnke, M., W. G. Kreyling, J. Lipka, S. Fertsch, A. Wenk, S. Takenaka, G. Schmid, and W. Brandau. 2008. "Biodistribution of 1.4- and 18-nm gold particles in rats." *Small* 4 (12):2108–11.

Song, K. S., J. H. Sung, J. H. Ji, J. H. Lee, J. S. Lee, H. R. Ryu, J. K. Lee, Y. H. Chung, H. M. Park, B. S. Shin, H. K. Chang, B. Kelman, and I. J. Yu. 2013. Recovery from silver-nanoparticle-exposure-induced lung inflammation and lung function changes in Sprague Dawley rats. *Nanotoxicology* 7 (2):169–80.

Stebounova, L. V., A. Adamcakova-Dodd, J. S. Kim, H. Park, P. T. O'Shaughnessy, V. H. Grassian, and P. S. Thorne. 2011. Nanosilver induces minimal lung toxicity or inflammation in a subacute murine inhalation model. *Part Fibre Toxicol* 8 (1):5.

Stern, J. M., J. Stanfield, Y. Lotan, S. Park, J. T. Hsieh, and J. A. Cadeddu. 2007. Efficacy of laser-activated gold nanoshells in ablating prostate cancer cells in vitro. *J Endourol* 21 (8):939–43.

Sun, L., D. Liu, and Z. Wang. 2008. Functional gold nanoparticle-peptide complexes as cell-targeting agents. *Langmuir* 24 (18):10293–7.

Sung, J. H., J. H. Ji, J. D. Park, J. U. Yoon, D. S. Kim, K. S. Jeon, M. Y. Song, J. Jeong, B. S. Han, J. H. Han, Y. H. Chung, H. K. Chang, J. H. Lee, M. H. Cho, B. J. Kelman, and I. J. Yu. 2009. Subchronic inhalation toxicity of silver nanoparticles. *Toxicol Sci* 108 (2):452–61.

Takahashi, H., Y. Niidome, T. Niidome, K. Kaneko, H. Kawasaki, and S. Yamada. 2006. Modification of gold nanorods using phosphatidylcholine to reduce cytotoxicity. *Langmuir* 22 (1):2–5.

Tiwari, D. K., T. Jin, and J. Behari. 2011. Dose-dependent in-vivo toxicity assessment of silver nanoparticle in Wistar rats. *Toxicol Mech Methods* 21 (1):13–24.

Trickler, W. J., S. M. Lantz, R. C. Murdock, A. M. Schrand, B. L. Robinson, G. D. Newport, J. J. Schlager, S. J. Oldenburg, M. G. Paule, W. Slikker, Jr., S. M. Hussain, and S. F. Ali. 2010. Silver nanoparticle induced blood-brain barrier inflammation and increased permeability in primary rat brain microvessel endothelial cells. *Toxicol Sci* 118 (1):160–70.

Trop, M., M. Novak, S. Rodl, B. Hellbom, W. Kroell, and W. Goessler. 2006. Silver-coated dressing acticoat caused raised liver enzymes and argyria-like symptoms in burn patient. *J Trauma* 60 (3):648–52.

Uboldi, C., D. Bonacchi, G. Lorenzi, M. I. Hermanns, C. Pohl, G. Baldi, R. E. Unger, and C. J. Kirkpatrick. 2009. Gold nanoparticles induce cytotoxicity in the alveolar type-II cell lines A549 and NCIH441. *Part Fibre Toxicol* 6:18.

Vlachou, E., E. Chipp, E. Shale, Y. T. Wilson, R. Papini, and N. S. Moiemen. 2007. The safety of nanocrystalline silver dressings on burns: A study of systemic silver absorption. *Burns* 33 (8):979–85.

Wang, L., X. Jiang, Y. Ji, R. Bai, Y. Zhao, X. Wu, and C. Chen. 2013a. Surface chemistry of gold nanorods: Origin of cell membrane damage and cytotoxicity. *Nanoscale* 5 (18):8384–91.

Wang, W., Q. Q. Wei, J. Wang, B. C. Wang, S. H. Zhang, and Z. Yuan. 2013b. Role of thiol-containing polyethylene glycol (thiol-PEG) in the modification process of gold nanoparticles (AuNPs): S tabilizer or coagulant? *J Colloid Interface Sci* 404:223–9.

Zhang, X. D., H. Y. Wu, D. Wu, Y. Y. Wang, J. H. Chang, Z. B. Zhai, A. M. Meng, P. X. Liu, L. A. Zhang, and F. Y. Fan. 2010. Toxicologic effects of gold nanoparticles in vivo by different administration routes. *Int J Nanomed* 5:771–81.

10 Uptake and Effects of Carbon Nanotubes

James C. Bonner

CONTENTS

10.1 INTRODUCTION

Nanotechnology is a rapidly developing industry that promises great societal and economic benefits. In particular, carbon nanotubes (CNTs) have enormous potential for innovation in the fields of engineering, electronics, and medicine. However, recent reviews of the scientific literature predict that CNTs will be a risk for lung diseases, particularly pulmonary fibrosis and granuloma formation, related to occupational and perhaps consumer exposure (Bonner 2010a, Bonner 2011). Inhalation studies using mice or rats demonstrate that CNTs deposit reach the alveolar regions in the lung and also migrate to the subpleural tissues beneath the mesothelial lining of the lung, where they remain embedded in the extracellular matrix or within resident lung cells for months following exposure (Ryman-Rasmussen et al. 2009a). CNTs also localize in lymphoid tissues (Ma-Hock et al. 2009, Pauluhn 2010) and stimulate systemic immune effects that influence extra-pulmonary tissues and organs such as the spleen and heart (Mitchell et al. 2007, 2009). In addition, genetic and environmental factors influence susceptibility to CNT-induced lung diseases in rodents. For example, CNTs exacerbate allergen-induced airway inflammation in mice (Ryman-Rasmussen et al. 2009b). The aim of this chapter is to summarize the uptake, fate, and effects of CNTs following pulmonary exposure to rodents to predict possible human health effects that could result as a consequence of occupational, consumer, or environmental exposure.

10.2 LUNG DEPOSITION AND TRANSLOCATION OF NANOTUBES

The deposition of inhaled CNTs is determined by a number of factors including particle size, shape, electrostatic charge, and agglomeration state. Published studies that document the fate and effects of inhaled CNTs are extremely valuable, because these studies much more closely model "real world" exposures and therefore more accurately model the true nature of deposition patterns for particles and fibers. Inhalation exposure results in deposition of CNTs in the distal regions; that is, alveolar duct bifurcations and alveolar epithelial surfaces, of the lungs of mice or rats (Ma-Hock et al. 2009; Pauluhn 2010; Mitchell et al. 2007; Shvedova et al. 2008; Ryman-Rasmussen et al. 2009a, 2009b). Most of the CNTs deposited in these regions are avidly taken up by alveolar macrophages. Although the majority of CNTs are engulfed by macrophages, some individual CNTs or small aggregates evade phagocytosis or uptake by macrophages and can be found within epithelial cells or mesenchymal cells (Ryman-Rasmussen et al. 2009a).

Aggregation of CNTs, alternatively referred to as state of dispersion, is an important factor that determines the pathologic response following exposure. The state of dispersion depends on electrostatic charge, functionalization, or suspension of CNTs in surfactant-containing media. As illustrated in Figure 10.1, dispersed CNTs are deposited throughout the lower lung with a relatively uniform distribution pattern, causing interstitial fibrosis that increases the thickness of airway and alveolar walls. Macrophages exposed to dispersed CNTs release growth factors [e.g., platelet-derived growth factor (PDGF) and TGF-β1], which stimulate fibroblast proliferation and collagen deposition that culminates in fibrosis and alveolar wall thickening. Agglomerated CNTs tend to cause granuloma formation, characterized by focal accumulation of macrophages and lymphocytes, followed by walling off of the CNT agglomerate with fibrotic scar tissue produced by fibroblasts. Figure 10.2 shows a lung fibroproliferative response to relatively well-dispersed multi-walled CNTs (MWCNTs) delivered by a single nose-only inhalation exposure (Ryman-Rasmussen et al. 2009b).

CNTs also reach the subpleural region of the lungs after inhalation exposure, either via macrophage-dependent or -independent processes, and remain embedded within the subpleural tissue or within macrophages for up to three months (Ryman-Rasmussen et al. 2009b). Some of these CNT-bearing macrophages exit the lung via the pleural lymphatic system and enter the pleural space or can be found in lung-associated lymph nodes (Ma-Hock et al. 2009; Pauluhn et al. 2010).

The majority of published studies on the pulmonary effects of CNTs have used intratracheal instillation (IT) or oropharyngeal aspiration (OPA) techniques for delivery to the lungs of rats or mice, respectively. Although studies using IT or OPA are important for determining biological effects, these studies rely on an aqueous bolus delivery and therefore do not precisely reproduce the deposition patterns that can be achieved by using inhalation exposure with dry aerosolized or nebulized suspensions of CNTs. Nevertheless, improvements have been made in dispersing CNTs in aqueous suspension, and the use of surfactant-containing media has

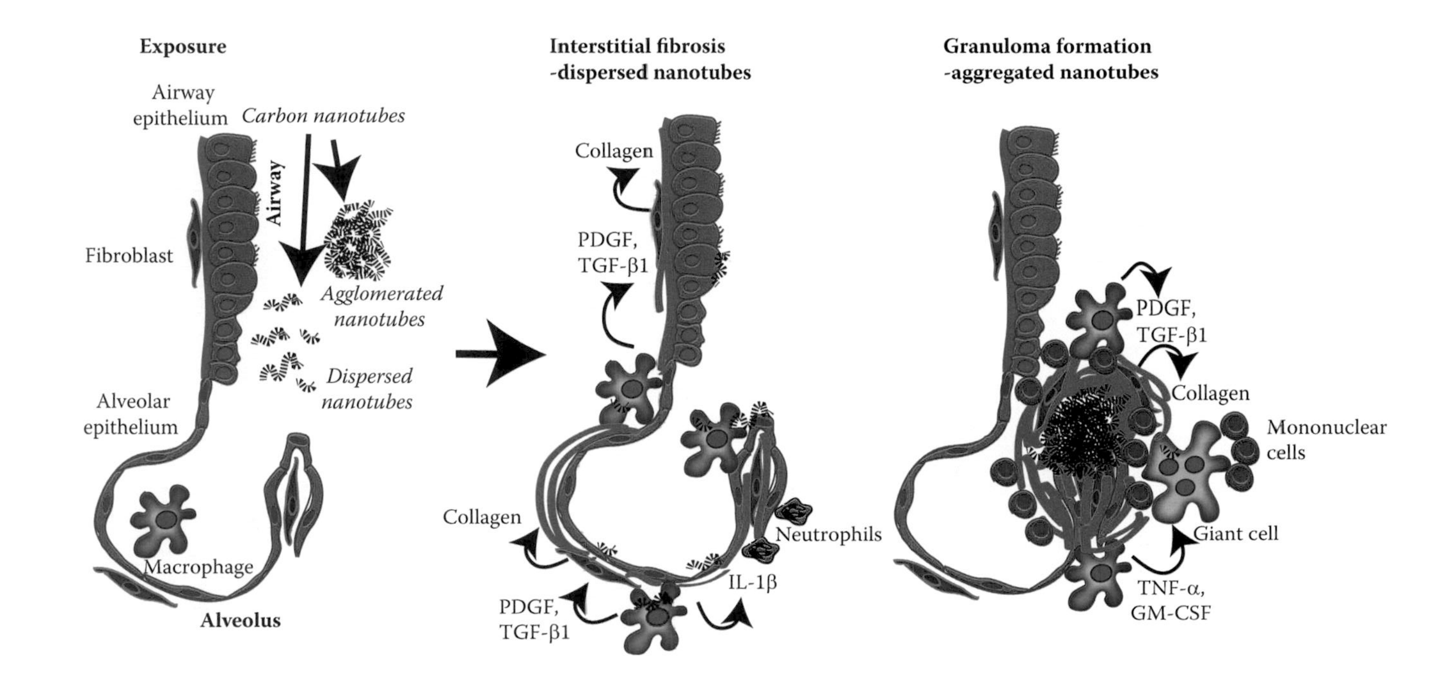

FIGURE 10.1 Illustration depicting the development of interstitial pulmonary fibrosis or granuloma in mice after exposure to dispersed or aggregated CNTs, respectively.

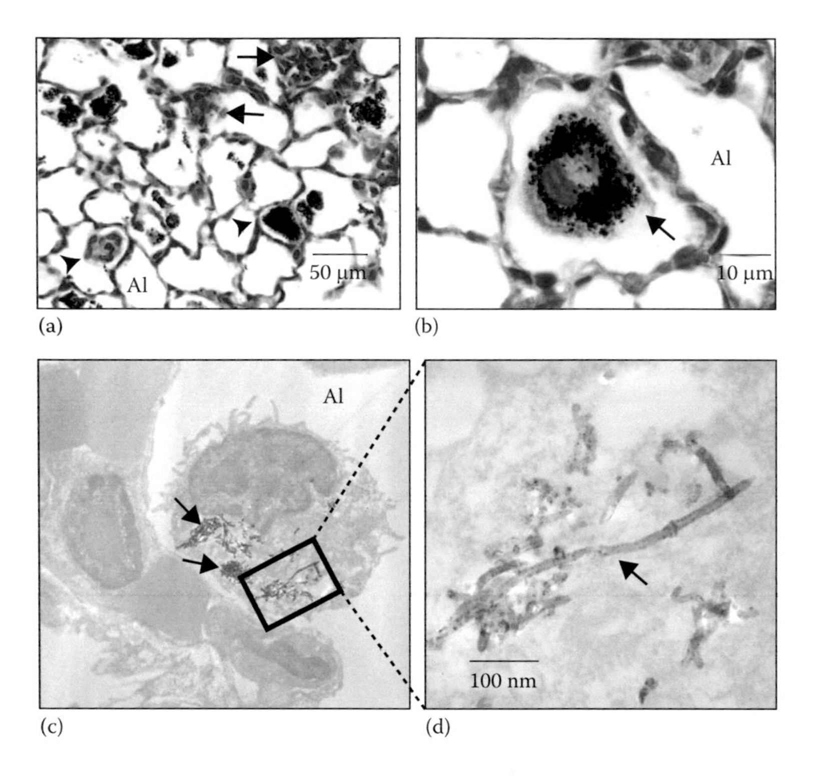

FIGURE 10.2 Light microscopic images and electron microscopic images showing uptake of multiwalled carbon nanotubes (MWCNTs) by macrophages in the lungs of mice following inhalation exposure. (a) and (b) Hematoxylin and eosin-stained mouse lung section 6 weeks after inhalation exposure to MWCNTs showing fibroproliferative lesions (arrows) and multinucleated giant cells (arrowheads). (c) and (d) TEM image showing MWCNTs (arrows) engulfed by an alveolar macrophage. (Images in (c) and (d) reprinted from Bonner, J.C., *Proc. Am. Thorac. Soc.*, 7, 138–141, 2013. With permission from American Thoracic Society.)

improved dispersal of CNT agglomerates for IT or OPA delivery to the lungs of rats and mice (Mercer et al. 2008).

10.3 ACUTE RESPONSES

10.3.1 Macrophage-Mediated Uptake and Clearance

Alveolar macrophages provide a primary immune defense against particles and fibers that enter the lungs. Macrophage recognition to CNTs may depend on tube width. For example, it has been reported that single-walled carbon nanotubes (SWCNT) delivered to the lungs of mice are not taken up by macrophages unless the SWCNTs are first treated with phosphatidylserine (Konduru et al. 2009). On the other hand, other work has shown that SWCNTs are engulfed by alveolar macrophages *in vivo* (Mangum et al. 2006). These seemingly conflicting reports could be due to differences in well-dispersed SWCNTs, which would likely not be readily taken up by phagocytosis, versus agglomerated SWCNTs, which would more likely be recognized as a larger particle and engulfed by roaming phagocytes. As discussed earlier, the aggregation status is also likely an important determinant toward evaluating the toxicity of CNTs and the type of resulting lung pathology (i.e., granuloma versus interstitial fibrosis).

CNTs taken up by macrophages via phagocytosis are cleared from the lungs through two primary mechanisms: (a) the mucociliary escalator and (b) the lymphatic drainage system. The mucociliary escalator is comprised of a coating of mucus on the surface of the airways that is constantly moving up the airways by the coordinated movement of cilia on the airway epithelium (Bonner 2008). Macrophages with engulfed particles or fibers migrate to the distal portion of small airways where they are transported by the escalator to larger airways and ultimately out of the trachea where they are swallowed or expelled through coughing. Migration of macrophages containing CNTs across the pleura could cause adverse effects (e.g., DNA damage to mesothelial cells) as CNTs could possess some physical characteristics (e.g., high aspect ratio, durability) similar to asbestos fibers (Bonner et al. 2010a; Donaldson 2010). However, the possible carcinogenicity of CNTs remains controversial.

The clearance of inhaled MWCNTs from the lungs and lung associated lymph nodes (LALNs) has been modeled in rats (Pauluhn et al. 2010). This was accomplished by measuring concentrations of residual cobalt (Co) catalyst in the lung and LALN 13 weeks after a nose-only inhalation exposure for 6 hours/day × 5 days/week, and a 6 month postexposure period. In this study, the authors observed the half-life for MWCNT elimination at 151, 350, 318, and 375 days to be 0.1, 0.4, 1.5, and 6 mg/m^3, respectively.

10.3.2 Cellular Responses to Nanotubes

Macrophages react to CNTs in a variety of different ways, some of which are dependent on the physical and chemical characteristics of the CNTs. Uptake of CNTs could have a variety of consequences related to macrophage biology and function. As illustrated in Figure 10.3, CNTs cause inflammasome activation in macrophages which results in the processing and secretion of the proinflammatory cytokines IL-1β and IL-18 (Palomäki et al. 2011; Meunier et al. 2012; Sun et al. 2012). Inflammasome activation by CNTs and other high aspect ratio materials (e.g., asbestos fibers) is mediated by lysosomal disruption and ROS production. Moreover, the level of residual nickel catalyst in MWCNTs has been shown to play an important role in inflammasome activation in macrophages (Hamilton et al. 2012). Inflammasome activation and consequent IL-1β production is an important innate immune response for recruiting neutrophils to the lung to participate in microbial killing (Martinon, Mayor, and Tschopp 2009). Also the dysregulation of inflammasomes has been implicated in a variety of disease states (Strowig et al. 2012). Macrophages can also be "alternatively activated" in the presence of Th2 cytokines IL-4 or IL-13 to produce growth factors (e.g., PDGF, TGF-β1) that are involved in the pathogenesis of fibrosis (Gordon and Martinez 2010). It is currently unknown whether CNTs stimulate alternative

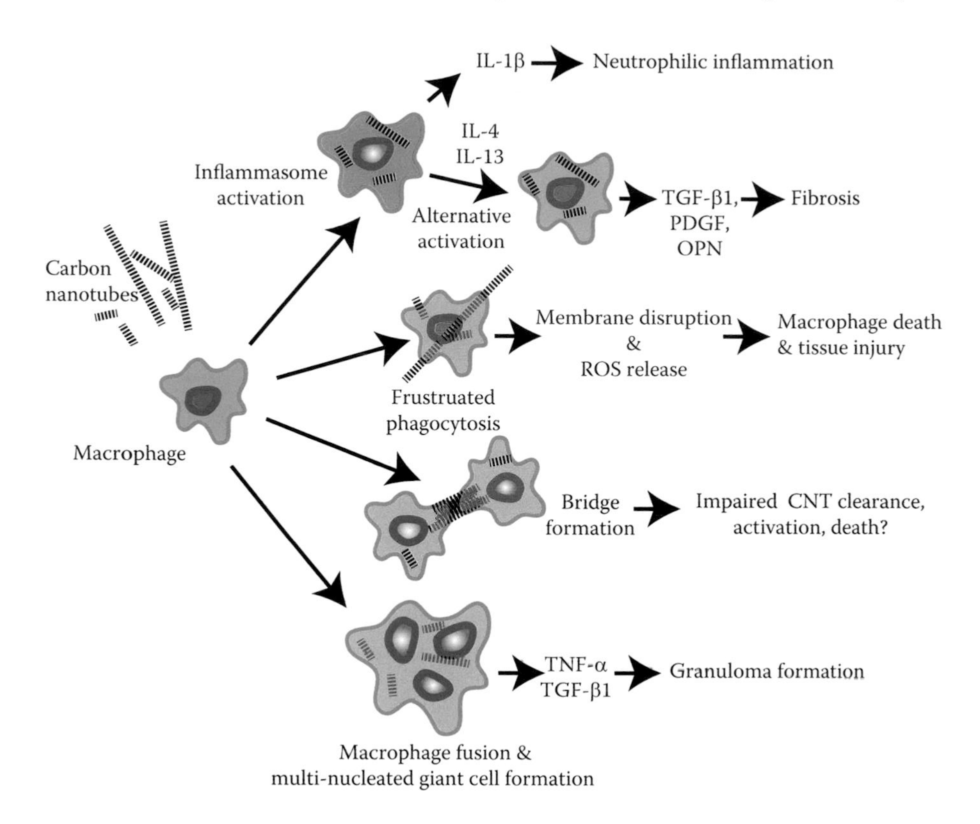

FIGURE 10.3 Illustration depicting the macrophage uptake of CNTs and possible outcome on macrophage biology and function.

activation of macrophages or whether Th2 cytokines influence CNT-induced inflammasome activation.

CNTs disruption of macrophage function is also determined by tube length. As shown in Figure 10.3, long, rigid CNTs cause frustrated phagocytosis, which subsequently causes cell death and release of ROS and cytokines, leading to tissue injury (Donaldson et al. 2010; Murphy et al. 2012). In addition to frustrated phagocytosis, CNTs align to form bridge-like structures that link two or more macrophages. These structures have been observed in rats exposed to SWCNTs by intratracheal instillation (Mangum et al. 2006). CNT bridge structures could represent a molecular interaction between the tube-like nanomaterial and the cytoskeleton and likely impair macrophage function.

The cellular signaling responses to CNTs are complex and at least some of these pathways are common to those activated by a variety of other particles and fibers. For example, CNTs increase levels of the COX-2 enzyme through a MAP kinase-dependent signaling pathway in mouse RAW267 cells *in vitro* (Lee et al. 2012). CNTs also activate the antioxidant mediator Nrf2 in cultured human THP-1 cells (Brown, Donaldson, and Stone 2010). Both COX-2 and Nrf2 are protective factors against lung diseases that are increased to counteract ROS-induced cellular stress initiated by CNTs.

10.4 PATHOLOGICAL EFFECTS

10.4.1 Fibrosis and Granuloma Formation

In some ways CNTs share features of asbestos fibers, mainly with regard to their fiber-like shape and aspect (length to width) ratio. Asbestos fibers are a known cause of fibrosis and mesothelioma in humans. However, CNTs also have uniquely different properties from asbestos, including nanoscale width and highly conformal structure. Also, different kinds of CNTs (e.g., multiwalled vs. single-walled) could produce different pathologic effects.

Rodent inhalation and instillation studies demonstrate that CNTs cause pulmonary inflammation (Li et al. 2007; Shvedova et al. 2008; Ryman-Rasmussen et al. 2009a, 2009b; Ma-Hock et al. 2009; Pauluhn et al. 2010). Pulmonary fibrosis is also a common pathologic feature observed in rodent studies of CNT exposure (Li et al. 2007; Shvedova et al. 2007; Ryman-Rasmussen et al. 2009a, 2009b; Pauluhn et al. 2010). CNT exposure also affects the upper respiratory tract, and goblet cell hyperplasia and inflammation have been observed in the nasal cavity and upper airways of rodents (Ma-Hock et al. 2009; Pauluhn et al. 2010). Pulmonary inflammation and fibrosis resulting from intratracheal instillation of CNTs generally produce granulomatous lesions, which is likely due to CNT aggregation into micron-sized bundles in aqueous media (Mangum et al. 2006; Lam et al. 2004; Muller et al. 2005; Shvedova et al. 2005; Warheit et al. 2004). Foci of granulomatous lesions and collagen deposition in mice are associated with dense particle-like SWCNT agglomerates (Murray et al. 2012). In contrast, interstitial pulmonary fibrosis is associated with dispersed CNTs that require transmission electron microscopy for detection (Ryman-Rasmussen et al. 2009a). Improved methods for CNT dispersion in aqueous media result in pathology that more closely resembles that caused by inhalation exposure. For example, dispersal of MWCNTs by bovine serum albumin (BSA) and dipalmitoylphosphatidylcholine has been reported to influence cytokine and growth factor production by lung cells *in vitro* and predict pulmonary fibrosis *in vivo* in mice. Well-dispersed MWCNTs were readily taken up by macrophages and induced more robust growth factor (PDGF-AA, TGF-β1) and IL-1β production than nondispersed MWCNTs. These results indicated that the dispersal state of MWCNTs affected profibrogenic cellular responses that correlate with the extent of pulmonary fibrosis.

Comparisons of the lung pathological effects between inhalation and other methods of pulmonary exposure have been performed. Comparison of instillation with inhalation methods for delivery of MWCNT in mice showed a more diffuse pattern of deposition with inhalation than that was observed after intratracheal instillation of a bolus dose of MWCNT and greater interstitial pulmonary fibrosis (Li et al. 2007). Also a study utilizing SWCNT that compared inhalation exposure with oropharyngeal aspiration (OPA) showed that the pulmonary effects were similar in both cases, but were more severe for the inhaled SWCNTs, which caused more pronounced alveolar wall thickening (i.e., interstitial fibrosis) as compared to instilled SWCNT, which produced more granuloma formation (Shvedova et al. 2008). These studies indicate that the degree of CNT dispersion, together with the influence of particle mass and aerodynamic properties, contributes to the differences in potency and physiological effects observed when comparing inhalation to other routes of pulmonary exposure such as intratracheal instillation or OPA. The agglomeration state of CNTs is an important determinant of pathologic outcome. As illustrated in Figure 10.1, agglomerated CNTs stimulate granuloma formation, whereas dispersed CNTs cause diffuse interstitial fibrosis in the alveolar region and around airways and blood vessels.

In addition to the method of pulmonary exposure and the degree of particle dispersion, two other major factors determine the toxicity of CNTs: (a) clearance and (b) contamination by metal catalysts. Longer CNTs impede clearance and structures longer than 10–15 μm (the approximate width of an alveolar macrophage) are difficult to clear from lung tissues via macrophage-mediated mechanisms. This is especially relevant to MWCNTs, which are more rigid than SWCNTs. Although rigidity and length are likely important for mediating frustrated phagocytosis of macrophages, CNT diameter has also shown to mediate toxicity. In particular, thinner (~10 nm diameter) MWCNTs are more toxic in the lungs of mice than thicker (~70 nm diameter) MWCNTs (Fenoglio et al. 2012). Long SWCNTs are more likely to fold and be taken up by macrophages, whereas long MWCNTs are more likely to cause frustrated phagocytosis and impede macrophage clearance. Second, the composition of CNTs must be carefully considered. Metals such as nickel, cobalt, and iron are commonly used as catalysts in the manufacture of CNTs and these same metals are well-known to cause pulmonary diseases in humans, including pulmonary fibrosis and asthma (Kelleher, 2000). For example, nickel is known to cause occupational asthma and contact dermatitis, whereas iron and cobalt cause interstitial pulmonary fibrosis in occupations related to mining and metallurgy. Metal catalysts can be removed to some extent from CNTs by acid washing, although this process usually does not completely remove metals.

The cellular and molecular mechanisms that cause CNT-induced lung disease have not been clearly elucidated, but it is expected that many lessons can be learned from studies on larger particles, fibers, and metals. However, CNTs and other engineered nanomaterials may also operate through unique and unexpected mechanisms to cause disease given that they can interact at the molecular level with cellular organelles, proteins, phospholipids, and nucleic acids. A variety of soluble mediators (growth factors, cytokines, and chemokines) that play important roles in fibrosis are induced in the lungs of rats or mice after exposure to CNTs (Bonner 2010b). Some of these mediators and their relationship to lung region and cell type are shown in Figure 10.5. Several studies have shown that either SWCNTs or MWCNTs delivered to the lung by intratracheal instillation in rats or inhalation in mice increase mRNA and protein levels of PDGF (Mangum et al. 2006; Ryman-Rasmussen et al. 2009a; Cesta et al. 2010). PDGF stimulates the replication, chemotaxis, and survival of lung fibroblasts to promote the pathogenesis of fibrosis (Bonner et al. 2004). SWCNTs delivered to the lungs of mice by OPA or inhalation cause induction of TGF-β1 (Shvedova et al. 2005; 2008), a central mediator of collagen production by mesenchymal cells (fibroblasts and myofibroblasts). However, MWCNTs containing nickel catalyst produce very little or no detectable TGF-β1 in the lungs of mice exposed to inhalation or in the lungs of rats exposed by instillation (Ryman-Rasmussen et al. 2009b, Cesta et al. 2010). Other factors such as osteopontin (OPN) could serve to stimulate collagen deposition and fibrosis in the absence of TGF-β1. OPN mRNA levels are highly induced in the lungs of rats exposed to SWCNTs (Mangum et al. 2006). Alveolar macrophages, as well as airway epithelial cells and fibroblasts, produce PDGF, TGF-β1, and OPN. Macrophages are also a rich source of IL-1β and IL-6 (Shvedova et al. 2008), and these two cytokines promote tissue fibrogenesis and are increased in the lungs of mice exposed to inhaled SWCNTs. Several chemokines are also induced by CNT exposure and drive the inflammatory response in the lung. CXCL8 (IL-8), a potent neutrophil chemoattractant, is produced by a human bronchial epithelial cell line *in vitro* after exposure to MWCNTs (Hirano et al. 2010). CCL2, also known as monocyte chemoattractant protein-1 (MCP-1), is elevated in the lavage fluid of mice exposed to MWCNT by inhalation (Ryman-Rasmussen et al. 2009a). CCL2 is also produced by pleural mesothelial cells (Visser et al. 1998), and is a candidate that mediates mononuclear cell accumulation that occurs at the pleura of mice after inhalation of MWCNTs (Ryman-Rasmussen et al. 2009a). In addition to soluble mediators, cell–cell interaction stimulated by CNTs could play a role in the pathogenesis of inflammatory or fibrogenic reactions. For example, understanding how CNTs affect the physical interaction between epithelial cells and underlying mesenchymal cells (termed the epithelial–mesenchymal cell trophic unit) could yield important insight into airway remodeling events. Likewise, since inhaled CNTs migrate to the subpleural tissues and remain there for months, it will be important to gain an understanding of how CNTs influence the interaction between mesothelial cells and subpleural mesenchymal cells.

10.4.2 Pleural Disease and Cancer

As mentioned earlier in this chapter, CNTs are cleared from the lungs via macrophage-mediated mechanisms and one route that macrophages take to remove particles from the lungs is across the pleural lining via the lymphatic drainage. Therefore, CNTs have the opportunity to interact with the mesothelial lining that makes up the pleura. The durable nature of CNTs along with their fiber-like shape could pose a problem of long-term lung persistence, which might result in asbestos-like behavior and carcinogencity (i.e., mesothelioma). As shown in Figure 10.4, inhaled CNTs migrate to the subpleural tissue in mice and produce transient proinflammatory lesions on the pleural surface that have been referred to as mononuclear cell aggregates (Ryman-Rasmussen et al. 2009a). As illustrated in Figure 10.5, the accumulation of mononuclear cell aggregates could be facilitated by increased levels of the chemokine CCL2, which is increased in the lungs of mice exposed to MWCNTs (Ryman-Rasmussen et al. 2009a). Moreover, CCL2 is produced by cultured pleural mesothelial cells stimulated with PDGF, which is produced by activated alveolar macrophages. In addition to mononuclear cell accumulation at the pleura, MWCNT inhalation exposure caused subpleural fibrosis

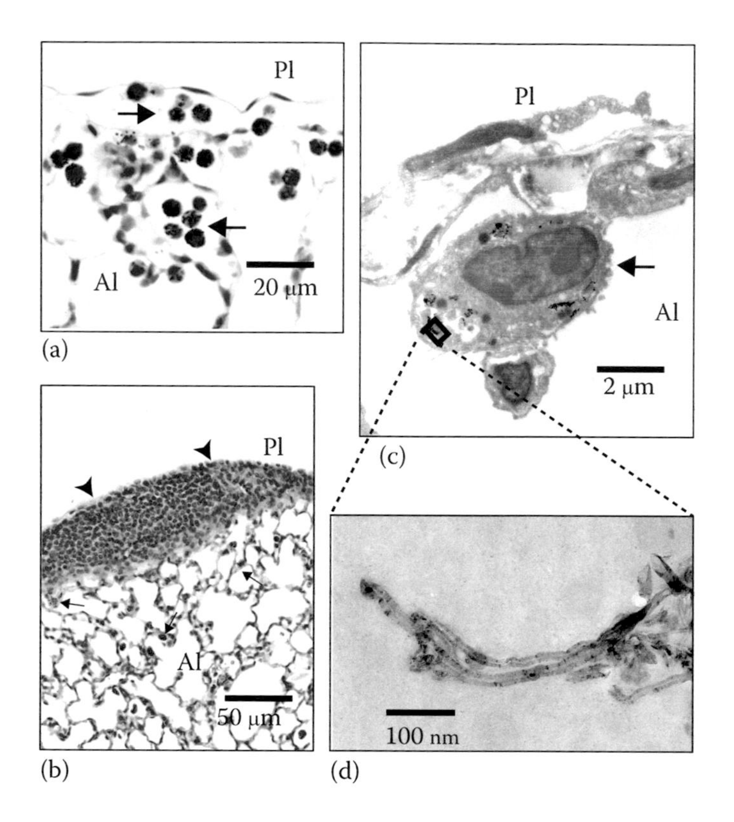

FIGURE 10.4 Multiwalled carbon nanotubes (MWCNTs) within the subpleural tissue after inhalation exposure in mice. (a) Light microscopy of macrophages with MWCNTs (arrows) abundant in the subpleural tissue and (b) associated with the formation of pleural (Pl) mononuclear cell aggregates (arrowheads). (c) and (d) Electron micrographs of MWCNTs within subpleural macrophage (arrow). (Images in (a), (b), (c) and (d) reprinted by permission from Macmillan Publishers Ltd., *Nat Nanotechnol*, Ryman-Rasmussen J.P. et al. Inhaled multi-walled carbon nanotubes reach the subpleural tissue in mice. 4: 747–751, copyright 2013.)

(Ryman-Rasmussen et al. 2009b). As shown in Figure 10.5, fibrosis at the subpleural region could be driven by growth factors, PDGF and TGF-β1, produced by MWCNT-stimulated macrophages.

Injection of long MWCNTs into the peritoneal cavity of mice (a surrogate for the pleura) induces inflammation and granuloma formation, suggesting that MWCNTs have asbestos-like pathogenicity (Poland et al. 2008). Also, p53 heterozygous null mice, which are susceptible to the development of mesothelioma, had increased incidence of mesothelioma in the abdominal cavity after intraperitoneal injection of MWCNTs (Takagi et al. 2008).

MWCNTs delivered to the lungs of mice by nose-only inhalation exposure (Ryman-Rasmussen et al. 2009a) or by OPA (Porter et al. 2010) demonstrated migration of CNTs to the pleura. The highest dose used in experiments by Ryman-Rasmussen

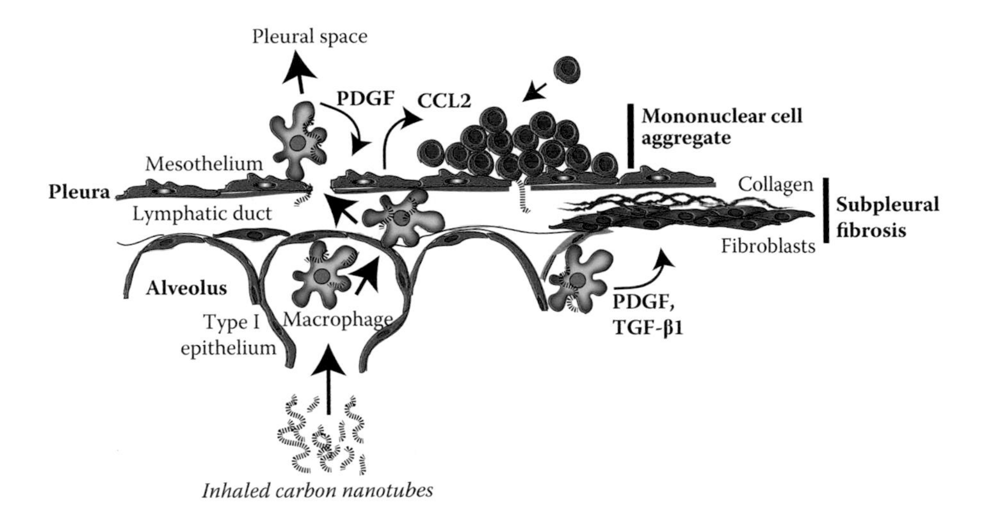

FIGURE 10.5 Illustration depicting migration of inhaled MWCNTs to the pleural lining surrounding the lungs and resulting inflammatory and fibrotic response.

et al. produced mononuclear cell accumulation at the pleural surface within 1 day after exposure and subpleural fibrotic lesions within 2 weeks, which resolved by 6 and 14 weeks, respectively. MWCNTs delivered to mice penetrate the pleural lining (Mercer et al. 2010). However, no evidence of mesothelioma has been observed in mice exposed to CNTs by any method, with the exception of genetically susceptible p53-deficient mice. This could be due to the fact that mice do not readily develop mesothelioma, even in response to known agents (e.g., asbestos) that cause mesothelioma in humans. The issue of whether CNTs are capable of causing mesothelioma in humans remains a key topic of research that will have important implications for responsible use of CNTs. So far it is unknown whether CNTs represent a new cancer risk factor for humans. However, a recent report by NIOSH set a recommended exposure limit for MWCNT based on rodent studies (Castranova et al. 2012). Although CNTs cause lung fibrotic reactions in the interstitium and pleura of mice, their carcinogenic potential has not been adequately addressed. Longer term, low dose studies with CNTs will need to be undertaken to adequately address the potential carcinogenicity of CNTs. Elucidating the carcinogenic potential of CNTs at the pleural lining in rodents using relevant inhalation exposures with appropriate positive controls (e.g., asbestos fibers) will have important implications for the future use and development of CNTs for a variety of applications.

Several studies show that CNTs have the potential to cause genotoxic effects in lung cell types and in rodents *in vivo*. Work by Sargent et al showed that SWCNTs caused fragmented centrosomes, mitotic spindle disruption, anaphase bridges, and aneuploid chromosome number in cultured primary or immortalized human airway epithelial cell types (Sargent et al. 2009, Sargent et al. 2012). These studies showed disruption of the mitotic spindle by SWCNT and the authors noted that the similar size and geometry of SWCNT bundles might account for physical interaction of SWCNT with the microtubules that form the mitotic spindle. Zhu et al assessed the DNA damage response to MWCNTs in mouse embryonic stem (ES) cells and found that MWCNTs can accumulate and induce apoptosis in mouse ES cells and activate the tumor suppressor protein p53, a marker of DNA damage (Zhu et al. 2007). Finally, Patlolla et al. showed that MWCNTs caused dose-dependent genotoxic effects of structural chromosome aberrations, DNA damage, and micronuclei formulation in mice when administered by intraperitoneal injection (Patlolla et al. 2010). Recently, persistent DNA damage was reported in rat lung cells after five days of inhalation exposure and one month postexposure to dispersed MWCNTs (Kim et al. 2012). Collectively, these studies show that SWCNTs and MWCNTs have genotoxic effects in cultured cells and can reach sensitive sites *in vivo* to cause genotoxic effects.

10.4.3 Effects of CNTs beyond Lungs

Inhaled CNTs can stimulate the immune system beyond the boundaries of the lung to impact distant organ systems, including the spleen and heart. Mitchell and coworkers showed that inhaled MWCNTs cause systemic immunosuppression and splenic oxidative stress (Mitchell 2007). The mechanism of splenic immunosuppression was further elucidated and involves the release of TGF-β1 from the lungs of MWCNT-exposed mice, which enters the bloodstream to signal COX-2-mediated increases in prostaglandin-E2 and IL-10 in the spleen, both of which play a role in suppressing T cell proliferation (Mitchell 2009). Erdely and coworkers also reported that SWCNTs or MWCNTs deposited into the lungs induced acute lung and systemic effects, suggesting that the systemic response, if chronic and persistent, could trigger or exacerbate cardiovascular dysfunction and disease (Erdely et al. 2009). Collectively, these studies suggest that inhaled CNTs could have profound effects on systemic targets.

10.4.4 Susceptibility Factors

Careful consideration should be given to the possibility that CNTs might have the most profound adverse health effects on individuals with pre-existing respiratory diseases such as asthma, bronchitis, or chronic obstructive pulmonary disease (COPD) (Bonner 2010a; Bonner 2011). Some susceptibility factors that might determine CNT toxicity are illustrated in Figure 10.6. MWCNTs have been shown to exacerbate pre-existing allergic airway diseases in mice that were first prechallenged with ovalbumin allergen (Ryman-Rasmussen et al. 2009b; Inoue et al. 2009). These studies suggest that CNTs pose a hazard to individuals with allergic asthma. SWCNTs have also been reported to exacerbate allergic airway inflammation in mice via enhanced activation of Th immunity and increased oxidative stress (Inoue et al. 2010; Nygaard et al. 2009). Repeated exposure to MWCNTs has also been shown to induce Th2 allergic responses in the absence of any allergen pre-exposure (Park et al. 2009). Although these studies suggest that individuals with allergic asthma are susceptible to lung and airway diseases caused by CNTs exposure, it remains unknown whether

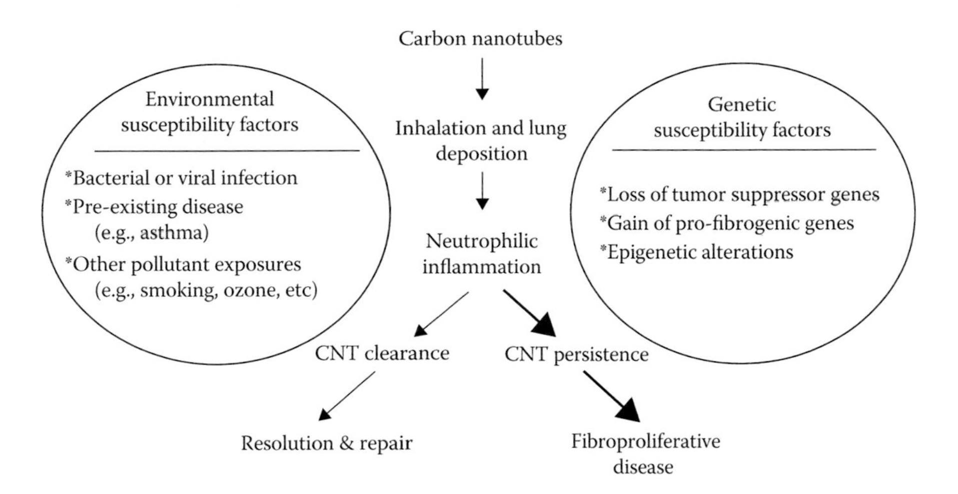

FIGURE 10.6 Environmental or genetic susceptibility factors (within circles) that are proposed or have been shown to increase inflammation and fibroproliferative lung disease in mice after exposure to CNTs.

CNTs will cause or exacerbate asthma in humans. However, the evidence in rodents suggests that CNTs would be a hazard to individuals with asthma.

The effects of CNTs in the lung could also be exacerbated by pre-existing bacterial or viral infection. For example, bacterial lipopolysaccharide (LPS), a potent proinflammatory agent, has been implicated in a number of occupational and environmental lung diseases in humans, including bronchitis, COPD, and asthma. SWCNTs or MWCNTs exacerbated LPS-induced lung inflammation, pulmonary vascular permeability, and production of proinflammatory cytokines in the lungs of mice (Inoue et al. 2008). LPS pre-exposure also enhanced MWCNT-induced pulmonary fibrosis in rats and synergistically enhanced MWCNT-induced production of PDGF by rat alveolar macrophages and lung epithelial cells (Cesta et al. 2010). These studies provide evidence that LPS-induced lung inflammation is a susceptibility factor that increases the severity of fibroproliferative lung disease caused by CNT exposure.

10.5 CONCLUSIONS AND SUMMARY

CNTs offer innovative structural platforms for a wide variety of novel applications in engineering, electronics, and medicine. However, careful consideration should be given to the potential human health risks that CNTs pose in occupational, consumer, or environmental exposure scenarios. Predictions that CNTs will have the potential to cause fibrosis and/or cancer in humans are based on physical similarities to toxic fibers and growing evidence that CNTs cause fibrosis and perhaps cancer in rodents. As with other engineered nanomaterials, the nanoscale of CNTs makes it difficult to predict how these structures will interact with intracellular structures such as DNA, cell membranes, and cytoskeletal proteins. Also, CNTs contain residual metals (e.g., nickel, cobalt, iron) introduced from the manufacturing process and many of these metals are known to cause fibrosis or cancer in humans. Therefore, multiple physical and chemical characteristics should be taken into consideration during the manufacturing process to ensure the safe design of products that contain CNTs. Human disease as a result of CNT exposure has not yet been documented and this is likely due to the infancy of the nanotechnology field and the lack of exposure or epidemiologic data. The evidence from studies with rodents indicates that there is the potential for CNTs to cause lung disease in humans and this possibility should be considered when weighing societal benefits of CNTs.

REFERENCES

Bonner, J.C. (2004) Regulation of PDGF and its receptors in fibrotic diseases. *Cytokine Growth Factor Rev* 15: 255–273.

Bonner, J.C. (2008) Respiratory Toxicology *in* Molecular and Biochemical Toxicology, 4th Edition. Eds: R.C. Smart and E. Hodgson. John Wiley & Sons, Inc. New York.

Bonner, J.C. (2010a) Nanoparticles as a potential cause of pleural and interstitial lung disease. *Proc Am Thorac Soc* 7: 138–141.

Bonner, J.C. (2010b) Mesenchymal cell survival in airway and interstitial pulmonary fibrosis. *Fibrogenesis Tissue Repair* 3: 15.

Bonner, J.C. (2011) Carbon nanotubes as delivery systems for respiratory disease: Do the dangers outweigh the potential benefits? *Expert Rev Respir Med* 5: 779–787.

Brown, D.M., Donaldson, K., Stone, V. (2010) Nuclear translocation of Nrf2 and expression of antioxidant defence genes in THP-1 cells exposed to carbon nanotubes. *J Biomed Nanotechnol* 6: 224–233.

Castranova, V., Schulte, P.A., Zumwalde, R.D. (2012) Occupational nanosafety considerations for carbon nanotubes and carbon nanofibers. *Acc Chem Res* 46: 642–649.

Cesta, M.F., Ryman-Rasmussen, J.P., Wallace, D.G., Masinde, T., Hurlburt, G., Taylor, A.J., Bonner, J.C. (2010) Bacterial lipo polysaccharide enhances PDGF signaling and pulmonary fibrosis in rats exposed to carbon nanotubes. *Am J Respir Cell Mol Biol* 43: 142–151.

Donaldson, K., Murphy. F.A., Duffin, R., Poland, C.A. (2010) Asbestos, carbon nanotubes and the pleuralmes othelium: Are view of the hypothesis regarding the role of long fibre retention in the parietalpleura, inflammation and mesothelioma. *Part Fibre Toxicol* 7: 5.

Erdely, A., Hulderman, T., Salmen, R., Liston, A., Zeidler-Erdely, P.C., Schwegler-Berry, D., Castranova, V., Koyama, S., Kim, Y.A., Endo, M., Simeonova, P.P. (2009) Cross-talk between lung and systemic circulation duringcarbonnanotube respiratory exposure. Potential biomarkers. *Nano Lett* 9: 36–43.

Fenoglio, I., Aldieri, E., Gazzano, E., Cesano, F., Colonna, M., Scarano, D., Mazzucco, G., Attanasio, A., Yakoub, Y., Lison, D., Fubini, B. (2012) Thickness of multiwalled carbon nanotubes affects their lung toxicity. *Chem Res Toxicol* 25: 74–82.

Gordon, S., Martinez, F.O. (2010) Alternative activation of macrophages: Mechanism and functions. *Immunity* 32: 593–604.

Hamilton, R.F Jr, Buford, M., Xiang, C., Wu, N., Holian, A. (2012) NLRP3 inflamma some activation in murine alveolar macrophages and related lung pathology is associated with MWCNT nickel contamination. *Inhal Toxicol* 24: 995–1008.

Hirano, S., Fujitani, Y., Furuyama, A., Kanno, S. (2010) Uptake and cytotoxic effects of multiwalled carbon nantobes in human bronchial epithelial cells. *Toxicol Appl Pharmacol* 249: 8–15.

Inoue, K., Koike, E., Yanagisawa, R., Hirano, S., Nishikawa, M., Takano, H. (2009) Effects of multi-walled carbonnanotube sonamurine allergic airway inflammation model. *Toxicol Appl Pharmacol* 237: 306–316.

Inoue, K., Takano, H., Koike, E., Yanagisawa, R., Sakurai, M., Tasaka, S., Ishizaka, A., Shimada, A. (2008) Effects of pulmonary exposure to carbonnanotubes on lung and systemic inflammation with coagulatory disturbance induced by lipopolysaccharide in mice. *Exp Biol Med* 233: 1583–1590.

Inoue, K., Yanagisawa, R., Koike, E., Nishikawa, M., Takano, H. (2010) Repeated pulmonary exposure to single-walled carbon nanotubes exacerbates allergic inflammation of the airway: Possible role of oxidative stress. *Free Radic Biol Med* 48: 924–934.

Kelleher, P. Pacheco, K., Newman, L.S. (2000) Inorganic dust pneumonia: The metal-related parenchymal disorders. *Environ Health Perspect* 108: 685–696.

Kim, J.S., Sung, J.H., Song, K.S., Lee, J.H., Kim, S.M., Lee, G.H., Ahn, K.H., Lee, J.S., Shin, J.H., Park, J.D., Yu, I.J. (2012) Persistent DNA damage measured by comet assay of Sprague dawley rat lung cells after five day of inhalation exposure and 1 month post-exposure to dispersed multi-wall carbon nanotubes (MWCNTs) generated by new MWCNT aerosol generation system. *Toxicol Sci* 128: 439–448.

Konduru, N.V., Tyurina, Y.Y., Feng, W., Basova, L.V., Belikova, N.A., Bayir, H., Clark, K., Rubin, M., Stolz, D., Vallhov, H., Scheynius, A., Witasp, E., Fadeel, B., Kichambare, P.D., Star, A., Kisin, E.R., Murray, A.R., Shvedova, A.A., Kagan, V.E. (2009) Phosphatidylserine targets single-walled carbon nanotubes to professional phagocytes *in vitro* and *in vivo*. *PLoS One* 4(2): e4398.

Lam, C.W., James, J.T., McCluskey, R., Hunter, R.L. (2004) Pulmonary toxicity of single-wall carbon nanotubes in mice 7 and 90 days after intra tracheal instillation. *Toxicol Sci* 77: 126–134.

Lee, J.K., Sayers, B.C., Chun, K.S., Lao, H.C., Shipley-Phillips, J.K., Bonner, J.C., Langenbach, R. (2012) Multi-walled carbon nanotubes induce COX-2 and iNOS expression via MAP kinase-dependent and -independent mechanisms in mouse RAW264.7macrophages. *Part Fibre Toxicol* 9: 14.

Li, J.G., Li, W.X., Xu, J.Y., Cai, X.Q., Liu, R.L., Li, Y.J., Zhao, Q.F., Li, Q.N. (2007) Comparative study of pathological lesions induced by multiwalled carbon nanotubes in lungs of mice by intratracheal instillation and inhalation. *Environ Toxicol* 22: 415–421.

Ma-Hock, L., Treumann, S., Strauss, V., Brill, S., Luizi, F., Mertler, M., Wiench, K., Gamer, A.O., van Ravenzwaay, B., Landsiedel, R. (2009) Inhalation toxicity of multiwall carbon nanotubes in rats exposed for 3 months. *Toxicol Sci*. 112(2): 468–481.

Mangum, J.B., Turpin, E.A., Antao-Menezes, A., Cesta, M.F., Bermudez, E., Bonner, J.C. (2006) Single-walled carbon nanotube (SWCNT)-induced interstitial fibrosis in the lungs of rats is associated with increased levels of PDGF mRNA and the formation of unique intercellular carbon structures that bridge alveolar macrophages *in situ*. *Part Fibre Toxicol* 3: 15.

Martinon, F., Mayor, A., Tschopp, J. (2009) The inflammasomes: Guardians of the body. *Annu Rev Immunol* 27: 229–265.

Mercer, R.R., Hubbs, A.F., Scabilloni, J.F., Wang, L., Battelli, L.A., Schwegler-Berry, D., Castranova, V., Porter, D.W. (2010) Distribution and persistence of pleural penetrations by multi-walled carbonnanotubes. *Part Fibre Toxicol* 7: 28.

Mercer, R.R., Scabilloni, J., Wang, L., Kisin, E., Murray, A.R., Schwegler-Berry, D., Shvedova, A.A, Castranova, V. (2008) Alteration of deposition pattern and pulmonary response as a result of improved dispersion of aspirated single-walled carbon nanotubes in a mouse model. *Am J Physiol Lung Cell Mol Physiol* 294: L87–L97.

Meunier, E., Coste, A., Olagnier, D., Authier, H., Lefèvre, L., Dardenne, C., Bernad, J., Béraud, M., Flahaut, E., Pipy, B. (2012) Double-walled carbon nanotubes trigger IL-1β release in human monocytes through Nlrp3 inflammasome activation. *Nanomedicine* 8: 987–995.

Mitchell, L.A., Gao, J., Wal, R.V., Gigliotti, A., Burchiel, S.W, McDonald, J.D. (2007) Pulmonary and systemic immune response to inhaled multiwalled carbon nanotubes. *Toxicol Sci* 100: 203–214.

Mitchell, L.A., Lauer, F.T., Burchiel, S.W., McDonald, J.D. (2009) Mechanisms for how inhaled multiwalled carbon nanotubes suppress systemic immune function in mice. *Nat Nanotechnol* 4: 451–456.

Muller, J., Huaux, F., Moreau, N., Misson, P., Heilier, J.F., Delos, M., Arras, M., Fonseca, A., Nagy, J.B., Lison, D. (2005) Respiratory toxicity of multi-wall carbon nanotubes. *Toxicol Appl Pharmacol* 207: 221–231.

Murphy, F.A., Schinwald, A., Poland, C.A., Donaldson, K. (2012) The mechanism of pleural inflammation by long carbon nanotubes: Interaction of longfibres with macrophages stimulates them to amplify pro-inflammatory responses in mesothelial cells. *Part Fibre Toxicol* 9: 8.

Murray, A.R., Kisin, E.R., Tkach, A.V., Yanamala, N., Mercer, R., Young, S.H., Fadeel, B., Kagan, V.E., Shvedova, A.A. (2012) Factoring-inagglomeration of carbon nanotubes and nanofibers for better prediction of their toxicity versus asbestos. *Part Fibre Toxicol* 9: 10.

Nygaard, U.C., Hansen, J.S., Samuelsen, M., Alberg T., Marioara, C.D., Løvik, M. (2009) Single-walled and multi-walled carbon nanotubes promote allergic immune responses in mice. *Toxicol Sci.* 109: 113–123.

Palomäki, J., Välimäki, E., Sund, J., Vippola, M., Clausen, P.A., Jensen, K.A., Savolainen, K., Matikainen, S., Alenius, H. (2011) Long, needle-like carbon nanotubes and asbestos activate the NLRP3 inflammasome through a similar mechanism. *ACS Nano* 5: 6861–6870.

Pauluhn, J. (2010) Subchronic 13-week inhalation exposure of rats to multiwalled carbon nanotubes: toxic effects are determined by density of agglomerate structures, not fibrillar structures. *Toxicol Sci.* 113(1): 226–242.

Park, E.J., Cho, W.S., Jeong, J., Yi, J., Choi, K., Park, K. (2009) Pro-inflammatory and potential allergic responses resulting from B cell activation in mice treated with multi-walled carbon nanotubes by intratracheal instillation. *Toxicology* 259: 113–121.

Patlolla, A.K., Hussain, S.M., Schlager, J.J., Patlolla, S., Tchounwou, P.B. (2010) Comparative study of the clastogenicity of functionalized and nonfunctionalized multiwalled carbon nanotubes in bone marrow cells of Swiss-Webster mice. *Environ Toxicol.* 25: 608–621.

Poland, C.A., Duffin, R., Kinloch, I., Maynard, A., Wallace, W.A.H., Seaton, A., Stone, V., Brown, S., MacNee, W., Donaldson, K. (2008) Carbon nanotubes introduced into the abdominal cavity of mice show asbestos-like pathogenicity in a pilot study. *Nat Nanotechnol* 3: 423–428.

Porter, D.W., Hubbs, A.F., Mercer, R.R., Wu, N., Wolfarth, M.G., Sriram, K., Leonard, S., Battelli, L., Schwegler-Berry, D., Friend, S., Andrew, M., Chen, B.T., Tsuruoka, S., Endo, M., Castranova, V. (2010) Mouse pulmonary dose- and time course-responses induced by exposure to multi-walled carbon nanotubes. Toxicology 269: 136–147.

Prato, M., Kostarelos, K., Bianco, A. (2008) Functionalized carbon nanotubes in drug design and discovery. *Acc Chem Res* 41: 60–68.

Ryman-Rasmussen, J.P., Cesta, M.F., Brody, A.R., Shipley-Phillips, J.K., Everitt, J., Tewksbury, E.W., Moss, O.R., Wong, B.A., Dodd, D.E., Andersen, M.E., Bonner, J.C. (2009a) Inhaled multi-walled carbon nanotubes reach the subpleural tissue in mice. *Nat Nanotechnol* 4: 747–751.

Ryman-Rasmussen, J.P., Tewksbury, E.W., Moss, O.R., Cesta, M.F., Wong, B.A., Bonner, J.C. (2009b) Inhaled multiwalled carbon nanotubes potentiate airway fibrosis in a murine model of allergic asthma. *Am J Resp Cell Mol Biol* 40: 349–358.

Sargent, L.M., Hubbs, A.F., Young, S.H., Kashon, M.L., Dinu, C.Z., Salisbury, J.L., Benkovic, S.A., Lowry, D.T., Murray, A.R., Kisin, E.R., Siegrist, K.J., Battelli, L., Mastovich, J., Sturgeon, J.L., Bunker, K.L., Shvedova, A.A., Reynolds, S.H. (2012) Single-walled carbon nanotube-induced mitotic disruption. *Mutat Res.* 745: 28–37.

Sargent, L.M., Shvedova, A.A., Hubbs, A.F., Salisbury, J.L., Benkovic, S.A., Kashon, M.L., Lowry, D.T., Murray, A.R., Kisin, E.R., Friend, S., McKinstry, K.T., Battelli, L., Reynolds, S.H. (2009) Induction of aneuploidy by single-walled carbon nanotubes. *Environ Mol Mutagen*. 50: 708–717.

Shvedova, A.A., Kisin, E.R., Mercer, R., Murray, A.R., Johnson, V.J., Potapovich, A.I., Tyurina, Y.Y., Gorelik, O., Arepalli, S., Schwegler-Berry, D., Hubbs, A.F., Antonini, J., Evans, D.E., Ku, B.K., Ramsey, D., Maynard, A., Kagan, V.E., Castranova, V., Baron, P. (2005) Unusual inflammatory and fibrogenic pulmonary responses to single-walled carbon nanotubes in mice. *Am J Physiol Lung Cell Mol Physiol* 289: L698–L708.

Shvedova, A.A., Kisin, E., Murray, A.R., Johnson, V.J., Gorelik, O., Arepalli, S., Hubbs, A.F., Mercer, R.R., Keohavong, P., Sussman, N., Jin, J., Yin, J., Stone, S., Chen, B.T., Deye, G., Maynard, A., Castranova, V., Baron, P.A., Kagan, V.E. (2008) Inhalation vs. aspiration of single-walled carbon nanotubes inC57BL/6mice: Inflammation, fibrosis, oxidative stress, and mutagenesis. *Am J Physiol Lung Cell Mol Physiol.* 295:L552–L565.

Strowig, T., Henao-Mejia, J., Elinav, E., Flavell, R. (2012) Inflammasomes in health and disease. *Nature* 481: 278–286.

Sun, B., Wang, X., Ji, Z., Li, R., Xia, T. (2012) NLRP3 inflammasome activation induced by engineered nanomaterials. *Small* 9: 1595–1607.

Takagi, A., Hirose, A., Nishimura, T., Fukumori, N., Ogata, A., Ohashi, N., Kitajima, S., Kanno, J. (2008) Induction of mesothelioma in p53 +/- mouse by intraperitoneal application of multi-wall carbon nanotube. *J Toxicol Sci* 33: 105–116.

Visser, C.E., Tekstra, J., Brouwer-Steenbergen, J.J., Tuk, C.W., Boorsma, D.M., Sampat-Sardjoepersad, S.C., Meijer, S., Krediet, R.T., Beelen, R.H. (1998) Chemokines produced by mesothelial cells: HuGRO-alpha,IP-10,MCP-1andRANTES. *Clin Exp Immunol* 112: 270–275.

Warheit, D.B., Laurence, B.R., Reed, K.L., Roach, D.H., Reynolds, G.A, Webb, T.R. (2004) Comparative pulmonary toxicity assessment of single-wall carbon nanotubes in rats. *Toxicol Sci* 77: 117–125.

Zhu, L., Chang, D.W., Dai, L., Hong, Y. (2007) DNA damage induced by multiwalled carbon nanotubes in mouse embryonic stem cells. *Nano Lett* 7: 3592–3597.

Section III

Emission and Exposure along the Lifecycle

11 Measurement and Monitoring Strategy for Assessing Workplace Exposure to Airborne Nanomaterials

Christof Asbach, Thomas A.J. Kuhlbusch, Burkhard Stahlmecke, Heinz Kaminski, Heinz J. Kiesling, Matthias Voetz, Dirk Dahmann, Uwe Götz, Nico Dziurowitz, and Sabine Plitzko

CONTENTS

11.1 BACKGROUND

Representative exposure measurement, following dedicated measurement strategies, is one of the foundations for epidemiological studies (Dahmann et al. 2008). Each exposure assessment can only be used in an international context if harmonized and accepted measurement strategies are applied. Although accepted protocols are available for assessing exposure to conventional materials, as of now, they are still lacking for nanomaterials. Over the recent years, several studies have been published that reported on exposure related measurements at nanotechnology workplaces (a review of these studies can be found in Kuhlbusch et al. 2011). These studies used their own measurement strategies, and most of them employed rather large sets of equipment, including instruments for measuring particle size distributions, number and mass concentrations, and samplers to collect particles for subsequent chemical and/or morphological analyses. The measurements often focused on particle number concentrations and their size distributions, because they are considered to be the most sensitive particle metrics when it comes to assessing exposure to nanomaterials. Although nanoscale particles usually occur in high number concentrations, they commonly only contribute negligibly to the total particle mass concentration, because the particle mass scales with the third power of the particle size. Some studies also suggested that health effects of inhaled particles may correlate better with their number rather than their mass (Peters et al. 1997) while other studies have reported that the particle surface area may play a more critical role (Oberdörster 2000; Tran et al. 2000).

Measurement instruments that are commonly used in exposure assessment are condensation particle counters (CPCs, McMurry 2000) for measuring number concentrations and scanning mobility particle sizers (SMPSs, Wang, and Flagan 1990) or similar electrical mobility spectrometers to determine the submicron particle number size distributions (see Chapter 2). Since nanoscale particles can occur as micron-sized agglomerates, concentrations of larger particles are often assessed by means of optical spectrometers or aerodynamic particle sizing. These instruments are rather bulky and mains operated. Simplified devices for measuring concentrations of airborne particles in workplaces are provided by different handheld, battery driven instruments (see Chapter 2), such as handheld CPC or diffusion charger based instruments (Fierz et al. 2011; Marra, Voetz, and Kiesling 2010; Asbach et al. 2012a; Kaminski et al. 2013). These instruments all measure the particle number and/or lung deposited surface area (LDSA) concentration and are hence usable for an easy screening of the workplace aerosol. Most diffusion chargers additionally measure the mean particle size. All the above-mentioned measurement techniques have in common that they only measure physical quantities, that is, particle size and/or concentration, but inherently they do not allow for the differentiation of the engineered nanomaterials from ubiquitous background particles, typically stemming from outdoor sources, combustion processes, or other human activity in the near and far field of the workplace (for a perspective on background sources and effects, compare Chapter 5). Workplace exposure measurements therefore always require knowledge about the particle background in the workplace, for example, by measuring the background concentration prior to and after the work process under investigation or simultaneously with a second set of equipment at a representative background site. Additionally, all these measurements can be accompanied by particle sampling for subsequent analysis of the particle morphology or the chemical composition. Sampling is typically done on filters or electrostatically onto flat surfaces, using an electrostatic precipitator (Dixkens and Fissan 1999). These analyses can then provide a more definitive proof for the presence or absence of the engineered nanomaterial in the workplace air. Still, the choice of instruments has to be carefully considered depending on the measurement task. Their use and data evaluation has to follow harmonized protocols concerning, for example, the placement of instruments, particle size and concentration ranges to be covered, averaging time intervals, and so on. It is obvious that such exhaustive measurement campaigns are very labor and cost intensive. Since there is no legal framework that outlines the measurements and their data evaluation, the results of such measurements are not comparable with similar measurements carried out elsewhere and also not suitable for storage in exposure databases, which eventually are the base for epidemiological studies. Furthermore, especially small and medium enterprises may not have the financial means to carry out such measurement campaigns. A pragmatic, clear, and harmonized approach is hence required for both measurement and data evaluation.

Several groups across the globe have been working on the development of such measurement strategies over the recent years (e.g., Methner, Hodson, and Geraci 2009a and 2009b; BAuA et al. 2011; Asbach et al. 2012b; Witschger et al. 2012, others are still unpublished) and efforts have been taken to harmonize these approaches (Brouwer et al. 2012). All these measurement strategies have many similarities. For example, all the aforementioned approaches are tiered to be pragmatically applicable. Tier by tier the measurement effort increases and the uncertainty hence decreases. Clear decision criteria for moving to a higher tier are defined. One of these measurement strategy frameworks was developed in the German project nanoGEM and is described in detail here. Advanced standard operation procedures (SOPs) are freely available for every measurement step of this approach (Asbach et al. 2012b) as well as for the use of the various measurement instruments applied.

11.2 THE NANOGEM TIERED APPROACH

The aforementioned challenges in assessing exposure to airborne engineered nanomaterials were considered in the nanoGEM approach, particularly taking into account that an exposure assessment strategy needs to be widely adopted, that is, in large industry, in small and medium enterprises, and in research laboratories. The approach therefore has to be pragmatic and also needs to offer means for a rather simple assessment. The nanoGEM strategy is based on a publication by the German Chemical Industry Association VCI (BAuA et al. 2011) which foresees three tiers. The goal of tier 1 is to gather as much information as possible about the workplace and the nanomaterial(s) considered. If a release of the nanomaterial into the workplace air cannot be excluded, a rather simple measurement of the particle concentrations in the workplace is done in tier 2. If these measurements reveal that the particle concentration in the workplace is significantly higher than the background concentration and this increase cannot be tracked back to other, not nanomaterial related processes in the workplace, an extensive measurement campaign is carried out in tier 3. If the result of a tier 3 measurement is that nanomaterials are released from the process leading to worker exposure, additional risk management steps have to be taken by the company. The overall three-tiered hierarchical structure of the concept was maintained in the nanoGEM strategy, as shown in Figure 11.1.

11.2.1 TIER 1

The task in tier 1 is to clarify, if and how nanomaterials are used in the respective workplace and whether there is a chance that they can be released from the corresponding processes. The latter usually requires an on-site inspection of the considered workplace(s). If a release cannot be excluded, the potential exposure has to be determined in tier 2. Furthermore, the type of nanomaterials used in the process has to be specified and information on its potential toxicity gathered. If the material

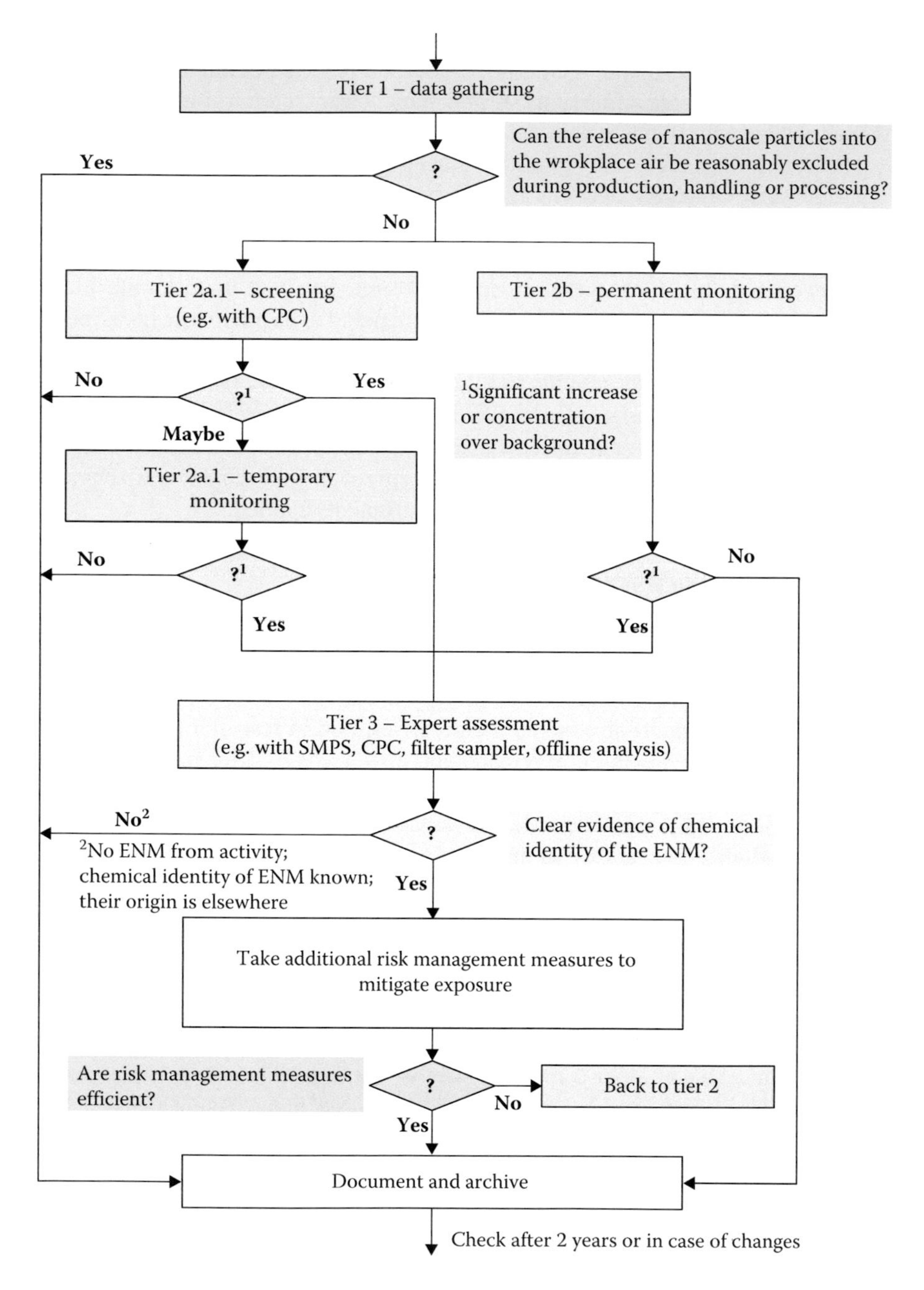

FIGURE 11.1 Flow scheme of the nanoGEM tiered approach for assessing workplace exposure to airborne engineered nanomaterials.

is of known high toxicity, then a more material or morphology specific detection technique is required than that used in tier 2. In this particular case, it has to be decided whether a tier 3 measurement is sufficiently specific or a completely different approach has to be applied.

11.2.2 Tier 2

As long as no health based limit values exist for engineered nanomaterials, measured concentrations need to be compared with the background concentration levels to determine a realistic value for the exposure to engineered nanomaterials. Measurements according to tier 2 foresee a simplified assessment of the workplace exposure by means of either a short-term screening (tier 2a.1) or long-term (tier 2a.2) or even permanent (tier 2b) monitoring. Measurements in tier 2 are conducted using particle size integrating, easy-to-use devices, measuring, for example, the total particle number concentration as shown in Figure 11.2. Typical instruments are handheld, battery-operated CPCs and diffusion charger based devices (for more details on instrumentation see Chapter 2). The most important outcome of tier 2 measurements is a decision whether an expanded measurement according to tier 3 is necessary or whether emission of nanoscale particles released from ENMs into workplace air during production, handling, or processing can be reasonably excluded. However, tier 2 may already give clear indication that additional (technical) measures are required to avoid exposure of workers. In that case, those measures may be realized first before tier 2 measurements are repeated for their verification.

If no permanent particle monitoring system is installed in the workplace under consideration, tier 2 measurements are conducted as a screening of the particle concentration. The main questions to be answered during screening measurements are

FIGURE 11.2 Typical screening measurement using a handheld CPC during cleaning operations.

What is the average background concentration taking into account the measurement characteristics of the measurement procedure involved (e.g., time characteristics) and what is its standard deviation?

Are defined sources of emissions identifiable/locatable (above background level)? If yes, what are the average emission concentration and its standard deviation?

Are nanoobjects in the breathing zone of workers (exposure concentrations) identifiable (above background level)? If yes, what are the average exposure concentration and its standard deviation?

Are the identified exposure or emission concentrations significantly above background level?

For the background determination, it has to be taken into account that other process-related particle sources, such as electrical motors or hot surfaces, may emit particles and therefore contribute to the particle concentration in the workplace. Often these particles are in a similar size range as the nanomaterial produced or handled and are therefore included in the particle concentration reading of the handheld particle concentration monitor. Ideally, the background is hence measured prior to the process step with all equipment and machinery running, only without the nanomaterial being applied. This is, however, often impossible. Instead, the background concentration can be measured either prior to and after the process step under investigation or during the investigated process step at a representative background site. If the workplace under consideration is enclosed and has a forced ventilation system, the background can best be measured by assessing the particle concentration directly in the supply air. The advantage of a parallel measurement of the background is that potential influences from outside (fork-lift, welding, smoking, etc.) can be excluded from the determination. This type of far-field measurement does, however, request the availability of a second measurement instrument.

The background determination measurements have to cover a time period of typically at least 45 min. During this period several measurements of equal lengths have to be performed (e.g., 9 times 5 min or 3 times 15 min). The results of these individual measurements have to be recorded, documented, and their arithmetic means as well as standard deviations have to be calculated and documented. The time series of the concentration measurements also needs to be taken into account and be critically evaluated, because they may reveal only short-lived particle releases which, depending on the local flow situation, can lead to very short spikes in the measured particle concentration. Depending on the averaging time, such short spikes may completely vanish and only have a small impact on the standard deviation. In such cases, the averaging times need to be lowered.

If a source of possible nanomaterial emissions could be identified, the concentration near that source has to be determined during application of the relevant process and the reason why this particular source of emission is considered relevant has to be documented. The measurements are repeated several times to cover the same measurement intervals as during the background determination (see the aforementioned). If the emission process does not last for such a time period, the complete duration of the process has to be covered as a minimum. Again, the arithmetic mean and standard deviation of the emission concentrations are calculated. If personal monitors are available, workers may be equipped with these instruments during the same measurement times, and the arithmetic mean and standard deviation of the individual exposure concentration in the breathing zone are determined as previously described..

The net emission or exposure concentrations are then determined by subtracting the mean background concentration from the mean emission or exposure concentration:

$$C_{(net-E)} = C_E - C_B \tag{11.1}$$

$$C_{(net-BR)} = C_{BR} - C_B \tag{11.2}$$

Here, C stands for the particle concentration and the indices E for emission, BR for breathing zone, and B for background.

To judge whether a potentially increased particle concentration in the workplace is significantly higher than in the background, the standard deviations of the background and emission or breathing zone measurements, respectively, are taken into account. The workplace concentration is considered to be significantly increased over the background if the net exposure or release concentration ($C_{(net-BR)}$ $C_{(net-E)}$) is larger than 3 times the standard deviation $s_{D,B}$ of the background concentration:

$$C_{(net-E)} > 3 \cdot s_{D,B} \tag{11.3}$$

$$C_{(net-BR)} > 3 \cdot s_{D,B} \tag{11.4}$$

It should be noted that this procedure is only applicable if the background concentration does not fluctuate strongly; that is, if the standard deviations of the background and the near-field measurements are of the same order of magnitude.

If a screening measurement according to tier 2a.1 reveals a significantly increased workplace concentration, this may be double-checked by a long-term monitoring (tier 2a.2). In tier 2a.2, small, easy-to-use devices are installed in the facility under surveillance, near the detected potential particle source or near the ventilation inlet (see Figure 11.3), if available. Such measurements can be conducted over several days to weeks. Since particle concentrations may vary significantly over such long time, the background particle concentration should frequently be measured, best by installing an additional long-term monitor in a representative location. Only instruments that do not require frequent attention or maintenance are suitable for the monitoring task; that is, diffusion charger based instruments are usable, whereas handheld CPCs are not an option, because they need a frequent refill of their alcohol reservoir.

If aerosol monitors are permanently installed in the facility, they are used to continuously measure the workplace concentration. As long as no legally binding limits exist, certain threshold concentration can be defined as a trigger for further investigations on "suspicious concentrations." Monitoring data are examined as described

FIGURE 11.3 Placement of diffusion charger based monitors (miniDiSC) during tier 2b measurements, left: at the ventilation inlet and right: above a valve that had been identified as a potential leak.

earlier for the screening measurements to judge whether the workplace concentration is significantly increased over the background concentration.

If tier 2a or 2b reveals that the particle concentration in the investigated workplace is significantly increased over the background concentration and the reason for this increase cannot unambiguously be tracked back to other, not-nanomaterial related sources, an expanded measurement according to tier 3 is recommended.

11.2.3 Tier 3

The aim of tier 3 measurement campaigns is to verify (or falsify) that nanomaterials have been released into the workplace under investigation, leading to an exposure of workers to nanomaterials. The tier 3 measurement strategy was developed for online measurements of particle size distributions and total concentrations, which are to be used in combination with particle sampling for consecutive chemical and morphological analyses. Tier 3 measurements are always accompanied by measurements of at least the total background particle concentration. On the basis of this combination it is possible to undertake an estimation of the inhalation exposure to nanoscale product materials at workplaces, and in particular to delimit them in relation to possible background concentrations.

Tier 3 typically includes the measurement of particle size distributions in the submicron and micron size range using electrical mobility spectrometers (SMPS or FMPS) and aerodynamic (APS) or optical (OPC) spectrometers, respectively. Those size distribution measurements are accompanied by the measurement of the total particle number concentration using CPC. Sometimes the LDSA concentration and—to comply with conventional exposure assessment—the particle mass concentration of, for example, the alveolar dust fraction (particles <10 μm) are also measured. The latter can be measured either time resolved by using Tapered Element Microbalances (Patashnick and Rupprecht 1991) or optical spectrometers or time integrated by sampling particles onto filters and weighing them before and after sampling. These filters can also be used for wet chemical analyses of the chemical composition of the collected particles. Other methods for determining the chemical composition of the particles in the workplace are by sampling them electrostatically (Dixkens and Fissan 1999) and by impaction (e.g., Marple et al. 1986) onto flat surfaces which are suitable for analysis by total reflection X-ray fluorescence (TXRF, John at el. 2001) or in electron microscopes with electron dispersive X-ray spectroscopy (EDX). The chemical analyses provide clear evidence for whether or not the nanomaterials produced or handled in the workplace were present in the airborne state. If the nanomaterial is a nonubiquitous material in the (workplace) atmosphere, the purely chemical analysis can already provide definitive proof, whereas in the case of materials that are commonly found in the atmosphere (e.g., silica), morphological analyses by electron microscopy are required. From the large variety of information that needs to be assessed in tier 3 measurements, it is obvious that the set of equipment needed for such measurements is quite extensive. A setup used during tier measurements at a pilot plant at IUTA (Huelser et al. 2011; Wang et al. 2012) is shown in Figure 11.4.

Similarly to tier 2, the particle background in the workplace can be determined by using a second set of measuring equipment to measure in a representative background site or by using the same set of equipment to measure before and after the considered process. For the extensive measurements in tier 3, it is desirable to have at least one instrument measuring the total number concentration (e.g., a CPC) in the background in parallel with high time resolution, because this can help to identify potential particle sources from outside the work area. If a complete second set of equipment is used for the background measurements, including size distribution measurements and particle sampling, more information on possible particle infiltration from outside can be derived, because it can be shown whether particles of certain sizes, composition, or morphology had been present in the background.

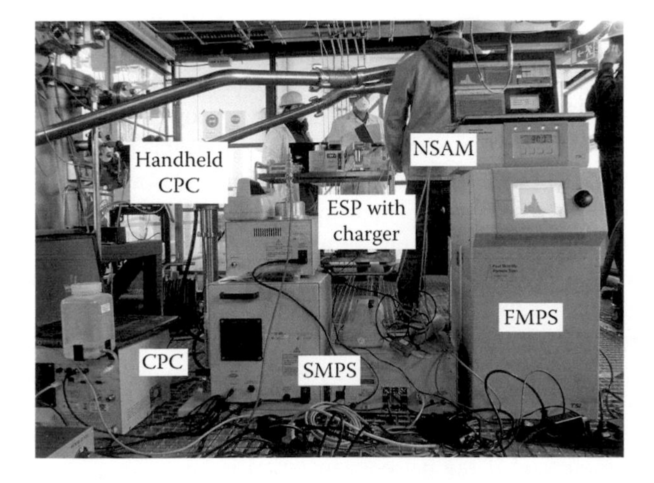

FIGURE 11.4 Setup of instruments for a tier 3 workplace measurement, including SMPS and FMPS for measuring submicron particle size distributions, a CPC for measuring particle number concentration, and an NSAM for LDSA concentration as well as an NAS for particle sampling for subsequent morphological and chemical analyses.

The sequence of a measurement campaign is such that all instruments to be used in the campaign are to be checked for plausibility of their results. Results from instruments measuring the same metrics must be intercompared. For the on-site measurements, the measurement staff commonly arrives a day before the actual start of the measurements to set up the instruments and synchronize the clocks. All instruments are left running overnight. The overnight measurements have two purposes. On the one hand they provide background data for a situation in the facility when no one is working (provided there is no night shift). On the other hand, a quick comparison of the instruments is used to detect possible damage to the instruments that may have occurred during their transport. For the measurements, the instrument(s) for determining the background are placed in a representative background location. If available, this may be the supply air from the local ventilation system. The instruments for measuring in the near field of the workplace are located near the location, where particle release is expected. All instruments should sample through a common inlet, which of course needs to be suitable for the resulting total flow rate, or at least very close to each other to avoid spatial variations affecting the measurements. Depending on the conditions in the workplace, the measuring instruments may need to be placed inside an air conditioned box. The near field measurements should be accompanied by personal exposure measurements, if appropriate personal samplers or monitors are available. Prior to the start of the measurements, all instrument flow rates need to be checked and documented.

The measurements are usually conducted over a full work shift or at least the length of the considered process. For data evaluation, in principle a similar procedure as in tier 2 is applied. The time series for the particle number concentration and the geometric mean of the particle size distribution at the workplace must be presented and every relevant event interpreted. The particle size resolved number concentration from the size distribution data in the workplace (and if available in the background) is divided into different particle size fractions which are then further used for comparison. Only the definition of such harmonized size ranges allows for a comparison of data gathered by different groups that may even use different types of instruments. The size ranges to be reported are: Lower size limit (LSL)—100 nm, 100–400 nm, 400 nm–1 μm, and 1–10 μm (only applicable to the size ranges that have been determined). Since aerosol measuring equipment commonly assumes particles to be spherical they usually report particle sizes as equivalent diameters of spheres, where the equivalency depends on the measurement method. All particle size data reported therefore need to include information on which equivalent diameter is used.

All data from workplace and background measurements are averaged over equal time steps (e.g., 5 min or 15 min) and the standard deviations calculated. For size distribution data, this needs to be done for each of the aforementioned size ranges. The net emission or exposure concentration is calculated according to Equation 11.1 or 11.2, respectively. The workplace or exposure concentration is considered to be significantly increased if the net concentration is larger than 3 times the standard deviation of the background concentration. In this case the particle samples are analyzed for their chemical composition and particle morphology. If the nanomaterial produced or handled in the workplace is found among the sampled particles, this proves that indeed the nanomaterial was released into the workplace and caused worker exposure. In this case, the operator of the facility will need to take additional risk management measures to mitigate exposure (see Figure 11.1). This may include closing leaks, adding an enclosure to the facility, or improving the local ventilation. The effectiveness of such measures will need to be verified by tier 2 measurements.

11.3 CONCLUSIONS

The tiered approach described above presents a pragmatic approach to assess exposure in workplaces, which is also feasible for small and medium enterprises. While in the past, most exposure related measurements for nanomaterials were quite cost and labor intensive, the possibility of doing a simplified screening or monitoring first simplifies the whole process. Now, such intensive measurements are only required if the screening revealed significantly increased concentrations. At the same time the approach and the corresponding SOPs for the first time suggested clear data evaluation procedures and clear decision criteria whether or not a measured particle concentration in the workplace is significantly increased. The approach therefore provides a first step towards a more harmonized and eventually standardized nanomaterial exposure assessment. The approach and SOPs have been widely distributed and are currently discussed within the framework of international harmonization activities.

It should, however, be noted that the presented tiered approach also has some limitations. Strictly, it is more focused on the release of nanomaterials from processes rather than the exposure itself. The assessment of exposure requires the use of personal samplers or monitors, taking samples in the breathing zone of a worker, that is, within a 30-cm hemisphere around mouth and nose. For assessing exposure to nanomaterials, such instruments have only very recently become available and are hence not yet included here. Furthermore, the approach is only usable for materials that have been shown to be not highly toxic. In the case of toxic materials, small amounts of such substances in the workplace air may have dramatic effects. If these amounts are within the typical fluctuations of the particle concentrations in the workplace, they cannot be detected with the means described earlier. In such a case a measurement technique that specifically detects the toxic material would need to be employed, similar to a gas alarm sensor. Since as of now there are no such online measurement techniques, only the use of particle sampling and subsequent offline analyses, as foreseen in tier 3 can be used.

ACKNOWLEDGMENTS

This research was conducted under the umbrella of the nanoGEM project, funded by the German Ministry for Education and Research (BMBF) under grant number 03X0105. The financial support is gratefully acknowledged.

REFERENCES

Asbach, C., Kaminski, H., von Barany, D., Kuhlbusch, T.A.J., Monz, C., Dziurowitz, N., Pelzer, J., Vossen, K., Berlin, K., Dietrich, S., Götz, U., Kiesling, H. J., Schierl, R., Dahmann, D. (2012a): Comparability of portable nanoparticle exposure monitors. *Ann Occup Hyg*, 56: 606–621.

Asbach, C., Kuhlbusch, T.A.J., Kaminski, H., Stahlmecke, B., Plitzko, S., Götz, U., Voetz, M., Kiesling, H. J., Dahmann, D. (2012b): NanoGEM Standard Operation Procedures for assessing exposure to nanomaterials, following a tiered approach, http://www.nanogem.de/cms/nanogem/upload/Veroeffentlichungen/nanoGEM_SOPs_Tiered_Approach.pdf (accessed on October 13th, 2013).

BAuA, BG RCI, IFA, IUTA, TUD, VCI (2011): Tiered Approach to an Exposure Measurement and Assessment of Nanoscale Aerosols Released from Engineered Nanomaterials in Workplace Operations, available online: https://www.vci.de/Downloads/Tiered-Approach.pdf (accessed on October 13th, 2013).

Brouwer, D., Berges, M., Virji, M.A., Fransman, W., Bello, D., Hodson, L., Gabriel, S., Tielemans, E. (2012): Harmonization of measurement strategies for exposure to manufactured nano-objects; Report of a workshop. *Ann Occup Hyg*, 56: 1–9.

Dahmann, D., Taeger, D., Kappler, M., Büchte, S., Morfeld, P., Brüning, T., Pesch, B. (2008): Assessment of exposure in epidemiological studies: the example of silica dust. *J Exposure Sci Environ Epidemiol*, 18: 452–461.

Dixkens, J., Fissan, H. (1999): Development of an electrostatic precipitator for off-line particle analysis. *Aerosol Sci Technol*, 30: 438–453.

Fierz, M., Houle, C., Steigmeier, P., Burtscher, H. (2011): Design, calibration, and field performance of a miniature diffusion size classifier. *Aerosol Sci Technol*, 45: 1–10.

Huelser, T., Schnurre, S.M., Wiggers, H., Schulz, C. (2011): Gas phase synthesis of nanoscale silicon as an economical route towards sustainable energy technology. *KONA Powder Part J*, 29: 191–207.

John, A.C., Kuhlbusch, T.A.J., Fissan, H., Schmidt, K.G. (2001): Size-fractionated sampling and chemical analysis by total-reflection X-ray fluorescence spectrometry of PMx in ambient air and emissions. *Spectrochim Acta Part B,* 56: 2137–2146.

Kaminski, H., Rath, S., Götz, U., Sprenger, M., Wels, D., Polloczek, J., Bachmann, V., Dziurowitz, N., Kiesling, H.J., Schwiegelshohn, A., Monz, C., Dahmann, D., Kuhlbusch, T.A.J., Asbach, C. (2013): Comparability of mobility particle sizers and diffusion chargers. *J Aerosol Sci*, 57: 156–178.

Kuhlbusch, T.A.J., Asbach, C., Fissan, H., Göhler, D., Stintz, M. (2011): Nanoparticle exposure at nanotechnology workplaces: a review. *Part Fibre Toxicol*, 8: 22.

Marple, V., Rubow, K., Ananth, G., Fissan, H. (1986): Micro-orifice uniform deposit impactor. *J Aerosol Sci*, 17: 489–494.

Marra, J., Voetz, M., Kiesling, H.J. (2010): Monitor for detecting and assessing exposure to airborne nanoparticles. *J Nanopart Res*, 12: 21–37.

McMurry, P.H. (2000): The history of condensation nucleus counters. *Aerosol Sci Technol*, 33: 297–322.

Methner, M., Hodson, L., Geraci, C. (2009a): Nanoparticle Emission Assessment Technique (NEAT) for the identification and Measurement of Potential Inhalation Exposure to Engineered nanomaterials–Part A. *J Occup Environ Hyg*, 7: 127–132.

Methner, M., Hodson, L., Geraci, C. (2009b): Nanoparticle Emission Assessment Technique (NEAT) for the identification and Measurement of Potential Inhalation Exposure to Engineered nanomaterials–Part B: Results from 12 field studies. *J Occup Environ Hyg*, 7: 163–176.

Oberdörster, G. (2000): Toxicology of ultrafine particles: in vivo studies. *Philos T Roy Soc A*, 358: 2719–2740.

Patashnick, H., Rupprecht, E.G. (1991): Continuous PM10 measurements using the tapered element oscillating microbalance. *J Air Waste Manage Assoc*, 41: 1079–1083.

Peters, A., Wichmann, H.E., Tuch, T., Heinrich, J., Heyder, J. (1997): Respiratory effects are associated with the number of ultrafine particles. *Am J Resp Crit Care*, 155: 1376–1383.

Tran, C.L., Buchanan, D., Cullen, R.T., Searl, A., Jones, A.D., Donaldson, K. (2000): Inhalation of poorly soluble particles. II. Influence of particle surface area on inflammation and clearance. *Inhal Toxicol*, 12: 1113–1126.

Wang, J., Asbach, C., Fissan, H., Hülser, T., Kaminski, H., Kuhlbusch, T.A.J., Pui, D.Y.H. (2012): Emission measurement and safety assessment for the production process of silicon nanoparticles in a pilot scale facility. *J Nanopart Res*, 14: 759.

Wang, S.C., Flagan, R.C. (1990): Scanning electrical mobility spectrometer. *Aerosol Sci Technol*, 13: 230–240.

Witschger, O., Le-Bihan, O., Reynier, M., Durand, C., Charpentier, D. (2012): Préconisation en matière de caractérisation et d'exposition des potentiels d'emission et d'exposition professionnelle aux aerosols lors d'operations nanomateriaux, INRS - *Hygiène et sécurité du travail - 1er trimestre 2012*, 226: 41–55.

12 Release from Composites by Mechanical and Thermal Treatment
Test Methods

Thomas A.J. Kuhlbusch and Heinz Kaminski

CONTENTS

12.1 INTRODUCTION

12.1.1 MOTIVATION

Nanotechnologies are increasingly being used in industries and by consumers. Currently, there are no major concerns linked to them in regards to their impact on health and the environment. This may change in the next few years if health or environmentally relevant harmful nano-objects become associated with nanotechnologies. One of the possible risks is the release of nano-objects, for example, during processing or use of nanomaterials with hazard potential for workers, consumers, and the environment. The most straightforward way to avoid risks is by minimizing exposure. Therefore, detailed knowledge is important to understand how and when a possible release of particles from products with nanoparticles or nanocomposites may occur. Harmonized approaches are needed for various processes along the lifecycle to enable risk assessors, regulators, and product developers to assess possible risks of exposures for humans and the environment. The focus of this chapter is on the methods of testing and characterizing various mechanical and thermal release processes along the lifecycles of nanocomposites. The possibility and need for a standardized release test method will be discussed. The review of the different test methods shows that release testing is feasible, but also that not all test setups can deliver the data needed to assess the exposure risk for a given release scenario.

12.1.2 Introduction to Release Testing

Nanocomposites have been used in commercial products for more than a century as can be seen by the use of carbon black and precipitated amorphous silica in rubber tires. The use of carbon black in tires, invented by the B.F. Goodrich Company in 1910, led to a more than 100-fold increase in strength and durability of rubber tires.* This nanofiller is still used today. But nowadays nanocomposites are also used more and more in other products such as construction and insulation materials (Van Broekhuizen et al. 2011; Lee et al. 2010) or as components in cars, air planes, and ships to, for example, strengthen the construction by keeping or even reducing their weight. Here, nanocomposites are understood as a solid material containing nanomaterials, phases, or structures with one to three dimensions smaller than 100 nm.

The increased use of nanocomposites has raised the concern on possible risks stemming from their production and use. One of the main concerns here is the use of nanomaterials as fillers in composites, which potentially leads to the release of nanomaterials into the environment. The risk that nanomaterials may pose for humans and nature was identified at the end of the last century and since then investigations have been conducted to assess the probability of release and exposure. It has to be noted that nanocomposites with pores in the nanoscale but no dispersed nanomaterial, as used in insulation materials, cannot lead to the release of original nano-objects added to the matrix.

To be able to assess potential exposure and consequently, when linked with the hazard potential, the risk of the use of a nanomaterial in a composite, detailed knowledge on the release is needed. Different approaches may be pursued to derive the information needed for exposure assessment. It is essential to recognize though that measurements allow the assessment of only one case study at a time while more general knowledge is needed to ensure safe use of nanocomposites. One tiered approach was introduced by Schneider et al. (2011), where source-related studies are used to derive information on release rates and the characteristics of the released material. Modeling of the transport and transformation of the released material can then be used to derive

* M. Bellis, History of tires, www.inventors.about.com, and D. Hiskey, Making tires black, www.todayifoundout.com, 1. Nov. 2013.

exposure concentrations at various locations, for example, in work environments. This approach is exemplarily discussed in Figure 12.1 (Schneider et al. 2011) but can be used in principle for all release, transport, and transformation processes.

Necessary input information for the modeling of transport and transformation processes of released nanomaterials includes particle number, surface area and/or mass concentrations, size distributions, release rates, and detailed information on the material and release process itself: Is the nano-object still embedded in the matrix, released as an agglomerate or a primary nanoscaled particle? Further information on the matrix material itself is also needed to enable the assessment of possible degradation of the matrix material, which ultimately would lead to the release of an embedded persistent nano-object.

The actual release characteristics and hence model input data are dependent on the release process itself. This means that, for example, nano-SiO_2 particles that were dispersed in a polyamide (PA) composite matrix will be released in a different form into the air during sanding compared to incineration. While the primary nanoparticles will most likely be embedded in the matrix when released during sanding, all carbon may be combusted and pure nano-SiO_2 released by incineration. But, also different input data may be needed for one single release process. Taking sanding of a nanocomposite surface with different sanding papers performed by different people as an example, possibly significant variations can be observed: coarse paper, fast and heavy weight compared to fine paper, and relatively slow and low weight sanding (Figure 12.2).

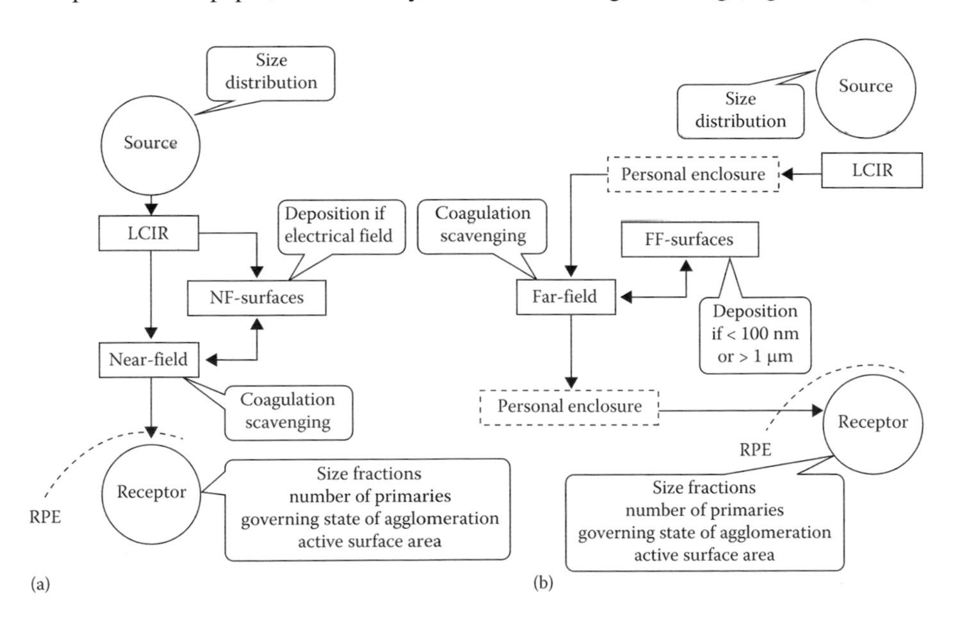

FIGURE 12.1 Conceptual model for inhalation exposure: (a) near-field (NF) source and (b) far-field (FF) source. The rectangles indicate the compartments, whereas the callouts indicate the transport process. LCIR, local influence region; RPE, respiratory protective device. (Adapted by permission from Macmillan Publishers Ltd., *J. Expos. Sci. Environ. Epidemiol.*, Schneider, T. et al., Conceptual model for assessment of inhalation exposure to manufactured nanoparticles, 21(5); 450–463, copyright 2011.)

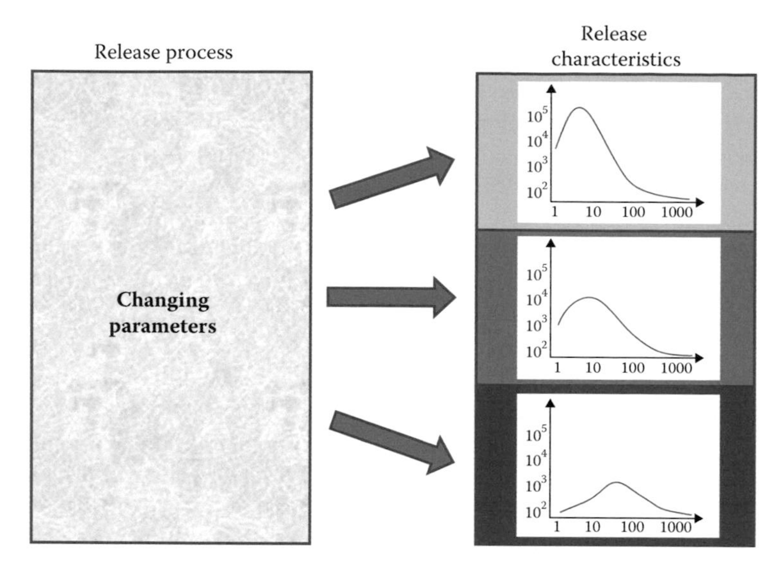

FIGURE 12.2 Influence of changing parameters during the release process on particle release.

This chapter focuses on the methods of testing and characterizing various mechanical and thermal release processes along the lifecycles of a nanocomposite. The possibility and need for standardized release test methods will be discussed. Other relevant environmental release processes are described by Nowak (2014) in Chapter 15 and Nguyen et al. (2014) in Chapter 14 of this book. The accompanying Chapter 13 by Brouwer et al. (2014) in this book presents the current knowledge on the results of release investigations linked to exposure-related measurements at workplaces. For further information on measurement techniques and analytical methods employed for the characterization of released particles the reader is referred to Chapters 1, 2, and 11 by Izak-Nau and Voetz (2014), Asbach (2014), and Asbach et al. (2014) all in this book.

12.2 RELEASE DUE TO MECHANICAL PROCESSES

Mechanical processes, mainly sheer forces, lead to stress, and release of nanomaterials may occur along the whole lifecycle of a nanocomposite. Some of the processes discussed as follows may take place in more than one step of the lifecycle, for example, abrasion. This slow and relatively soft process may occur during processing, use, and recycling or during environmental transport of nanocomposites (fragments) in natural waters; other processes such as drilling occur only during further processing of nanocomposite materials. Overall, all of the processes described in this section are potentially relevant for the release of nanomaterials into the environment either suspended in air or in liquids, mainly water. The definitions of the processes are based on those discussed in a report by Canady et al. (2013).

12.2.1 Stirring and Sonication

Mixing by stirring and sonication are mechanical processes where the nanomaterial is brought into or is already in a liquid phase. This process is required to disperse different nano-objects in suspensions. This handling step should normally be conducted with local exhaust ventilation along with a glove box to minimize exposure (Dahm et al. 2012).

Two general different release processes can be differentiated for this handling step: (a) the emission of nano-objects during the handling of the powder and (b) emissions due to the activity of stirring and sonication.

Dustiness tests of powders have already been standardized (EN 15051) but normally several hundred grams of powder are needed. Therefore, developments have been made in the commonly established single or continuous drop methods to reduce the amount of powder needed for testing as well as the use of additional measurement techniques such as the fast mobility particle sizer (FMPS) to allow for the quantification of nano-object release (Schneider and Jensen, 2008; Evans et al. 2013). Some methods using stronger shear forces to test agglomerate stability have also been developed allowing release assessments for shear forces larger than those present during dropping (e.g., Stahlmecke et al. 2009). Overall, dustiness tests with rotating drums, single or continuous drop methods are quite well developed but comparisons of different methods showed significant differences in absolute values as well as in ranking of different nanopowders (Tsai et al. 2012). Anyhow, prenormative tests are currently (2013) organized within Comité Européen de Normalisation (CEN).

Emissions during stirring and sonication were, for example, studied by Johnson et al. (2010). They used defined sonication protocols and determined elevated particle number concentrations in the submicrometer range with droplets containing nanomaterials. The overall release was not quantified. Fleury et al. (2013) used a commercial batch mixer equipped with roller rotors not further described. They reported no significant release and no free nano-objects during mixing but elevated concentrations during cleaning processes. The current status for international harmonization of release testing for stirring and sonication is viewed to be limited due to the lack of detailed further studies.

12.2.2 Sanding

Sanding is a high energy type of mechanical stress where shear forces, due to processing of a material with a rough surface, act on the matrix. Cracks can be propagated through the nanocomposite matrix and thus nano-objects can be pulled out of the matrix due to shear forces. As with all high speed processes sanding may also produce heat influencing a possible release.

Göhler et al. (2010) developed a sanding test setup in a particle-free environment for the quantification of the nanoparticle release into air from surface coatings (Figure 12.3). For a defined sanding of surface coatings, a miniature sander was used. The sanding machine consists of a small abrasion wheel with a diameter of 7 mm that rotates with an adjustable peripheral speed in the range of 1.8–24 m/s.

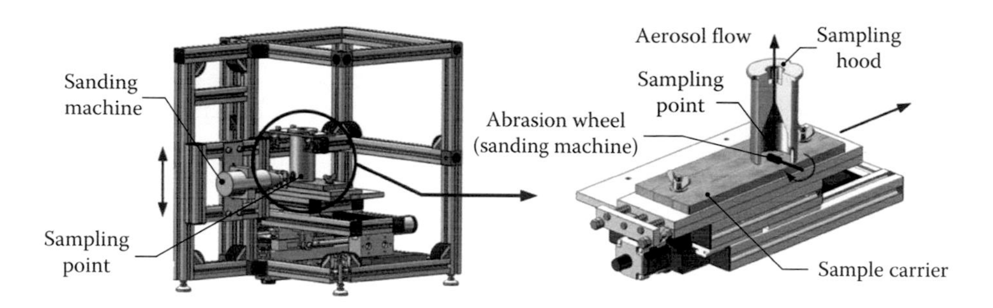

FIGURE 12.3 3D-CAD model of the test device employed for sanding with marked sampling hood. (Adapted from Göhler, D. et al., Characterization of nanoparticles release from surface coatings by the simulation of a sanding process, 2010, by permission of Oxford University Press.)

TABLE 12.1
Comparison of Characteristic Values between the Employed Miniature Sander and Professional Sanding Processes

	Sander (Dremel)	Typical Sanding Process
Contact force (N)	0.2–1.0	400–40,000
Contact pressure (Pa)	10,000–50,000	3,000–20,000
Peripheral speed (m/s)	1.8–24	3.0–17.0

Source: Adapted from Göhler, D. et al., *Ann. Occup. Hyg.*, 54, 615–624, 2010.

The sanded surface was a rectangular area of 13 cm², the contact pressure 10 kPa, the relative peripheral speed 1.83 m/s, and the sanding paper was P600 with a grit size of 25.8 μm. The abraded material during sanding was continuously sucked through the sampling hood (see aerosol flow in Figure 12.3). The employed measurement instruments, an FMPS, a laser aerosol particle sizer, and a condensation particle counter (CPC), received the sample flows through a flow splitter installed directly behind the sampling hood. A comparison of the characteristic parameters between the used sander and typical industrial sanding processes is given in Table 12.1.

Five different coatings (without nanoparticle [NP] and with ZnO or Fe_2O_3-NP) were examined in this sanding study in five repetitions. The results show considerable generation of nanoparticles by the sanding process, but with no significant difference between coatings containing nanoparticle additives and without additives. Furthermore, the measurements exhibit a large spread in the released particle concentrations even though the swarf mass data were in a reasonable agreement.

Cena and Peters (2011) conducted a study in a facility that produces test samples composed of epoxy reinforced with carbon nanotubes (CNTs). The test samples of different formulations and hence different properties of the nanocomposites were rectangular sticks. After production the nanocomposite test samples that contained 2 mass% CNTs were broken apart manually and sanded with P220 sanding paper (68 μm grit size) to remove excess material until final dimensions were achieved. Other parameters with regard to the sanding process are not given. During the sanding process airborne particle number concentrations were measured using a CPC and an optical particle counter (OPC) at two locations (adjacent to the sanding process and in the breathing zone of the operator). Cena and Peters (2011) used the ratios of the geometric mean concentration measured during the sanding process to the measured background concentrations (P/B ratio) as indices of the impact of the process. They found that during the sanding process the nanoparticle number concentration was negligibly elevated compared to the background concentration with a P/B ratio of 1.04. Furthermore, they found that sanding epoxy containing CNTs may generate micrometer-sized particles with embedded CNTs as shown by the corresponding P/B ratio of 5.90 and electron microscopy analyses.

Huang et al. (2012) characterized the emission of the airborne particles from epoxy resin test sticks with different CNT loadings and two commercial products during a sanding process. An experimental setup consisting of a system to simulate sanding and a system for sampling and monitoring airborne particles was developed (Figure 12.4).

The experimental setup consisted of a sand blasting cabinet, which was housed in a secondary clear plastic enclosure, both supplied with high efficiency particulate airfilter (HEPA)-filtered air to maintain a low background concentration. A terminal pump was used to extract a portion of the cabinet exhaust through a sampling manifold from which the measurement devices, CPC, OPC, and scanning mobility particle sizer (SMPS), received their sample flows. Inside the sand blasting cabinet a commercial lathe was used where a disk plate with sanding paper was mounted to the spindle

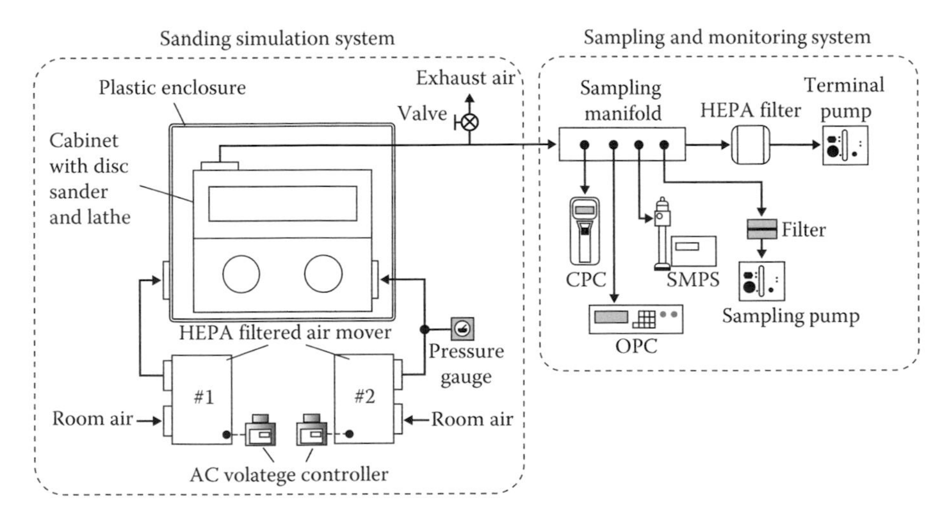

FIGURE 12.4 Experimental setup used. (With kind permission from Springer Science+Business Media: *J. Nanoparticle. Res.*, Evaluation of airborne particle emissions from commercial products containing carbon nanotubes, 14, 2012, 1–13, Huang, G. et al.)

of the lathe. The test materials with 0–4 mass% CNTs (1% gradation) were fed by a carriage, which was driven towards the sanding disk with a speed of 0.0025 cm/s by a turning screw beneath the lathe. Three levels of sanding paper roughness were used (P80 with 201 μm grit size, P150 with 100 μm grit size, and P320 with 46.1 μm grit size) with three disk sander rotation rates (586; 1,425; and 2,167 rpm).

The results showed that the highest particle number concentrations (4,670 #/cm³, mainly <100 nm diameter) were produced with coarse sanding paper, 2% CNT, and medium disk sander speed. The lowest concentration (92 #/cm³) was produced with medium sanding paper, 2% CNT, and slow disk sander speed. No free CNTs were observed by electron microscopy in airborne samples, except for tests with 4% CNTs. Also, heating and thermal decomposition were observed due to the friction between the samples and the sanding paper leading to the formation of particles <50 nm diameter.

Hirth et al. (2013) used the same experimental setup as Huang et al. (2012). The test samples were prepared by mixing 2 mass% multiwalled CNTs with epoxy resin or neat epoxy resin. The automated sanding simulation system was equipped with sanding paper (68 μm grit size). Other sanding parameters were not given. Hirth et al. (2013) found that mechanically released fragments in some cases show tubular protrusions on their surface. With the help of chemically resolved microscopy and a suitable preparation protocol, they identified these protrusions unambiguously as naked CNTs. Size-selective quantification of the fragments revealed that as a lower limit at least 95% of the CNTs remain embedded.

Koponen et al. (2011) investigated the particle size distribution and the total number of dust particles generated during sanding of paints, lacquers, and fillers doped with nano-objects as compared to their conventional counterparts. The sanding experiments were conducted inside a human exposure chamber of 20.6 m³ with a HEPA filtered air supply and an air exchange rate of 9.2 per hour. A commercial handheld orbital sander (Metabo Model FSR 200 Intec) with sanding paper P240 (58.5 μm grit size) was used as the sander. The sander has an internal fan for dust removal and normally a filter bag attached to the exhaust air. Instead, for this study the outlet was connected with a flexible tube to the aerosol sampling chamber. The measurement devices used were an aerodynamic particle sizer (APS), FMPS, and a modified commercial electrostatic precipitator (ESP) for particle sampling. All sanding experiments were performed by the same person using the same protocol. The test objects were 13 paints, lacquers, and fillers with nano-objects (TiO_2, SiO_2, kaolinite, carbon black, and perlite) applied on wooden plates produced by several manufacturers and, additionally, reference plates without nano-objects for each product type.

The investigation results show that the particle number size distribution of sanding dust (corrected for the emissions from the sanding machine) was dominated by 100–300 nm sized particles, whereas the mass spectra were dominated by micrometer particles. The added nano-objects had no or only very little influence on the particle modes in the sanding dust when compared to the reference products. Nevertheless, Koponen et al. (2011) observed considerable differences in the release rates as expressed in number concentrations of the different size modes. Still no significant link to the investigated nano-objects could be made.

Within the nanoGEM project, IUTA developed an experimental setup to investigate possible nanoparticle release during sanding of various composite materials using defined but adjustable test settings. Therefore, a test rig was constructed where the electrical engine was placed outside of an isolated chamber, simultaneously allowing particle measurements during the sanding procedure (Figure 12.5). The chamber is flushed with HEPA filtered air through two inlets and sealed dustproof to ensure low background particle concentrations. The test object is mounted to a rotating disk in the middle of the flow channel. The external motor moves the sanding plate by means of a magnetic clutch. The rotational speed and thus the velocity of the sample are adjustable. A stationary stamp fixes the sanding paper.

The sampling of the airborne sanding dust was conducted using conductive tubes. The background measurement in the chamber served as reference for each examined material. Two different sandpapers (201 and 46.2 μm grit size) were used, the

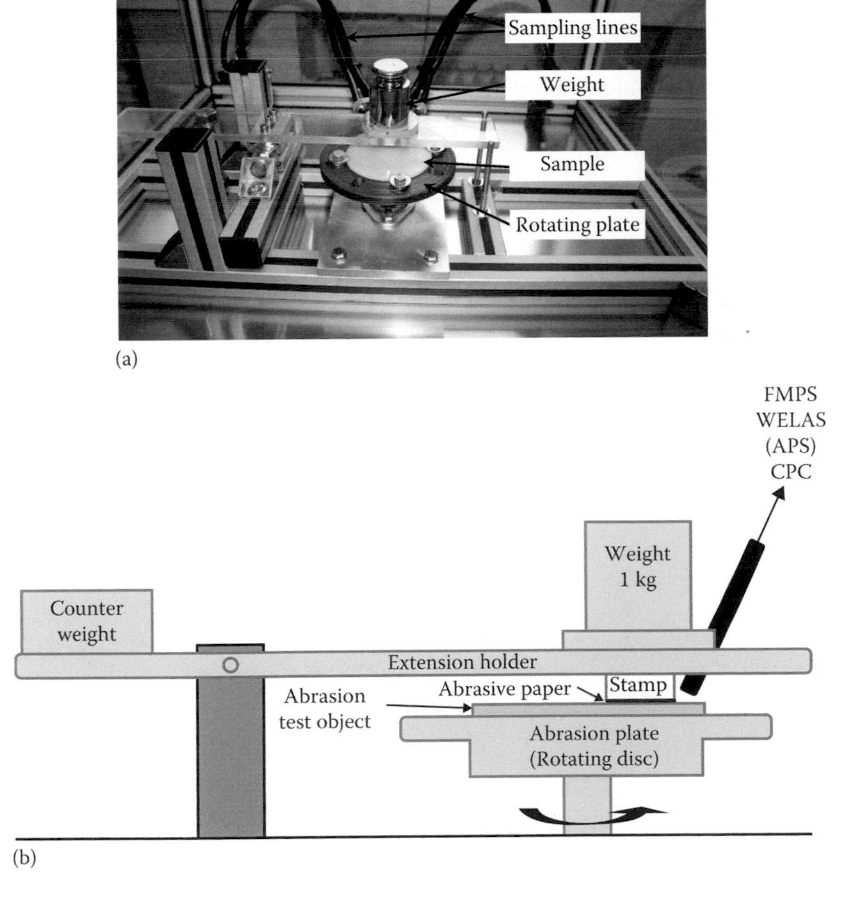

FIGURE 12.5 Example of a test setup to investigate particle release by sanding. (Courtesy of IUTA, nanoGEM.)

velocities between sanding paper and sample were 0.5 and 1.8 m/s, and the contact pressure was approximately 12,500 Pa. The test objects were epoxy resin and PA disks, the latter without and with 4 mass% nanoSiO_2. The released particles during the sanding process were measured with FMPS and APS. Additionally, ESP and contact samples were taken by carbonaceous adhesive pads for offline scanning electron microscope (SEM) analysis.

The ratios of the geometric mean concentration measured during the sanding process to the concentration measured in the background (P/B ratio) are shown in Table 12.2. Epoxy resin emitted significantly more particles (especially coarse particles) than PA. Furthermore, it was found that during the sanding process of PA the nanoparticle number concentration measured with the FMPS (with about 90% particles <100 nm) was negligible compared with the background concentration with a mean P/B ratio of 1.11 for PA with nanoSiO_2. The micrometer-ranged particles (measured with APS) showed in almost all cases very high P/B ratios indicating a high particle release in this size range.

Table 12.3 summarizes the test parameters and results of the selected test methods. The presented studies show that in most cases enclosed boxes were employed to (a) reduce the background aerosol concentration and (b) to ensure the safety of the operators. Additionally, it is evident that particle samples had to be taken to find out if the nanofillers or other nanoscale particles were released.

A comparison of all sanding test methods employed shows a general trend to testing in enclosures with possibilities to adjust and define parameters and test conditions. Still, some of the published test setups do not allow easily for quantitative release measurements and standardization. Anyhow, basic conditions for harmonized

TABLE 12.2
Results of the Sanding Tests from IUTA

Test Object	Grit Level	Speed, m/s	P/B FMPS 5–550 nm Diameter	P/B APS 700 nm–5 μm Diameter
Epoxy resin	80	0.5	9.8	772.1
Epoxy resin	80	1.8	17.4	507.8
Epoxy resin	320	0.5	3.3	97.3
Epoxy resin	320	1.8	12.7	131.5
PA neat	80	0.5	1.1	59.8
PA with SiO_2	80	0.5	1.09	12.0
PA neat	80	1.8	5.7	45.3
PA with SiO_2	80	1.8	1.32	61.4
PA neat	320	0.5	1.4	15.1
PA with SiO_2	320	0.5	0.96	2.9
PA neat	320	1.8	4.9	20.1
PA with SiO_2	320	1.8	1.07	10.9

Note: P/B is the ratio of sanding process/background.
Abbreviations: APS, aerodynamic particle sizer; FMPS, fast mobility particle sizer; PA, polyamide.

TABLE 12.3
Test Method Parameters and Results

	Contact Pressure [Pa]	Peripheral Speed [m/s]	Grit Level	Added NP	Particle Release		Comment
					Nano	Coarse	
Göhler et al. (2010)	10,000	1.83	600	ZnO, Fe_2O_3	Yes		No significant difference if NP were used or not
Cena and Peters (2011)			200	CNT	No	Yes	CNT embedded in the matrix
Huang et al. (2012)			80, 150, 320	CNT	Yes		No free CNTs were observed, exception 4 mass% CNT
Hirth et al. (2013)			220	CNT			Released fragments may show tubular protrusions
Koponen et al. (2011)			240	TiO_2, SiO_2, kaolinite, carbon black, perlite	Yes	Yes	No clear effect of ENPs on dust emissions from sanding
IUTA	12,500	0.8 and 1.8	80, 320	SiO_2	No	Yes	No clear effect of NPs on dust emissions from sanding

Note: CNT stands for carbon nanotube.

testing have been developed and the next step is the international collaboration on defining a test setup and testing conditions.

12.2.3 Abrasion

Abrasion is a mechanical process describing the dynamic friction between two surfaces, one of the two sometimes having a rough surface. It simulates, for example, use of consumer products like cleaning, sliding, walking, and scratching. Hence, it is a low energy type of mechanical stress with a lower likeliness of heat production and is not the same as sanding. One of the most commonly used methods for simulating abrasive damage is the Taber Abraser test, which is described in national and international standards (e.g., DIN 53754:1977, DIN 68861-2:1981, ISO 5470-1:1999, and ASTM D 4060-95:2007). Briefly, the Taber Abraser rotates the sample against the abrasion wheels and wear rate is often measured as a function of mass loss by the sample.

Vorbau et al. (2009) developed a method based on the Taber Abraser (Model 5131, Taber Industries, USA) for the characterization of abrasion-induced nanoparticle release into air from surface coatings due to its wide spread use and description in many national and international standards. The stress of the Taber test corresponds to the typical stress applied to surface coatings in a domestic setting, for example, when walking with sandy shoes on a floor surface. The Taber Abraser consists of two abrasion wheels, which are moved by the rotation of the sample (Figure 12.6). The abrasion is caused by the friction at the contact line between the sample surface and the abrasion wheels. The samples are rotated at 60 rpm with a normal force of 2.5 N and the stressing cycles were 3 × 100 revolutions. The abraded zone is an annular area and the abraded material is removed continuously with a stationary exhaust device. The wear rate is determined by weighing the clean sample at the beginning and at the end of the Taber test.

A sampling hood was realized as a small box, which almost completely encapsulated the sample suction zone behind the abrasion wheel on the surface coating (Figure 12.6, right). A CPC (Model 3022, TSI, Inc.) and an SMPS (Model 3934, TSI, Inc.) were connected to the sampling aerosol flow from the sampling hood. The test objects were sample carriers with three different coatings that contained nanoparticles (ZnO) or no nanoparticles. For each coating four different samples were examined. The results showed that during the abrasion tests the released particle mass depends on substrate and coating, but there is no significant correlation to nanoparticle content. Furthermore, the transmission electron microscope (TEM) and energy dispersive X-ray spectroscopy (EDX) analysis showed that the particles <100 nm remain embedded in the coarse wear particles.

Schlagenhauf et al. (2012) investigated the released particles and the possibility of releasing CNTs from an epoxy-based nanocomposite (with 0, 0.1, and 1 mass% CNT) by simulating a sanding process with a commercial Taber Abraser. A setup was developed where the abraded particles are characterized by aerosol measurement devices APS, FMPS, and SMPS directly after the abrasion. After collection on filters or grids, the abraded particles were analyzed by SEM, TEM, and EDX. For the particle measurements and collection only one of the two abrasive wheels was used whereon a small chamber was designed (Figure 12.7). During the tests the samples rotated at 60 rpm with a normal force of 7.5 N.

In this study, release of particles smaller than 100 nm was not observed. The detected smallest particle sizes were 300–400 nm. A trend of increasing size with increasing nanofiller content was observed. Imaging by TEM revealed that free individual CNTs and agglomerates were emitted during abrasion.

A similar setup was chosen by Guiot et al. (2009) (Figure 12.7, right), which was also tested for its measurement efficiency depending on the abrasion speed (0.05–0.2 m/s) using a CPC. They injected 6 nm Pt particles into the abrasion zone and defined 100% as the number concentration at a rotation speed of 0. The rotation speed was stepwise increased to 0.2 m/s and they found a decrease in the concentration to about 80% of the initial concentration. This study reveals that evaluations of test setups for different conditions are needed in all cases to ensure that comparability is achieved.

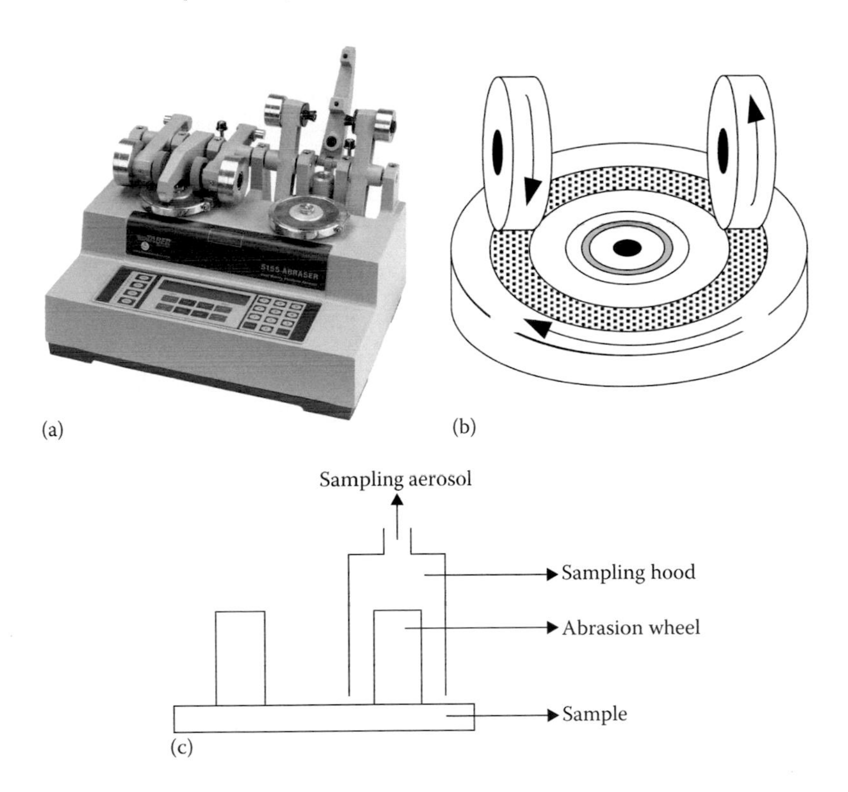

FIGURE 12.6 (a) Taber Abraser. (b) Abrasion scheme of a Taber Abraser with test piece (sample), abraded area, abrasion wheels, and the direction of rotation (marked with arrows), and (c) schematic illustration of the modified Taber Abraser with sampling hood. ((a) Courtesy of Taber Industries, USA. (c) Reprinted from *Aerosol. Sci.*, 40, Vorbau, M. et al., Methods for the characterization of the abrasion induced nanoparticles release into air from surface coatings, 209–217, Copyright 2009, with permission from Elsevier.)

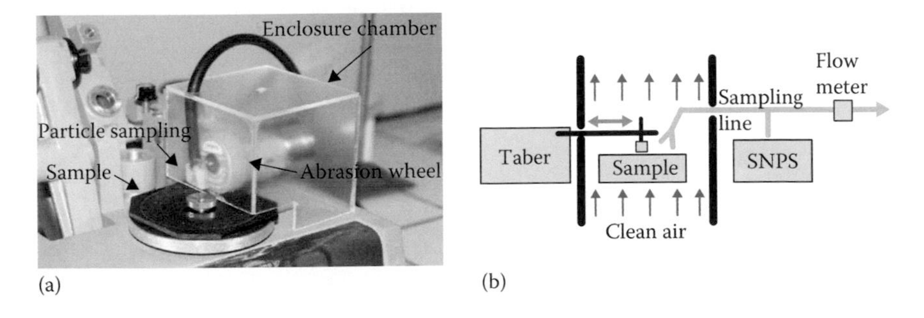

FIGURE 12.7 Abrasion area of the Taber Abraser with the enclosure chamber from Schlagenhauf et al. (2012) (a) and (b) principle test setup of a Taber Abraser to test nanoparticle release. (Reprinted with permission from Guiot, A. et al., Measurement of nanoparticleremoval by abrasion. J Phys Conf Ser 170:012014. Copyright 2009 American Chemical Society. Reprinted with permission from Schlagenhauf, L. et al., Release of carbon nanotubes from an epoxy-based nanocomposite during an abrasion process. *Environ Sci Technol* 46: 7366–7372. Copyright 2012 American Chemical Society.)

Overall, the general test setups for abrasion testing are comparable to each other and machines for testing are readily available due to the existing standards. But the evaluation study by Guiot et al. (2009) stresses the point that small changes in the setup and parameters, for example, rotations per minute, may have significant influence on the test results. Some further adjustments of the current setups are also needed to allow the quantification of the total release rates, improve the reproducibility, and ensure safety for the personnel.

12.2.4 Scratching

Scratching is a special case of low speed mechanical processes with a limited contact area to the material. This process can be simulated by using, for example, a metallic comb being pushed over a surface with a specific weight (Figure 12.8; Golanski et al. 2012). Golanski et al. (2012) developed a scratching tool inducing nonstandardized stresses and tested on model polymers with defined CNT content. Therefore, a homemade stainless steel rake on Taber equipment was used. The rake had a width of 26 mm and a teeth distance of 3 mm. The aspiration of abraded or scratched particles was measured with SMPS and/or electrical low pressure impactor (ELPI). Additionally, samples were collected for subsequent SEM or TEM analysis.

The scratching results showed a release of nanoscaled particles (mainly <30 nm) from a polyvinylchlorid (PVC) fabric containing NP with negligible release from the corresponding NO-free reference material.

Le Bihan et al. (2013) investigated the release of the nano-objects during different operations related to the handling and processing of an automotive part. The experiments were carried out in a sealed glove box testing surface scratching in three directions, surface sawing, and surface sanding (19,000 rpm). An iron alloy shaft cap with a fullerene coating (thickness 0.1 μm) was tested. During the tests all measurement devices (CPC, SMPS, and OPC) were fed from a unique sampling point (Figure 12.9). The clean air inlet flow (4.5 lpm) allowed working in a nearly particle-free atmosphere.

The particles emitted by scratching were mainly in the size range of 10–500 nm with total number concentrations of approximately 10 #/cm³. Peak concentrations during sawing reached 1,000 #/cm³ and 900 #/cm³ during sanding (Le Bihan et al. 2013). The sanding results showed that 50% of the released particles were smaller than 40 nm.

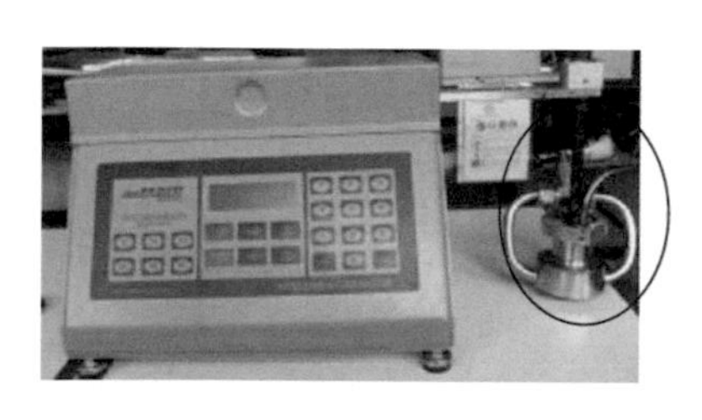

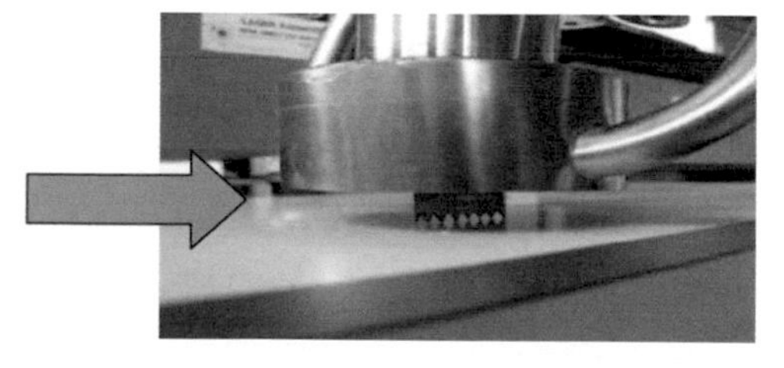

FIGURE 12.8 Solicitation tool used in the study: a stainless steel rake integrated into a linear Taber test. (With kind permission from Springer Science+Business Media: *J. Nanopart. Res.*, Release-ability of nanofillers from different nanomaterials (toward the acceptability of nanoproduct), 14, 2012, 1–9, Golanski L. et al.)

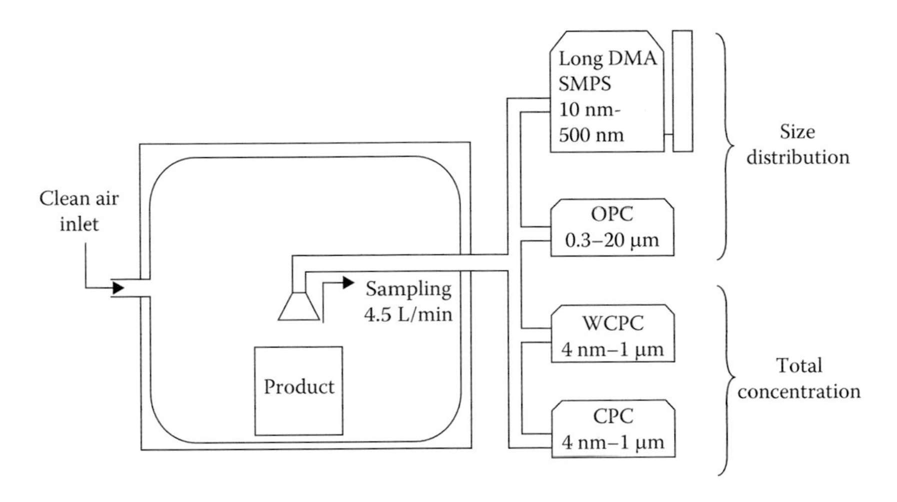

FIGURE 12.9 Experimental setup. (From Le Bihan, O. et al., *Adv. Nanopart.*, 2, 39–44, 2013.)

Only limited publication and test results are currently available for scratching of nanomaterial, even though commercial instruments are available for testing. Hence, the potential of international harmonization is given. Since no or very limited release was seen in the aforementioned tests, the need for international harmonization for this type of release may be of lower priority.

12.2.5 Shredding and Milling (Grinding)

Grinding is understood as a mixed process of milling and cutting of materials. It can occur upstream or downstream of the manufacturing process. Cutting, for example, in the case of shredding, is often a short-term process while milling may continue for some period of time. Anyhow, both processes can be linked to heat and hence to thermal alterations of the material. At the end-of-life recycling stage, composite waste can be ground or cut to generate new pellets for recycling.

Fleury et al. (2013) used a commercial grinder in a clean room to test possible release of nano-objects during grinding activities. Blank measurements had to be conducted since the electrical motor emissions were not decoupled from the sampled aerosol. They found that the grinding process can generate significant amounts of airborne MWCNTs still embedded in its matrix (Fleury et al. 2013). Release may especially occur during the act of opening the grinder lid and product removal hence leading also to noncontinuous emission.

Stahlmecke et al. (2014b) pursued a similar approach by placing a commercial small-scale shredder in a glove box with forced, particle-free air. They sampled close to the shredder area before and during shredding (Figure 12.10). Significant amounts of nanoscale particles were already released during the idling period prior to shredding, and also load-dependent electrical engine emission was observed. Particle size distributions were determined with an FMPS and APS; furthermore, samples were

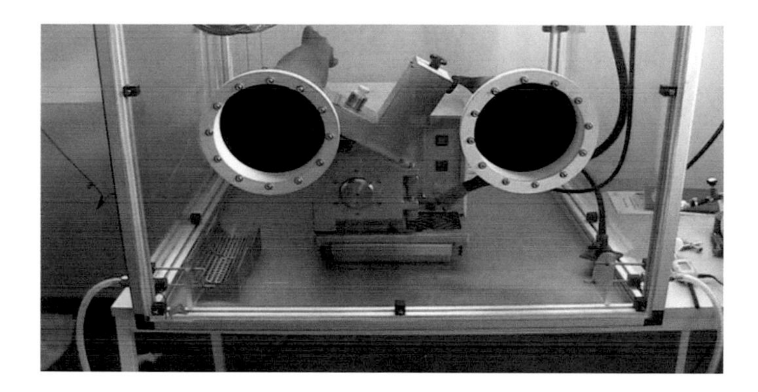

FIGURE 12.10 A small scale shredder with aerosol sampling lines in a glove box. (Courtesy of IUTA.)

collected by electrostatical precipitation for subsequent SEM analysis. No release of free CNT or CNT-agglomerates was observed but matrix material (PA, polycarbonate, and polyethylene) with embedded CNT was observed.

Park et al. (2009) conducted the investigations of nanoparticle release during grinding directly at a production facility, here for silver nanoparticles. The silver particles were produced in liquid and dried before grinding. The grinder is not further described in the paper. Significant release of nanosilver particles in the size range around 40 nm was observed when the grinder hatch was opened but not at other time periods. Hence, it is concluded that grinding should be conducted inside airtight systems that are placed in well ventilated rooms.

A few studies investigating shredding and grinding as possible sources of nano-objects have been investigated but mainly in an exploratory way. Some process-related studies and basic agreements are needed before starting any international harmonization efforts .

12.2.6 Drilling

Drilling is understood as a mechanical process where high speed mechanical shear forces are used to produce a hole. Release from drilling is effected by composite type, speed (rpm), sample thickness, and dry versus wet conditions (Bello et al. 2010). Wet drilling can be accomplished by continuously spraying the sample composite with distilled water leading to reduced heat.

Bello et al. (2009b) investigated in a study the airborne exposures to nanoscale particles and fibers during solid core drilling of base alumina and CNT–alumina composites. They considered the composite type (base alumina and CNT–alumina), diamond drill diameter (1/4" and 3/8"), drill speed (725 and 1,355 rpm), and dry and wet drilling. A schematic layout of the research laboratory is shown in Figure 12.11, above. Drilling exposure data were generated during a single session with each five replications. The instrument inlets were positioned (a) 10 cm from the potential emission source or (b) in the breathing zone of the operator (Figure 12.11b). Measurements of airborne particles were accomplished by FMPS, APS, CPC, thermal precipitator

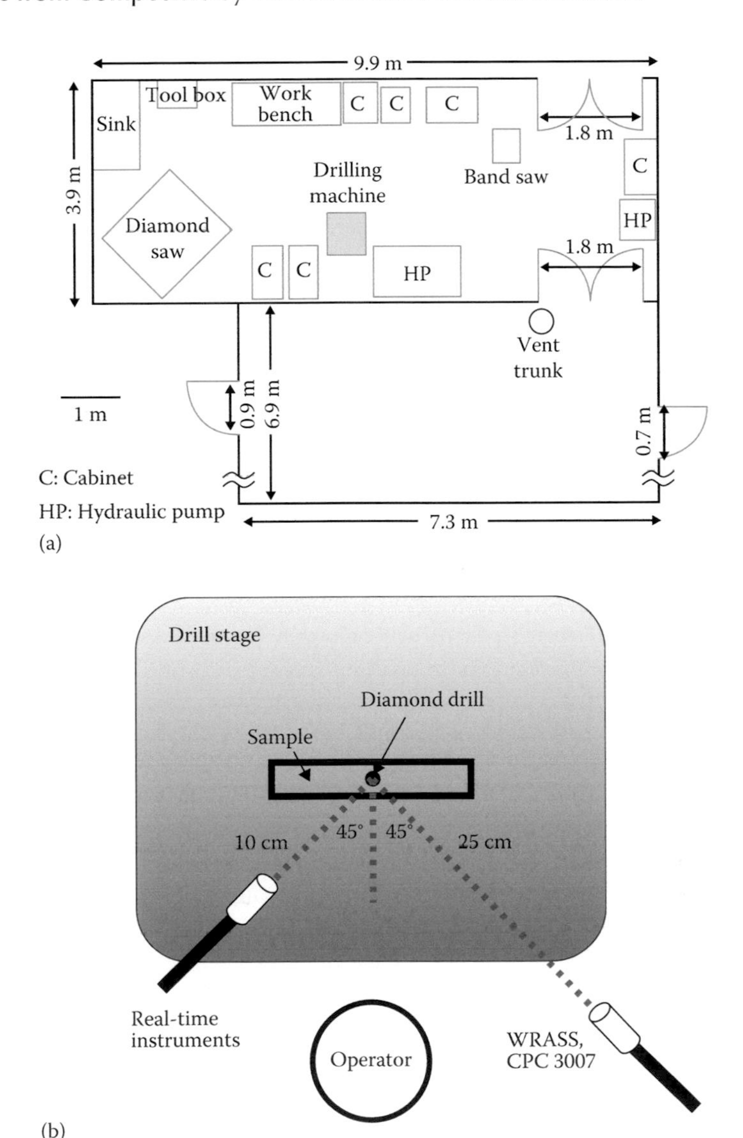

FIGURE 12.11 Schematic of the laboratory depicting the location of the drilling machine (a) and schematics of sampler arrangement for drilling and the drill stage (b). (Adapted from Bello, D. et al. (2009b), Exposures to nanoscale particles and fibers during handling, processing, and machining of nanocomposites and nanoengineered composites reinforced with aligned carbon nanotubes, International Committee on Composite Materials 17, Edinburgh, Scotland.)

(TP), and a wide range aerosol spectrometer system (WRASS). Additionally, particles were sampled using a precipitator and subsequently analyzed by electron microscopy.

The results showed that drilling may generate substantial exposure to nanoscale and submicron particles, in which higher exposures tend to be generated at higher

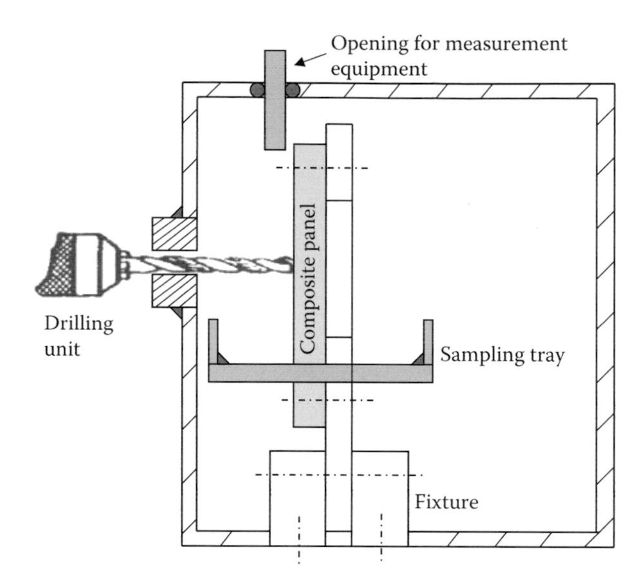

FIGURE 12.12 Drill chamber for particle collection. (From Sachse, S. et al., *NanoProduct Rev.* J., 1, 11, 2013a.)

speeds, with larger drills, and thicker composites. The CNT–alumina composite tends to generate less airborne particles than the pure alumina sample of the same thickness. On the TEM grids no CNTs could be found (individual or in bundles) in the air samples and the dust left behind by drilling. Dry drilling of composites did generate a visible smoke plume (potential exposures to organic thermal decomposition products of the epoxy components) and wet drilling generated considerably less particulate exposures.

Sachse et al. (2013a) studied the release of particles during drilling of glass fiber–polymer composites panels (PA6 and PP) reinforced with nanosilica and microsilica. The drilling tests of the nano-reinforced panels were conducted in a specially designed drilling chamber (Figure 12.12). Inside the chamber the panels were mounted on a fixture. An angle drill (Makita BDA351Z) with a drill diameter of 10 mm and a speed of 1,800 rpm was used during the tests. Background particle size distribution was measured before drilling over a period of 1 h. The following drilling was performed for 14 min. Afterwards, the drill bit was removed and the opening sealed. The actual measurements were continued for 2.5 h after drilling.

The results showed that drilling generated airborne nanosized particles regardless of whether the composite contains nanofillers or not. Furthermore, it could be seen that the total particle concentration varied significantly depending on the type of filler and matrix: PA6 panels generated approximately 10 times higher concentrations than that of PP. In another study, Sachse et al. (2012) found that the integration of nanofillers in the matrix (nanoclay) decreased release rates during drilling. The test chamber was similar to the one shown in Figure 12.12.

Stahlmecke et al. (2014a) investigated typical handling processes, here drilling, of novel building insulation materials. The materials studied contained typical

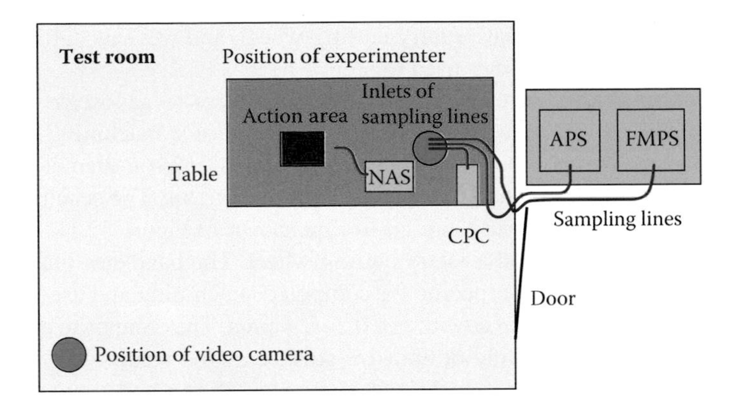

FIGURE 12.13 Schematic setup of test room and instruments. (Courtesy of IUTA.)

insulation stone wool without and with aerogel as a nanoporous additive to improve the insulation capabilities. The test setup chosen was similar to the one of Bello et al. (2009a) by using a test room of 25 m³ without forced ventilation instead of an enclosure. This was done to simulate working with building materials in a private home (Figure 12.13).

To assess the number size distribution of the particles in the test room an FMPS, APS, and a CPC were employed, if possible outside of the test chamber to avoid particle emission from the measurement devices. Particles were also collected with an ESP for subsequent identification of released particles.

Five holes were drilled into the composites with a 10-mm drill bit for wooden material within a 30-s time interval. To determine the particle contribution from the drilling machine itself, a blank test, that is, running the drill for 30 s without load, was conducted. The results showed that particle emissions by drilling are dominated by nanoscale particles. These particles appear independent of the material under investigation and can be attributed to motor emissions. A significant influence of the nanoadditive on the emission in this size class was not observed and the total number of additional particles in the investigated size range is low when compared to ambient or workplace particle number concentrations. Nevertheless, fragments of the aerogel were identified on SEM samples.

Drilling generates airborne particles in nanoscale range, but often stemming from the drilling machine. They also appear to be not influenced by the presence of nanomaterial in the material.

The general test setups for drilling testing are not completely comparable to each other and no comparisons between them have been made. Still basic investigations on the process itself and the correct test setup seem to be necessary.

12.2.7 Cutting and Sawing

Cutting/sawing is a relatively low speed mechanical process with a limited contact area to the material. It is used to derive specific forms and pieces from composite materials (e.g., with band-saw, rotary cutting wheel, and wet saw cutting). Wet cutting and sawing are sometimes used to reduce heat.

Bello et al. (2009a) investigated airborne exposures to nanoscale particles and fibers, which were generated during dry and wet abrasive machining of two three-phase advanced composite systems containing CNTs, micron-diameter continuous fibers (carbon or alumina), and thermoset polymer matrices. The schematic layout of the research laboratory is very similar to that shown in Figure 12.11. Two methods were applied: band-saw and a rotary cutting wheel. The band-saw blade was a diamond grit (220 grit, 60 μm) specific for composite machining and used with a blade speed of 20 m/s and a rate of advance of 0.2–0.4 cm/s. The composite cutting wheels employed used the same kind of abrasive surface as the band-saw, but with water to flush dust particles during machining, a speed of 15 m/s and a rate of advance of 0.4 cm/s. The cutting wheel was covered with guards on all sides, which helped to reduce aerosol emissions.

The duration of one cut was on average 5–15 s, so several cuts were performed on each composite to improve reliability of the data. The whole cutting cycle of 4–5 cuts lasted 1–3 min. Measurements of airborne particles during the cutting were detected by FMPS, APS, and CPC. Additionally, particles were collected for subsequent electron microscopy characterization.

Wet cutting, the common approach, did not significantly increase the particle number concentration above background whereas dry cutting, without any emission controls, resulted in high airborne concentrations of nanoscale and fine particles, as well as submicron and respirable fibers. The results showed that particle release depended on composite thickness and type, but the concentration of nanoscale, fine particles, and fibers did not vary between composites without and with nanofiller.

Stahlmecke et al. (2014a) investigated the sawing process of novel building insulation materials. For the schematic setup of the test room see Figure 12.13. To assess the number size distribution of the particles in the test room an FMPS, APS, and a CPC were employed. An ESP was used for sampling and subsequent SEM analysis. Sawing of the materials into smaller pieces was simulated using an electrical jigsaw operated with 1500 strokes per minute. A piece of material with a width of 15 cm was cut into four smaller pieces. Therefore, three lines, corresponding to an operation time of approximately 30 s of the jigsaw, were cut during each experiment. For an MDF-board as the comparison material and the experimental nanomaterial composite board, a blade intended for wood was used. In the case of the hard Rockpanel®, a blade intended for metal had to be used. Five repetitions for each material and each task were conducted. The results showed for all materials significantly higher concentrations of nanoscale particles as well as micrometer-range particles but without a difference between the materials without and with nanoporous additive.

Even though sawing was employed and tested in several studies no single defined method can currently be identified. Research on the test procedure, setup, and parameters still has to be conducted to be able to suggest a reliable test method for nanomaterial release by cutting and sawing.

12.2.8 Mechanical Shock

Mechanical shock is another special case of low energy mechanical processes. This process is seen to simulate short hits as they may occur when the product falls down to the ground or experiences sudden mechanical hits by accident. This process can be simulated by a vibrating plate engraver with a round tip as was used by Golanski et al. (2012). Release of CNTs from the matrix is reported by Golanski et al. (2012) using the engraver but no further details on the method and CNT identification are given.

Another method employed is the one used by Le Bihan et al. (2013) and Stahlmecke et al. (2014a). They dropped the material under investigation from a defined height, for example, 20–80 cm. Standard fast reading aerosol measurement tools such as an FMPS, APS, or OPS were employed with these test methods to be able to determine any release. Le Bihan et al. (2013) determined very small concentrations of about 10 #/cm³ in the clean box used in the test. Stahlmecke et al. (2014a) conducted the test in a "normal" room simulating homeworker conditions (Figure 12.13). No significant particle release could be determined in these conditions.

A third approach to test nano-objects release from nanomaterials is the one pursued by Sachse et al. (2013b, Figure 12.14). They produced the so-called crash cones of the nanomaterial of interest to be placed directly under a falling impactor with a defined weight. The aerosol sampling lines were placed close to the crashing zone in the test chamber. The SMPS measurements were conducted for extended periods prior and after the crash along with sampling with an electrostatical precipitator. Significant release of nanoscale particles was observed for all materials, those with silica nanofiller and without, but no significant differences in particle size and number concentrations for the different materials were observed.

Three different approaches testing nano-object release from nanomaterials were investigated, all of them simulating different conditions and hence release scenarios.

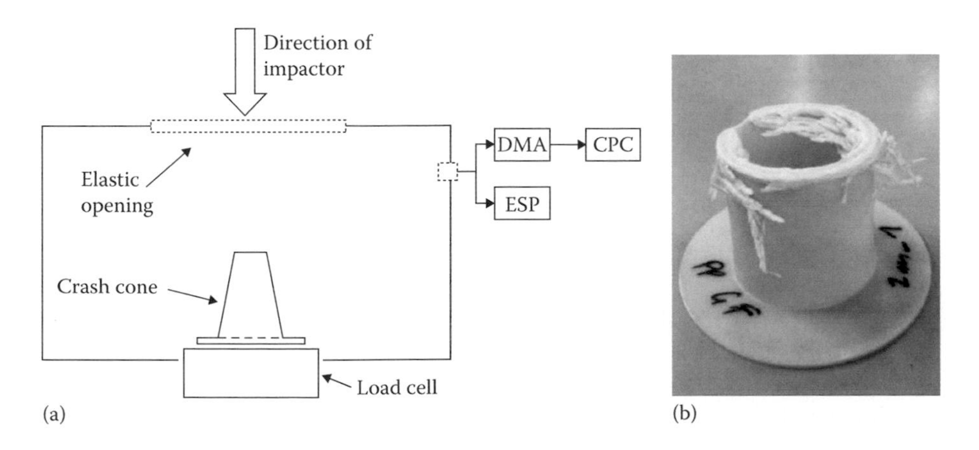

FIGURE 12.14 Experimental setup for testing of nano-objects release by mechanical shock (a) and a crashed cone (b). (From Sachse, S. et al., *NanoProduct Rev.* J., 1, 11, 2013a.)

In no case mechanistic studies investigating the influence of the variables on release were conducted. Overall, the studies show that nanoscale particles can be released by mechanical shock and that the release rate most likely depends on the energy input. Anyhow detailed studies still have to be conducted and release scenarios to be defined before pursuing international harmonization efforts.

12.3 RELEASE FROM THERMAL PROCESSES

12.3.1 Thermal Degradation

Thermal degradation is another mechanism selectively removing the matrix from a nano-object containing polymer. The mechanisms involved here are evaporation of matrix components as well as changes in the chemical composition by elimination and cracking, as well as some oxidation reactions. All studies investigating thermal degradation in the following presentation used commercially available thermogravimetric analyzers.

Orhan et al. (2012) examined thermal stability of nanocomposites containing nanoclay (NC) and MWCNT using a thermogravimetric analyzer (TGA). They found that the onset of degradation was delayed by the presence of the nanomaterials. In the presence of NC better flame retarding characteristics were observed, explained by anhydride formation.

Peng et al. (2007) investigated the thermal degradation mechanism of polyvinyl alcohol (PVA) with SiO_2 nanocomposites and found that the PVA/SiO_2 nanocomposite showed a significantly improved thermal resistance. The degradation products identified by Fourier transform infrared/thermogravimetric analysis (FTIR/TGA) and pyrolysis-gas chromatography/mass spectrometric analysis (Py-GC/MS) suggest that the first degradation step (ca. 350°C) of the nanocomposite mainly involves the elimination reactions of H_2O and residual acetate groups, as well as quite a few chain-scission reactions. The second degradation step (>450°C) is dominated by chain-scission reactions and cyclization reactions, and continual elimination of residual acetate groups is also found in this step.

Costache et al. (2006) studied the thermal degradation of polymethyl methacrylate (PMMA) and its nanocomposite to determine if the presence of clays (anionic and cationic) or CNT has an effect on the degradation pathway. The thermal degradation has been investigated by cone calorimetry and TGA, and the products of degradation have been studied with FTIR/TGA and gas chromatography/mass spectrometry (GC/MS). They found that the presence of clay or CNTs has no qualitative effect on the degradation mechanisms of PMMA, but the degradation of the nanocomposite occurs at higher temperatures. Since PMMA undergoes thermal degradation by a single process, the presence of filler cannot change its degradation pathway, in contrast to other systems where fillers can promote one thermal decomposition pathway at the expense of another.

Pramoda et al. (2003) focused on the study of thermal degradation and evolved gas analysis using TGA coupled to FTIR spectroscopy to study PA6 with and without clay nanocomposites prepared by melt compounding. PA6 with 2.5 mass% of clay loading showed a 12 K higher onset temperature for degradation compared to that for neat PA6. The onset temperature for degradation remained almost unchanged for samples with higher clay loading (5, 7.5, and 10 mass% clay). These results are related to the morphological observations that showed an optimal exfoliated structure only for the nanocomposite with 2.5 mass% clay, and distinct clay agglomeration in those with higher clay loadings.

Zanetti et al. (2004) found that during the thermal degradation of polyethylene (PE)/clay nanocomposites by means of TGA in the oxidant atmosphere, the formation of a protective layer on the polymer surface was observed caused by a charring process of PE. But PE is normally a nonchar-forming polymer.

Chrissafis et al. (2007) studied the thermal degradation mechanism of polycaprolactone (PCL) and its nanocomposites containing different nanoparticles (two layered silicates, SiO_2, and MWNT) through TGA using nonisothermal conditions at different heating rates. The results showed that modified montmorillonite and fumed silica accelerate the decomposition of PCL due to respective aminolysis and hydrolytic reactions that the reactive groups on the surface of these materials can induce. On the other hand, CNTs and unmodified montmorillonite can decelerate the thermal degradation of PCL due to a shielding effect.

Overall, quite a few studies investigating the thermal degradation and release of nano-object containing material were conducted. The test setups for thermal degradation are comparable to each other because of the general use of thermogravimetric analyzer and are therefore advanced for harmonization and standardization.

12.3.2 Combustion

Combustion is an exothermal process not entirely different from thermal degradation. The main difference is the uncontrolled high energy release in the form of heat during explosion or fire leading to mainly oxidized compounds as the final product. This also means that different from thermal degradation, combustion may lead to partial or total oxidation and hence release of the embedded nano-objects.

Bouillard et al. (2013) used a home-made demonstrator combustion system to assess nano-object release during combustion. Therefore, a high-performance nanocomposite polymer (acrylonitrile–butadiene–styrene, ABS) mixed with 3 mass% of MWCNTs was tested. An oven was developed to simulate specific combustion regimes (mainly in low-temperature substoichiometric conditions) and conceived with proper mass and heat-transfer controls to assess favorable conditions for the release of nano-objects. To detect the potential release of CNTs during the combustion, the particle size distribution in the fumes was measured using an ELPI. Additionally, particles were sampled using an aspiration-based TEM grid sampler. The study showed release of MWCNT at about 400°C, the combustion temperature of ABS with 3 mass% of MWCNTs. CNTs of about 12 nm diameter and 600 nm length were observed in the combustion emissions.

Calogine et al. (2011) focused on the combustion of nanocomposite samples under various ventilation conditions. The tests were performed with the fire propagation apparatus according to ISO 12136:2011, and PMMA with various nanofillers (CNTs,

Al_2O_3, and SiO_2) was employed. In the well-ventilated cases all tested PMMA samples with fillers and/or fire retardants showed lower effective heat of combustion values and a lower average mass loss rate compared to PMMA without any additive. The most significant decrease in heat of combustion was 6% for PMMA with 15 mass% Al_2O_3. In the underventilated case the highest effective heat of combustion was observed for PMMA with 15 mass% SiO_2, which is an increase of about 48% compared to neat PMMA. The highest mass loss rates were determined for samples containing CNT.

Motzkus et al. (2012) investigated the fire behavior and the characterization of solid and gaseous fire effluents of PMMA and PA6 filled with NPs (silica, alumina, and CNTs) used to improve their flame retardancy. The impact of these composites on the emission of airborne particles produced during their combustion in accidental fire scenarios was determined with an experimental setup, which was developed to measure the mass distribution in the size range of 30 nm–10 μm, and the concentrations of submicrometer particles in the aerosol using an ELPI, CPC, and cone calorimeter. Comparisons were made between unfilled and filled polymers. The influence of filler surface treatments (silane-based) as well as combinations with a flame retardant (APP) was also investigated. The results showed that for all samples of PMMA or PA6 modified or not with and without nanofillers of SiO_2, MWCNT, or Al_2O_3 high mass fractions were seen in the submicrometer particles (close to 80%). A 10% decrease in the mass fractions of ultrafine particles, due to addition of flame retardant (APP), was noticed.

12.3.3 Incineration

Incineration differs from thermal degradation and combustion in that it aims at full combustion to CO_2 and H_2O by high temperatures in an oxygen sufficient system with a minimum residence time of the compounds in the incineration zone.

Stahlmecke et al. (2014b) conducted such a study within the German Innovation Alliance CNT initiative. Therefore, an incinerator was simulated in the laboratory and two tube furnaces, each with a heated tube length of 550 mm, were used in series (see experimental setup in Figure 12.15). The test material was polycarbonat (PC), PA, and PE, each pure and with 5% and 7.5% CNT. Released particle concentration was measured with an FMPS and particle samples were taken with a NAS for consecutive morphological SEM analysis.

The results showed that there was no release of free CNTs and no major differences in the resulting particle quantities, sizes, and morphologies were found for the same materials with different CNT contents. During incineration experiments, only residue of PC materials that remained in the crucible after incineration showed CNT-like structures embedded in the slag, which was caused by an incomplete combustion of the material due to experimental deficiencies. Overall, the incineration experiments showed that burning of the different polymers always released very high particle number concentrations, independent of the CNT content. However, since none of the measurements revealed any evidence for airborne free CNTs, it can be concluded that under the given experimental conditions, the CNTs in the materials decomposed completely.

Derrough et al. (2013) investigated in a study the behavior of nanoparticles during high temperature with a setup of an incineration mounting at a laboratory scale (Figure 12.16). They selected pure silver, tin, and nickel NP as samples, and

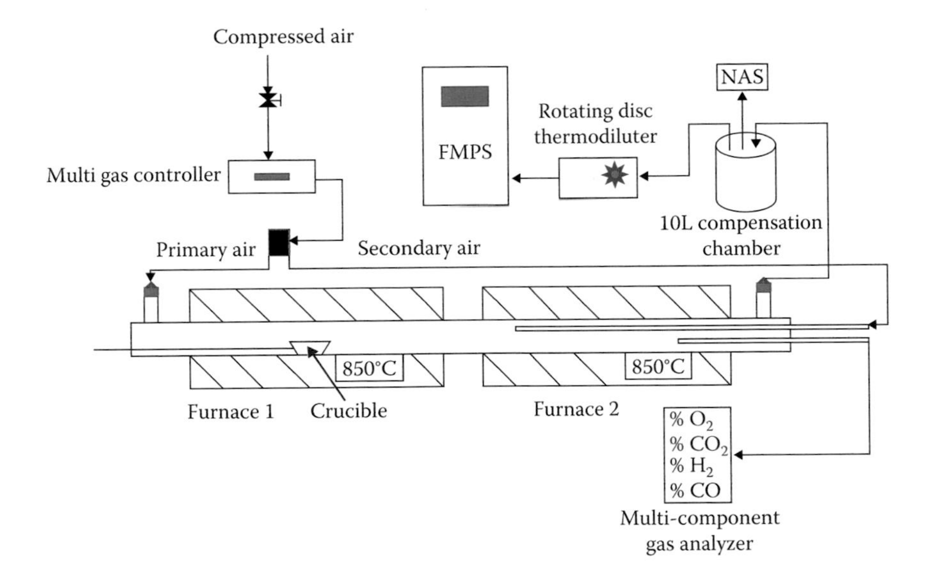

FIGURE 12.15 Schematic of the experimental setup for mimicking incineration of composite materials. (Reprinted from *Handbook of Nanosafety: Particle measurement, exposure assessment and risk management of engineered nanomaterials*, Stahlmecke, B. et al., 242–255, Copyright 2014, with permission from Elsevier.)

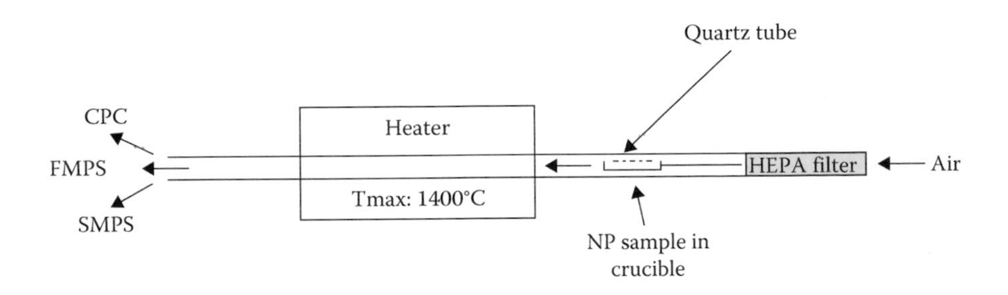

FIGURE 12.16 Schematic of the furnace. (Adapted from Derrough, S. et al. Behaviour of nanoparticles during high temperature treatment (Incineration type), 2013, *J. Phys. Conf Ser.*)

for each incineration trial, around 600 mg of NP were used. The incineration tests were performed using a bench top tubular furnace and were conducted at 850°C and 1,100°C.

Nickel and tin NP exhibited maximum concentration around seven times higher at 1,100°C compared to 850°C, whereas for silver NP only a low temperature influence was observed. The released particle size for nickel and tin NP was <50 nm at 850°C and significantly higher at 1,100°C (>100 nm). The silver NP behaves similar to those of tin and nickel at 850°C but different at 1,100°C with a mode close to 100 nm. This specific behavior of the silver sample is most likely linked to the melting point of 850°C for silver. SEM pictures of the silver residue, collected from the crucible after an 850°C heating, showed that melting already occurred.

Walser et al. (2012) demonstrated the persistence of nano-objects in a municipal solid-waste incineration plant. Walser et al. (2012) used nano-CeO_2 particles due to low cerium background concentrations in the waste and the environment. The nanoparticles were introduced (a) directly onto the waste before incineration or (b) into the gas stream exiting the furnace. Samples were taken from the flue gas and the residues of the combustion process and analyzed for cerium content. Nano-objects were shown to survive the waste incineration process and were detected in the fly ash and slag but not in the stack emissions. This suggests that no significant nano-CeO_2 emissions can be expected from thermal waste treatment plants provided that up-to-date flue gas cleaning systems are installed.

12.4 CONCLUSIONS AND RECOMMENDATIONS

Harmonized approaches are needed for various processes along the lifecycle to enable risk assessors, regulators, and product developers to assess possible risks of exposures for humans and the environment. The review of the different test methods shows that release testing is feasible and also that not all test setups can deliver the data needed to assess the exposure risk for a given release scenario. From the review of aforementioned release studies, several general conclusions can be derived.

Conditions for harmonized testing: All general criteria for harmonized or standardized approaches certainly apply also for nano-objects release testing. They are high accuracy, reproducibility, and comparability. Also all parameters of significant influence for the results should be known and well defined. An additional requirement is the safety for the testing personnel. Affordability of the testing equipment and easiness of use and cleaning are further criteria.

Data requirements for testing: The purpose of the release testing is to obtain information for specific exposure scenarios, modeling of these exposures, and to enable or at least facilitate exposure assessments linked to internal doses, for example, for humans. This means that the test methods should allow the determination of release rates, particle size distributions, and concentrations, as well as particle morphologies, for those particles stemming from the nanomaterial.

General setup conditions for release testing: From the requirements previously listed, it can be deduced that the use of enclosures facilitates the discrimination of nano-objects from background particles. The placing of any testing equipment outside of the enclosure, especially electrical engines, also facilitates the discrimination as well as determination of release rates. For most test setups some variability in test parameters influencing, for example, the wear energy is advantageous to allow for a better assessment of release probabilities. Another important asset of using enclosures is the safety of the testing personnel, who is not directly exposed to the released particles with possible hazardous properties.

Grouping of release processes: Release processes may be grouped as was done in this chapter for mechanical and thermal processes. Sometimes these cannot be clearly separated since mechanical treatments may also cause heat, like sanding or milling, and hence those simulated release processes will cover both thermal and mechanical mechanisms. Further release processes not tackled in this chapter are exposures of nanomaterials to reactive gases or liquids, for example, ozone or hydrogen peroxide in water. Simulations of weathering combine thermal processes and exposure to reactive chemical compounds. Yet, going another step further is the combination of weathering followed by mechanical stress like sonication as reviewed in Hirth et al. (2013). Especially, this combination of matrix degradation by weathering and mechanical stress by sanding and abrasion mechanisms is seen as a possible mechanism releasing single or agglomerated nano-objects.

Another methodology for grouping was, for example, discussed by Le Bihan et al. (2013). They list different test methods, the release scenario simulated by this method and the wear energy linked to it. In a first assessment some correlation between released concentrations and the energy input was identified by Le Bihan et al. (2013).

Current status in view of harmonization: The review presented here shows that some test methods, for example, dustiness, abrasion, and sanding, can be viewed as quite advanced and ready for standardization while others are more in their infancy. A simple grouping of release test scenarios into three levels for harmonization maybe (see Table 12.4): Level 1—ready for standardization; Level 2—basic information on mechanisms and release rates is available but several test setups are available or some basic information is missing; and Level 3—nearly no information and test methods are available. This type of evaluation linked to the need of standardization, for example, due to exposure risk is a feasible way to prioritize future developments.

Comparison of material behavior to different stress conditions: First tests of the same nanomaterial using different test methods have been conducted (e.g., Wohlleben et al. 2011 and 2013). This type of tests, once evaluated and established, will allow complete assessment of release probabilities facilitating safer design and thus a reduction in likeliness of exposure.

Release to exposure: The ultimate aim for release testing is to ensure the safety of workers, consumers, the population, and the environment. Hence, a good robust link from release to exposure or even dose is needed. Due to the high physical and chemical variability, the interaction with the environment during transport, the homogeneous interaction at high concentrations, and uptake probabilities can vary significantly so that models are needed to simulate the processes and interactions well enough for, for example, exposure assessments. Some computational fluid dynamic (CVD) models are available for specific workplace scenarios. Extensions to other exposure scenarios and the environment are urgently needed. This will facilitate the safe implementation of the use of nanomaterials for the benefit of humans and nature.

TABLE 12.4
Readiness for Standardization

Level	Release Scenario
1	Dustiness, abrasion, sanding, and drilling
2	Milling, grinding, sawing scratching, dropping, cutting/shredding, thermal stress, incineration, and mixed processes (e.g., weathering)
3	Reactive liquids, reactive gases, and fire/explosion

REFERENCES

Bello, D., Wardle, B. L., Yamamoto, N., de Villoria, R. G., Garcia, E. J., Hart, A. J., Ahn, K., Ellenbecker, M. J., Hallock, M. (2009a). Exposure to nanoscale particles and fibers during machining of hybrid advanced composites containing carbon nanotubes. *J Nanopart Res* 11: 231–249.

Bello, D., Wardle, B. L., Yamamoto, N., de Villoria, R. G., Hallock, M. (2009b). Exposures to nanoscale particles and fibers during handling, processing, and machining of nanocomposites and nanoengineered composites reinforced with aligned carbon nanoturbes. International Committee on Composite Materials 17, Edinburgh, Scotland.

Bello, D., Wardle, B. L., Zhang, J., Yamamoto, N., Santeufemio, C., Hallock, M., Virji, M. A. (2010). Characterization of exposures to nanoscale particles and fibers during solid core drilling of hybrid carbon nanotube advanced composites. *Int J Occup Environ Health* 16(4): 434–450.

Bouillard, J. X., R'Mili, B., Moranviller, D., Vignes, A., Le Bihan, O., Ustache, A., Bomfim, J. A. S., Frejafon, E., Fleury, D. (2013). Nanosafety by design: Risks from nanocomposite/nanowaste combustion. *J Nanopart Res* 15:1519.

Calogine, D., Marlair, G., Bertrand, J-P., Duplantier, S., Lopez-Cuesta, J-M., Sonnier, R., Longuet, C., Minisini, C., Chivas-Joly, C., Guillaume, E., Parisse, D. (2011). Gaseous effluents from the combustion of nanocomposites in controlled-ventilation conditions. *J Phys Conf Ser* 304: 012019.

Canady, R., Kuhlbusch, T., Renker, M., Lee, E., Tsytsikova, L. (2013). NanoReleasePhase 2.5 Report: Comparison of existing studies of release measurements of MWCNT-polymer composites. ILSI, USA: pp. 44.

Cena, L., Peters, T. (2011). Characterization and control of airborne particles emitted during production of epoxy/carbon nanotube nanocomposites. *J Occup Environ Hyg* 8: 86–92.

Chrissafis, K., Antoniadis, G., Paraskevopoulos, K. M., Vassiliou, A., Bikiaris, D. N. (2007). Comparative study of the effect of different nanoparticles on the mechanical properties and thermal degradation mechanism of in situ prepared poly(caprolactone) nanocomposites. *Composites Sci Technol* 67: 2165–2174.

Costache, M. C., Wang, D., Heidecker, M. J., Manias, E., Wilkie, C. A. (2006). The thermal degradation of pol(methyl methacrylate) nanocomposites with montmorillonite, layered double hydroxides and carbon Nanotubes. *Polym Adv Technol* 17: 272–280.

Dahm, M., Evans, D., Schubauer-Berigan, M., Birch, M., Fernback, J. (2012). Occupational exposure assessment in carbon nanotube and nanofiber primary and secondary manufacturers. *Ann Occup Hyg* 56: 542–556.

Derrough, S., Raffin, G., Locatelli, D., Nobile, P., Durand, C. (2013). Behaviour of nanoparticles during high temperature treatment (Incineration type), *J Phys Conf Ser* 429:012047.

EN 15051 2006. Workplace atmospheres— measurement of the dustiness of bulk materials—requirements and reference test methods. Brussels, Belgium: European Committee for Standardization.

Evans, D. E., Turkevich, L. A., Roettgers, C. T., Deye, G. J., Baron, P. A. (2013). Dustiness of Fine and Nanoscale Powders. *Ann Occup Hyg* 57 (2): 261–277.

Fleury, D., Bomfim, J. A. S., Vignes, A., Girard, C., Metz, S., Muñoz, F., R'Mili, B., Ustache, A., Guiot, A., Bouillard, J. (2013). Identification of the main exposure scenarios in the production of CNT-polymer nanocomposites by melt-mouldingprocess. *J Clean Prod* 53: 22–36.

Göhler, D., Stintz, M., Hillemann, L., Vorbau, M. (2010). Characterization of nanoparticle release from surface coatings by the simulation of a sanding process. *Ann Occup Hyg* 54:615–624.

Golanski L., Guiot A., Pras M., Malarde M., Tardif F. (2012). Release-ability of nano fillers from different nanomaterials (toward the acceptability of nanoproduct). *J Nanopart Res* 14: 1–9.

Guiot, A., Golanski, L., Tardif, F. (2009). Measurement of nanoparticle removal by abrasion. *J Phys Conf Ser* 170:012014.

Hirth, S., Cena, L., Cox, G., Tomovic, Z., Peters, T., Wohlleben, W. (2013). Scenarios and methods that induce protruding or released CNTs after degradation of composite materials. *J Nanopart Res* 15: 1504.

Huang, G., Park, J., Cena, L., Shelton, B., Peters, T. (2012). Evaluation of airborne particle emissions from commercial products containing carbon nanotubes. *J Nanoparticle Res* 14: 1–13.

ISO 12136:2011. Reaction to fire tests–Measurement of fundamental material properties using a fire propagation apparatus.

Johnson, D. R., Methner, M. M., Kennedy, A. J., Steevens, J. A. (2010). Potential for Occupational Exposure to Engineered Carbon-Based Nanomaterials in Environmental Laboratory Studies. *Environ Health Perspect* 118(1):49–54.

Koponen, I. K., Jensen, K. A., Schneider, T. (2011). Comparison of dust released from sanding conventional and nanoparticle-doped wall and wood coatings. *J Exposure Sci Environ Epidemiol* 21: 408–418.

Le Bihan, O., Shandilya, N., Gheerardyn, L., Guillon, O., Dore, E., Morgeneyer, M. (2013). Investigation of the release of particles from nanocoated product. *Adv Nanopart* 2: 39–44.

Lee, J., Mahendra, S., Alvarez, P. J. J. (2010). Nanomaterials in the construction industry: A review of their application and environmental health and safety considerations. *ACS Nano* 4(7): 3580–3590.

Motzkus, C., Chivas-Joly, C., Guillaume, E., Ducourtieux, S., Saragoza, L., Lesenechal, D., Mace, T., Lopez-Cuesta, J.-M., Longuet, C. (2012). Aerosol emitted by the combustion of polymers containing nanoparticles. *J Nanoprt Res* 14: 687.

Orhan, T., Isitman, N. A., Hacaloglu, J., Kaynak, C. (2012). Thermal degradation of organophosphorus flame-retardant poly(methyl-methacrylate) nanocomposites containing nanoclay and carbon nanotubes. *Polym Degrad Stabil* 97: 273–280.

Park, J., Kwak, B. K., Bae, E., Lee, J., Kim, Y., Choi, J. Y. (2009). Characterization of exposure to silver nanoparticles in a manufacturing facility. *J Nanoprt Res* 11: 1705–1712.

Peng, Z., Kong, L. X. (2007). A thermal degradation mechanism of polyvinyl alcohol/silica nanocomposites. *Polym Degrad Stabil* 92: 1061–1071.

Pramoda, K. P., Liu, T., Liu, Z., He, C., Sue, HJ. (2003). Thermal degradation behavior of polyamide 6/clay nanocomposites. *Polym Degrad Stabil* 81: 47–56.

Sachse, S., Gendre, L., Silva, F., Zhu, H., Leszczyńska, A., Pielichowski, K., Ermini, V., Njuguna, J. (2013b). On Nanoparticles Release from Polymer Nanocomposites for Applications in Lightweight Automotive Components. *J Phys Conf Ser* 429: 012046.

Sachse, S., Njugana, J., Michaowski, S., Leszczynska, A., Pielichowski, K., Zhu, H. (2013a). The release of nanofillers from polymer matrix: Case studies on polypropylene, polyamide and polyurethane nanocomposites. *NanoProduct Rev J* 1: 11.

Sachse, S., Silva, F., Zhu, H., Irfan, A., Leszczynska, A., Pielichowski, K., Ermini, V., Blazquez, M., Kuzmenko, O., Njuguna, J. (2012). The effect of nanoclay on dust generation during drilling of PA6 nanocomposites. *J Nanomaterials* 189386.

Schlagenhauf, L., Chu, B. T. T., Buha, J., Nüesch, F., Wang, J. (2012). Release of carbon nanotubes from an epoxy-based nanocomposite during an abrasion process. *Environ Sci Technol* 46: 7366–7372.

Schneider, T., Brouwer, D. H., Koponen, I. K., Jensen, K. A., Fransman, W., Van Duuren-Stuurman, B., Van Tongeren, M., Tielemans, E. (2011). Conceptual model for assessment of inhalation exposure to manufactured nanoparticles. *J Expos Sci Environ Epidemiol* 21(5): 450–463.

Schneider, T., Jensen, K. A. (2008). Combined single-drop and rotating drum dustiness test of fine to nanosize powders using a small drum. *Ann Occup Hyg* 52(1): 23–34.

Stahlmecke, B., Asbach, C., Donaldson, K., Andersen, J. R., Jorgensen, K. S., Schneider, T., Kuhlbusch, T. A. J. (2014a). Investigations on the release of nanoscale particles during typical handling processes of improved insulation materials (submitted JOEH).

Stahlmecke, B., Asbach, C., Todea, A. M., Kaminski, H., Kuhlbusch, T. A. J. (2014b). Investigations on CNT release from composite materials during end of life, Chapter 7.4 in *Handbook of Nanosafety: Particle measurement, exposure assessment and risk management of engineered nanomaterials*. Ed. U. Vogel, ISBN 13:9780124166042, Elsevier (in press).

Stahlmecke, B., Wagener, S., Asbach, C., Kaminski, H., Fissan, H., Kuhlbusch, T. A. J. (2009). Investigation of airborne nanopowder agglomerate stability in an orifice under various differential pressure conditions. *J Nanopart Res* 11(7):1625–1635.

Tsai C.-J., Lin, G.-Y., Liu, C.-N., He, C.-E., Chen, C.-W. (2012). Characteristic of nanoparticles generated from different nano-powders by using different dispersion methods. *J Nanopart Res* 14:777.

Van Broeckhuizen, P., van Broeckhuizen, F., Cornelissen, R., Reijnders, L. (2011). Use of nanomaterials in the European construction industry and some occupational health aspects thereof. *J Nanopart Res* 13: 447–462.

Vorbau, M., Hillemann, L., Stintz, M. (2009). Method for the characterization of the abrasion induced nanoparticle release into air from surface coatings. *Aerosol Science* 40: 209–217.

Walser, T., Limbach, L. K., Brogioli, R., Erismann, E., Flamigni, L., Hattendorf, B., Juchli, M., Krumeich, F., Ludwig, C., Prikopsky, C., Rossier, M., Saner, D., Sigg, A., Hellweg, S., Günther, D., Stark, W. J. (2012). Persistence of engineered nanoparticles in a municipal solid-waste incineration plant. *Nat Nanotechnol* 7:520–524.

Wohlleben, W., Brill, S., Meier, M. W., Mertler, M., Cox, G., Hirth, S., von Vacano, B., Strauss, V., Treumann, S., Wiench, K., Ma-Hock, L., and Landsiedel, R. (2011) On the life cycle of nanocomposites: Comparing released fragments and their in-vivo hazards from three release mechanisms and four nanocomposites. *Small* 7(16):2384–2395.

Wohlleben, W., Meier, M. W., Vogel, S., Landsiedel, R., Cox, G., Hirth, S., Tomović, Ž., (2013). Elastic CNT–polyurethane nanocomposite: Synthesis, performance and assessment of fragments released during use. *Nanoscale* 5(1):369–380

Zanetti, M., Bracco, P., Costa, L. (2004). Thermal degradation behaviour of PE/clay nanocomposites. *Polym Degrad Stabil* 85:657–665.

13 Field and Laboratory Measurements Related to Occupational and Consumer Exposures

Derk Brouwer, Eelco Kuijpers, Cindy Bekker, Christof Asbach, and Thomas A.J. Kuhlbusch

CONTENTS

13.1 INTRODUCTION

As part of the nano-safety research, exposure assessment has evolved over the last couple of years from explorative research toward more comprehensive exposure assessment to provide data for an appropriate risk assessment. Several review papers, for example, Kuhlbusch et al. (2011) and Clark et al. (2012), concluded that there is an urgent need for a systematic approach to harmonize and standardize both, workplace exposure assessment and test procedures simulating release during workplace activities and processes. More specifically, agreement on harmonization of measurement metric, strategy, data processing, including statistical analysis, and reporting were indicated as key issues to enable future data pooling and building REACH-compliance Exposure Scenarios. Experts in the area of exposure assessment anticipated this challenge and established an informal group focusing on Global Harmonization of Measurement Strategies for Exposure to Manufactured Nano-Objects. A first report listed recommendations with respect to all aspects of exposure assessment (Brouwer et al. 2012) and formed the basis for ongoing activities to establish a nano exposure database. This database Nano Exposure and Contextual

Information Database (NECID) is currently constructed under the umbrella of the European partnership of Occupational Health and Safety research (PEROSH*). The NECID template is currently been used in all EU projects where exposure data are generated and facilitates the population and use of the database.

The NECID database requires that data were determined according to measurement strategies that support a harmonized comprehensive exposure assessment. Several measurements strategies have been developed as described in Chapter 11 by Asbach et al. Such strategies are pragmatic decision schemes in order to avoid comprehensive, time-consuming, and expensive exposure measurements. In the case of nano-objects, their agglomerates, and aggregates (NOAA, EN ISO 2012), it reflects the lack of analytical capabilities and the costs of employing scientific instruments in the field. In these tiered approaches, information is collected in each successive tier at a more detailed level in order to reduce the uncertainty in the exposure estimates. The NEAT approach as developed by NIOSH (Methner et al. 2010) was one of the first tiered approaches applied for nano exposure scenarios. Presently, a number of tiered-approach strategies have been published: the one proposed by a number of institutions in Germany on which the German nanoGEM project based their closely related proposal (Asbach et al. 2012), a French approach (Witschger et al. 2012), and the tiered approach proposed by McGarry and coauthors (McGarry et al. 2012). All approaches start with a relatively simple and limited set of measurements and/or gathering basic information on processes, jobs, and materials in a first tier followed by extended assessment in subsequent tiers. The decision criteria to enter the next tier are key factors for such approaches; for example, nanoGEM uses a situational-driven criterion: An emission or exposure concentration at a workplace is considered to be significantly increased above background if it exceeds the background concentration plus its threefold standard deviation (Asbach et al. 2012). A slightly different approach has been proposed by The British Standards Institution (BSI 2010), and adopted by van Broekhuizen et al. (2012), respectively, that makes use of benchmark levels or nano reference values. If the exposure exceeds an "action level," actions are required, for example, for more detailed measurements or control measures. The nonsubstance specific reference values are expressed as particle size integrated total particle number concentration and could as well be used in the framework of a tiered approach. Recently, some projects have started to evaluate the decision criteria proposed in the various tiered approaches, for example, the Business and Industry Advisory Committee within OECD WPMN SG8 and CEN project on Guidance and so on, see also Chapter 11 (Asbach et al., this book).

Occupational exposure limits have not yet been derived for any NOAA, neither by the European SCOEL (the Scientific Committee on Occupational Exposure Limits) nor by any national OEL-setting authority. However, NIOSH (2011, 2013) has proposed mass-based recommended time-weighted exposure limits for two materials, that is, nanoTiO_2 and nanocarbon nanotubes (CNTs). The German MAK (maximale Arbeitsplatzkonzentrationen) Committee is currently discussing a general limit value for airborne nano-objects and nanomaterials as described in the BegKS 527.†

* http://www.perosh.eu/research-projects/perosh-projects/exposure-measurements-and-risk-assessment-of-manufactured-materials-nanoparticles-devices/.

† http://www.baua.de/en/Topics-from-A-to-Z/Hazardous-Substances/TRGS/Announcement-527.html;jsessionid=26C257784F8C77E286DD99E15D507CFD.1_cid389.

Exposure modeling has been paid much attention in the last few years. The conceptual model for inhalation exposure to nanoparticles was proposed by Schneider et al. (2011), built on previously proposed models, for example, by Maynard and Zimmer (2003), who already addressed specific issues related to size distribution evolution through coagulation, settling, and diffusion. Several other research groups have addressed the specific aerosol dynamics for nano aerosols especially with regard to the evaluation of size distribution over time by coagulation in indoor air, for example, Seipenbusch et al. (2008), Rim et al. (2011), Anand et al. (2012), and Asbach et al. (2014). Walser et al. (2012) used a one-box model to predict temporal fluctuations of particle number concentration due to accidental events.

A number of pragmatic tools for risk management or risk prioritization, for example, control banding tools, have been developed (Brouwer 2012a) and a few use the conceptual model as the basis for the exposure assessment band of the tool. However, actual exposure data will still be needed to provide solid estimates for exposure and concomitant risk assessment. These, the real or simulated exposure studies are the focus of this chapter. Recent developments with respect to measurement devices and methods will be discussed here only briefly.

13.2 MEASUREMENT DEVICES

In general, devices used to assess exposure to nanomaterial or nano-size aerosols can be subdivided into devices that monitor (on-line) a chemical substance or aerosol by "near or quasi" real-time detection and devices that sample (time-aggregated) chemical substances or aerosols on a substrate, followed by off-line analysis.

A suite of real-time devices are already available (Kuhlbusch et al. 2011), and quite recently Asbach et al. (2012a) compared the performance of five different real-time portable and battery-operated nanoparticle monitors measuring particle number and surface area concentrations. Furthermore, Kaminski et al. (2013) investigated the comparability of mobility particle sizers and diffusion chargers. These investigations are needed to allow the assessment of exposure related measurements and assessments. Also, new devices have been developed and are likely to become available on the market in the near future. For example, additionally to the established real-time handheld devices, a device has become commercially available that provides estimates of the surface area of the aerosol fraction deposited in the gas-exchange region (Fierz et al. 2011). Another example of new developments is the extension of the usability domain of existing devices. In addition to the currently available stationary monitors using a DMA with a radioactive source as a bipolar charger for size-resolved measurement of nanoscale particles, a number of devices came up with a nonradioactive charger, Naneum Nano-ID NPS500 a proprietary Corona unipolar charger, whereas TSI model 3087 uses bipolar diffusion charging using soft X-ray. Within the recently concluded NANODEVICE project* interesting near-market prototype real-time devices have been developed that (a) reduce size and weight of existing device concepts that afford portability (and could be used in a position closer to the receptor) and (b) increase the size ranges for both the

* www.nano-device.eu.

size-resolved real-time monitors and size-resolved samplers, which increase the ease of use, since combination of devices can be avoided. Examples are (a) a wide-range real-time classifier (10 nm–27 μm) (mobility + optical detection), (b) a nano aerosol classifier and short/long term sampler system (20–450 nm), and (c) a low-cost total (active) surface area monitor.

The mentioned portable online devices all determine physical parameters like number, surface area, or mass concentrations. No portable devices detecting (size resolved) chemical composition or particle morphology are yet available. So far only transportable devices like aerosol mass spectrometer exist but have not been employed at workplaces yet. Instruments being capable of delivering this information are important to facilitate the fast discrimination of product nano-objects/NOAA from background particles. Therefore, sampling devices for personal and areal measurements are currently often employed to allow such discrimination.

(Personal) samplers collect the aerosol fraction that can penetrate into only one or all compartments of the human respiratory tract. The samplers try to emulate one or more of the sampling conventions which are based on the entry efficiency of particles in the respiratory tract, for example, EN ISO 13138. With respect to nanoparticle sampling in workplaces, Tsai et al. (2012) have developed an active Personal Nanoparticle Sampler (PENS) that consists of an impactor that is mounted downstream of an existing cyclone for the respirable fraction. The impactor has an aerodynamic cut-size of 100 nm and is followed by a back-up filter. Slightly earlier, Furuuchi and coworkers (Furuuchi et al. 2010) have developed a personal nanosampler based on another principle, that is, impaction in a layered metal fibre mesh filter.

Very interesting are the developments that match the ICRP/ISO criteria (EN ISO 2012) for respiratory tract compartment specific lung-deposited fractions, which enable interpretation of the results with respect to the lung-deposited dose. Cena et al. 2011 have developed an add-on device to the SKC personal aluminum cyclone for the respirable fraction. This sampler is called the Nanoparticle Respiratory Deposition (NRD) sampler and consists of two stages. The sampler is inserted between the cyclone outlet and the usual 37-mm filter cassette. Its first stage is an impactor with a cutoff of 0.3 μm, and the total pressure drop is low enough to allow a normal personal sampling pump. The final stage emulates the total deposition by diffusion in all compartments of the respiratory tract.

Within the NANODEVICE projects, a number of prototype instruments were developed aiming to meet these conventions; for example, (a) a wide-range size resolving personal sampler (2 nm–5 μm) up to 8 size fractions; (b) a sampler for aerosol fraction deposited in the gas-exchange region (20 nm–5 μm), 8 size classes; and (c) a sampler for aerosol fraction deposited by diffusion in the anterior nasal region 5–400 nm (www.nano-device.eu).

In the case of sampling directly through a substrate suitable for electron microscope techniques (EM), for example, silicon substrates or Transmission Electron Microscopy (TEM) grids, three different collection principles can be observed: (a) particles are deposited by diffusion directly from the air stream passing through the grid openings, (b) particles are deposited onto a substrate by an electric field, or (c) particles are deposited by thermophoresis. Many versions of the first principle are also available as personal samplers, for example, the Aspiration Electron Microscopy Sampler designed by VTT Technical Research Centre of Finland (from where it is commercially available) (Lyyränen et al. 2009) and the Mini Particle Sampler developed by INERIS and distributed by EcoMesure (R'mili et al. 2013). An electrostatic precipitator (EP) collects the charged particles on a sample carrier, for example, TEM grid, by employing a strong electric field (5–10 kV) between two parallel plates (electrodes). Either the natural particle charge distribution or a corona discharger, to enhance the charge level, is used in these devices. Within the NANODEVICE project, a "plug and play" EP has been developed for relatively short sampling periods in the range of 20–400 nm (www.nano-device.eu). In the thermophoretic sampler (TP), a strong temperature gradient exists between a heated plate and a temperature sink on which the TEM grid is mounted. Azong-Wara et al. (2013) describe the development and testing of a personal TP, which shows very good results for homogeneous loading of a substrate in the range of a few nanometers to approximately 300 nm. This type of TP has been applied for the development of a so called cyto-TP where the substrate is a well with a cell culture for in vitro testing of real workplace nano aerosols (Broßell et al. 2013). A wider review on aerosol characterization methods is given by Asbach in Chapter 2 of this book.

Even though workplace exposure measurements are focused on personal exposure, it becomes evident that this is not fully possible yet for nanoparticle exposure measurements. Very few online and only a limited set of off-line sampling devices for personal exposure measurements exist. All of them have only been developed within the recent years so that only a very limited data set on exposure (related) measurements exists.

13.3 REAL AND SIMULATED WORKPLACE OR RELEASE STUDIES

As mentioned earlier, a comprehensive review of nano exposure relevant publications was published by Kuhlbusch et al. (2011). Since then a number of new studies have been published and this chapter will give an update of the review; however, we have restructured the presentation of the summary of studies. The conceptual model of inhalation exposure to nanoparticles (Schneider et al. 2011) has formed the backbone of recent exposure modeling. Therefore the papers have been categorized according to the so called source domains that include the vast majority of current and near-future exposure situations for manufactured nano-objects. The rationale for this categorization is that the source domains reflect different mechanisms of release and consequently possible different forms of released aerosols. Moreover, they are associated with the lifecycle stages of the nanomaterials, that is, production of the nanomaterial, downstream use/incorporation in a matrix/a nano-enabled product, the application of the product, the use phase, and activities related to the end-of-use.

1. *Point source or fugitive emission* during the production phase (synthesis) prior to harvesting the bulk material; for example, emissions from the reactor, leaks through seals and connections, and incidental releases (Figure 13.1).
2. *Handling and transfer* of bulk manufactured nanomaterial powders with *relatively low energy*; for example, collection, harvesting, bagging/bag dumping, bag emptying, scooping, weighing, and dispersion/compounding in composites (Figure 13.2).

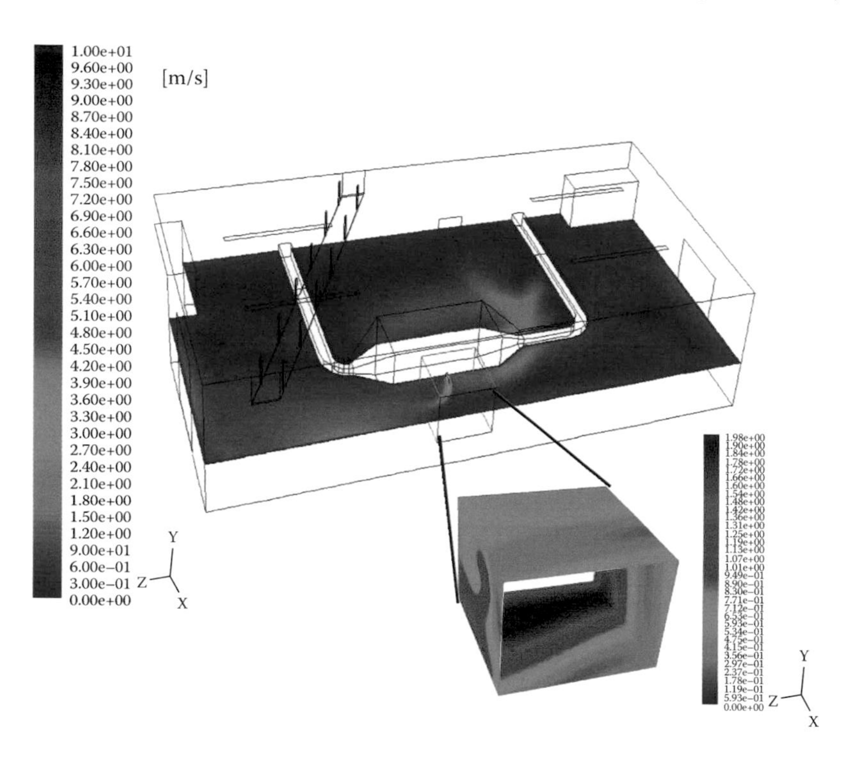

FIGURE 13.1 Modeling results of nanoparticle release from a leak in the reactor of a nanoparticle production facility. (Reprinted from *Handbook of Nanosafety,* Seipenbusch, M. et al. From Source to Dose: Emission, Transport, Aerosol Dynamics and Dose Assessment for WP Aerosol Exposure, Copyright 2014, with permission from Elsevier.)

3. *Relatively high energy dispersion* of either (a) nanopowders or (liquid) intermediates containing highly concentrated (>25%) nanoparticles, for example, pouring/injection molding, (jet) milling, stirring/mixing, or (b) application of (relatively low concentrated <5%) ready-to-use products; for example, application of coatings or spraying of solutions that will form nanosized aerosols after condensation (Figure 13.3).
4. *Activities resulting in fracturing and abrasion* of manufactured *nanoparticles-enabled end-products* at work sites; for example, (a) low energy abrasion, manual sanding, or (b) high energy machining, for example, sanding, grinding, drilling, cutting, and so on. High temperature processes like burning, and so on are included (Figure 13.4).

The focus of the review is to collect evidence of either release of or exposure to NOAA in real and simulated workplaces, whereas the data can be used to calibrate exposure models and to feed exposure scenario libraries, for example, the NANEX Exposure library.*

* http://nanex-project.eu/mainpages/exposure-scenarios-db.html.

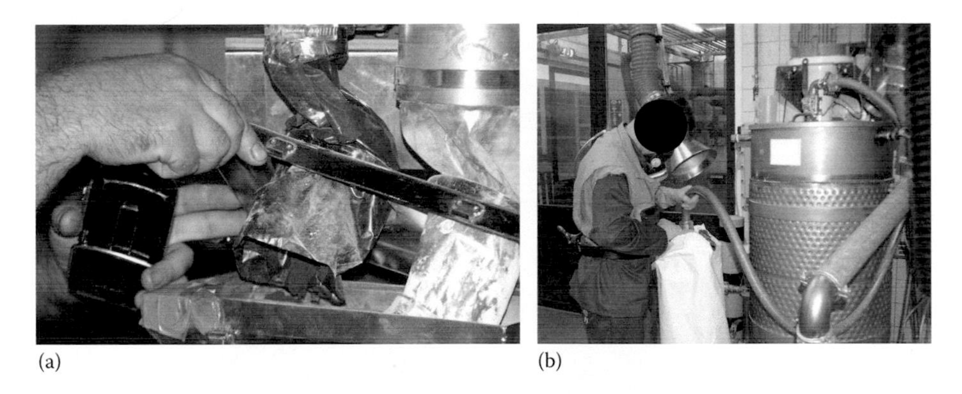

FIGURE 13.2 Examples of handling of nanopowders with low dispersion energy transfer, here bagging: (a) bag exchange and (b) bag filling. (Photos courtesy of TNO/C. Bekker.)

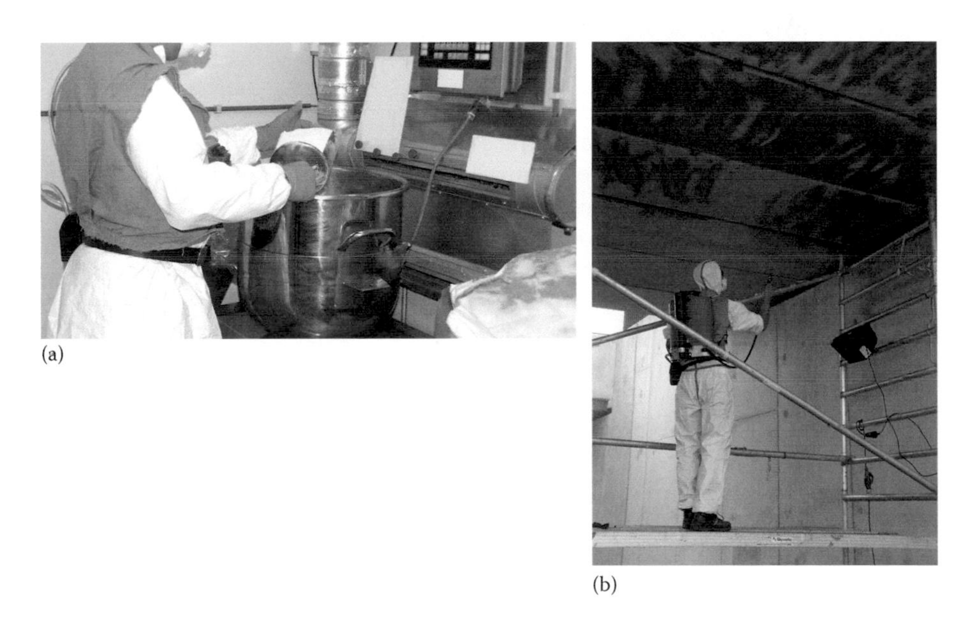

FIGURE 13.3 Examples of handling of nanopowders with high dispersion energy transfer, here (a) mixing and (b) spraying of a liquid coating. (Photos courtesy of TNO/C. Bekker.)

13.3.1 Source Domain 1: Synthesis of Nanoparticles

Gas-phase based aerosol processes have, in general, high production rates and low volume requirements. Examples are the hot wall (tubular) reactor process, laser-based aerosol production (e.g., laser pyrolysis), and plasma (arc/sputtering-based aerosol production) and (liquid) flame synthesis. Reasons why these types of synthesis are more often used over wet-based methods (e.g., sol–gel) are the often more complex nature and low yields of the latter. However, an excellent control of particle size distribution is achievable with the wet-based methods. Another method, the

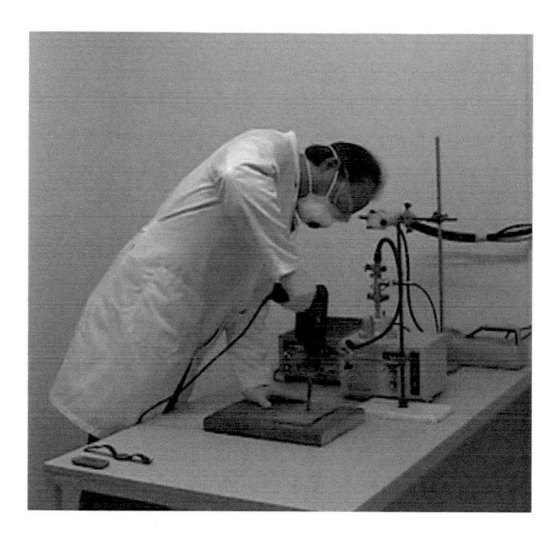

FIGURE 13.4 Investigations of nano-object release from nanomaterials during drilling.

top-down approach (attrition/milling) has low energy efficiency and produces inhomogeneous particle sizes. Aerosol spray methods atomize a solution or suspension and heat the droplets (in furnace or flame) to produce solid particles, which are then collected by precipitation. In addition, inert gas condensation processes exist with or without catalytic interaction, that is, chemical vapor deposition (CVD) and physical vapor deposition, respectively. All these processes have different potentials of releasing NOAA into the workplace environment, here especially as aerosols.

A few workplace or simulated workplace studies that fell into this source domain have been published after the Kuhlbusch et al. (2011) review or were not included (Table 13.1). In general, the most frequently studied process of synthesis is flame spraying (Leppänen et al. 2012; Walser et al. 2012; Koivisto et al. 2011; Koponen et al. 2013), which is a relatively open process where synthesis and collection take place in a fume hood. The spray process is a well-controlled process that may generate small sized primary particles, for example, 10–20 nm with a narrow distribution, for example, geometric standard deviation (GSD) < 1.8. However, due to high concentration, that is, above 10^6 cm^{-3}, coagulation may cause a shift of size mode. During normal production conditions some synthesized nanoparticles could be emitted, usually resulting in low breathing zone concentrations. Anyhow, it has been demonstrated that these particles will dominate the deposition in the alveolar region. Other processes like nano-objects production by CVD (Ogura et al. 2011; Lee et al. 2011) show a release of NOAAs (here CNT) only during the opening of the reactor. In addition, a relatively high elevation of the concentration compared to the background was reported during the preparation of the catalyst for CNT production using the CVD process. In more industrial scale operations the reactor room can be separated from other workplace, thus preventing these emissions from the reactor into workplace areas (Wang et al. 2012).

Not all studies assessed size distribution over a wide range of particle sizes, that is, from 10 nm to 10,000 nm, but usually limited this to the submicron range.

TABLE 13.1
Recent Publications on Point Source or Fugitive Emission (Source Domain 1)

Reference	Synthesis Process	Nanomaterial	Characterization	Metric	Summary/Conclusions
Ogura et al. (2011)	CVD	CNT	SEM	PNC, PSD (14–740; 500–20,000)	In on-site aerosol measurements at a research facility manufacturing and handling CNTs, airborne CNTs were only found inside a fume hood and glove box, except for a small amount of CNTs released from the glove box when it was opened. The airborne CNTs released during the synthesis of CNTs were considered by the authors to be practically negligible
Lee et al. (2011)	CVD (various workplaces)	CNT	STEM	PNC, PSD (14–500 nm) PMC (25–3,200 nm)	Nanoparticles and fine particles were most frequently released after the CVD (chemical vapor deposition) cover was opened, followed by catalyst preparation
Lee et al. (2012)	Induced coupled plasma with electric atomizer	Ag	SMPS, SEM-EDX, TEM	PNC, PSD (14–750 nm) PMC	Spatial and personal measurements were conducted at this pilot plant producing nanoAg-particles. Significantly increased PNC and PMC for Ag were observed during reactor operation but no clear quantification of nanoAg-particles was conducted. The number of particles ranged from 224–2,300 × 10^5 #/cm³ with the main fraction in the particle size range <100 nm (measured with SMPS)
Leppänen et al. (2012)	(enclosed) Flame spray process	CeO_2	SEM/TEM	PNC, PSD (4–120; 10–11,000 nm), PSD/PMD 30–10,000), PMC	The average particle number concentration varied from 4.7 × 10^3 to 2.1 × 10^5 #/cm³ inside the enclosure, and from 4.6 × 10^3 to 1.4 10^4 #/cm³ outside the enclosure. The average mass concentrations inside and outside the enclosure were 320 and 66 mg/m³, respectively. A batch-type process caused significant variation in the concentrations, especially inside the enclosure. CeO_2 NOAAs were mainly chain-like aggregates, consisting of spherical 20–40 nm primary particles having crystalline structures

(Continued)

TABLE 13.1 (*Continued*)

Recent Publications on Point Source or Fugitive Emission (Source Domain 1)

Reference	Synthesis Process	Nanomaterial	Characterization	Metric	Summary/Conclusions
Wang et al. (2012)	Hot-wall	SiO_2	SEM	PNC, PSD (15–750 nm), PSAC	Emission of silicon nanoparticles was not detected during the processes of synthesis, collection, and bagging. This was attributed to the completely closed production system and other safety measures against particle release. Emission of silicon nanoparticles significantly above the detection limit was only observed during the cleaning process when the production system was open and manually cleaned
Walser et al. (2012)	Flame spray pyrolysis	$CaCO_3$	No	PNC	Three scenarios of equipment failure were simulated during gas phase production of nanoparticles in a laboratory. While under normal production conditions, no elevated NOAA concentrations were observed, worst case scenarios led to homogeneous indoor NOAA concentrations of up to 10^6 #/cm^3 in the production room after only 60 s. The fast dispersal in the room was followed by an exponential decrease in number concentration after the emission event. The worst case emission rate at the production zone was also estimated at 2×10^{13} #/s with a stoichiometric calculation based on the precursor input, density, and particle size. NOAA intake fractions were 3.8–5.1 $\times 10^{-4}$ #/cm³ inhaled NOAA per produced NOAA in the investigated setting

Koponen et al. (2013)	Liquid flame spraying (LFS)	Fe*x*O*y*		PNC, PSD	The results for the experimental part focused on agglomeration showed that over time (>3 times 440 s) for low concentrations, that is, below 10^5 #/cm^3, the concentration close to the LFS generator increases, however, the mode of the primary aerosol size distribution did not shift to a larger mode, which indicates that no homogeneous coagulation occurred. For higher concentrations, that is, 2. × 10^6 #/cm^3, the FMPS data close to the LFS show a shift to a 20 nm mode, indicating that homogeneous coagulation may have occurred. Further away are slightly lower concentrations than at 1 m indicating that scavenging, agglomeration, and diffusion losses could be decisive factors when determining far-field concentrations. The results of the CFD modeling and the agglomeration experiments confirm the commonly assumed limit for homogeneous coagulation of 10^6 #/cm^3, which is an important finding with respect to exposure modeling
Koivisto et al. (2012a)	LFS	TiO_2	TEM, EDX	PNC, MC, PSD	PNC in the process room was on average 101 × 10^3 #/cm^3

Source: Göhler, D. et al. (2010), Characterization of nanoparticle release from surface coatings by the simulation of a sanding process, *Ann Occup Hyg* 54: 615–624.

Abbreviations: CVD, chemical vapor deposition; PSD, particle size distribution; PNC, particle number concentration; SA, surface area; M, mass; NAPD, number-averaged particle diameter.

The most commonly reported size modes are between 20 and 80 nm. Koivisto et al. (2011) reported in the area of a laminar flow reactor a GMD of 80 nm (1.8), and D_{AED} GMD of 120 nm (1.7). During liquid flame spraying synthesis a D_{mob}GMD of 14 nm (1.8) was reported by Koivisto et al. (2012b).

13.3.2 Source Domain 2: Handling of Powders, Low Energy

Powder handling activities are related to the actual production of nanomaterials, for example, collection and harvesting during (gas-phase) synthesis and packing, for example, bagging, as well as to downstream processes using nanomaterials for the manufacturing of nano-enabled products, for example, bag dumping, weighing, scoping and so on, and low energy dispersion into composites. A substantial number of workplace studies reported the resulting (combined) exposure from several activities (Table 13.2).

Especially in the assessment of powder handling scenario, it has been observed that exposure mostly results from agglomerates and aggregates, which contribute to relatively high mass concentrations, whereas their contribution to the elevation of particle number concentration is very limited. Curwin and Bertke (2011) and Tsai et al. (2011) reported MMAD of 4 μm and larger. Several authors, for example, Koivisto et al. (2012a) and Dahm et al. (2013), propose for these scenarios to focus on mass concentration or include PSD up to 10,000 nm in the measurement setup. Dahm et al. (2013) clearly demonstrated that activity-based measurements by direct reading instruments (DRIs) have limited value with respect to quantification of exposure. In their workplace study on exposure to CNTs and CNFs, they used both personal and workplace sampling followed by off-line detection of elemental carbon (EC) and CNT structure counts by TEM and DRIs. They observed no correlation between the results from sampling and off-line analysis and time-weighted averages of DRI results for corresponding processes. This is partly due to the high correlation of time series, as has been reported earlier by Klein Entink et al. (2011).

In general, clear indications on which powder handling scenario will result in the highest exposure concentrations cannot be extracted from these studies. Recently published results of a survey amongst 19 workplaces revealed increased particle number concentrations in the size range between 10 and 100 nm, compared to bagging of powders; however, the opposite was observed for particles with size ranges above 100 nm (Brouwer et al. 2013). Currently, data are lacking that confirm that amount of powder handled and level of energy, for example, fall-height, are major modifying factors of exposure.

13.3.3 Source Domain 3: Handling of Nanomaterials, High Energy

A number of (experimental) studies have been published (Table 13.3) that report the dispersion of nano-enabled products for personal care or surface coating by spray cans or hand pumps (Nørgaard et al. 2009; Chen et al. 2010; Lorenz et al. 2011; Nazarenko et al. 2011; Quadros and Marr 2011; Bekker et al. 2013). The use of nanosprays is relevant for both consumers, personal care (hair, facial, and antiperspirant spray), and professional use (hair products, leather impregnation, disinfections,

TABLE 13.2
Publications on Relatively Low Energy Handling and Transfer of (Source Domain 2)

Reference	Nanomaterials	Metric	Information	Conclusion
Cena and Peters (2011)	CNT	PNC (0.01–20 μm), PSD MC	For test sample production, bulk multiwall CNTs of 10–50 nm diameter and 1,000–20,000 nm length were weighed and mixed.	Weighing of CNT PNC ($N = 300$: GM 166, GSD 1.08), no difference to background, indicating no release, MC ($n = 51$: GM 0.03 μg/m^3 GSD 3.50)
van Broekhuizen et al. (2011)	Silica	PNC, SA (10–300 nm)	Two measurements regarding the mixing of Nanocrete (silica), 25 kg, 11 L mortar.	For one experiment a high increase in particle number concentration compared to the background was detected (199,508 vs. 20,763 #/cm^3) and an increase in particle mean diameter (62 nm vs. 41 nm), the other measurement showed a small increase in both variables, no chemical characterization was conducted
Dylla and Hassan (2012)	TiO_2	TEM EDX, PNC + PSD (2–150 nm)	Powder TiO_2, suspended TiO_2, and a control material were weighed and mixed into cement, sand and/or water.	The highest particle concentrations were found for dry mixing and mixing in general. The particle diameter varied from 29–53 nm. No good characterization
Fleury et al. (2013)	CNT	PSD, PNC (7 nm–10 μm)	Three sub-operations, CNT container opening, CNT transfer from the container to a first recipient for weighing, and transfer of weighed CNT to final container.	Opening the CNT container seems to provoke a sudden increase in nanoscale particles (2,200 #/cm^3) and a decrease in particle size, probably due to small air flow. Other activities showed concentration close to background level
Wang et al. (2012)	Silicon (57 nm)	PSD (10–750 nm), SA, PNC	Production cycle: generation from the reactor, collection by filters, bagging, packaging and cleaning of the system	Emission of nanoparticles was not detected during the processes of synthesis, collection, and bagging; only observed exposure during open cleaning process, 17,000 (#/cm^3), twice as high as during other processes and background. Surface area concentration 174 μm^2/cm^3, four times higher than before cleaning, mainly agglomerated particles (100–400 nm)

(Continued)

TABLE 13.2 (*Continued*)
Publications on Relatively Low Energy Handling and Transfer of (Source Domain 2)

Reference	Nanomaterials	Metric	Information	Conclusion
Zimmermann et al. (2012)	For example, Si, Pt, Co, Cu, Zn, Ti	PNC, PSD	Fifteen workplaces with different reactor cleanout methods, different materials were investigated by stationary measurements	Variation is found in the cleanout method and exposure ranges from 10 #/cm^3 to 10^6 #/cm^3, wet cleaning lowers exposure compared to dry clean-out, whereas most aggressive cleanout methods, that is, the use of a heat gun showed the highest emission. Other parameters that affect the emission of particles are the chemical element involved, the amount of matter used, and the periodicity of the maintenance and cleanout
Ham et al. (2012)	TiO_2, Ag, Al, Cu	PSD, PNC, SA (10–1,000 nm), M	Two workplaces are measured, manufacturing different nanoparticles, semi-closed processes, stationary measurements	Cleaning floor next to reactor resulted in PNC of 45,000 #/cm^3, with a size mode of 33.4 nm
Tsai (2013)	Al	PSD, PNC	Manual handling of NPs (transferring, pouring), variation in evaluated hoods, face velocity, and sash	Average number concentration in breathing zone, outside enclosure was 1,400 #/cm^3, estimated mass concentration 83 μg/m^3
Koivisto et al. (2012a)	TiO_2	PNC, PSD (5.5 nm–30 μm)	Four packaging areas, around 100 kg/h	Workers' average exposure varied from 225–700 μg/m^3, and from 1.15×10^4 to 20.1×10^4 #/cm^3. Over 90% of the particles were smaller than 100 nm. These were mainly soot and particles formed from process chemicals. Mass concentration originated primarily from the packing of p TiO_2 (pigment grade) and n TiO_2 (nano grade) agglomerates. The n TiO_2 exposure resulted in a calculated dose rate of 3.6×10^6 # min^{-1} and 32 μg min^{-1} where 70% of the particles and 85% of the mass was deposited in head airways

Evans et al. (2010)	CNF	PNC, SA, M	Manufacturing and processes CNF facility: production, mixing, drying, thermal treatment	Elevated PNC and MC indicate release of significant amounts (up to 1.15×10^6 #/cm^3, not due to CNF) of nanoscale particles and their agglomerates; no definite indication of release of single and agglomerated carbon nanofibers
Curwin and Bertke (2011)	Metal oxides	PSD, PNC, M, SA	Exposure assessment survey at seven facilities, small, medium, and large manufacturers and end users, only description of task is handling or production	TiO_2 mass concentration varied from GM 10.56 (1.88) to 78.3 (5.6) μg/m^3, PNC varied from GM 7,214 (1.27)#/cm^3 to 28,991 (2.69)#/cm^3, SA from 32 μm^2/cm^3 (2.4) to 145 μm^2/cm^3 (7.98). Generally, the greater mass of particles is found in the larger particle sizes. However, for production processes, the predominant mass of particles is found in the 0.1 to 1.0 μm particle diameter range
Dahm et al. (2012)	CNT/CNF	M (EC) respirable	Six sites assessed, primary or secondary manufacturers of CNT/CNF, personal breathing zone samples in occupational setting	Dry powder handling, mass concentration respirable EC 0.25–8 μg/m^3 (mean 4.5), production/harvesting 0.75–5.5 μg/m^3 (mean 2.5). During harvesting individual fibres were identified
Dahm et al. (2013)	CNT/CNF	PNC, respirable mass, SA	Six sites assessed, primary or secondary manufacturers of CNT/CNF, direct-reading instrument in occupational setting	Differences were not observed among the various sampled processes compared with concurrent indoor or outdoor background samples using different direct reading instruments (DRIs). These data are inconsistent with results for filter-based samples collected concurrently at the same sites (Dahm et al. 2012). Significant variability was seen between these processes as well as the indoor and outdoor backgrounds. However, no clear pattern emerged linking the DRI's results to the EC or the microscopy data (CNT and CNF structure counts). With respect to PNC, on average harvesting showed the highest concentrations, whereas, dry powder handling (weighing, mixing, extrusion, transferring) showed the highest mass concentration. Background concentrations showed the highest SA concentration

(Continued)

TABLE 13.2 (*Continued*)
Publications on Relatively Low Energy Handling and Transfer of (Source Domain 2)

Reference	Nanomaterials	Metric	Information	Conclusion
Tsai et al. (2010)	Al	NC, PSD	Three hood designs (constant-flow, constant-velocity, and air curtain hoods), manual handling of nano aluminum, low amounts of 100 g	Constant-flow hood: high variability in exposure; constant-velocity varied by operating conditions: very low exposure, new air curtain hood: exposure barely detected
Tsai et al. (2011)	SiO_2, CB, $CaCO_3$	PM, NC, PSD	Mixing of SiO_2 (40 kg) 2–5 times a day, dumping CB 600 kg per bag, 100 bags a day, dumping $CaCO_3$ 25 kg per bag, 900 bags a day	Mixing SiO_2: total mass 4,653 μg/m^3, dumping CB 732 μg/m^3, dumping $CaCO_3$ 510 μg/m^3. In addition, powders were tested with rotating drum MMAD (μm) 4.61 (2.4), 6.15 (2.3), 5.23 (2.7) for SiO_2, CB, $CaCO_3$, respectively

Abbreviations: EC, elemental carbon; GSD, geometric standard deviation.

TABLE 13.3
Publications on Relatively High Energy Dispersion (Source Domain 3)

Reference	Scenario	Nanomaterial	Characterization	Metric	Summary/Conclusions
Nørgaard et al. (2009)	Pump and propellant gas spray can	Silane, siloxane, TiO_2	GC/MS and GC/FID	PNC + PSD (6 nm–18.4 μm)	The number of generated particles was in the order of 3×10^8 to 2×10^{10} #/m³ per g sprayed NFP and was dominated by nanosized particles
Chen et al. (2010)	Propellant gas spray can	TiO_2	SEM/EDX	Gravimetric mass, PSD (5 nm–20 μm) NC (calculated from mass and density)	Results indicated that, while aerosol droplets were large with a CMD of 22 μm during spraying, the final aerosol contained primarily solid TiO_2 particles with a diameter of 75 nm. This size reduction was due to the surface deposition of the droplets and the rapid evaporation of the aerosol propellant. In the breathing zone, the aerosol, containing primarily individual particles (>90%), had a mass concentration of 3.4 mg/m³, or 1.6×10^5 #/cm³, with a nanoparticle fraction limited to 170 μg/m³ or 1.2×10^5 #/cm³
Lorenz et al. (2011)	Pump and propellant gas spray	Ag, ZnO	TEM	PNC + PSD (10–500 nm)	NOAA were identified in the dispersions of two products (shoe impregnation and plant spray). Nanosized aerosols were observed in three products that contained propellant gas. The aerosol number concentration increased linearly with the sprayed amount, with the highest concentration resulting from the antiperspirant. Modeled aerosol exposure levels were in the range of 10^{10} nanosized aerosols cumulated per person and application event for the antiperspirant and the impregnation sprays, with the largest fraction of nanosized aerosol depositing in the alveolar region. Negligible exposure from the application of the plant spray (pump spray) was observed.

(*Continued*)

TABLE 13.3 (*Continued*)
Publications on Relatively High Energy Dispersion (Source Domain 3)

Reference	Scenario	Nanomaterial	Characterization	Metric	Summary/Conclusions
Nazarenko et al. (2011)	Pump and nebulizer	Ag, Cu, Ca, Mg, Zn	TEM	PNC + PSD (13 nm–20 μm)	Realistic application of the spray products near the human breathing zone characterized airborne particles that are released during use of the sprays. Aerosolization of sprays with standard nebulizers was used to determine their potential for inhalation exposure. Electron microscopy detected the presence of nanoparticles in some nanotechnology-based sprays as well as in several regular products, whereas the photon correlation spectroscopy indicated the presence of particles of 100 nm in all investigated products. During the use of most nanotechnology-based and regular sprays, particles ranging from 13 nm to 20 μm were released, indicating that they could be inhaled and consequently deposited in all regions of the respiratory system. The results indicate that exposures to nanoparticles as well as micrometer-sized particles can be encountered owing to the use of nanotechnology-based sprays as well as regular spray products
Quadros and Marr (2011)	Pump spray	Ag	TEM, SEM/EDX	PNC + PSD (10 nm–10 μm), surface area	Three products were investigated: an anti-odour spray for hunters, a surface disinfectant, and a throat spray. Products emitted 0.2456 ng of silver in aerosols per spray action. The plurality of silver was found in aerosols with 12.5 μm in diameter for two products. Both, the products' liquid characteristics and the bottles' spray mechanisms played roles in determining the size distribution of total aerosols, and the size of silver-containing aerosols emitted by the products was largely independent of the silver size distributions in the liquid phase.

					Silver was associated with chlorine in most samples. Results demonstrate that the normal use of silver-containing spray products carries the potential for inhalation of silver containing aerosols. Exposure modeling suggests that up to 70 ng of silver may deposit in the respiratory tract during product use.
Bekker et al. (2013)	Propellant gas spray	SiO_2, Al	TEM, SEM/EDX	PNC + PSD (14 nm–20 μm), surface area	For all four spray products, the maximum number and surface area concentrations in the "near field" exceeded the maximum concentrations reached in the "far field." At 2 min after the emission occurred, the concentration in both the "near field" and "far field" reached a comparable steady-state level above background level. These steady-state concentrations remained elevated above background concentration throughout the entire measurement period (12 min). The results of the real-time measurement devices mainly reflect the liquid aerosols emitted by the spray process itself rather than only the MNO, which hampers the interpretation of the results. However, the combination of the off-line analysis and the results of the real-time devices indicates that after the use of nano-spray products, personal exposure to MNOs can occur not only in the near field but also at a greater distance than the immediate proximity of the source and at a period after emission occurred
Van Broekhuizen et al. (2011)	Spray coating (of surfaces)	TiO_2	—	PNC (20–300 nm)	Workplace measurements suggest a modest exposure to nanoparticles (NPs) above background of construction workers associated with the use of nanoproducts. The measured particles were within a size range of 20–300 nm, with the median diameter below 53 nm. Positive assignment of this exposure to the nanoproduct or to additional sources of ultrafine particles, like the electrical equipment used was not possible within the scope of this study.

(*Continued*)

TABLE 13.3 (*Continued*)
Publications on Relatively High Energy Dispersion (Source Domain 3)

Reference	Scenario	Nanomaterial	Characterization	Metric	Summary/Conclusions
Dylla and Hassan (2012)	Spray coating (of surfaces)	TiO_2	TEM EDX	PNC + PSD (2–150 nm)	The actual construction activities for spray-coat application resulted in higher nanoparticle concentrations compared to the laboratory-simulated construction activities totalling 2.39×10^8 #/cm³ released. The base coat released significantly smaller nanoparticles compared to the top coat. Both nanoparticle distributions correlated to the corresponding nanoparticle sizes comprised in each suspension. The nanoparticles collected were spherical with some agglomerating and none matching the shape of the photocatalytic nanorods in the top coat suspension.
Nazarenko et al. (2012)	Brush application of cosmetic powders	Various	TEM EDX	PNC + PSD (14 nm–20 μm	TEM observations and aerosol measurements suggest that exposure to nanomaterial(s) due to the use of cosmetic powders will be predominantly in the form of agglomerates or nanomaterials attached to larger particles that would deposit in the upper airways of the human respiratory system rather than in the alveolar and tracheobronchial regions of the lung, as would be expected based on the size of the primary nanoparticles.
Yang et al. (2012)	Dispersion of powder by compressed air	TiO_2	FESEM	PSD, PNC	Two peaks of nano-TiO_2 aerosol in the diameter range of 10–200 nm and 500–900 nm in the workplace are presented in the distribution profiling of number concentration curves with the number concentration of more than 4,000 #/cm³ and 1,000–2,000 #/cm³, respectively. Owing to the spray force of the aerosol source and the inertia force of flying aerosol particles, the number concentration at a distance of 3 m shows higher concentrations than in another sampling distance at the height of breathing zone

Yang et al. (2011)	High speed jet milling	TiO_2	FESEM	PSD, PNC	The flight and agglomeration behavior of aerosol nanoparticles influenced and limited the number concentration distribution of nano-TiO_2 aerosols in the workplace, while the dispersive forces on the aerosol nanoparticles from the discharging port had a profound influence on the flight and agglomeration of the particles. Agglomerated particles were the most frequently found form of aerosol particles, with a high number concentration in the workplace. Primary particles of the nano-TiO_2 aerosol (70 nm) reached a relatively high concentration at a height of 1.5 m (in the breathing zone), after a working time of 2 h, but exhibited lower concentrations at other working times. With the aerosol enclosure, a distinct reduction in the aerosol concentrations was observed for all diameter ranges. Furthermore, with the help of the aerosol enclosure, the diameter of agglomerated aerosol particles increased from the mode diameter of 124 nm to 175 nm

Abbreviations: PSD, particle size distribution; PNC, particle number concentration; SA, surface area; M, mass; NAPD, number-averaged particle diameter.

and cleaners). Most of the nanosprays contain nanosized metals or metal oxides. In general, it can be concluded that the release method, for example, pressure and nozzle size, determines the initial size distribution. For example, gas pressured spray cans generate large fractions of aerosols with particle sizes below 100 nm. However, the solute concentration will also significantly influence the number concentrations of the released aerosols. There is no indication that the presence of NOAA in spray solution affects the size distribution (Nørgaard et al. 2009; Nazarenko et al. 2011). Most studies report that individual ENPs have been observed in the breathing zone and could be deposited in the alveolar region of the lungs.

No evidence for exposure to NOAAs has been given by the few (field) studies on occupational spray coating of nano-enabled products in outdoor conditions (Table 13.3), e.g. (van Broekhuizen et al. 2011; Dylla and Hassan 2012); however, decisive conclusions cannot be drawn from the limited number of studies.

High energy dispersion of powders (Yang et al. 2011) will result in relatively high concentrations of agglomerates in the breathing zone; however, the presence of primary particles has been shown as well. The latter will be much depending on the stability of the agglomerates of the powder and the shear forces applied to the powder (Stahlmecke et al. 2009). A relatively low energy dispersion process, that is, manual brushing, has been investigated for personal care consumer products (Nazarenko et al. 2012). In this study mostly agglomerates were observed.

Most of the studies conducted for spray-can use were experimental studies in small size experimental rooms (often glove boxes). Only a few studies characterized the particle size distribution over a large size range (up to 10,000 nm). Chen et al. (2010) reported a CMD of 75 nm (2.3) and a NMD of 395 nm (1.6) and MMAD of 836 nm (1.6). Other studies reported size modes below and above 100 nm. Quadros and Marr (2011) reported NMD of 167 nm (±9) and 217 nm (±23). Bekker et al. (2013) reported for various spray cans size modes (D_{mob}) of 50 and 90 nm, and modes between 140–210 D_{AED} (electrical low pressure impactor data) and 500–600 nm (aerodynamic particle sizer data), which were similar to background. Delmaar (2013, personal communication) found number mean diameters between 90–112 nm and 160 nm for various spray cans, with MMAD (D_{AED}) between 2.07 μm (±0.34) and 4.2 μm (±0.41).

Nazarenko et al. (2012) studied the brush application of nano cosmetic powders and reported size modes <100 nm and MMADs of 1.44 μm (GSD 1.7) to 1.65 μm (GSD 1.7), whereas the regular powders showed a higher MMAD, that is, 2.86 (1.9) μm to 3.12 (1.6) μm.

For high speed milling of powders simulating workplace exposure, Yang et al. (2011) reported size modes of 37–525 nm (D_{mob}).

13.3.4 Source Domain 4: Nanoparticle-Enabled End Products

It is assumed that the liberation of NOAA from a matrix, for example, nanocomposites, and eventual discharge into the air (emission) will often result from a combination of energy input, for example, by mechanical or thermal processes. These processes and the composite material properties will determine the type of particles being released as discussed in the chapter by Kuhlbusch and Kaminski (Chapter 16).

Below we will summarize the results of simulated release and (workplace) exposure scenarios as listed in Table 13.4. The simulated scenarios investigated are low energy abrasion (Golanski et al. 2011, 2012; Schlagenhauf et al. 2012; Wohlleben et al. 2013) and high energy abrasion (Golanski et al. 2012; Hirth et al. 2013), sanding (Cena and Peters 2011; Huang et al. 2012; Wohlleben et al. 2013), grinding, polishing, cutting/drilling (van Broekhuizen et al. 2011; van Landuyt et al. 2012, 2013; Methner et al. 2012; Fleury et al. 2013), and other high-energy impacts on nanocomposites (Sachse et al. 2013; Golanski et al. 2012; Raynor et al. 2012).

With respect to CNT fragments released from machining from polymer matrices, so far there is a consensus that the debris mass is dominated by micron-sized composite fragments of matrix with bound CNTs, not by freely released CNTs, neither individually nor in bundles, for example, CNT-epoxy (Cena and Peters 2011; Golanski et al. 2012; Huang et al. 2012, Hirth et al. 2013), CNT-polyurethane (Schlagenhauf et al. 2012; Wohlleben et al. 2013; Hirth et al. 2013), and CNT-cement (Hirth et al. 2013). Similar observations were made for release from nanoplatelet composites (Raynor et al. 2012; Sachse et al. 2013), pigment composites with various polymers (Golanski et al. 2011), and dental composites (van Landuyt et al. 2013). In exception, Schlagenhauf et al. (2012) and Methner et al. (2012) identified free-standing CNTs and nonagglomerated CNFs, respectively, due to machining (cutting, grinding, and abrasion) of CNT or CNF containing composites. In addition, protrusions of CNTs at the composite fragments surface were observed (Cena and Peters 2011; Schlagenhauf et al. 2012; Hirth et al. 2013). Hirth et al. (2013) hypothesized that the phenomenon of protrusion is material-depending, that is, brittle versus tough matrix scenarios.

In general, it can be concluded that the number of particles released depends very much on the level of input energy, for example, high shear wear by, for example, turning speed, granular size of the sander paper, and the rigidity/hardness of the matrix. Release of nanofiller has been associated with inhomogeneity of dispersion of the nanofiller in the matrix (Schlagenhauf et al. 2012; Golanski et al. 2012), extreme high shear forces (Golanksi et al. 2012), or degradation of the matrix (Hirth et al. 2013).

In the few cases where a shift of mode or mean diameter was observed between nano- and non-nano control matrices, the tendency was that nano matrices shifted to a slightly larger size mode.

13.4 DISCUSSION AND CONCLUSION

Over the last few years a tremendous progress in the area of exposure assessment can be observed, both with respect to the development of devices and measurement strategies, and the number of field exposure and laboratory release studies.

Recent developments underline the usefulness of tiered approaches and the use of control banding tools based on exposure models, as first tier estimates of the potential for release. Furthermore, the usefulness of dose-estimating measurement devices that mimic deposition curves has been identified (Fissan et al. 2007), and (a) preprototypes of a wide-range size resolving personal sampler (2 nm–5 μm) up to 8 size fractions, (b) a sampler for aerosol fraction deposited in the gas-exchange region

TABLE 13.4
Publications Describing Scenarios Related to Source Domain 4

Reference	Scenario	Nanomaterial	Observed NM Release	Metric	Summary/Conclusions
Golanski et al. (2011)	Abrasion	TiO_2 (from coated surfaces)	Yes, no free or agglomerated nanoparticles	PSD (7 nm to 10 μm), PNC (7 nm to 10 μm)	Wet abrasion resulted in no release of nanoparticles, dry abrasion showed very low release of super- and submicron particles. No free or agglomerated TiO_2 nanoparticles were observed: TiO_2 nanoparticles (~30 nm) seem to remain embedded in the paint matrix. Agglomerates of matrix particles having sizes around 110 nm are observed. The sizes vary between 100 nm and 300 nm. Under wet abrasion, submicron and micrometer particles were released, but no nanoscale particles. Dry abrasion using different tools resulted in very low release of micron and submicron particles while no nanoscale particles were detected. SEM images showed no free nor agglomerated nanoparticles released during abrasion in the air and liquid: nanoparticles seem to remain embedded in the paint matrix
Cena and Peters (2011)	Sanding	CNT (epoxy/CNT composite)	Yes, large particles, commonly protuberances of CNT (attached to matrix)	PNC (0.01–20 μm), PSD	Processing of CNT-epoxy nanocomposite materials released respirable size airborne particles (mass concentration process/background ratio: weighing = 1.79; sanding = 5.90) but generally no nanoparticles (P/B ratio ~1). The particles generated during sanding were predominantly micron sized with protruding CNTs and very different from bulk CNTs that tended to remain in large (>1 μm) tangled clusters. Respirable mass concentrations in the operator's breathing zone were lower when sanding was performed in the biological safety cabinet (GM = 0.20 μg/m^3) compared with those with no LEV (GM = 2.68 μg/m^3) or those when sanding was performed inside the fume hood (GM = 21.4 μg/m^3)

van Broekhuizen et al. (2011)	Drilling	Silica	No	PNC, NAPD, SA (10–300 nm)	Measurements were carried out during drilling in conventional concrete as well as in a wall that was constructed with NanoCrete mortar. For drilling in cured NanoCrete concrete the arithmetic mean NP concentration for the down wind position exceeds the concentration in the up-wind position by 40,000 #/cm^3. For drilling in "normal" concrete this difference is ca. 16,000 #/cm^3 The median values for these situations differ by 20,000 #/cm^3 and 6,000 #/cm^3, respectively. The NPs emission generated by drilling in NanoCrete concrete is 2–3 times higher than the emission of drilling in "traditional" concrete, suggesting a higher release of NPs from the NanoCrete concrete. However, the emission of NPs from the idle-running drill indicates that the higher emission during the NanoCrete-concrete drilling may as well be caused by engine-generated NPs from the higher drilling intensity in the denser NanoCrete concrete
Methner et al. (2012)	Wet saw cutting, grinding, hand sawing	CNF (composites)	Yes, free fibers, bundles, and agglomerates	PNC (20 nm to 1,000 nm) PSD of fragments (by TEM)	Wet cutting and hand sanding of CNF composite without control revealed the highest particle number concentrations in the breathing zone
Schlagenhauf et al. (2012)	Abrasion	CNT (epoxy/CNT composite)	Yes, free-standing CNTs and agglomerates were emitted during abrasion	PSD	The mode corresponding to the smallest particle sizes of 300–400 nm showed a trend of increasing size with increasing nanofiller content. The three measured modes, with particle sizes from 0.6 to 2.5 µm, were similar for all samples. The measured particle concentrations were between 8,000 and 20,000 #/cm^3 for measurements with the SMPS and between 1,000 and 3,000 #/cm^3 for measurements with the APS. Imaging by TEM revealed that free-standing individual CNTs and agglomerates were emitted during abrasion

(*Continued*)

TABLE 13.4 (*Continued*)
Publications Describing Scenarios Related to Source Domain 4

Reference	Scenario	Nanomaterial	Observed NM Release	Metric	Summary/Conclusions
Sachse et al. (2013)	High energy impact	Nanoclays Silica	No	PNC, PSD (5.6–1,083 nm)	Particulate emissions were evaluated based on size-resolved from various silica based composites during impacting process. Physical characterization of the number concentration and size distribution of sub-micron particles from 5.6 to 512 nm was carried out, for the different composites. In general, nano and ultrafine airborne particles were emitted from all investigated materials. However, composite filled with nanoclay emitted higher amounts of particles than those filled with nano and microsilica. One reason for the increase of particle emission of the nanoclay filled composites was the change of the failure behaviour of the matrix. Nanoclay induced a transition from ductile to brittle fracture. Brittle material behavior results in fracture of material in many pieces and a low deformation, and hence more particles were generated. However similar results of particle emission were obtained for both nano and microsilica fillers, which in general did not vary significantly from the results obtained from traditionally reinforced glass fiber polymer composites

van Landuyt et al. (2012)	Polishing, grinding	Silica from dental composites	Yes, single nano-filler particles frequently observed	MC	Polymerized blocks of contemporary composites (Nano-Filler 54–70% total volume) were ground with a diamond bur according to a clinically relevant protocol. All composites released respirable dust in experimental setting. These observations were corroborated by the clinical measurements; however only short episodes of high concentrations of respirable dust upon polishing composites could be observed. Electron microscopic analysis showed that the size of the dust varied widely with particles larger than 10 μm, but submicron and even nano-sized particles could also be observed. The dust particles often consisted of multiple filler particles contained in resin, but single nano-filler particles could also frequently be distinguished
van Landuyt et al. (2013)	Grinding	Commercial dental composites	SEM/TEM/EDX	PSD (SMPS)	Exposure measurements of dust in a dental clinic revealed high peak concentrations of nanoparticles in the breathing zone of both dentist and patient, especially during aesthetic treatments or treatments of worn teeth with composite build-ups. Further laboratory assessment confirmed that all tested composites released very high concentrations of airborne particles in the nano-range (>10^6 #/cm^3). The median diameter of composite dust varied between 38 and 70 nm. Electron microscopic and energy dispersive X-ray analysis confirmed that the airborne particles originated from the composite, and revealed that the dust particles consisted of either filler particles, resin, or both

(*Continued*)

TABLE 13.4 (*Continued*)
Publications Describing Scenarios Related to Source Domain 4

Reference	Scenario	Nanomaterial	Observed NM Release	Metric	Summary/Conclusions
Huang et al. (2012)	Sanding, composite material	CNT filler			The quantity and size distribution of airborne particles emitted when CNT-containing materials are sanded depend on characteristics of the material being sanded and the conditions under which sanding occurs The determinants of particle concentration include the brittleness of test samples, which is a function of the loading of nanomaterial, and the abrasion energy, which is associated with sandpaper grit and disk sander speed during the sanding process. In addition to the abrasion, the friction originated by the contact of test samples and sandpaper may cause the sandpaper or adhesive backing to thermally decompose and produce sub-50-nm particles. No free CNTs were observed on air filters, except for tests conducted with 4% by weight CNT test sticks
Wohlleben et al. (2013)	Sanding, abrasion	CNT	Yes, a release of free CNT was not detected	PSD (SMPS) PSD of fragment	Three degradation scenarios were investigated that may lead to the release of CNTs from the composite: normal use, machining, and outdoor weathering. Unexpectedly, it was found that the relative softness of the material actually enhances the embedding of CNTs also in its degradation fragments. A release of free CNTs was not detected under any conditions using several detection methods. At very low rates over years, weathering degrades the polymer matrix as expected for polyurethanes, thus exposing a network of entangled CNTs.

					The absolute numbers of aerosol during high speed sanding are clearly above clean-air background, and comparable to the environmental background. Importantly, the numbers are not significantly different between the nanomaterial-free comparison material and the nanocomposite. In mass metrics, only larger fragments with 10–200 mm diameters are generated. Free CNTs were not observed: not by morphology, not by quantitative surface chemistry, not in the quantitative size distribution. Protrusions of CNTs on the polymer fragments were not observed either in morphology or in surface chemistry
Golanski et al. (2012)	Abrasion, scratching, mechanical shocks, sanding	CNT composite SiO_2 paint	High energy released CNTs only in the case of bad dispersion, only release of SiO_2 in the case of bad dispersion	PSD, PNC, TEM EDX	Abrasion was performed on polycarbonate, epoxy, and PA11 polymers containing carbon nanotubes (CNT) up to 4%wt. Release under mechanical shocks and hard abrasion was observed but only when nanomaterials present a bad dispersion of CNT within the epoxy matrix. Under the same conditions no release was obtained from the same material presenting a good dispersion. The CNT used in this study had an external diameter of 12 nm and a mean length of 1 µm. Release from paints under hard abrasion using a standard rotative taber tool was observed from an intentionally non-optimized paint containing SiO_2 nanoparticles up to 35%wt. The primary diameter of the SiO_2 was around 12 nm

(Continued)

TABLE 13.4 (*Continued*)
Publications Describing Scenarios Related to Source Domain 4

Reference	Scenario	Nanomaterial	Observed NM Release	Metric	Summary/Conclusions
Fleury et al. (2013)	Grinding, cutting	CNT	Yes, no isolated CNT and strange aggregates	PSD, PNC (7 nm–10 μm)	The grinding process generates a large amount of airborne particles, mostly small pieces of polymer containing CNT, which seems roughly proportional to the quantity of polymer introduced into the grinder. As this process does not involve heating, the CNTs stay stuck into the matrix and the probability to aerosolize fibers might be very small. A blank test was performed by running the grinder empty. It reveals an emission leading to very small-particle concentrations probably resulting from friction of rotating elements and blades, plus the natural emission of the electric motor. 10 min after the introduction of nanocomposites into the grinder, the measurement devices show a particle concentration of 3×10^3 #/cm^3 mainly composed of diameter below 50 nm (average size is close to 20 nm)
Hirth et al. (2013)	High energy abrasion/ sanding, nanocomposites	CNT	STEM images, elemental mapping, and photoelectron line shape analysis	PSD (SMPS) PSD of fragment	Scenarios of high-energy input (sanding) and of dry and wet weathering were simulated to the polymer matrix to investigate the airborne particles released from polymer nanocomposites. Protrusions of CNTs from fragments of CNT-epoxy and CNT-cement composites after sanding were observed, whereas such protrusions were not observed for CNT-polyoxymethylene and CNT-polyurethane fragments. There is no indication of freely released CNTs from mechanical forces alone. Based on size characterization with validated methods, at least 95 wt% of the CNT nanofillers remain embedded

Raynor et al. (2012)	Shredding, nanocomposite	CNT	SEM	NC, PSA, PSD	Tests conducted to measure the potential of nanoclay-reinforced polypropylene parts to generate airborne nanoparticles when they are shredded for recycling indicated that test plaques made from plain polypropylene resin, the same resin reinforced with talc, and the resin reinforced with montmorillonite nanoclay, all generated airborne nanoparticles with a count median diameter of approximately 10–13 nm. The highest particle concentrations were generated during shredding of plain polypropylene test plaques. Fewer particles were generated by shredding the nanoclay-filled test plaques, but not as few as were generated during the shredding of the talc-filled test plaques. Because the nanoclay and talc reinforcing materials were larger than the majority of particles measured, most of the particles that were observed must have been generated by a mechanism other than the direct release of the nanoclay or talc. SEM imaging did not identify any nanoparticles that appeared to be nanoclay liberated from the nanocomposite

Abbreviations: PSD, Particle size distribution; PNC, Particle number concentration; SA, Surface area; M, Mass; NAPD, Number-averaged particle diameter.

(20 nm–5 μm) 8 size classes, and (c) a sampler for aerosol fraction deposited by diffusion in the anterior nasal region 5–400 nm (Cena et al. 2011) have been developed. From another angle the combination of near-real time size-resolved measurement results with deposition models will also result in dose estimates. This has been applied for consumer exposure scenarios, for example, Chen et al. (2010), Lorenz et al. (2011), Quadros and Marr (2011), and has recently been introduced for worker exposure scenarios (Koivisto et al. 2011, 2012b).

Studies conducted to assess exposure in source domains 1 and 2, that is, during the production of nanomaterials, which include the synthesis and activities related to harvesting, milling/grinding, packaging, and the downstream use of these materials (usually in powder form) present a lot of data mostly from DRIs. This limits the quantification of the exposure to the NOAA since no clear quantification is possible. Survey-type of studies, for example, those conducted by NIOSH (Curwin and Bertke 2011; Dahm et al. 2012), tend to focus on risk assessment and risk management (comparison with recommended reference value), and provide shift-based (mass) concentrations. Most studies demonstrate potential for exposure to NOAA, however, the amount of data is still limited to get a clear picture of the range of exposure and its within (day-to-day) or between worker variation.

With a few exceptions, studies conducted to assess exposure in source domains 3 and 4, that is, scenarios with nano-enabled products such as sprays, or potential for release during the use phase of products, are workplace (or consumer) scenario simulated studies, which provide good information on relevant processes and determinants of release. This type of data is important for exposure modeling.

Laboratory release measurements are here considered as complementary to field exposure measurements since laboratory release studies indicate possible workplace and production processes which may lead to exposure and flag the need for systematic assessments which should be linked to appropriate safety measures. The conceptual model as developed by Schneider et al. (2011) is considered useful for a systematic evaluation of the various emission and release processes over the lifecycle of a manufactured nanomaterial. The first two source domains cover more or less the potential for release during processes and activities related to the production and downstream use of nanomaterials, whereas the last two source domains reflect actual application of nano-enabled products and use scenarios (and partly end-of-life scenarios), so they also include consumer exposure scenarios. Field studies have focused on the first two source domains, whereas the vast majority of data for the last two source domains are generated in experimental studies.

Despite the increased number of data it is still difficult to draw general conclusions from individual (and sometimes small scaled) studies whether there is significant workplace exposure to NOAA. Larger surveys, for example, the NIOSH studies by Curwin and Betke (2011) and Dahm et al. (2012, 2013), however, enable more robust conclusions, as in these studies the exposure levels were evaluated with respect to a (health-based) recommended exposure level (REL). These studies show that for TiO_2 the shift averages did not exceed the REL, whereas they did (for some job titles) for CNT/CNFs. The recently published NANOSH study, which included 19 enterprises in Europe, evaluated the results of activity-based measurements using a decision logic, where the results of all types of measurements were combined (Brouwer et al.

2013). The authors concluded that only 1 out of 54 (2%) fully characterized exposure scenarios was very likely to result in significant exposure to NOAA, however, for 42% other scenarios exposure to NOAA could not be excluded.

Specific scenarios in source domain 3, more specifically spray-can use, show high evidence of release (and potential for exposure) of NOAA, which especially for consumers might be (health)-relevant. However, for the occupational applicator of nano-enabled products there are fewer indications for exposure, though the number of available data is still scarce.

The interpretation of the results from experimental studies in source domain 4 with respect to the potential for exposure is not unambiguous, since only releases were studied. Again, some very specific processes can be flagged for their potential to result in exposure to NOAA.

In conclusion it can be stated that since 2011 many studies have been reported and the amount of data relevant for (worker) exposure assessment is increasing rapidly. However, in view of the number of workplaces and exposure scenarios it is still limited. Meta-analyses of data from field (and experimental) studies are needed to derive a better insight into quantitative exposure assessment and exposure modeling. Data pooling is an important condition to achieve relevant data sets for these purposes, thus activities related to harmonization of data collection, analysis, and reporting have to be continued.

REFERENCES

Anand, S., Mayya, Y.S., Yu, M., Seipenbusch, M., Kasper, G (2012) A numerical study of coagulation of nanoparticle aerosols injected continuously into a large, well stirred chamber. *J Aerosol Sci* 52: 18–32.

Asbach, C., Aguerre, O.,Bressot, C., Brouwer, D., Gommel, U., Gorbunov, B., Le Bihan, O., Jensen, K.A., Kaminski, H., Keller, M., Koponen, I.K., Kuhlbusch, T.A.J., Lecloux, A., Morgeneyer, M., Muir, R., Shandilya, N., Stahlmecke, B., Todea, A.M (2014) *Examples and Case Studies,* in *Handbook of Nanosafety*. Ed. Vogel et al., Elsevier.

Asbach, C., Kaminski, H., Von Barany, D., Kuhlbusch, T.A.J., Monz, C., Dziurowitz, N., Pelzer, J. Berlin, K., Dietrich, S., Götz, U., Kiesling, H.-J., Schierl, R., Dahmann, D (2012a) Comparability of portable nanoparticle exposure monitors. *Ann. Occup. Hyg.* 56: 606–621.

Asbach, C., Kuhlbusch, T.A.J., Kaminski, H., Stahlmecke, B., Plitzko, S., Götz, U., Voetz, M., Kiesling, H.-J., Dahmann, D (2012) "Standard Operation Procedures For assessing exposure to nanomaterials, following a tiered approach." www.nanoGEM.de.

Azong-Wara, N., Asbach, C., Stahlmecke, B., Fissan, H., Kaminski, H., Plitzko, S., Bathen, D., Kuhlbusch, T.A.J (2013) Design and experimental evaluation of a new nanoparticle thermophoretic personal sampler. *J Nanopart Res* 15 (4): art. no. 1530 1.

Bekker C., Brouwer D.H., van Duuren-Stuurman B., Tuinman I.L., Tromp P., Fransman, W (2013) Airborne manufactured nano-objects released from commercially available spray products: Temporal and spatial influences. *J Expo SciEnvEpid*, pp. 1–8.

Broßell, D., Tröller, S., Dziurowitz, N., Plitzko, S., Linsel, G., Asbach, C., Azong-Wara, N., Fissan, H., Schmidt-Ott, A (2013) A thermal precipitator for the deposition of airborne nanoparticles onto living cells-Rationale and development. *J Aerosol Sci* 63: 75–86.

Brouwer, D.H (2012a) Control banding approaches for nanomaterials. *Ann OccupHyg* 56 (5): 506–514.

Brouwer, D., Berges, M., Virji, M.A., Fransman, W., Bello, D., Hodson, L., Gabriel, S., Tielemans, E (2012b) Harmonization of measurement strategies for exposure to manufactured nano-objects; Report of a workshop 2012. *Ann OccupHyg* 56 (1): 1–9 10.

Brouwer, D.H., vanDuuren-Stuurman, B., Berges, M., Bard, D., Jankowska, E., Moehlmann, C., Pelzer, J., Mark, D (2013) Work place air measurements and likelihood of exposure to manufactured nano-objects, agglomerates, and aggregates. *J Nanopart Res* 15 (11): DOI 10.1007/s11051-013-2090-7.

BSI (2010) "Nanotechnologies—part 3: guide to assessing airborne exposure in occupational settings relevant to nanomaterials." British Standards Institution: London, UK. (BSI PD 6699-3:2010).

Cena, L.G., Anthony, T.R., and Peters, T.M (2011) A Personal Nanoparticle Respiratory Deposition (NRD) Sampler. *Environ SciTechnol* 45(15): 6483–6490.

Cena, L.G., Peters, T.M (2011) Characterization and Control of Airborne Particles Emitted During Production of Epoxy/Carbon Nanotube Nanocomposites. *J Occup Environ Hyg* 8: 86–92.

Chen, B.T., Afshari, A., Stone, S., Jackson, M., Schwegler-Berry, D., Frazer, D.G., Castranova, V., Thomas, T.A (2010) Nanoparticles-containing spray can aerosol: Characterization, exposure assessment, and generator design. *InhalToxicol* 22: 1072–1082.

Clark, K., Van Tongeren, M., Christensen, F.M., Brouwer, D., Nowack, B., Gottschalk, F., Micheletti, C., Riediker, M (2012) Limitations and information needs for engineered nanomaterial-Specific exposure estimation and scenarios: Recommendations for improved reporting practices. *J Nanopart Res* 14 (9): art. no. 970 1.

Curwin, B., Bertke, S (2011) Exposure characterization of metal oxide nanoparticles in the workplace. *J Occup Environ Hyg*8: 580–587.

Dahm, M.M., Evans, D.E., Schubauer-Berigan, M.K., Birch, M.E., Deddens, J.A (2013) Occupational exposure assessment in carbon nanotube and nanofiber primary and secondary manufacturers: Mobile direct-reading sampling. *Ann OccupHyg*, 57 (3): 328–344.

Dahm, M.M., Evans, D.E., Schubauer-Berigan, M.K., Birch, M.E., Fernback, J.E (2012) Occupational exposure assessment in carbon nanotube and nanofiber primary and secondary manufacturers. *Ann OccupHyg*, 56(5): 542–556.

Dylla, H., Hassan, M.M (2012) Characterization of nanoparticles released during construction of photocatalytic pavements using engineered nanoparticles. *J Nanopart Res* 14 (4):, art. no. 825.

EN ISO 13138 2012. "Air Quality—Sampling conventions for airborne particle deposition in the human respiratory system." Comité Européen de Normalisation, Brussels, Belgium.

Evans, D.E., Ku, B.K., Birch, M.E., Dunn, K.H (2010) Aerosol monitoring during carbon nanofiber production: Mobile direct-reading sampling. *Ann Occup Hyg*, 54(5): 514–531.

Fierz, M., Houle, C., Steigmeier, P., Burtscher, H (2011) Design, calibration, and field performance of a miniature diffusion size classifier. *Aerosol SciTechnol* 45(1): 1–10.

Fissan, H., Neumann, S., Trampe, A., Pui, D.Y.H., Shin, W.G (2007) Rationale and principle of an instrument measuring lung deposited nanoparticle surface area. *J Nanopart Res* 9: 53–59. doi: 10.1007/s11051-006-9156-8.

Fleury, D., Bomfim, J.A.S., Vignes, A., Girard, C., Metz, S., Munoz, F., R'Mili, B., Ustache, A., Guiot, A., Bouillard, J.X (2013) Identification of the main exposure scenario's in the production of CNT-polymer nanocomposites by melt-moulding process. *J Cleaner Production*, 1–15.

Furuuchi, M., Choosong, T., Hata, M., Otani, Y., Tekasakul, P., Takizawa, M., and Nagura, M (2010) Development of a personal sampler for evaluating exposure to ultrafine particles. *Aerosol Air Qual. Res.* 10(1): 30–37.

Golanski, L., Gaborieau, A., Guiot, A., Uzu, G., Chatenet, J., Tardif, F (2011) Characterization of abrasion-induced nanoparticle release from paints into liquids and air. *J Phys: Conf Ser* 304 012062.

Golanski, L., Guiot, A., Pras, M., Malarde, M., Tardif, F (2012) Release-ability of nano fillers from different nanomaterials (toward the acceptability of nanoproduct). *J Nanopart Res* 14 (7): art. no. 962.

Ham, S., Yoon, C., Lee, E., Lee, K., Park, D., Chung, E., Kim, P., Lee, B (2012) Task-based exposure assessment of nanoparticles in the workplace. *J Nanopart Res* 14:1126.

Hirth, S., Cena, L., Cox, G., Tomović, Z., Peters, T., Wohlleben, W (2013) Scenarios and methods that induce protruding or released CNTs after degradation of nanocomposite materials: Technology transfer and commercialization of nanotechnology. *J Nanopart Res*, 10.1007/s11051-013-1504-x.

Huang, G., Jae, H.P., Lorenzo, G.C., Betsy, L.S., Peters, T.M (2012) Evaluation of airborne particle emissions from commercial products containing carbon Nanotubes. *J Nanopart Res* 14: 1231.

Kaminski, H., Kuhlbusch, T.A.J., Rath, S., Götz, U., Sprenger, M., Wels, D., Polloczek, J., Bachmann, V., Dziurowitz, N., Kiesling, H. J., Schwiegelshohn, A., Monz, C., Dahmann, D., Asbach, C (2013) Comparability of Mobility Particle Sizers and Diffusion Chargers. *J Aerosol Sci* 57: 156–178.

Klein Entink, R.H., Fransman, W., Brouwer, D.H (2011) How to statistically analyse nano exposure measurement results: using an ARIMA time series approach. *J Nanopart Res* 13: 6991–7004.

Koivisto, A.J., Aromaa, M., Mäkelä, J.M., Pasanen, P., Hussein, T., Hämeri, K. (2012a) Concept to estimate regional inhalation dose of industrially synthesized nanoparticles. *ACS Nano* 6(2): 1195–1203 1.

Koivisto, A.J., Lyyränen, J., Auvinen, A., Vanhala, E., Hämeri, K., Tuomi, T., Jokiniemi, J. (2012b) Industrial worker exposure to airborne particles during the packing of pigment and nanoscale titanium dioxide. *Inhal Toxicol*, 24(12): 839–849.

Koivisto, A.J., Mäkinen, M., Rossi, E.M., Lindberg, H.K., Miettinen, M., Falck, G.C.-M., Norppa, H., Alenius, H., Korpi, A., Riikonen, J., Vanhala, E., Vippola, M., Pasanen, P., Lehto, V.-P., Savolainen, K., Jokiniemi, J., Hämeri, K (2011) Aerosol characterization and lung deposition of synthesized TiO_2 nanoparticles for murine inhalation studies. *J Nanopart Res* 13(7): 2949–2961.

Koponen, I.K., Asbach, C., Brouwer, D.H (2013) Study on nanoparticle aerosol emission and evolution using laboratory scale liquid flame spray nanoparticle generation system. Deliverable 5.2. Nanodevice.

Kuhlbusch, T.A.J., Asbach, C., Fissan, H., Göhler, D., Stintz, M (2011) Nanoparticle exposure at nanotechnology workplaces: A review. *Part Fibre Toxicol* 8: art. no. 22.

Lee, J.H., Ahn, K., Kim, S. M., Jeon, K. S., Lee, J. S., Yu, J.Il (2012) Continuous 3-day exposure assessment of workplace manufacturing silver nanoparticles, *J Nanopart Res* 14:1134.

Lee, J.H., Kwon, M., Ji, J.H., Kang, C.S., Ahn, K.H., Han, J.H., Yu, I.J (2011) Exposure assessment of workplaces manufacturing nanosized TiO_2 and silver. *Inhal Toxicol* 23(4): 226–236.

Leppänen, M., Lyyränen, J., Järvelä, M., Auvinen, A., Jokiniemi, J., Pimenoff, J., Tuomi, T (2012) Exposure to CeO2 nanoparticles during flame spray process. *Nanotoxicology*, 6(6): 643–651.

Lorenz, C. Hagendorfer, H., von Goetz, N., Kaegi, R.,Gehrig, R., Ulrich, A., Scheringer, M., Hungerbühler, K. (2011) Nanosized aerosols from consumer sprays: Experimental analysis and exposure modeling for four commercial products. *J Nanopart Res* 13: 3377–3391.

Lyyränen, J., Backman, U., Tapper, U., Auvinen, A., Jokiniemi, J (2009) A size selective nanoparticle collection device based on diffusion and thermophoresis. *J Phys Conf Ser* 170: 012011.

Maynard, A.D., Zimmer, A.T (2003) Development and validation of a simple numerical model for estimating workplace aerosol size distribution evolution through coagulation, settling, and diffusion. *Aerosol Sci Technol* 37(10): 804–817.

McGarry, P., Morawska, L., Morris, H., Knibbs, L., Capasso, A (2012) "Measurements of particle emissions from nanotechnology processes, with assessment of measuring techniques and workplace controls." Safe Work Australia.

Methner, M., Crawford, C., Geraci, C (2012) Evaluation of the potential airborne release of carbon nanofibers during the preparation, grinding, and cutting of epoxy-based nanocomposite material. *J Occup Environ Hyg* 9: 308–318.

Methner, M., Hodson, L., Geraci, C (2010) Nanoparticle emission assessment technique (NEAT) for the identification and measurement of potential inhalation exposure to engineered nanomaterials part a. *J Occup Environ Hyg* 7(3): 127–132.

Nazarenko, Y., Han, T.W., Lioy, P.J., Mainelis, G (2011) Potential for exposure to engineered nanoparticles from nanotechnology-based consumer spray products. *Journal of Exposure Sci Environ Epidemiol* 21: 515–528.

Nazarenko, Y., Zhen, H., Han, T., Lioy, P.J., Mainelis, G (2012) Potential for inhalation exposure to engineered nanoparticles from nanotechnology-based cosmetic powders. *Environ Health Perspect* 120(6): 885–892.

NIOSH (2011) "Current intelligence bulletin 63: occupational exposure to titanium dioxide." Department of Health and Human Services, Public health Service, Center for Disease Control and Prevention. Cincinnati, OH. Publication No. 2011–160.

NIOSH (2013) "Draft current intelligence bulletin: occupational exposure to carbon nanotubes and nanofibers. " US Department of Health and Human Services, Centers for Disease Control, National Institute for Occupational safety and Health, Cincinnati, OH. DHHS (NIOSH), NIOSH Docket Number: NIOSH No. 2013–145; Available at http://www.cdc.gov/niosh/docs/2013-145/pdfs/2013-145.pdf.

Nørgaard, A.W., JensenK. A., Janfelt C., Lauritsen F.R., Clausen, P.A.,WolkoffP (2009) Release of VOCs and particles during use of nanofilm spray products. *Environ Sci Technol* 43: 7824–7830.

Ogura, I., Hiromu, S., Kohei, M., Masashi, G (2011) Release potential of single-wall carbon nanotubes produced by super-growth method during manufacturing and handling. *J Nanopart Res* 13: 1265–1280.

Quadros, M.E., Marr, L.C (2011) Silver nanoparticles and total aerosols emitted by nanotechnology-related consumer spray products. *Environ SciTechnol* 45: 10713–10719.

R'mili, B., Le Bihan, O.L.C., Dutouquet, C., Aguerre-Charriol, O., Frejafon, E (2013) Particle Sampling by TEM Grid Filtration. *Aerosol Sci Technol* 47: 767–775. dx.doi.org/10.1021/es204580f.

Raynor, P.C., Ingraham Cebula, J., Spangenberger, J.S., Olson, B.A., Dasch, J.M., D'Arcy, J.B (2012) Assessing potential nanoparticle release during nanocomposite shredding using direct-reading instruments. *J Occup Environ Hyg* 9:1: 1–13.

Rim, D., Green, M., Wallace, L., Persily, A., Choi, J.-I (2011) Evolution of ultrafine particle size distributions following indoor episodic releases: Relative importance of coagulation, deposition and ventilation. *Aerosol Sci* 46: 494–503.

Sachse, S., Gendre, L., Silva, F., Zhu, H., Leszczyńska, A., Pielichowski, K., Ermini, V. and Njuguna, J (2013) On Nanoparticles release from polymer nanocomposites for applications in lightweight automotive components. *J Phys Conf Ser*. 429: 012046.

Schlagenhauf, L., Chu, B.T.T., Buha, J., Nuesch, F., Wang, J (2012) Release of carbon nanotubes from an epoxy-based nanocomposite during an abrasion process. *Environ Sci Technol* 46(13): 7366–7372.

Schneider, T., Brouwer, D.H., Koponen, I.K., Jensen, K.A., Fransman, W., Van Duuren-Stuurman, B., Van Tongeren, M., Tielemans, E (2011) Conceptual model for assessment of inhalation exposure to manufactured nanoparticles. *J Exposure Sci Environ Epidemiol* 21(5): 450–463.

Seipenbusch, M., Binder, A., Kasper, G (2008) Temporal distribution of nanoparticle aerosols in workplace exposure. *Ann Occ Hyg* 52(8): 707–716.

Seipenbusch, M., Yu, M., Asbach, C., Rating, U., Kuhlbusch, T.A.J., Lidén, G (2014). From Source to Dose: Emission, Transport, Aerosol Dynamics and Dose Assessment for WP Aerosol Exposure, in *Handbook of Nanosafety*. Ed. Vogel et al., Elsevier.

Stahlmecke, B., Wagener, S., Asbach, C., Kaminski, H., Fissan, H., Kuhlbusch, T.A.J (2009) Investigation of airborne nanopowder agglomerate stability in an orifice under various differential pressure conditions. *J Nanopart Res* 11(7): 1625–1635.

Tsai, C.J., Huang, C.Y., Chen, S.C., Ho, C.E., Huang, C.H., Chen, C.W., Chang, C.P., Tsai, S.J., Ellenbecker, M.J (2011) Exposure assessment of nano-sized and respirable particles at different workplaces. *J Nanopart Res* 13: 4161–4172.

Tsai, C.J., Liu, C.N., Hung, S.M., Chen, S.C., Uang, S.N., Cheng, Y.S., and Zhou, Y (2012) Novel active personal nanoparticle sampler for the exposure assessment of nanoparticles in workplaces. *Environ Sci Technol* 46(8): 4546–4552.

Tsai, C.S.J (2013) Potential inhalation exposure and containment efficiency when using hoods for handling nanoparticles. *J Nanopart Res* 15: 1880.

Tsai, S.J.C., Huang, R.F., Ellenbecker, M.J (2010) Airborne nanoparticle exposures while using constant-flow, constant-velocity, and air-curtain-isolated fume hoods. *Ann Occup Hyg* 54(1): 78–87.

Van Broekhuizen, P., van Broekhuizen, F., Cornelissen, R., Reijnders, L (2011) Use of nanomaterials in the European construction industry and some occupational health aspects thereof. *J Nanopart Res* 13(2): 447–462.

Van Broekhuizen, P., Van Broekhuizen, F., Cornelissen, R., Reijnders, L (2012) Workplace exposure to nanoparticles and the application of provisional nanoreference values in times of uncertain risks. *J Nanopart Res* 14(4): 770.

Van Landuyt, K.L., Hellack, B., Van Meerbeek, B., Peumans, M., Hoet, P., Wiemann, M., Kuhlbusch, T.A.J., Asbach, C (2013) Nanoparticle release from dental composites. *Acta Biomaterialia* (accepted for publication).

Van Landuyt, K.L., Yoshihara, K., Geebelen, B., Peumans, M., Godderis, L., Hoet, P., van Meerbeek, B (2012) Should we be concerned about composite (nano-)dust? *Dental Materials* 28: 1162–1170.

Walser, T., Hellweg, S., Juraske, R., Leuchinger, A., Wang, J., Fierz, M (2012) Exposure to engineered nanoparticles: Model and measurements for accidental situations in laboratories. *Sci Total Environ* 420: 119–126.

Wang, J., Asbach, C., Fissan, H., Hülser, T., Kaminski, H., Kuhlbusch, T.A.J., Pui, D.Y.H (2012) Emission measurement and safety assessment for the production process of silicon nanoparticles in a pilot-scale facility. *J Nanopart Res* 14: 759.

Witschger, O., Le Bihan, O., Reynier, M., Durand, C., Marchetto A., Zimmerman E., Charpentier, D (2012) Recommendations for characterizing potential emissions and exposure to aerosols released from nanomaterials in workplace operations. *Hyg. Secur. Trav.* - 1er trimestre 226: 41–55.

Wohlleben, W., Meier, M.W., Vogel, S., Landsiedel, R., Cox, G., Hirth, S., Tomovic, Z (2013) Elastic CNT-polyurethane nanocomposite: syntheses, performance and assessment of fragments released during use. *Nanoscale* 5: 369.

Yang, Y., Mao, P., Wang, Z.P., Zhang, J.H (2012) Distribution of Nanoparticle Number Concentrations at a Nano-TiO_2 Plant. *Aerosol Air Quality Res* 12934–940.

Yang, Y., Mao, P., Xu, C.I., Chen, S.W., Zhang, J.H., Wang, Z.P (2011) Distribution Characteristics of nano-TiO_2 Aerosol in the Workplace. *Aerosol Air Quality Res* 11: 466–472.

Zimmermann, E., Derrough, S., Locatelli, D., Durand, C., Fromaget, J.L., Lefranc, E., Ravanel, X., Garrione, J (2012) Results of potential exposure assessments during the maintenance and cleanout of deposition equipment. *J Nanopart Res* 14: 1209.

14 Mechanisms of Aging and Release from Weathered Nanocomposites

Tinh Nguyen, Wendel Wohlleben, and Lipiin Sung

CONTENTS

14.1 INTRODUCTION

From wood and tar to today's synthetic plastics, polymers provide indispensable materials for a wide range of applications. By adding fillers, pigments, and additives, a variety of polymeric materials with unique properties can be produced. Nanofillers (i.e., fillers having a dimension below 100 nm), such as spherical nanoparticles, layered platelets, tubes, and rods, have exceptional properties and high surface area. Incorporating a small amount (<5 mass%) of the high surface area nanofillers into a polymer (i.e., polymer nanocomposites) can substantially enhance many properties of the host matrix, as evidenced in numerous reviews on polymer nanocomposites for a variety of nanofillers (McNally and Pötschke 2011; Ma et al. 2010; Potts et al. 2011; Li and Zhong 2011; Pavlidou and Papaspyrides 2008; Zou et al. 2008). Today, polymer nanocomposite applications have gained a strong commercial footing, due to the outstanding performance of these advanced materials and the efforts of resin manufacturers and compounders who offer user-friendly products (Future Markets 2012). In 2006, only about 300 commercial products claimed to contain nanomaterials (McIntire 2012), but in 2013, more than 1,300 manufacturers around the world produced commercial nano-enabled products with a market value of 1.6 trillion

US dollars and an annual growth of 49% between 2009 and 2013 (Research and Markets 2009).

Polymer nanocomposites differ from traditional plastic composites in that they provide greatly enhanced properties with a minimum effect on mass and without major processing modifications. Depending on the type of nanofiller used, advantages of polymer nanocomposites over traditional polymer products include their being stronger, harder, tougher, lighter, more dimensionally stable, less permeable, and more durable. In coming years, polymer nanocomposites will enter the consumer markets in large quantities, and they will be increasingly used in essentially every segment of the industry from textiles, electronics, constructions, sporting goods to aerospace. Whatever the application, both the long-term performance of the composite and the fate of the nanofillers in the matrix during the product's lifecycle play a key role in the commercialization and uses of these nanocomposite products (Kingston et al. 2013). The reason for that is most common polymers tend to undergo degradation during exposure to weathering environments. The serious consequence of the degradation of the host matrix is that the embedded nanofillers could be exposed to the surface and released to the environment via the effects of rain, condensed water, wind, and mechanical vibrations. For those nanosize materials that have shown potential risks to the environment, health, and safety (Nel et al. 2006; Poland et al. 2008; Maynard 2006; Aschberger et al. 2010; Handy et al. 2008; Lee et al. 2010), we need to know rates and probabilities of release of the nanofillers during the lifecycle of a polymer nanocomposite to select products and scenarios of safe use.

Weather-induced degradation of the host matrix in a polymer nanocomposite is inherently a complex process controlled by the exposure environment, materials properties, and processing. In addition, nanoscale materials possess unique electronic structure, create a large polymer/particle interfacial volume fraction in the nanocomposites, and have a tendency to agglomerate. These distinctive characteristics of nanofillers likely affect the rate and mechanism of degradation of the polymer matrices. Yet, little data is available on (1) the behavior and mechanisms of polymer nanocomposite degradation under different weathering environments, (2) the fate of the imbedded nanofillers, (3) how the nanofillers may be released, and (4) the methodology to estimate the release rate as a function of weathering. The lack of this set of information hinders our ability to understand the nanofiller release mechanisms, to predict the potential release during the nanocomposite lifecycles and to develop strategies to mitigate this potentially serious problem. Due to the lack of such data, the long-term performance of polymer nanocomposites and the potentially harmful effects of the nanofillers incorporated in polymers on the environment and human health cannot be correctly assessed.

This chapter will briefly present and discuss recent advances in the weathering of polymer nanocomposites and release of nanofillers caused by weather-induced degradation of the polymer matrix.

14.2 WEATHERING OF POLYMER NANOCOMPOSITES

Polymer nanocomposites provide a range of appealing properties such that, in many applications, they are exposed to harsh conditions. Whatever the application, there is a real concern regarding their susceptibility to attack by weathering environments. This is because (1) the bond energies of most organic polymers are in the same range of the photon energies of the solar radiation (Rabek 1995), (2) the nanofillers have unique electronic structure and filler/matrix interfacial volume is high whose effects on the degradation and stabilization of the matrix are unknown, and (3) there is lack of data on weathering performance of polymer nanocomposites.

Weathering or weatherability of polymeric materials, for example, plastics, coatings, fiber-reinforced polymer composites, and polymer nanocomposites, is generally defined as the ability to withstand impacts from complex and variable atmospheric conditions similar to those in nature (Rabek 1995; Davis and Sims 1983). Testing the weathering of polymeric materials is generally carried out by exposing samples to natural outdoor environments and/or accelerated laboratory conditions. The primary environmental factors responsible for degradation of polymeric materials during weathering are solar radiation, temperature, moisture, atmospheric contaminants, oxygen content in air, and mechanical forces (Rabek 1995; Davis and Sims 1983; Hamid and Hussain 2000). These factors act synergistically, and their effects are not constant but always changing with time and location. Among these factors, the ultraviolet (UV) portion of the solar radiation inflicts the most severe damage to polymeric materials through a series of photochemical reactions. Other factors play only a minor role in the weathering of most common polymeric materials by promoting the photodegradation caused by UV light. Therefore, this section only reviews recent studies on the weathering of polymer nanocomposites caused by the UV radiation, that is, photodegradation, but following sections include further stressors such as hydrolysis. The performance of polymer nanocomposites subjected to weathering environments can be evaluated by a variety of techniques given in Figure 14.1.

Photodegradation depends strongly on the chemical composition and molecular structure of the polymers. Although most pure polymers do not absorb terrestrial sunlight to a great extent, the fact that most polymers undergo photodegradation during exposure to solar radiation (>290 nm) indicates that some chromophores, such as residual initiators, catalysts, antioxidants, fillers, and other additives, must be present in these polymer matrices (Rabek 1995; Hamid and Hussain 2000). Chromophores can absorb radiation at long wavelengths and initiate oxidative reactions of polymers, which is responsible for chain scission, mass loss, physical changes, and mechanical property reduction. Further, UV-induced degradation of polymeric materials is not a uniform progression, but a highly spatially inhomogeneous process (Nguyen et al. 2012). The degradation initiates at isolated nano- or microdomains, resulting in the formation of pits having nanoscale dimensions. Over time, these pits deepen and enlarge. Such inhomogeneous photodegradation of polymeric materials has a strong implication on the fate and release of nanofillers from polymer nanocomposites used outdoors. Nanofillers are likely to be exposed on the surface locations where degradation of the matrix occurs. On prolonged weathering, the degradation of the matrix will spread from these sites, resulting in locally exposing nanofillers on the nanocomposite surface.

Recent advances in the weathering of polymer nanocomposites containing multiwalled carbon nanotubes (MWCNTs), metal oxide nanoparticles, and nanoclays are briefly reviewed here, because these are the three most common nanofillers

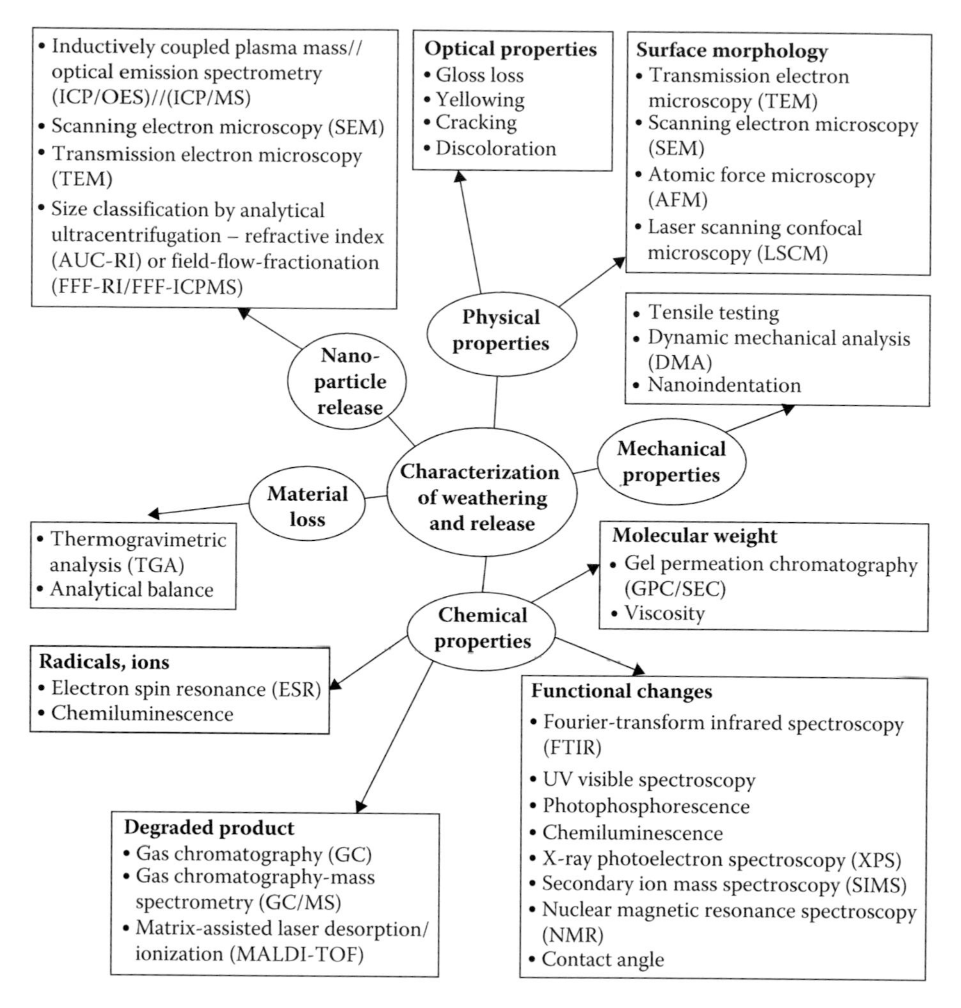

FIGURE 14.1 Techniques for characterizing weathering of polymer nanocomposites and nanofiller release.

that are being or potentially used in polymer nanocomposites for a variety of applications. MWCNTs are technically suitable nanofillers for enhancing the performance of polymers where high strength, exceptional electrical conductivity, light weight, and high aspect ratio can distinguish them from other nanofillers (McNally and Pötschke 2011; Ma et al. 2010), although only niche applications tolerate today's high prices. Degradation behavior of polymer–MWCNT nanocomposites subjected to UV radiation has been reported. Measured release from polymer–MWCNT and release probabilities extrapolated from degradation of the pure polymer have been reviewed (Kingston et al. 2013), identifying matrix degradation as the only relevant release scenario. An early study by Najafi and Shin (2005) found that the degradation rate of poly(methyl methacrylate)/MWCNT nanocomposite irradiated by UV in ozone is substantially reduced as the loading of MWCNTs increased from 0.02%

mass fraction to 0.5% mass fraction. This study quantified the nanocomposite degradation by measuring the etching depth with an AFM (see Figure 14.1 for all methods and their abbreviations). The enhanced UV resistance of the nanocomposite was attributed to the electron ring of the MWCNT network, which can disperse and filter radiation energy, and the strong interaction between free radicals (generated during irradiation) and MWCNTs. Similarly, the photo-oxidation of vinyl acetate (EVA)/MWCNT nanocomposites was observed to be lower than that of neat EVA, and the effect was attributed to MWCNTs acting as filters as well as an antioxidant (Morlat-Therias 2007). This study was conducted using UV radiation >300 nm at 60°C in the presence of oxygen, and photo-oxidation was characterized by FTIR. Similar MWCNT photostabilization effect was also observed for high density polyethylene (HDPE) (Grigoriadouet al. 2011).

However, chemical analysis by Wohlleben et al. (2011) suggested that an order of magnitude higher MWCNT content in polyoxymethylene accelerated the degradation of this photolabile polymer after irradiating it with an UV radiation dose equivalent to nine-month outdoor exposure. These authors also observed that a MWCNTs-containing layer having a thickness of approximately 3.0 μm was formed on the sample surface, and the surface-exposed MWCNTs were devoid of polymer matrix. Kumar et al. (2009) reported that, in the presence of singlet oxygen, MWCNT also catalyze the photodegradation of ethylene propylene diene monomer (EPDM). The enhanced degradation was attributed to the photocycloaddition reaction between singlet oxygen and double bonds on the composites, followed by cleavage. A study on graphene oxide (GO)/polyurethane (PU) nanocomposite under 295–400 nm UV radiation revealed that this carbon nanofiller did not seem to affect the rate of photodegradation or mass loss as compared to neat polymer (Bernard et al. 2011). However, both AFM and SEM images of this nanocomposite clearly showed the presence of GO particles accumulated on the composite surface with irradiation time.

Probably the most extensive study on the aging of polymer/MWCNT nanocomposites was performed by Nguyen and coworkers (Nguyen et al. 2008, 2009, 2011; Petersen et al. 2013) of the National Institute of Standards and Technology (NIST). Using the NIST Simulated Photodegradation via High Energy Radiant Exposure (SPHERE) and a variety of analytical techniques, they investigated the photodegradation of epoxy and PU nanocomposites containing between 0.72% and 3.5% mass fraction of pristine and isocyanate functionalized MWCNTs. This UV chamber provides a highly uniform UV radiation of 290–400 nm (Chin et al. 2004). Their spectroscopic (XPS and FTIR) results demonstrated that the degradation rates of the MWCNT nanocomposites were lower than those of the corresponding neat polymers. One example of their results is displayed in Figure 14.2 for an epoxy/0.72% MWCNT nanocomposite. Figures 14.2a and b show changes in the transmission FTIR intensity with UV irradiation time at 50°C and 75% RH for the C–O band at 1245 cm^{-1} of the epoxy chains and the C=O band at 1714 cm^{-1} formed during irradiation. (It is worth noting that the degradation behavior and degradation mechanism of this epoxy at 75% RH are similar to those exposed to outdoor during the summer in the state of Maryland.) These two FTIR bands represent chain scission (a) and oxidation (b), respectively. Figures 14.2a and b also include FTIR intensity changes of an epoxy/5% nanosilica

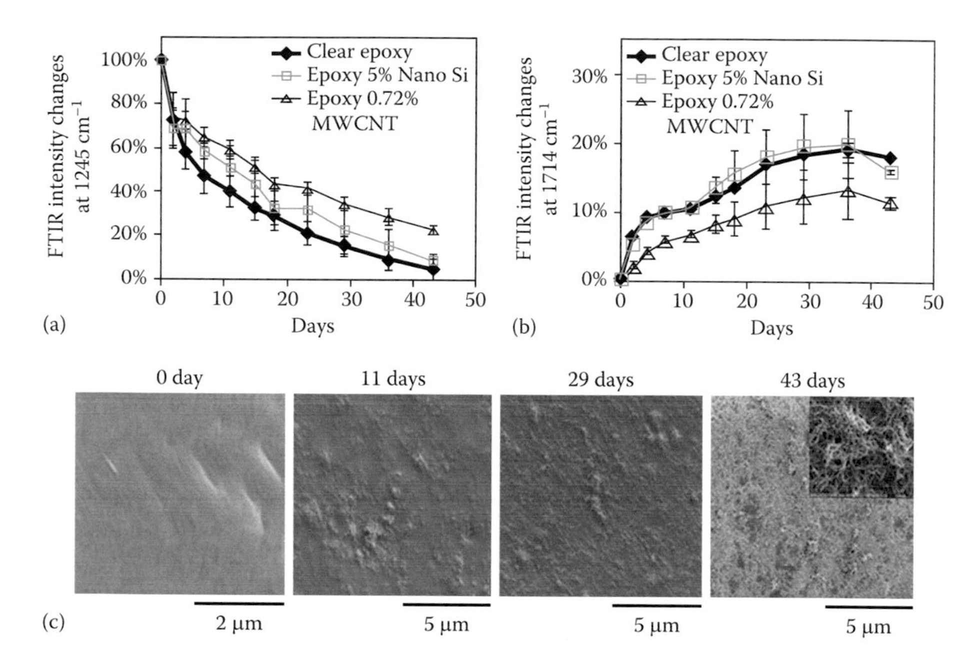

FIGURE 14.2 Transmission FTIR intensity changes of (a) 1245 cm^{-1} and (b) 1714 cm^{-1} bands with exposure time for a 6-μm film of neat epoxy, epoxy/0.72% MWCNT composite, and epoxy/5% nanosilica composite exposed to UV/50°C/75% RH at 18.5 MJ/m^2/day in the NIST SPHERE. Each data point was the average of four specimens and error bars represent one standard deviation. (c) SEM images of epoxy/0.72% MWCNT composite surface for several exposure times, showing the dense entangled network of CNTs.

composite for comparison. These data were obtained from nanocomposite and neat epoxy films having a thickness of approximately 6 μm cast on CaF_2 prisms. (It should be mentioned that transmission FTIR provides more accurate quantitative data on polymer degradation than those obtained by FTIR in the attenuated total reflection [ATR] mode, which are often reported in the polymer degradation literature.) The results of Figures 14.2a and b clearly show that the rates of both chain scission and photo-oxidation of the epoxy/MWCNT nanocomposites were lower than those of the neat epoxy or epoxy/nanosilica composites, consistent with data for epoxy/3.5% MWCNT composite (Petersen et al. 2013), PU/MWCNT composite (Nguyen 2008), and the photostabilization effect of the aforementioned reviewed MWCNTs.

SEM images of the same epoxy/MWCNT nanocomposite irradiated under the same UV conditions for several time intervals are displayed in Figure 14.2c. The surface before exposure appeared smooth with little evidence of MWCNTs. The absence of nanofillers and the smooth appearance, which is similar to that of epoxy/nanosilica composite (Figure 14.3), suggested that a thin layer of epoxy material probably covered the nanocomposite surface. The presence of such a layer potentially influences the rate of matrix degradation of a polymer nanocomposite. As seen in Figure 14.2c, bundles of MWCNTs became visible on the surface after 11 days, and their concentration increased with increasing exposure time. After 43 days, MWCNTs

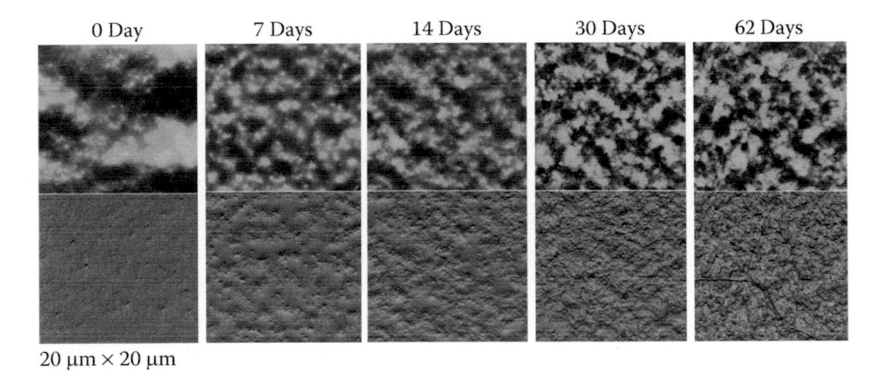

FIGURE 14.3 Height (top) and phase (bottom) 20 micron AFM images of epoxy/5% silane-treated nanosilica composite surface before and after UV irradiation for different times, showing a gradual accumulation of silica nanoparticles on the composite surface (at 18.5 MJ/m²/day).

have formed a dense entangled layer on the composite surface, which is clearly seen in the high magnification image (inset). Similar surface evolution of MWCNTs was observed for epoxy nanocomposite containing 3.5% mass fraction of MWCNTs in the NIST SPHERE (Petersen et al. 2013) and for PU with 3% MWCNTs under ISO 4892 irradiation (Wohlleben et al. 2013). From the microscopic and spectroscopic results, these studies suggested that the increased MWCNT concentration at the nanocomposite surface with UV irradiation was a result of the matrix photodegradation. As the epoxy matrix degraded and was removed, MWCNTs were increasingly exposed on the surface and formed a dense entangled structure with no evidence of release, even after long exposure time. The thickness of the MWCNT layer on the composite surface was approximately 400 nm (by cross-section SEM imaging) after 43 days of exposure and remained essentially unchanged between 45-day and 9-month exposures, suggesting that MWCNT layer on the composite surface probably shielded the UV radiation from attacking the epoxy matrix underneath.

Metal oxide nanoparticles, such as alumina (Al_2O_3) and titania (TiO_2), and nonmetal oxides, such as silicon dioxide (SiO_2), are currently the largest volume nanofillers used for polymer nanocomposites (Future Markets 2012). The metal oxide fillers serve many functions in the plastics and coatings industries. TiO_2 and ZnO are traditionally used as pigments to enhance the appearance and improve the durability of polymeric products. However, due to their ability to absorb broadband UV, these materials at nanosize have been exploited in applications such as self-cleaning coatings, UV-resistant coatings, and sunscreens. These nanofillers are also used for modifying optical properties, for example, increasing the refractive index of coatings. The degradation of several polymer/metal oxide nanoparticle composites irradiated with UV radiation has been investigated. For example, a photo-oxidation study of a PU coating containing different amounts of two types of TiO_2 nanoparticles showed that both photocatalysis of PU/anatase TiO_2 nanocomposite and photostability of PU/rutile TiO_2 nanocomposite increased with an increase in the TiO_2 nanoparticle content (Chen et al. 2007).

The photodegradation improvement by TiO_2 nanoparticles was also demonstrated in a PE-oxidized wax–nanoTiO_2 system, which revealed a higher degradation efficiency than either PE/TiO_2 nanocomposite or neat PE materials (Fa et al. 2010).

Similar to TiO_2, ZnO nanoparticles also exhibited both photocatalytic and photostabilizing behavior in polymer nanocomposites. The enhanced photostability due to ZnO was observed for composites made with polypropylene (PP) (Chandramouleeswaran et al. 2007; Zhao et al. 2006), linear-low-density PE (Yang et al. 2005), PU (Hegedus et al. 2008), and wood polymer composite (Dekaand Maji 2012). The photostability was attributed to the superior UV radiation screening effects of the ZnO nanoparticles. However, other studies reported the catalytic effect of ZnO nanoparticles. For example, the degradation of poly(vinyl chloride) (PVC)/ZnO nanoparticle composite was found to be greater than that of neat PVC (Sil et al. 2010). Similar accelerated degradation was observed by Gu et al. (2012a, 2012b) for PU/ZnO nanoparticle composite during its exposure to 295–400 nm UV light in the NIST SPHERE under dry and humid conditions. The rate of acceleration depended strongly on ZnO concentration and relative humidity of the exposure. This study also noted that AFM in the phase imaging mode is useful for revealing the preferential degradation of the polymer layer near the ZnO nanoparticle surface. Using the same UV source, Nguyen and coworkers (Nguyen 2012; Nguyen et al. 2012; Nguyen et al. 2011; Gorham et al. 2012) showed that the epoxy composite containing 5% mass fraction of silane-coated nanosilica degraded at a higher rate than the neat epoxy (e.g., Figure 14.2a). However, there is little difference in the oxidation rate (Figure 14.2a) between the nanocomposite and the neat epoxy. Further, the matrix degradation resulted in an increasing accumulation of silica nanoparticles on irradiated surfaces with exposure time, as demonstrated by AFM imaging displayed in Figure 14.3. Before exposure (day 0), the epoxy/nanosilica composite surface appeared smooth and nearly featureless, with little evidence of nanofillers, similar to that observed for MWCNT composites. After seven days of UV exposure, silica nanoparticles started to appear on the surface, and after 62 days, silica nanoparticles had covered almost the entire composite surface. The presence of silica nanoparticles on the surface was also supported by XPS and IR results (Nguyen et al. 2012; Gorham et al. 2012).

Clay nanofillers are currently incorporated in many commercial polymers to improve their mechanical strength, fire retardancy, gas barrier, and dimensional stability (Pavlidou and Papaspyrides 2008). The weathering performance of polymer/clay nanocomposites has received some attention (Kumar et al. 2009). Depending on material type, nanoclays displayed both catalytic and stabilizing effects on photodegradation of polymers. For example, nanocomposites made with montmorillonite (MMT) and PP(Morlat-Therias et al. 2004 and Mailhot et al. 2003), PE (Quin et al. 2003), and EPDM (Morlat-Therias et al. 2005b) were found to have a higher photo-oxidation rate than their neat respective polymers. The acceleration of photo-oxidation was attributed to the degradation of the alkyl-ammonium cation exchanged in natural MMT and the catalytic effect of iron impurities in the chemically modified MMT. The catalytic active sites in the nanoclays can accept electrons from donor molecules of the polymer matrices and induce the formation of free radicals in the polymer chains upon UV irradiation (Qin et al. 2004), leading to degradation of the polymers. In the presence of clay nanofillers, the efficiency of UV stabilizers and antioxidants for photodegradation of polyolefins was also substantially reduced (Chmela 2005; Mailhot et al. 2003). Further, clay nanofillers and their compatibilizers did not change the photo-oxidation mechanism, but they strongly affected the oxidation rates of these polymers (Morlat 2004; Hrdlovic 2004). Similar enhanced photodegradation was also observed for PP/TiO_2-organoclay, polycarbonate/layered silicate (Sloan et al. 2003), and polystyrene/layered double hydroxide (LDH) nanocomposites (Peng and Qu 2005). However, the rate of PP photodegradation was reduced when LDH (Ding and Qu 2006) or clay nanofiller modified with dopamine (Phua et al. 2013) was used.

14.3 RELEASE OF FRAGMENTS FROM MATRIX DEGRADATION: SPONTANEOUS OR INDUCED?

Once the matrix is degraded, the forces occurring in certain lifecycle scenarios (Le Bihan et al. 2013) may be sufficiently strong to overcome the remaining cohesion within the matrix and/or the adhesion between the matrix and nanofiller. As a consequence, fragments of the nanocomposite and/or individual nanofillers may be released into the atmosphere, soil, or liquids. If the physicochemical interactions with external media dominate, we designate the release as "spontaneous"; if external forces such as mechanical stress contribute substantially, we designate the release as "induced." Both can occur in real life and in laboratory-scale simulations.

Spontaneous release of nanofillers into run-off water was investigated by the NanoPolyTox project, and first results on the degradation of the nanocomposites (Vilar et al. 2013) and the released fragments in run-off waters (Busquets-Fite et al. 2013) have been reported. The tests were performed for 1,000 h in a climatic chamber (Suntest XXL+, Atlas) according to ISO 4892/06 with 0.5 W/m^2 nm at 340 nm (60 W/m^2 integrated from 300 nm to 400 nm). In the adapted protocol, after 29 min dry irradiation, rain was simulated for 1 min and the run-off waters were collected. The characterization was performed using a multitude of techniques, but specifically the solid content of fragments released into the run-off was assessed after lyophilization using gravimetry (Figure 14.4), and further identified using TEM and energy-dispersive X-ray spectroscopy (EDXS). For many other established colloidal techniques the total amount of fragments (on the order of 10 mg), dispersed in several liters of run-off, is prohibitively low.

By comparison of 12 nanocomposites (MWCNT, SiO_2, TiO_2, and ZnO nanofillers compounded in polyamide (PA), PP, and EVA matrices), Busquets-Fité et al. correlated the compatibility between nanofiller and matrix, as evidenced by the degree of nanofiller dispersion in the composite, with a significantly reduced probability of release during weathering. For good matrix/filler compatibility systems, for example, pristine MWCNTs in PP, the total release remained as low as 0.02% of the total specimen mass, corresponding roughly to 200 mg/m^2 release per irradiated surface at an UV dose of 216 MJ/m^2. No evidence of free MWCNTs was observed. In contrast, a sample with substantial aggregation of nanofillers in the matrix such as propyl-functionalized SiO_2 in PP led to 10-fold higher release, and an over-representative accumulation of SiO_2 nanofiller in the run-off waters (Figure 14.5h) (Busquets-Fite et al. 2013).

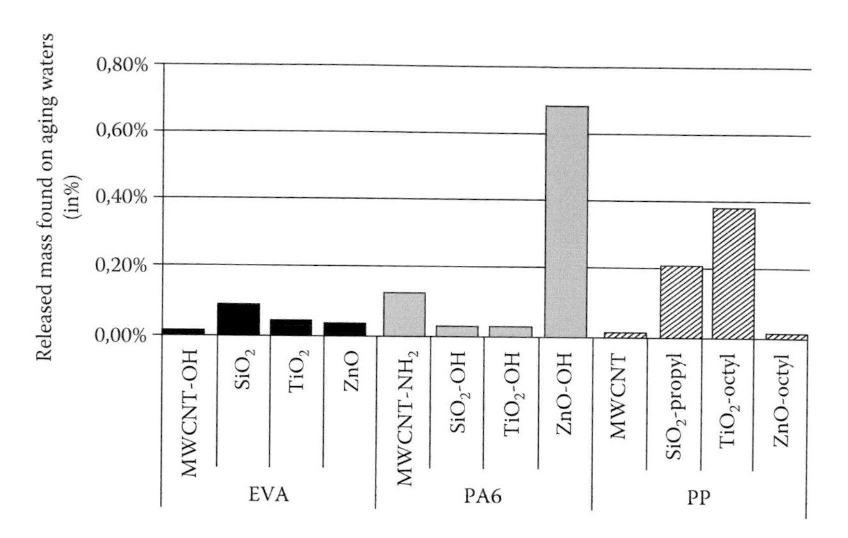

FIGURE 14.4 Spontaneous release during weathering with rain cycles: the solid content of collected run-off water increases for nanocomposites with low compatibility between nanofiller and matrix. (Adapted from Busquets-Fite, M. et al., Exploring release and recovery of nanomaterials from commercial polymeric nanocomposites, Institute of Physics.)

To explore the relation between spontaneous release and induced release, and to assess the origins of the variation, a pilot interlaboratory comparison was performed in the frame of the nanoGEM project between Leitat, NIST, and BASF, to simulate and detect aging and release of 4% hydrophobized SiO_2 from PA nanocomposites (Wohlleben et al. 2014). Both induced release by mechanical shear after dry weathering and spontaneous release during wet weathering were investigated at different UV intensities. For induced release, irradiated specimens between 10 and 50 cm^2 were immersed in a minimal volume of water (below 10 ml) to maximize signal-to-noise from released fragments. The forces to induce release were gradually increased from 24 h immersion (interfacial forces only) to 24 h shaking (assumed to be closest to reality), to 1 h ultrasound bath (as worst-case estimate). The SEM and TEM evaluation of fragments in run-off water (spontaneous release, identical protocol to the NanoPolyTox results reported earlier) and of fragments in immersion fluids (induced release) were qualitatively the same with respect to complex geometrical shapes and chemical compositions, ranging from polymer fragments with embedded or bound SiO_2 to occasional observations of pure polymer or pure nanofiller (Figures 14.5e through 14.5g). Given the coexisting morphologies and high polydispersity, a statistically significant quantification of fragments is difficult. Centrifugation methods were developed (Wohlleben 2012) and applied to size-selective quantification of release. By reference to the irradiated surface area that was immersed, values of mg fragments released from m^2 irradiated surface can be extracted. The amount of fragments with less than 150 nm diameter, where free SiO_2 would appear, grew with the UV irradiance, and was grossly consistent between laboratories, (Wohlleben et al. 2014). However,

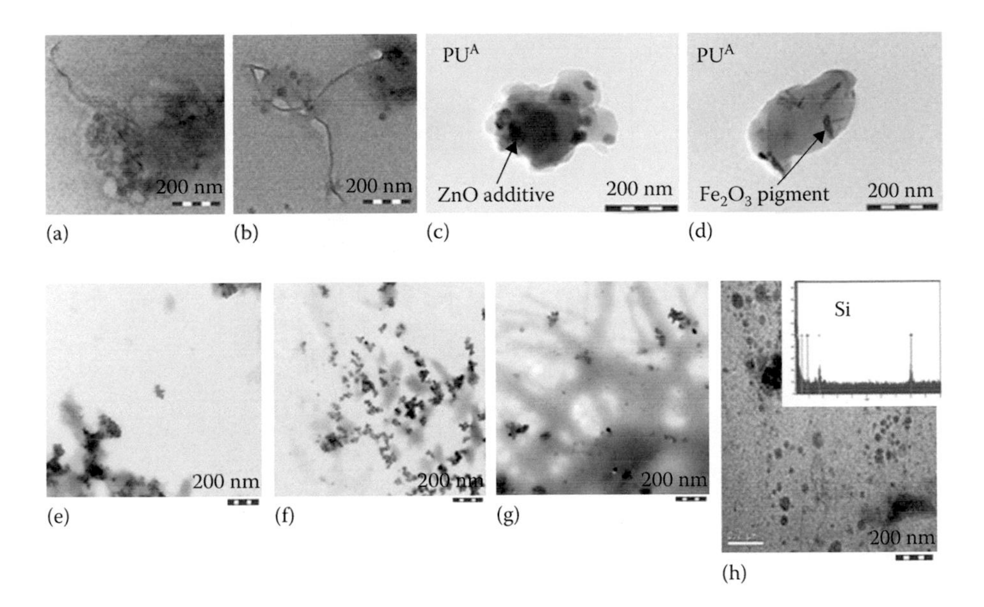

FIGURE 14.5 TEM image gallery of released fragments after simulated weathering. (a) Fragment from MWCNT–TPU after UV + immersion, using shaker; (b) ditto using ultrasound; (c) PU coating with ZnO pigment, irradiated and sanded, sampled from atmosphere; (d) ditto, with Fe_2O_3 pigment. (e) Fragment from nanoGEM. PA.SiO2 after UV + immersion, irradiated with NIST SPHERE dry 85 MJ/m²; (f) ditto, irradiated with ISO 4892 rain cycles 157 MJ/m²; (g) ditto, irradiated with reduced rain cycles 234 MJ/m²; (h) same protocol as (g), but from PP–SiO_2–propyl composite. ((a), (b) From Hirth, S. et al., *J. Nanopart. Res.*, 15, 1504, 2013. (c), (d) With permission from Göhler D. et al. (2013b), Nanoparticle release quantification of sanding-induced airborne particle emissions from artificially aged/weathered nanocomposites, Proceedings of Nanotech 2013, Boston, USA. (h) Adapted from Busquets-Fite M. et al. Exploring release and recovery of nanomaterials from commercial polymeric nanocomposites, 2013, Institute of Physics.)

within the error bars, the quantity of release (Figure 14.6a) was the same from the neat PA reference and from the PA SiO_2 nanocomposite material, confirming that SiO_2 has no significant potential to modify the matrix degradation. Further, accelerated irradiation at 140 W/m² over-proportionally increased release compared to ISO-4892 irradiation at 60 W/m² (Figure 14.6a). Aging and sampling conditions were critical for the release rates, not for release characteristics (Wohlleben et al. 2014). The immersion after rain + UV weathering released less fragments, indicating that fragments from SiO_2 nanocomposites have already been released spontaneously, without shear forces, into run-off waters during weathering (Wohlleben et al. 2014).

Using the same "UV + immersion" protocol of irradiation of ISO 4892 (Suntest XLS+) and induced release, Hirth and coworkers (2013) investigated the release probability of MWCNT from a thermoplastic PU (TPU) matrix. Again, TEM analysis of aliquots from the immersion fluids showed a coexistence of micron-sized

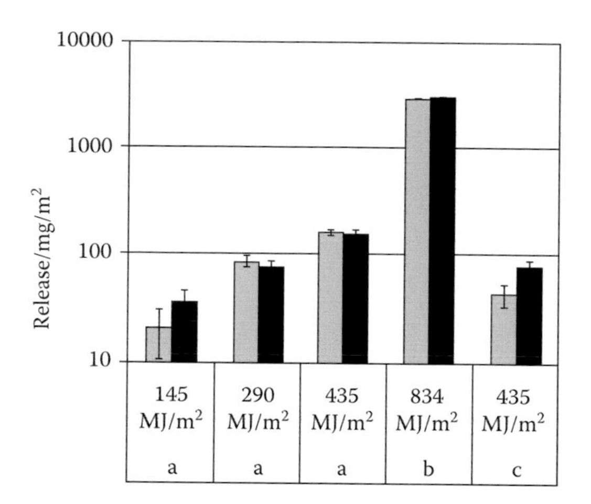

FIGURE 14.6 Quantity of release induced by immersion after weathering. PA released fragments in the size range below 150 nm by immersion and shaking with increasing UV irradiance, in different labs collapsed in one graph. Gray: neat PA, black: PA with 4% SiO_2 with weathering conditions a: dry_60W/m²_immersed; b: dry_140W/m²_immersed; c: 15%rain_60W/m²_immersed (ISO4892-2).

fragments and small fragments with diameters below 150 nm (Figure 14.5a and b). Their ratio shifts towards smaller sizes with increasing shear energies, suggesting not only increased release but also possibly further reduction of fragment diameters in the immersion fluids after detachment, thus changing the released entities. Only at the worst-case shear forces (ultrasound, bath, or probe), structures appeared that were rather individualized MWCNTs with attached polymer than a polymer fragment with protruding MWCNTs. Identical results were obtained for induced release after wet weathering.

Size-selective results showed that fragments from PA with SiO_2 released spontaneously with only 50% increase for ultrasound versus unagitated immersion, whereas fragments from TPU with MWCNTs required strong shear with 160% increase for sonication (Hirth et al. 2013). In contrast, the total release rates are primarily determined by the photodegradation rate of the polymer matrices, as evidenced by the well-differentiated release levels on the order of 100 mg/m² from PA versus 10 mg/m² from TPU (from neat polymers and nanocomposites alike) after 435 MJ/m² UV energy.

The UV + immersion + centrifugation method is hence capable of ranking materials with regard to their probability of release under UV irradiation, where particles release spontaneously, whereas MWCNTs do not, and where PA releases more ultrafine fragments (incl. nanofillers) than TPU. To verify that the ultrasound bath is truly a worst case, the energy input was further increased by ultrasonic probes, resulting in no significant increase of released amounts, but qualitatively decreasing the entanglement of the released MWCNTs (Hirth et al. 2013). Considering that MWCNTs were found to collapse into a dense layer after degradation of the polymer matrix

(Figure 14.2c), the finding that strong shear is required for the release of MWCNTs matches the experience from the formulation of as-produced MWCNTs into aqueous suspensions, where ultrasound is critical for dispersing MWCNTs.

A closely related approach to facilitate the collection of spontaneous release from run-off waters by instead performing a UV + immersion induced release was investigated at CEREGE. They subjected acrylic coatings with 7% nano-CeO_2 to an adaptation of standard UNE EN 927-6 (realized by Suntest XLS+, UV 300–400 nm at 65 W/m²), performing rain only four times per week and collecting the run-off waters for ICP-MS analysis of Ce content (neglecting structural information in a first step). Alternatively, they irradiated by UV 105 W/cm² (300–400 nm) and immersed specimen for 1.5 h four times a week (Scifo et al. 2012). The UV apparatus was identical and the irradiated area of the repeatedly immersed specimens (12 cm²) at CEREGE was in the same range as used for single immersion after irradiation as used at BASF (Hirth et al. 2013). After an initial lag time for the immersion protocol, the Ce amounts released grew faster than for the modified spontaneous protocol, but were in the exact same range of 350–800 μg/m² (at 400 kJ/m²). Only the Ce content is known, but if one assumes that the fragments are of the same 7% CeO_2 + polymer composition as the original specimens, this range corresponds to 5–11 mg/m² total release, which is more than an order of magnitude below the values obtained for SiO_2-PA (Figure 14.6) or MWCNT-PP (Figure 14.4), but in the same range as the values obtained for MWCNT-TPU (Hirth et al. 2013). Related UV + immersion protocols determined by ICP-OES the release of TiO_2 from photocatalytic glass, finding 25 mg/m² release after 1.1 MJ/m² UV energy (Olabarrieta et al. 2012). The same setup was later upgraded with mechanical shear by brushing and rolling wheels (Zorita 2013).

Similar ranges of release are obtained by another variant of induced release, namely gently spraying water onto vertically held large (71.2 cm²) samples for 10 min between or after irradiations. This protocol was proposed by Nguyen on the example of nanoSiO_2-epoxy, and the collected run-off was measured by ICP-OES (Nguyen 2012). These authors also observed a lag time and then measured a total 14–56 mg/m² Si release after 800 kJ/m² UV radiation, which one may again extrapolate to 280–1,123 mg/m² release of assumed SiO_2/epoxy fragments. For the same samples, the order of magnitude of release (50 mg/m² Si after 800 kJ/m²) was confirmed by another complementary approach, which unfortunately is applicable to SiO_2 composites only. Degraded nanocomposites (shown in Figure 14.3) were immersed in a 5% HF solution to dissolve the SiO_2 nanofillers that are accessible for the acid from the surface (Nguyen et al. 2012). The extracted solution was then analyzed by ICP-OES. Sampling in this protocol was limited to extraction and quantification, and the characterization of morphology was necessarily lost. The extracted Si amounts scaled with the 5% or 10% SiO_2 content as expected and confirmed that more Si (roughly doubled amounts) can be dissolved after UV irradiation as compared to the nonirradiated nanocomposites, from which significant Si was extracted, too, possibly by diffusion of the acid through percolating nanofiller networks. The HF protocol for induced release is not a realistic measure of release probability, but a quick method to estimate for a given nanocomposite the total SiO_2 mass with vanishing coverage by the polymer matrix. It may serve as worst-case benchmark to put the SiO_2 amounts released by simulations of real-world scenarios into context. Nguyen and coworkers finally also screened the spontaneous release via the airborne phase. A polytetrafluoroethylene (PTFE) sample collector was placed below the vertically fixed nanoSiO_2–epoxy nanocomposite during irradiation (Nguyen et al. 2011; Nguyen et al. 2012). Microscopic analysis (SEM) of the PTFE collector after irradiation revealed the presence of fragments, which can only have fallen off the nanocomposite. These fragments contained Si and O elements (determined by EDXS), allowing identification of spontaneous release, albeit with unknown polymer/nanofiller ratio and unknown quantification.

The complementary study of induced release into the atmosphere was performed by Stintz and coworkers, who first irradiated PU and polyacrylate coatings containing ZnO or Fe_2O_3 nanoparticles (following EN 927-6:2006), then released by a sanding process (Göhler et al. 2010). The number of aerosol particles with diameters below 100 nm systematically increased as determined by engine exhaust particle sizer (EEPS), typically by a factor of 2–8 compared to the nonirradiated samples. However, the SEM and TEM analysis of morphology and EDX analyses of material identified these airborne particles as fragments of the matrix with bound nanofillers, not as individual nanofillers. Results of a further study confirmed these findings. Herein, investigations on the sanding-induced particle release were performed based on artificially weathered (ISO 11341:2004) polyacrylate coatings and PP plastics containing different kinds of nanopigments. In contrast to the coatings, the weathering-caused change in the number of released particles showed a systematic decrease for the analyzed PP samples (Göhler et al. 2013a).

14.4 CHARACTERIZATION OF RELEASED FRAGMENTS: DETECTION, QUANTIFICATION, IDENTIFICATION

After either spontaneous or induced release, all studies used electron microscopy, mostly SEM, some TEM, to detect if release occurred. Sampled fragments on collectors, air filters, or liquid aliquots were in many cases identified by elemental information (EDXS, ICP-MS) or further by chemical information (TGA, XPS, FTIR). In the airborne state, the same instruments as for release from sanding (compare Chapter 12) were employed for quantification (often CPC), which can additionally be coupled to classification steps to derive also size-classified quantification, with different detection limits of fast mobility particle sizer (FMPS), EEPS, and scanning mobility particle sizer (SMPS) (Göhler et al. 2013a). Especially for low forces and hence low release rates, the background dominates even under clean-room conditions with less than 1,000 particles/cm^3, so that no size information can be extracted without sampling and identification of released fragments against the background.

In liquids (immersion or run-off waters), several complementary methods can quantify the released fragments: ICP-MS and ICP-OES to quantify all released nanofillers regardless of fragment morphology, gravimetry to quantify all released fragments together (Busquets-Fite et al. 2013), and ultracentrifugation with spectroscopy of supernatants to quantify free nanofillers selectively. If coupled with synchronized refractive index detection, the ultracentrifugation approach (AUC-RI) has the advantage (just like SMPS for atmospheres) to provide both size distributions and size-classified quantification in mass metrics (Wohlleben 2012). Calcination/thermogravimetry can be used to identify the nanofiller content in run-off waters at the expense of losing information on fragment morphology (Vilar et al. 2013). Experimental techniques to characterize nanofiller release from polymer nanocomposite weathering are included in Figure 14.1. Standardized methods for larger fragments such as laser diffraction are not applicable to immersion or run-off waters due to very limited amounts of particles.

14.5 CONCLUDING REMARKS

Polymer nanocomposites are stronger, tougher, lighter, more dimensionally stable, less permeable, and more durable than neat polymers or traditional polymer composites. In coming years, polymer nanocomposite products will enter the consumer markets in large quantities, and they will be increasingly used in essentially every segment of the industry from textiles, buildings, sporting goods, electronics to aerospace. Whatever the application, both the long-term weathering performance of the nanocomposites and the fate of the nanofillers in the matrices during the products' lifecycle play a key role in the widespread uses of these products. In this chapter, we have briefly presented recent advances in the weathering of polymer nanocomposites and the weather-induced release of nanofillers. Extensive literature data indicated that the presence of nanofiller either decreases or increases the degradation rate of the host matrix. Regardless of the effect, the overwhelming conclusion is that UV radiation of the weathering environments will degrade the matrix and the nanofillers will be exposed on the composite surface, and eventually released to the environment spontaneously or induced by other external factors such as mechanical forces, rain, hails, and snow.

The process of nanofiller surface accumulation and release is summarily illustrated in Figure 14.7 for, as two examples, polymer/nanosilica and polymer/MWCNT composites exposed to solar UV. Under this weathering environment, the matrix layer near the surface will first undergo photodegradation and is removed. The loss of the matrix surface layer will result in a gradual increase in the number of nanofillers on the composite surface with increasing exposure time or increasing radiation dose. For some nanofillers, at a critical thickness/concentration, they will release spontaneously, whereas for others such as MWCNTs (Kingston et al. 2013), where a entangled network is formed on the surface (Figure 14.7b), additional forces such as mechanical forces will be required to liberate the surface exposed nanofillers from the nanocomposite surface.

The eroded surface layer of polymers without UV stabilizers was found to release fragments on the order of 1 g/m²/year (corresponding to a 1-μm layer thickness), depending primarily on the degradation of the polymer matrix and the nanofiller shape, and also on the filler–matrix compatibility and on shear during use. If we assume a mm-to-cm thickness of typical polymer structural parts in cars of consumers products, the total nanoparticle release measured throughout a 10-year use phase corresponds to 0.001% of the entire polymer product (Wohlleben et al. 2014).

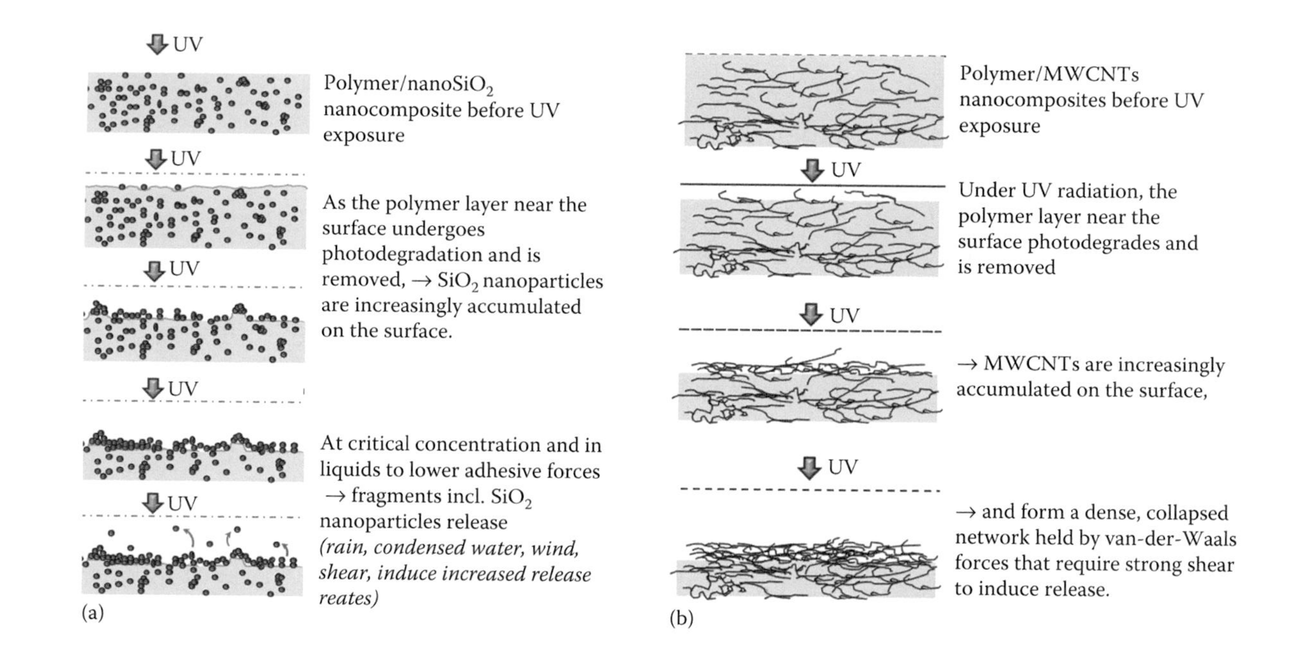

FIGURE 14.7 Conceptual models for (a) surface accumulation and release of SiO_2 nanoparticles from epoxy/nanosilica composites. (b) The formation of a dense collapsed MWCNT network on the nanocomposite surface during exposure to UV radiation.

Because most common polymers are susceptible to photodegradation under weathering environments, even those containing UV stabilizers, these nanofiller release models are equally applicable to other polymers. Further, matrix degradation will cause extensive oxidation, chain scission, and crosslinking of the polymer chains, which results in an increase in brittleness, loss of mechanical properties, and cracking of the polymer nanocomposites. This loss of mechanical integrity will promote nanofiller migration and release. However, appropriate measures, such as adding UV stabilizers to the formulation, are routine praxis in polymers for outdoor applications. Addition of UV stabilizers or protective coatings delays the nanofiller release to the disposal phase, where landfilling or waste combustion could be emission sources, and are discussed in Chapters 12 and 15. During the use phase, release of polydisperse fragments including nanofillers occurs at nonzero rates, but limited to micrometer-thin surface layers, with emerging mechanistic understanding. For environmental mass flow estimates, the use phase is hence a relatively small source compared to production, recycling, and disposal phases.

REFERENCES

Aschberger K, Johnston HJ, Stone V, Aitken RJ, Hankin SM, Peters SA, Tran CL, Christensen FM (2010) Review of carbon nanotubes toxicity and exposure–Appraisal of human health risk assessment based on open literature. *Crit. Rev. Toxicology* 40:759–790.

Bernard C, Nguyen TL, Pellegrin B, Holbrook RD, Zhao M, Chin J (2011) Fate of graphene in polymer nanocomposite exposed to UV radiation. *J. Phys.: Conf. Ser.* 304:012063.

Busquets-Fite M, Fernandez E, Janer G, Vilar G, Vazquez-Campos S, Zanasca R, Citterio C, Mercante L, Puntes V (2013) Exploring release and recovery of nanomaterials from commercial polymeric nanocomposites. *J. Phys.: Conf. Ser.* 429:012048.

Chandramouleeswaran S, Mhaske ST, Kathe AA, Varadarajan PV, Prasad V and Vigneshwaran N (2007) Functional behaviour of polypropylene/ZnO-soluble starchnanocomposites. *Nanotechnology* 18:385702.

Chen XD, Wang Z, Liao ZF, Mai YL, Zhang MQ (2007) Roles of anatase and rutile TiO_2 nanoparticles in photooxidation of polyurethane. *Polym. Test.* 26:202–208.

Chin J, Byrd E, Embree N, Garver J, Dickens B, Fin T, and Martin JW (2004) Accelerated UV weathering device based on integrating sphere technology. *Rev. Sci. Instrum.*75:4951.

Chmela S (2005) Photo-oxidation of sPP/organoclay nanocomposites. *J Macromol Sci A: Pure ApplChem* 42:821–9.

Davis and Sims D (1983) Weathering of Polymers. Appl. Sci. Publishers. New York.

Deka BK and Maji TK (2012) Effect of nanoclay and ZnO on the physical and chemical properties of wood polymer nanocomposite. *J. Appl. Polym. Sci.* 124:2919–2929.

Ding P, Qu B (2006) Synthesis of exfoliated PP/LDH nanocomposites via melt-intercalation: Structure, thermal properties, and photooxidative behavior in comparison with PP/MMT nanocomposites. *Polym. Eng. Sci.* 46:1153–9.

Fa WJ, Yang CJ, Gong CQ, Peng TY, Zan L (2010) Enhanced photodegradation efficiency of polyethylene-TiO2 nanocomposite film with oxidized polyethylene wax. *J. Appl. Polym. Sci.*118:378–384.

Future Markets, Inc. (2012) Nanomaterials in plastics and advanced polymers, Market Report # 52, April 2012.

Göhler D, Nogowski A, Fiala P, Stintz M (2013a) Nanoparticle release from nanocomposites due to mechanical treatment at two stages of the life-cycle. *J. Phys.: Conf. Ser.* 429:012045.

Göhler D, Nogowski A, Stintz M (2013b) Nanoparticle release quantification of sanding-induced airborne particle emissions from artificially aged/weathered nanocomposites. Proceedings of Nanotech 2013, Boston, USA.

Göhler D, Stintz M, Hillemann L, Vorbau M (2010) Characterization of nanoparticle release from surface coatings by the simulation of a sanding process. *Ann. Occup. Hyg.* 54:6.

Gorham J, Nguyen T, Bernard C, Stanley D, Holbrook D (2012) Photo-induced surface transformations of silica nanocomposites. *Surf. Interface Anal.* 44: 1572–1581.

Grigoriadou I, Paraskevopoulos KM, Chrissafis K, Pavlidou E, Stamkopoulos TG, Bikiaris D (2011) Effect of different nanoparticles on HDPE UV stability. *Polym. Deg. Stab.* 96:151–163.

Gu X, Chen G, Zhao M, Watson SS, Nguyen T, Chin JW, Martin JW, Gu X (2012a) Critical role of particle/polymer interface in photostability of nano-filled polymeric coatings. *J. Coat. Technol. Res.* 9:251–267.

Gu X, Zhe D, Zhao M, Chen G, Watson SS, Stutzman PE, Nguyen T, Chin JW and Martin JW (2012b) More durable or more vulnerable?–Effect of nanoparticles on long-term performance of polymeric nanocomposites during UV exposure. *Int. J. Sustainable Materials and Structural Systems*, 1:1-1015-624.

Hamid SH and Hussain I (2000) Life Time Prediction of Plastics, in Handbook of Polymer Degradation, S. Hamid, Ed., Marcel Dekker, N.Y. 2nd Ed. 699–726.

Handy RD, Henry TB, Scown TM, Johnston BD and Tyler CR (2008) Manufactured nanoparticles: Their uptake and effects on fish-A mechanistic analysis, *Ecotoxicology* 17:396–409.

Hirth S, Cena LG, Cox G, Tomovic Z, Peters TM, Wohlleben W (2013) Scenarios and methods that induce protruding or released CNTs after degradation of composite materials. *J. Nanopart. Res.* 15:1504.

Hrdlovic P (2004) Photochemical reactions and photophysical processes: photo-oxidative degradation of polyolefin nanocomposites. *Polym. News* 29:368–70.

Kingston C, Zepp R, Andrady A, Boverhof D, Fehir R, Hawkins D, Roberts J, Sayre P, Shelton B, Sultan Y, Vejins V, Wohlleben W (2013), Release Characteristics of Selected Carbon Nanotube Polymer Composites, *Carbon*, doi: 10.1016/j.carbon.2013.11.042

Kumar AP, Depan D, Singh Tomer N, Singh RP (2009) Nanoscale particles for polymer degradation and stabilization—trends and future perspectives. *Prog. Polym. Sci.* 34 (6): 479–515.

Le Bihan O, Shandilya N, Gheerardyn L, Guillon O, Dore E, Morgeneyer M (2013) Investigation of the Release of Particles from a Nanocoated Product. *Advances in Nanoparticles* 2:39–44.

Li B and Zhong WH (2011) Review on polymer/graphite nanoplateletnanocomposites, *J. Mat. Sci.* 46:5595–5614.

Li R and Zhao H (2006) A study of photo-degradation of zinc oxide (ZnO) filled polypropylene nanocomposites. *Polymer* 47:3207–3217.

Ma PC, Siddiqui NA, Marom G and Kim JK (2010) Dispersion and functionalization of carbon nanotubes for polymer-based nanocomposites: A review. *Composites A–Appl. Sci. Manufact.* 41:1345–1367.

Mailhot B, Morlat S, Gardette J-L, Boucard S, Duchet J, Gerard J-F (2003) Photodegradation of polypropylene nanocomposites. *Polym. Degrad. Stab.* 82:163–7.

Maynard AD (2006) Nanotechnology: Assessing the risks. *Nanotoday* 2:22–33.

McIntire R (2012) Common nano-materials and their uses in real world applications, *Science Prog.*95:1–22.

McNally T and Pötschke P Eds. (2011) Polymer-Carbon Nanotube Composites, Preparation, Properties, and Applications, Woodhead Publishing, Philadelphia.

Morlat-Therias S, Fanton E, Gardette JL, Peeterbroeck S, Alexandre M, Dubois P (2007) Polymer/carbon nanotube nanocomposites: Influence of carbon nanotubes on EVA photodegradation. *Polym. Degrad. Stab.* 92:1873–1882.

Morlat-Therias S, Mailhot B, Gardette J-L, Da Silva C, Haidar B, Vidal A (2005b) Photooxidation of ethylene–propylene-diene/montmorillonitenanocomposites. *Polym. Degrad. Stab*. 90:78–85.

Morlat-Therias S, Mailhot B, Gonzalez D, Gardette J-L (2004) Photo-oxidation of polypropylene/montmorillonite nanocomposites. 1. Influence of nanoclay and compatibilizing agent. *Chem. Mater.* 16:377–83.

Morlat-Therias S, Mailhot B, Gonzalez D, Gardette J-L (2005a) Photooxidationof polypropylene/montmorillonite nanocomposites. 2. Interactions with antioxidants. *Chem. Mater.* 17:1072–8.

Najafi E and Shin K (2005) Radiation resistant polymer-carbon nanotube nanocomposite thin films. *Coll. Surf. A: Phys. Eng. Aspects*, 257–258: 333.

Nel, Xia T, Mädler L, Li N (2006) Toxic potential of materials at the nanolevel. *Science* 311:622–627.

Nguyen T (2012) Quantitative Studies of Photo-induced Surface Accumulation and Release of Nanoparticles in Polymer Nanocomposites. Presented at NanoSafe 2012, Grenoble.

Nguyen T., Granier A, Shapiro A, Martin JW (2008) Mechanical properties and UV resistance of acrylic polyurethane coatings containing unmodified and NCO-functionalized carbon nanofillers, European PU Conference.

Nguyen T, Gu X, Vanlandingham M, Byrd E, Ryntz R and Martin JW (2013) Degradation modes of crosslinked coatings exposed to Photolytic Environment. *J. Coat. Technol. Res.* 10:1–14.

Nguyen T, Pelligrin B et al. (2009) Degradation and nanofiller release of polymer nanocomposites exposed to ultraviolet radiation, in Natural and Artificial Ageing of Polymers. T. Reichert Ed. 149–162.

Nguyen T, Pellegrin B, Bernard C, Gu X, Gorham JM, Stutzman P, Stanley D, Shapiro A, Byrd E, Hettenhouser R, Chin J (2011) Fate of nanoparticles during life cycle of polymer nanocomposites. *J Physics: Conf Series* 304:012060.

Nguyen T, Pellegrin B, Bernard C, Rabb S, Stuztman P, Gorham J, Gu X, Yu L and Chin JW (2012) Characterization of surface accumulation and release of nanosilica during irradiation of polymer nanocomposites by ultraviolet light. *J. Nanosci. & Nanotech.* 12:6202–6215.

Olabarrieta J, Zorita S, Pena I, Rioja N, Monzon O, Benguria P, Scifo L (2012) Aging of photocatalytic coatings under a water flow: Long run performance and TiO_2 nanoparticles release. *Appl. Catalysis B: Environmental* 123:182–192.

Pavlidou S and Papaspyrides CD (2008) A review on polymer-layered silicate nanocomposites. *Prog. Polym. Sci.* 33:1119–1198.

Petersen EJ, Lam T, Gorham, JM, Scott, KC, Long, CJ, Stanley D, Sharma R, Liddle JA, Pellegrin B, Nguyen T (2013) Impact of UV Irradiation on the Surface Chemistry and Structure of Multiwall Carbon Nanotube Epoxy Nanocomposites; *Carbon*. DOI: 10.1016/j.carbon.2013.12.016.

Phua SL, Yang LP, Toh CL, Ding GQ, Lau SK, Dasari A and Lu XH (2013) Simultaneous enhancements of UV resistance and mechanical properties of polypropylene by incorporation of dopamine-modified clay. *ACS Appl. Mater. Interf.*, 5:1302–1309.

Poland CA, Duffin R, Kinloch I, Maynard A, Wallace WAH, Seaton A, Stone V, Brown S, MacNee W, Donaldson K (2008) Carbon nanotubes introduced into the abdominal cavity of mice show asbestos-like pathogenicity in a pilot study. *Nature Nanotech*. 3:423–428.

Potts JR, Dreyer DR Bielawski, C.W., Ruoff, R.S (2011) Graphene-based polymer nanocomposites. *Polymer* 52:5–25.

Qin H, Zhang S, Liu H, Xie S, Yang M, Shen D (2005) Photo-oxidative degradation of polypropylene/montmorillonitenanocomposites. *Polymer* 46:3149–56.

Qin H, Zhang Z, Feng M, Gong F, Zhang S, Yang M (2004) The influence of interlayer cations on the photo-oxidative degradation of polyethylene/montmorillonite composites. *J. Polym. Sci. B: Polym. Phys.* 42:3006–12.

Qin H, Zhao C, Zhang S, Chen G, Yang M (2003) Photo-oxidative degradation of polyethylene/montmorillonitenanocomposites. *Polym. Degrad. Stab.* 81:497–500.

Rabek JF (1995) Polymer Photodegradation: Mechanism and Experimental Methods, Chapman & Hall, NY.

Research and Markets (2009) Nanotechnology Market forecast to 2013.

Scifo L, Chaurand P, Masion A, Auffan M, Diot M.-A, Labille J, Bottero JY and Rose J (2012) Release of CeO_2 nanoparticles upon aging of acrylic wood coatings. presented at NanoSafe2012, Grenoble.

Sil D, Chakrabarti S (2010) Photocatalytic degradation of PVC -ZnO composite film under tropical sunlight and artificial UV radiation: A comparative study. Solar Energy 84:476–485.

Sloan JM, Patterson PH, Hsieh A (2003) Mechanisms of photo degradation for layered silicate-polycarbonate nanocomposites. *Polym Mater Sci Eng* 88:354–5.

Vilar G, Fernandez-Rosas E, Puntes V, Jamier V, Aubouy L, Vazquez-Campos S (2013) Monitoring migration and transformation of nanomaterials in polymeric composites during accelerated aging. *J. Phys.: Conf. Ser.* 429:012044.

Wohlleben W (2012) Validity range of centrifuges for the regulation of nanomaterials: from classification to as-tested coronas. *J. Nanopart. Res.* 14:1300.

Wohlleben W, Brill S, Meier MW, Mertler M, Cox G, Hirth S, von Vacano B, Strauss V, Treumann S, Wiench K, Ma-Hock L, Landsiedel R (2011) On the life cycle of nanocomposites: Comparing released fragments and their vivo hazard from three release mechanisms and four nanocomposites. *Small* 7, 2384–2395.

Wohlleben W, Meier MW, Vogel S, Landsiedel R, Cox G, Hirth S, Tomovic Z (2013) Elastic CNT-polyurethane nanocomposite: Synthesis, performance and assessment of fragments released during use. *Nanoscale* 5:369–380.

Wohlleben W, Vilar G, Fernández-Rosas E, González-Galvez D, Gabriel C, Hirth S, Frechen T, Stanley D, Gorham J, Sung L, Hsueh H-C, Chuang Y-F, Nguyen T, Vázquez-Campos S (2014) A pilot interlaboratory comparison of protocols that simulate aging of nanocomposites and detect released fragments. *Environ Chemistry* doi: 10.1071/EN14072.

Yang R, Li Y, Yu J (2005) Photo-stabilization of linear low density polyethylene by inorganic nanoparticles. *Polym. Degrad. Stab.* 88:168–174.

Zorita S, Presentation held at Env. Effects of Nanomaterials, Aix-en-Provence 2013.

Zou H, Wu SS, Shen J (2008) Polymer/silica nanocomposites: Preparation, characterization, properties, and applications. *Chem. Rev.* 108:893–3957.

15 Emissions from Consumer Products Containing Engineered Nanomaterials over Their Lifecycle

Bernd Nowack

CONTENTS

15.1 INTRODUCTION

In the last years we have obtained a lot of knowledge about the behavior and effects of engineered nanomaterials (ENM) in the environment. Many studies have investigated processes such as agglomeration, sedimentation, dissolution, removal during wastewater treatment, or transformation reactions (Nowack and Bucheli 2007; Wiesner et al. 2009; Farre et al. 2011; Kaegi et al. 2011; Peralta-Videa et al. 2011; Klaine et al. 2012; Lombi et al. 2012; Lowry et al. 2012). Also the possible effects of ENM on environmental organisms have received a lot of attention and a vast body of research has been performed so far (Baun et al. 2008; Handy et al. 2008; Klaine et al. 2008; Scown et al. 2010; Menard et al. 2011).

However, in contrast to all the fate and effect studies, very little is actually known about the release of ENM into the environment (Gottschalk and Nowack 2011). Currently, techniques are not yet available for most ENM to monitor ENM release and quantify their trace concentrations in the environment (Hassellov and Kaegi 2009; von der Kammer et al. 2012), and thus reports on release and environmental concentrations are very rare.

Besides knowing the amounts of ENM released into the environment, it is equally important to investigate in what form ENM are released (Nowack et al. 2012). First results show that most of the ENM released from products are present in matrix-bound form, of which also some fraction is released as single, dispersed nanoparticles. Almost all fate and effect studies so far have been performed with pristine ENM, those materials that are synthesized and produced in the laboratory or by companies. However, these ENM undergo transformation and aging reaction when incorporated into products and when those products are used or disposed of (Nowack et al. 2012). The materials that actually reach the environment may be completely different from the materials originally produced by the industry. We therefore urgently need information on the amount and characteristics of the materials that are actually released under real-world conditions.

It is the aim of this chapter to present the available data about release of materials from consumer products containing nanomaterials. Only those studies that report results on release during normal use or from test mimicking normal use are included. Release during manufacturing of the products is not covered, as this is the topic of Chapter 12 in this book. Also data from occupational studies Kuhlbusch et al. (2011) (Chapter 13) and the release from disposed (Chapter 14) or incinerated product residues (Chapter 12) are included elsewhere. The available studies are reviewed here with emphasis on the characterization of the materials released.

15.2 ANALYSIS OF EMISSION

If we would like to quantify the release of ENM from commercial products in the real world, we are faced with problems of how to study the relevant release mechanisms and especially how to identify the presence of released ENM and how to quantify them.

15.2.1 Type of Release Studies

Release studies can be performed at different levels of complexity: batch tests with simplified and standardized test materials allow controlling the physical and chemical conditions to a large extent and are therefore suitable to investigate the mechanisms of release and to develop new experimental and analytical methods. The results from these studies may have only a limited value in estimating the amount released into the real world.

In the next level of complexity, studies are performed that mimic the real world but are still performed under controlled conditions. Examples include test washing machines to study release from textiles or weathering chambers to investigate release from paints. In these tests more parameters are fixed but still a sufficient control over the chemical and physical conditions is possible. However, because the tests are performed on a larger scale the number of replicates or treatments is more limited. Very often these tests are based on some established standard norms and thus certain parameters are fixed according to the standard. An example is the use of ISO washing tests for color fastness that were adapted to investigate release of ENM from textiles (Geranio et al. 2009). Adaptations had to be made to be able to collect the released materials and to guarantee sufficiently low levels of silver background. Although these studies are not directly transferable to the real world, they still allow at least estimating the order of magnitude of a certain release.

The highest level of relevance for real-world conditions are of course studies performed under normal use of products. This could be to follow weathering and release of materials from facades painted with nano-paints or washing of clothes in normal washing machines. However, the main problems are that the exposure conditions are much less controlled and that especially the collection of the released materials is very difficult if not even impossible. Consider washing a single nano-T-shirt together with 4 kg of other textiles and then quantifying and characterizing the released materials in several liters of washing and rinsing liquid, where any eventually released ENM are diluted to a large extent and occur together with a large variety of other materials released from the various textiles. In the lab washing machine the solid:water ratio and the amount of textiles can be optimized to allow detection of released materials under the chosen analytical methods. However, if performed well, real-world studies provide of course data that can be directly used for release and exposure scenarios and are in fact the prime source of data for modeling or risk assessment studies.

Figure 15.1 compares the different types of studies presented here and lists the possibilities and advantages of them. All studies have their merits and ideally we would have data from all three levels, allowing us to understand the mechanisms and have data that can be used for exposure scenarios. In the following sections that describe release data we will see that for the different nano-products that have been investigated so far, studies at all three levels have been performed.

Type of release study	To be used for	Advatages/ disadvantages
Mechanistic study – Batch tests – Simplified systems	– Method development – Understanding mechanisms	– Conditions can be completely controlled – Fast and simple experiments
Model study – Climate chamber – Test washing	– Method development – Understanding mechanisms – Release scenarios	– Conditions are to some extent controlled
Real-world study – Washing machine – Outdoor weathering	– Release scenarios – Exposure scenarios – Risk assessment	– Specific trace analytical methods needed – Very little control of conditions

FIGURE 15.1 Classification of release studies.

15.2.2 Analytical Issues

A second subject that is relevant when studying release processes of nano-products is the use of analytical methods to detect the released materials. In the simplest case only total element concentrations are measured and thus no size-specific information is obtained. Several studies investigating release of Ag from products used this approach. These measurements do not distinguish between release of dissolved Ag (after dissolution of the nano-Ag), release of nano-Ag, and release of larger agglomerates or other transformation products. Other studies used size-fractionation techniques before analyzing the total concentration, and therefore additional information on the size of the released materials could be obtained. Preferably the smallest filter size is small enough that only dissolved compounds can pass through so that information on dissolution can be obtained. This process is very important for some ENM, for example, Ag or ZnO, because it may result in complete elimination of the particulate nature of the material. In certain cases, for example, for dissolved Ag^+, methods for analysis using ion-selective electrodes are available and thus direct determination of the truly dissolved metals is possible. The direct coupling of size-fractionation techniques with bulk-analytical techniques, for example, field flow fractionation with ICPMS (Hagendorfer et al. 2010; Ulrich et al. 2012), is expected to provide additional characterization data when this method is applied in the future to more release studies.

Many studies complement the bulk chemical analyses using electron microscopic investigations which not only allow imaging the released particles but also enable the element-specific characterization of single particles. This technique allows gaining direct information on possible transformation reactions, a task that the other methods are currently not able to carry out. All comprehensive release studies need to combine a suite of different analytical techniques to quantify and characterize the release. However, many of the few published studies did not employ the full spectrum of possible methods and thus in many cases only fragmentary information on release was obtained. The next section that presents the release studies in detail does not discriminate any studies, mainly because in the present situation with only a few release investigations each single study, even if just with minimal analytical methods, is able to provide some much-needed information.

15.3 RELEASE STUDIES

15.3.1 Textiles

Nano-textiles are the nano-products that have been studied most often and therefore a detailed comparison between different studies is possible. Most of the studies targeted nano-Ag but one study also looked at TiO_2-containing textiles (Windler et al. 2012). Because different studies have analyzed various commercially available textiles, it is possible to get an overview on the concentrations of nano-Ag in the textiles (Figure 15.2).

Surprisingly there are many "nano-Ag textiles" that do not contain any measurable Ag, a fact that is observed in most studies. The majority of the textiles have Ag concentrations below 100 mg/kg, however, a few have even concentrations in

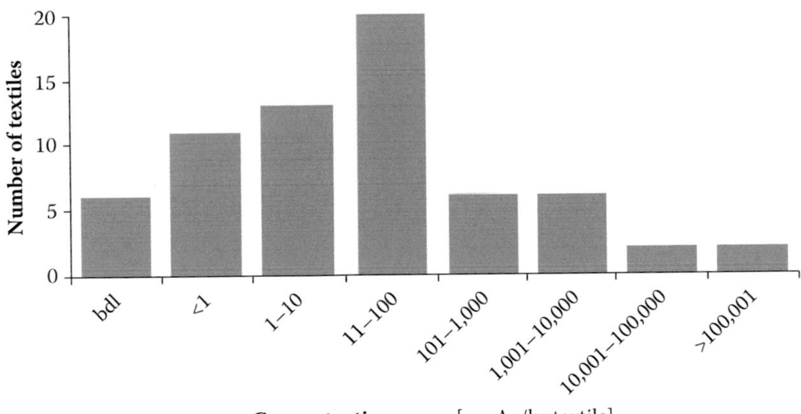

FIGURE 15.2 Ag concentrations in textile consumer products. bdl: below detection limit. (Data extracted from Benn, T.M., Westerhoff, P. *Environ. Sci. Technol.*, 42, 4133–4139, 2008; Geranio, L. et al., *Environ. Sci. Technol.*, 43, 8113–8118, 2009 GSM , 8, 2009; Benn, T. et al. *J. Environ. Quality*, 39, 1875–1882, 2010; Kulthong, K. et al., *Part Fibre Toxicol.* 7, 8, 2010; DEPA (2012), Assessment of nanosilver in textiles on the Danish market, Environmental Project No. 1432, Copenhagen, Danish Environmental Protection Agency; KEMI (2012), Antibacterial substances leaking out with the washing water: analyses of silver, triclosan and triclocarban in textiles before and after washing, Sundbyberg, Swedish Chemicals Agency; Lorenz, C. et al., *Chemosphere* 89, 817–824, 2012.)

the g/kg range or even in the %-range. Nano-textiles usually have Ag concentrations that are lower than textiles using conventional Ag or AgCl as biocide (Windler et al. 2013). The designation as "nanotextiles" is based in most cases on the labeling of the textiles. Analysis of ENM in complex matrices is notoriously difficult (von der Kammer et al. 2012) and therefore almost no studies have tried to verify the presence of nanomaterials in the textiles. Most studies about nano-textiles equalize the presence of Ag in the textiles to the presence of nano-Ag, although this is of course not correct because many conventional uses of Ag are on the market. Some studies have used textiles that were directly obtained from companies producing nanomaterials and first-hand information on the type of materials was obtained.

In Figure 15.3, the main results of the release experiments are compiled. It shows how much of the total Ag in the textiles is released in the different studies. The studies mimicking washing conditions report most often releases above 51% of the total Ag. 64% of all investigated textiles released more than 10% of the Ag in a single washing. However, 26% of the textiles released less than 1% of silver.

There are less experiments performed in artificial sweat but it seems that lower amounts of Ag are released into this medium, most often a release below the detection limit was reported.

Several studies also looked at the characterization of the released materials. Benn and Westerhoff (2008) showed that in distilled water the majority of the released Ag was in dissolved form (up to 86%). Geranio et al. (2009) and Lorenz et al. (2012) found that in washing solution most of the released Ag was present in particulate

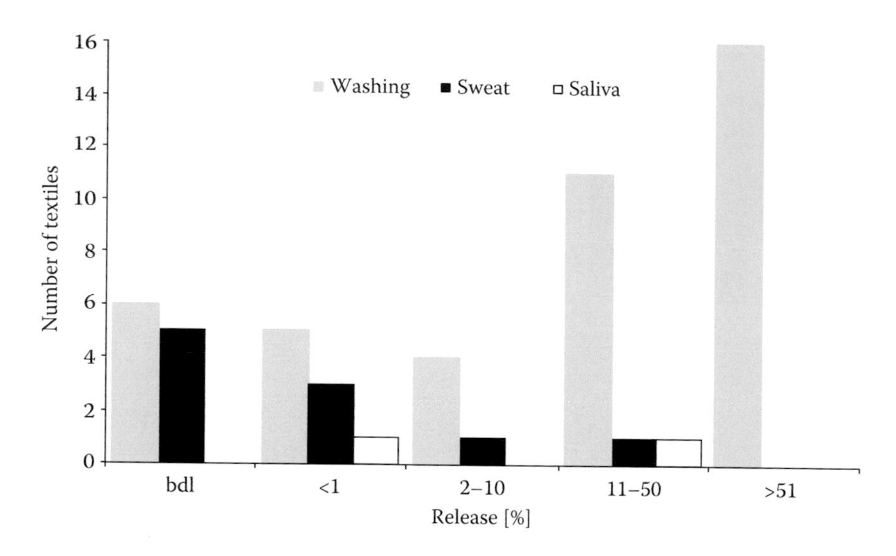

FIGURE 15.3 Percent of the total Ag released from nano-textiles. (Data extracted from Benn, T.M., Westerhoff, P. *Environ. Sci. Technol.,* 42, 4133–4139, 2008; Geranio, L. et al., *Environ. Sci. Technol.,* 43, 8113–8118, 2009 GSM, 8, 2009; Benn, T. et al. *J. Environ. Quality,* 39, 1875–1882, 2010; Kulthong, K. et al., *Part Fibre Toxicol.* 7, 8, 2010; DEPA (2012), Assessment of nanosilver in textiles on the Danish market, Environmental Project No. 1432, Copenhagen, Danish Environmental Protection Agency; KEMI (2012), Antibacterial substances leaking out with the washing water: analyses of silver, triclosan and triclocarban in textiles before and after washing, Sundbyberg, Swedish Chemicals Agency; Lorenz, C. et al., *Chemosphere* 89, 817–824, 2012.)

fraction and that the fraction >0.45 μm was the most important one. The dissolved fraction was for most textiles only of very minor importance. The fraction >0.45 μm, which includes the nanoparticulate fraction, was 75%–100% for the various textiles (except for one textile where it was only 40%) (Geranio et al. 2009), indicating that only a small fraction of the silver is released as single dispersed nanoparticles. Electron microscopy investigations revealed that AgCl was the most often observed form of Ag in the washing solution, Figure 15.4 (Lorenz et al. 2012).

Because textiles are in direct contact with the human skin several authors have also studied the release of Ag into sweat (Kulthong et al. 2010; Yan et al. 2012).

One study targeted the release of Ag from the washing machine itself, using a commercially available washing machine that supplies Ag to the wash water during the wash cycle (Farkas et al. 2011). The washing machine released silver into the washing solution with an average concentration of 11 μg/L. Less than 2% of this Ag was in the dissolved form. The presence of Ag nanoparticles was confirmed by a single particle ICP-MS, an average size of 10 nm was measured with a transmission electron microscopy.

Cleveland et al. (2012) exposed different Ag textiles to an estuarine mesocosm (with a salinity of 20%), mimicking release from textiles that are disposed of directly in the environment. After 60 days of exposure about 95% of the total silver content of the sock had been released. The wound dressing had even lost 99% of the silver content, the toy bear 82%.

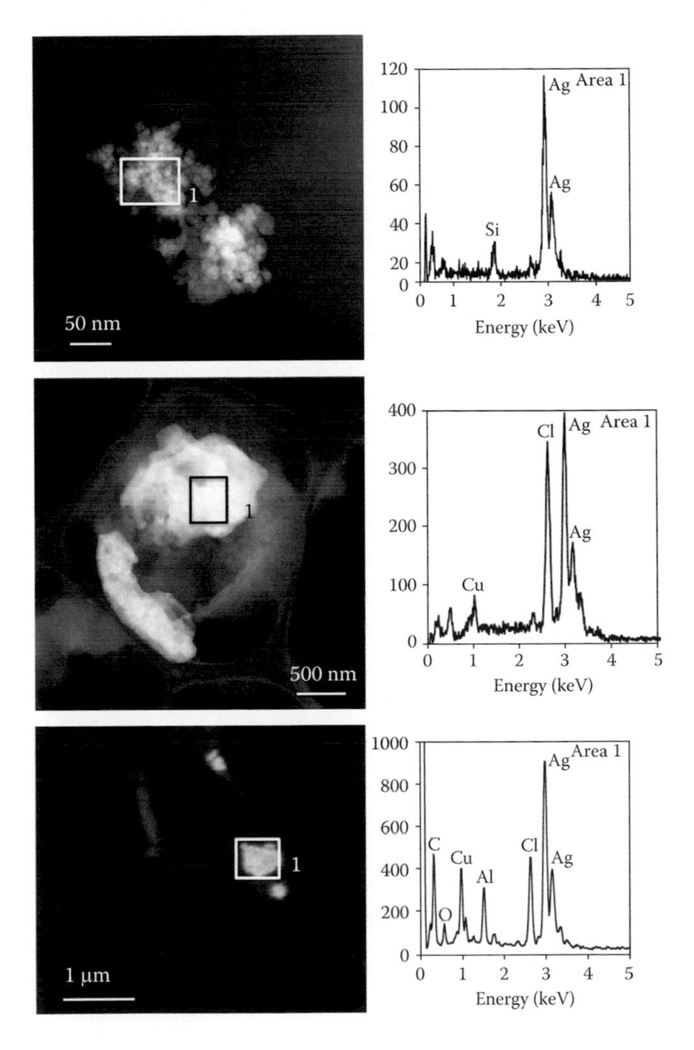

FIGURE 15.4 Electron microscopy images of particles found in the washing solution of nano-Ag containing textiles. Top: agglomerate of metallic Ag NP; middle: large AgCl crystals; bottom: agglomerates of small AgCl particles. (Reprinted from Chemosphere, 89, Lorenz, C. et al., Characterization of silver release from commercially available functional (nano)textiles, 817–824. Copyright 2012, with permission from Elsevier.)

All the studies presented so far targeted Ag-containing textiles. However, textiles also contain a variety of other nanoparticles, for example, TiO_2 or SiO_2 (Som et al. 2010). Windler et al. (2012) have investigated the release of TiO_2 from six different textiles. Analysis of fiber cross-sections revealed that the TiO_2 was contained within the fiber matrix. Five of the textiles (with sun-protection functionality) released low amounts of Ti (0.01–0.06 wt% of total Ti) in one washing cycle. One textile (with antimicrobial functionality) released much higher amounts of Ti (5 mg/L, corresponding to 3.4 wt% of total Ti in one washing cycle). Size fractionation of the Ti in the washing solution showed that about equal amounts were released as particles smaller and larger than 0.45 μm. Electron microscopy revealed that the TiO_2 particles released into the washing solution had primary particle sizes between 60 and 350 nm and formed small aggregates with up to 20 particles. These results indicate that functional textiles release only small amounts of TiO_2 particles and that the released particles are mostly not in the nanoparticulate range.

If we summarize all data obtained so far with textiles we can state that clearly from many textiles significant release is observed but that the majority of the released materials is not present as single nanoparticles but that a large variety of agglomerates, transformation products, or precipitates are present in the washing solution. Depending on the conditions also high concentrations of dissolved Ag can be found. These results show that in most cases the nano-properties of the original material in the textiles are lost and that the materials released are similar for nano-textiles and conventional textiles containing the same material in non-nano form.

15.3.2 Paints

Paints containing nanoparticles represent an important use of ENM in the construction industry (Broekhuizen et al. 2011). Many paints for indoor and outdoor applications contain biocides and additives for protection against microbial, physical, and chemical deterioration (Kaiser et al. 2013). Several studies have already investigated the release of materials from paints containing ENM. The first study—and in fact the first report of an ENM in the environment—was performed by Kaegi et al. (2008), who investigated the release of TiO_2 from outdoor paints. The authors presented direct evidence of the release of nanomaterials from buildings into the aquatic environment. TiO_2 particles were traced from painted facades through runoff to surface waters. The concentration of Ti in runoffs from new facades was as high as 600 μg/L. A centrifugation based sample preparation technique was used which extracted TiO_2 particles between 20 and 300 nm and it was found that 85%–90% of the total Ti in the samples occurred in this size range. Electron microscopy revealed that the released Ti was present as TiO_2 particles. Figure 15.5 shows examples of particles observed in the runoff. Some of the particles are still embedded in the organic binder (light gray zone around the particles). The same type of particles were also found in a small stream into which the runoff was discharged. This finding marks the first analytical verification of the presence of an ENM in the environment. However, these nano-TiO_2 particles were released from a facade that did NOT contain any nano-TiO_2 but only pigment TiO_2. It has been shown that pigment TiO_2 has a size distribution that extends into the nanoparticulate range (Weir et al. 2012). It is therefore not completely correct to call the observation of nano-TiO_2 in the environment in the study of Kaegi et al. (2008) a verification of the presence of ENM in the environment. It is rather a report about the presence of an incidental nanoparticle, which is of course of utmost relevance when evaluating the potential presence of real ENM. Further studies need to investigate how such engineered nano-TiO_2 particles behave in paints and how their behavior is different from that of the pigment TiO_2.

The same authors also studied the behavior of a nano-Ag containing paint (Kaegi et al. 2010). A painted panel was exposed to natural weather conditions

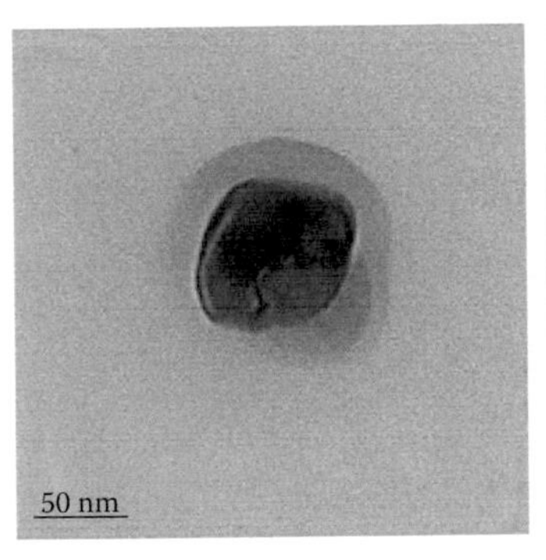

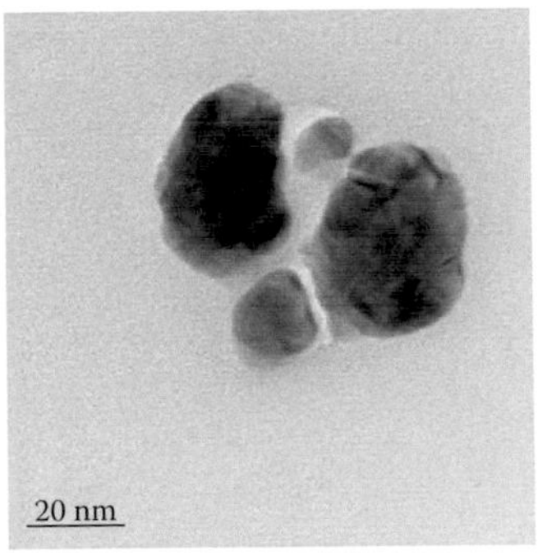

FIGURE 15.5 Transmission electron microscopy image of TiO_2 particles found in the runoff of the aged facade. (Reprinted from *Environ Pollut,* 156, Kaegi, R. et al., Synthetic TiO_2 nanoparticle emission from exterior facades into the aquatic environment, 233–239. Copyright 2008, with permission from Elsevier.)

on a small model house for one year. The Ag and Ti concentrations in the collected runoff samples from the facades were measured. A high leaching of Ag was observed during the first runoff events with a maximum Ag concentration of 145 μg/L (Figure 15.6). After 1 year the Ag concentrations were below 1 μg/L. A mass balance calculation showed that after 1 year more than 30% of the Ag was leached out of the facade. Also very high concentrations of Ti were measured in the initial leachates (Figure 15.6), but the values dropped below the detection limit after a few months.

The particles from selected runoff events were collected on transmission electron microscopy (TEM) grids and imaged under the electron microscope. Individual Ag nanoparticles with a size smaller than 15 nm were found to be attached to the organic binder. Also large agglomerates of such Ag particles were observed. The authors could not exclude the presence of single Ag NP in the runoff, but clearly the vast majority of all nano-Ag particles was incorporated in an organic matrix. The size of these particles, which formed a composite of Ag-NP and organic binder, varied between a few 100 nm and a few μm. energy dispersive X-ray spectroscopy analyses of the Ag particles showed some evidence for the formation of Ag_2S but not for AgCl (Figure 15.7).

The studies by Kaegi and coworkers looked at release during weathering under ambient conditions. Koponen et al. (2009) investigated the formation of dust particles during sanding of ENM-containing paint. The particle size distribution of the sanding dust was characterized by five size modes: three modes were under 1 μm and two modes around 1 and 2 μm. The sander used to generate the dust was the only source of particles smaller than 50 nm and these particles dominated the particle number concentration. Addition of nanoparticles to the paints caused only minor changes in the mean diameters of the particle modes compared to control paint without nanoparticles. However, the number concentrations varied considerably between the nano-paints and the nano-free reference paints. No further characterization of the released particles was made in this study.

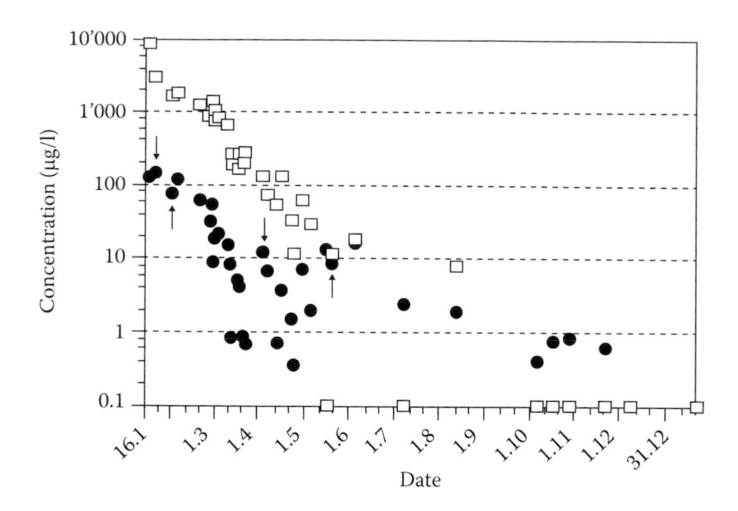

FIGURE 15.6 Ag (filled circles) and Ti (open squares) concentrations of runoff events from a nano-Ag containing paint applied to a small model house. (Reprinted from *Environ. Pollut.*, 158, Kaegi, R. et al., Release of silver nanoparticles from outdoor facades, 2900–2905. Copyright 2010, with permission from Elsevier.)

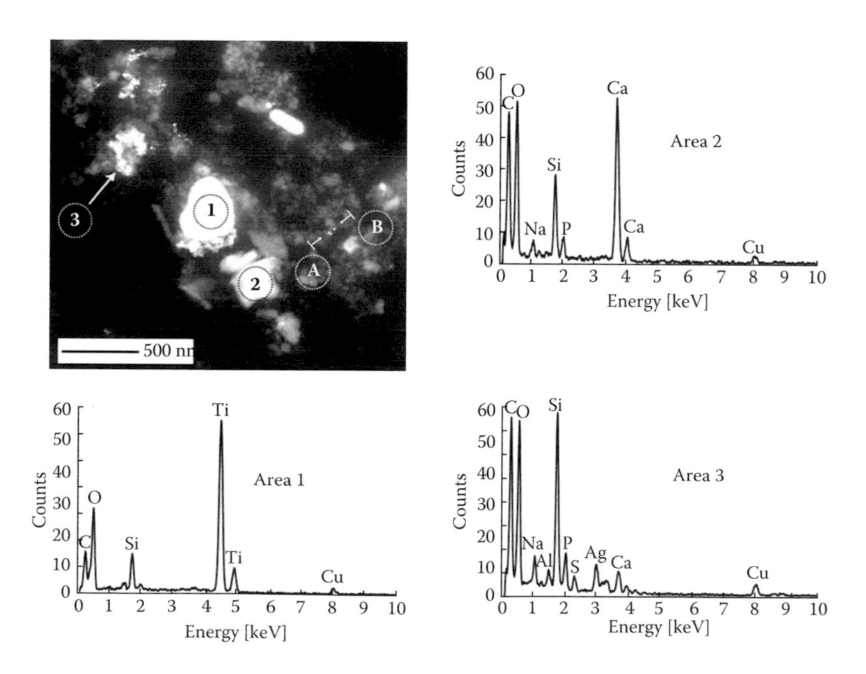

FIGURE 15.7 Electron microscopy image of Ag-containing particles released from paint and energy dispersive X-ray spectroscopy spectra of three areas. (Reprinted from *Environ. Pollut.*, 158, Kaegi, R. et al., Release of silver nanoparticles from outdoor facades, 2900–2905. Copyright 2010, with permission from Elsevier.)

15.3.3 Coatings

Coatings containing ENM represent an important use of ENM in the construction industry (Lee et al. 2010; Broekhuizen et al. 2011). These coatings can be applied to glass, metal, or other surfaces. Chabas et al. (2008) have investigated the durability of self-cleaning glass coated with nano-TiO_2. SEM investigations and chemical analyses on the samples exposed to ambient conditions for 6, 12, and 24 months showed no evidence of significant weathering of the surfaces or loss of Ti.

Olabarrieta et al. (2012) developed an experimental protocol to simulate the accelerated aging of photocatalytic coatings under water flow. The influence of environmental parameters such as the composition of the water matrix or the presence of UV light was investigated and the release of TiO_2 nanoparticles into water was studied. TiO_2 emissions as high as 150 μg/L were observed. The release was enhanced by the presence of NaCl and illumination with UV light. A commercially available coated glass released in 4 weeks up to 4 mg/m^2 TiO_2. The authors postulate three mechanisms that are involved in TiO_2 emissions (compare also Chapter 14): (1) TiO_2 agglomerates can be stripped off of the coatings by the water flow, resulting in the release of big TiO_2 agglomerates; (2) dispersion of TiO_2 agglomerates in the presence of NaCl and UV light, resulting in suspended TiO_2 nanoparticles; (3) direct release of TiO_2 nanoparticles from the matrix, also resulting in suspended TiO_2 nanoparticles.

15.3.4 Food Containers

Food contact materials containing nano-Ag have been widely used in the last years (Simon et al. 2008). Theoretical calculations show that detectable migration of ENM will only take place from polymers with low viscosity for ENM with a radius in the order of 1 nm (Simon et al. 2008). It was also predicted that no appreciable migration occurs in the case of bigger ENM contained within a polymer with high viscosity. Several studies are available that experimentally investigated the release and migration of Ag into water or matrices mimicking foods. Hauri and Niece (2011) measured Ag release from two types of plastic storage containers into distilled water, tap water, and acetic acid for up to 2 weeks. The highest observed concentration was 2.6 μg/L in the acetic acid container. No characterization or size fractionation of the released Ag was done. Huang et al. (2011) measured release into pure water, acetic acid, alcohol, and oil simulating solution (hexane) and found releases of up to 4 μg/dm^2, with an increase over time and with increasing temperature. The highest release was observed for alcohol, followed by acetic acid and water. Electron microscopy revealed the presence of silver particles with a diameter of 300 nm in the pure water leachate. Song et al. (2011) performed a similar study and measured Ag release from plastic material into ethanol and acetic acid at temperatures between 20 and 70°C. In acetic acid up to 0.6% of the Ag was released after 9 hours, for ethanol up to 0.2%. No further characterization of the released Ag was performed.

The studies published so far give therefore only very limited evidence for the release of particulate Ag, they all measured just total release of Ag. Because all studies found the highest release in acetic acid, it is very likely that the majority of the released Ag was present as dissolved Ag^+. It is known that the dissolution of Ag is enhanced at low pH (Elzey and Grassian 2010; Liu and Hurt 2010).

15.3.5 Sprays

Sprays containing ENM, mainly nano-Ag, are on the market and are thus used by many people (Quadros and Marr 2010). Sprays result in the direct aerosolization of particles, and therefore complete release from the bottle to the environment occurs. However, even sprays not containing any ENM form nanosized aerosols (Norgaard et al. 2009), but the chemical identity of the nano-sized fraction was not studied by these authors. However, for understanding the further fate and the effects of the aerosolized particles their detailed characterization after release is necessary. Hagendorfer et al. (2010) developed a method for quantification of the size and the composition of ENM released from sprays. Controlled spray experiments were performed in a glove box, and the time-dependent particle size distribution was studied using a scanning mobility particle sizer (SMPS). The aerosol was also transferred to an electrostatic TEM sampler. The particles deposited on the TEM grid were investigated using electron microscopy. The solid ENM were differentiated from the solvent droplets by applying a thermo-desorbing unit. This setup was used to study various sprays. Pump sprays resulted in no significant release of nanosized materials whereas gas-pressurized cans resulted in significant release of nanosized materials. Figure 15.8 shows an agglomerate of nano-NP from a propellant gas spray dispenser. The particles in the original dispersions seem to be smaller in diameter than the released particles after spraying, especially after a certain aging period of the aerosol.

Lorenz et al. (2011) investigated four commercially available nano-sprays and identified ENM in two of them and nanosized aerosols in the three products that used propellant gas. These nanosized aerosols consisted of a mixture of the spray ingredients. Depending on the products and the mixture of the ingredients, different

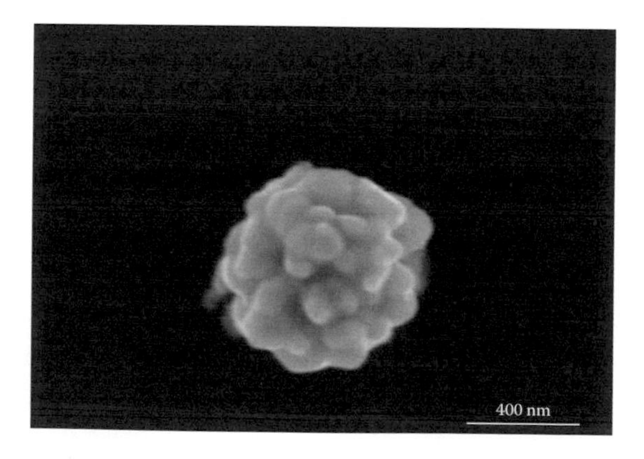

FIGURE 15.8 SEM image of an agglomerated Ag particle released from a spray. (With kind permission from Springer+Business Media: *J. Nanopart. Res.* 12, Size-fractionated characterization and quantification of nanoparticle release rates from a consumer spray product containing engineered nanoparticles, 2010, 2481–2494, Hagendorfer, H. et al.)

types of particles were detected in the aerosol. A spray containing nano-ZnO for example contained agglomerated ZnO surrounded by an acrylate polymer matrix.

Quadros and Marr (2011) also found that three commercially available sprays containing Ag released the Ag mainly as particles larger than 0.5 μm in the aerosol. Collection of the droplets and drying under vacuum revealed spherical particles that were mostly below 100 nm in size.

15.3.6 Sunscreens

Among the wide variety of nanomaterials that are used in consumer products, the UV absorbing nanoparticles used in sunscreens are the ENM category with the highest priority for further elucidating their exposure and effects (Wijnhoven et al. 2009; Dhanirama et al. 2011). Sunscreens are used up during use and thus almost complete release into the environment can be expected. For a further understanding of the fate and effects of these released materials it is therefore important to study the fate and behavior of the materials released into water. The fate of a specific hydrophobic nanocomposite used in sunscreens after release into water has been studied extensively by French researchers (Auffan et al. 2010; Labille et al. 2010; Botta et al. 2011). Figure 15.9 shows the nanocomposite consisting of a rutile (photocatalytic TiO_2) core, an aluminum oxide shell, and a PDMS coating (polydimethylsiloxane).

When this hydrophobic material comes into contact with water, it becomes hydrophilic and forms aggregates (Auffan et al. 2010). The organic Si layer is desorbed to more than 90% and the PDMS is degraded. The aluminum oxide layer is only affected to a very small extent. A stable suspension of particles with sizes from 50 to 700 nm was formed (Labille et al. 2010). About one fourth of the total mass of the added material was found in this stable suspension after 48 h of exposure and 48 h of sedimentation,

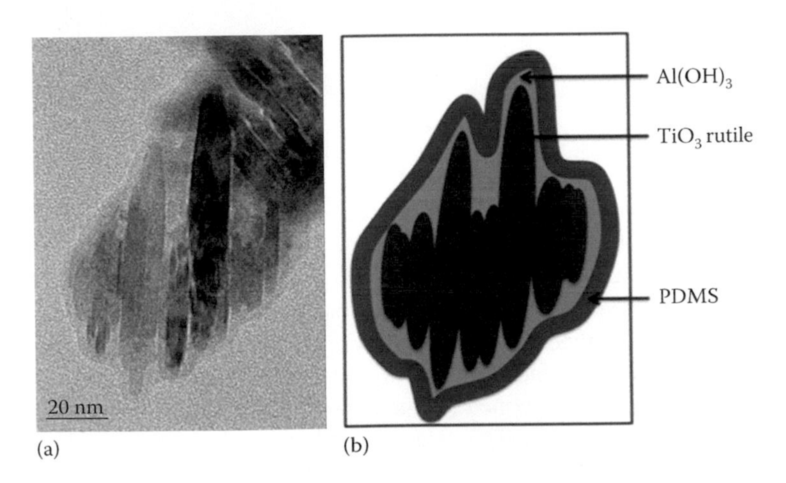

FIGURE 15.9 Transmission electron microscopy image of TiO_2–AlOOH nanocomposite used in sunscreen (a) and schematic view of the nanocomposite formulation (b), including the PDMS layer. (Reprinted from *Environ. Pollut.*, Labille, J. et al., Aging of TiO_2 nanocomposites used in sunscreen. Dispersion and fate of the degradation products in aqueous environment, 3482–3489, Copyright 2010, with permission from Elsevier.)

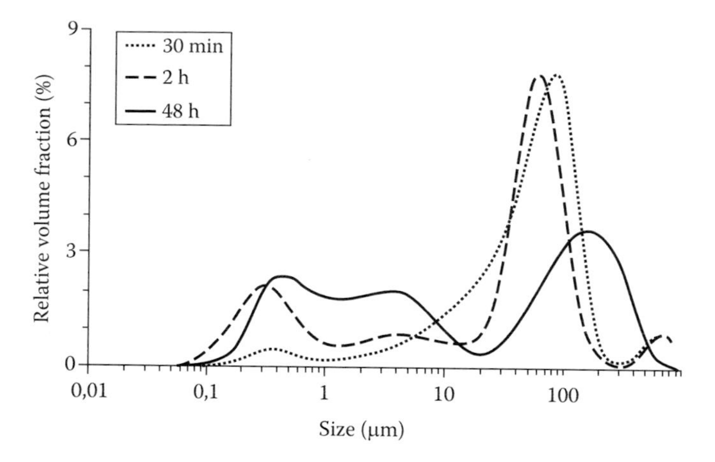

FIGURE 15.10 Size distribution of a nanocomposite used in sunscreen (Figure 15.9) aged in water for different times. (Reprinted from *Environ. Pollut.*, Labille, J. et al., Aging of TiO_2 nanocomposites used in sunscreen. Dispersion and fate of the degradation products in aqueous environment, 3482–3489, Copyright 2010, with permission from Elsevier.)

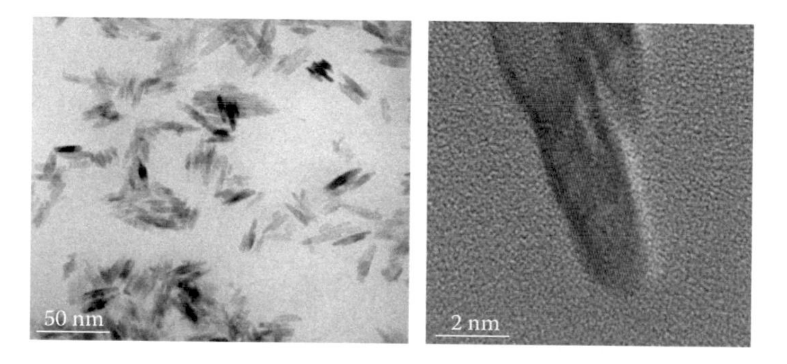

FIGURE 15.11 Transmission electron microscopy image of the stable suspension of a sunscreen aged in water. (Reprinted from *Environ. Pollut.*, Botta, C. et al., TiO_2-based nanoparticles released in water from commercialized sunscreens in a life-cycle perspective: Structures and quantities, 159, 1543–1548, Copyright 2011, with permission from Elsevier.)

the rest was agglomerated and sedimented rapidly (Figure 15.10). The stability of the suspension was determined by the properties of the aluminum oxide coating.

Botta et al. (2011) extended this work and studied not the pure nanocomposites but the release from four commercially available sunscreens. The results showed that a significant fraction of nano-TiO_2 was released from all sunscreens when they were put into water. A stable colloidal phase was formed which contained between 16 and 38% of the mass of the sunscreen and 1%–30% of the total TiO_2. About 10%–30% of the particles in this suspension were smaller than 1 µm, however, the volumetric fraction below 100 nm was almost negligible. A TEM investigation of the stable suspension showed rods of nano-TiO_2 with approximately 10 × 50 nm size that were agglomerated to larger particles (Figure 15.11).

These studies about the behavior of the original sunscreen TiO_2 and the release from commercial sunscreens show that the behavior of actually used ENM and their behavior in their matrix (the sunscreen) are much more complicated than the study of, for example, pure TiO_2 in water, see, for example, Ottofuelling et al. (2011).

15.4 MODELING RELEASE AND ENVIRONMENTAL FATE

A handful of modeling studies have so far incorporated ENM release into the model (Gottschalk and Nowack 2011). These studies modeled either the release of ENM from products during the consumer use phase or the release throughout the whole lifecycle of nano-products. The models vary how they conceptualize the release of ENM. The studies can be categorized into top-down and bottom-up approaches based on the release assessment frameworks used (Gottschalk and Nowack 2011). The top-down approach is based on information on the use of certain products, and uses a market fraction of nano-products and knowledge on the use of the product. In the bottom-up approaches, data on the production of ENM is combined with information on the distribution of the ENM to product groups. Some of the top-down models consider only the ENM release from a very small set of nano-products (Boxall et al. 2007; Blaser et al. 2008; Park et al. 2008; O'Brien and Cummins 2010). Some of the models using a bottom-up approach include the whole spectrum of nano-products and an exhaustive assessment of all possible applications of ENM (Mueller and Nowack 2008; Gottschalk et al. 2009; 2010b). In this approach an inventory of all known commercially relevant nano-products using a specific ENM is made and all these nano-products are categorized according to their potential to release ENM.

Mueller and Nowack (2008) published the first modeling of ENM considering ENM release from a whole lifecycle perspective. For nano-Ag, nano-TiO_2, and carbon nanotubes (CNT) direct releases from the nano-products to air, water, soil, waste incineration, sewage treatment plants, and landfills were modeled. Release processes that were considered include, for example, abrasion during use and washing of textiles. Gottschalk and coworkers (Gottschalk et al. 2009; Gottschalk et al. 2010a; Gottschalk et al. 2010b) extended this type of release modeling by replacing the scenario analysis used by Mueller and Nowack (2008) with a probabilistic and stochastic approach (Gottschalk et al. 2010a). The uncertainty and variability of the release of ENM were considered by means of Monte Carlo algorithms. Release was calculated for nano-TiO_2, nano-ZnO, nano-Ag, CNT, and fullerenes for the three regions, United States, Europe, and Switzerland (Gottschalk et al. 2009). Figure 15.12 shows an example of the material-flow diagrams which are the result of the modeling by Gottschalk et al. (2009). It depicts the flows of nano-Ag in the EU, considering all known applications of nano-Ag in products and incorporating release during production, manufacturing, use, and disposal. The major release pathways were from products into wastewater, caused by a high percentage of nano-Ag used in applications that have contact with water and the relatively high release of nano-Ag from these nano-products.

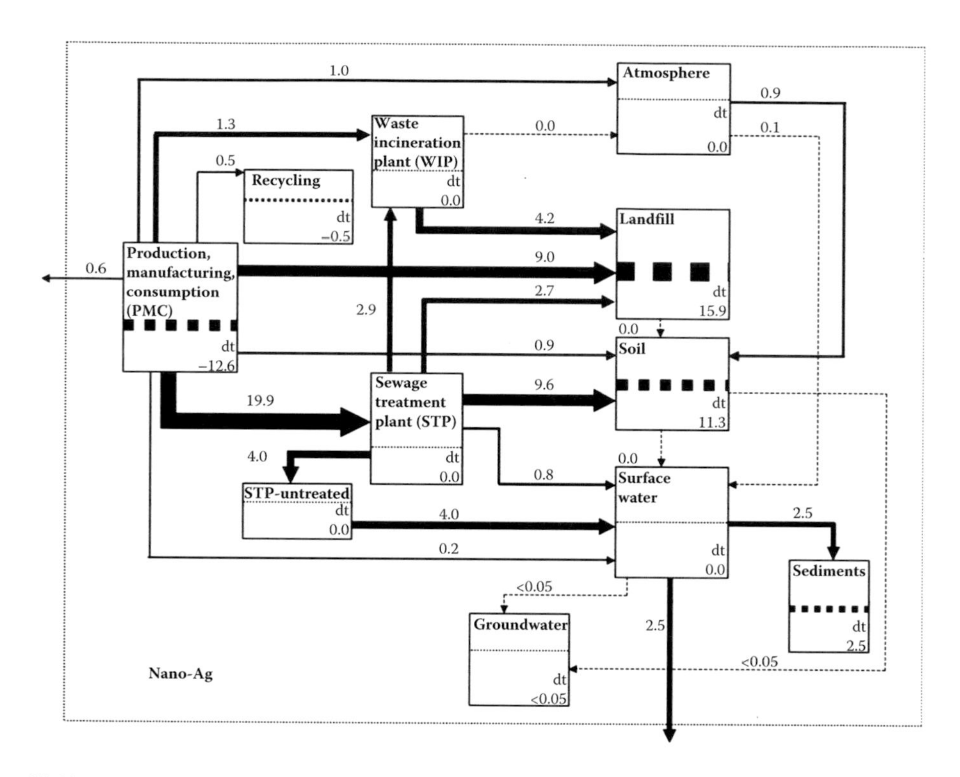

FIGURE 15.12 Material-flow model for nano-Ag in the EU. (Based on data reported from Gottsehalk, F. et al., *Environ. Sci. Technol.*, 43, 9216–9222, 2009.)

15.5 CONCLUSIONS

This chapter shows that there is already some information available about release processes from commercial products containing ENM. However, in the majority of the studies no nano-specific methods were used. The presence and identity of nanoscaled materials was in some studies confirmed using electron microscopy, whereas the magnitude of release was determined by bulk chemical methods after some separation of particles and dissolved metals. By combining both methods, some conclusions regarding the release of nanomaterials can be made, but in many cases the exact identity of the released nanomaterials remains vague. It is clear, however, that the majority of the released materials is not in the form of single dispersed ENM and that a multitude of different matrix fragments, agglomerates, transformation products, and dissolved metals are found in the released fraction. Only in a few studies single ENM were detected in the leachates. More studies using advanced analytical methods are clearly needed to obtain more information on the identity and mass-fractionated concentrations of the released materials. Also more studies under real-world conditions are needed that are able to provide sufficient information for modeling and environmental risk assessment.

REFERENCES

Auffan, M., Pedeutour, M., Rose, J., Masion, A., Ziarelli, F., Borschneck, D., Chaneac, C., Botta, C., Chaurand, P., Labille, J., Bottero, J.Y. (2010) Structural degradation at the surface of a TiO_2-based nanomaterial used in cosmetics. *Environ Sci Technol* 44, 2689–2694.

Baun, A., Hartmann, N.B., Grieger, K., Kusk, K.O. (2008) Ecotoxicity of engineered nanoparticles to aquatic invertebrates: a brief review and recommendations for future toxicity testing. *Ecotoxicology* 17: 387–395.

Benn, T., Cavanagh, B., Hristovski, K., Posner, J.D., Westerhoff, P. (2010) The release of nanosilver from consumer products used in the home. *J Environ Quality* 39: 1875–1882.

Benn, T.M., Westerhoff, P. (2008) Nanoparticle silver released into water from commercially available sock fabrics. *Environ Sci Technol* 42: 4133–4139.

Blaser, S.A., Scheringer, M., MacLeod, M., Hungerbuhler, K. (2008) Estimation of cumulative aquatic exposure and risk due to silver: Contribution of nano-functionalized plastics and textiles. *Sci Total Environ* 390: 396–409.

Botta, C., Labille, J., Auffan, M., Borschneck, D., Miche, H., Cabie, M., Masion, A., Rose, J., Bottero, J.Y. (2011) TiO_2-based nanoparticles released in water from commercialized sunscreens in a life-cycle perspective: Structures and quantities. *Environmental Pollution* 159: 1543–1548.

Boxall, A.B.A., Chaudhry, Q., Sinclair, C., Jones, A.D., Aitken, R., Jefferson, B., Watts, C. (2007) Current and future predicted environmental exposure to engineered nanoparticles. Central Science Laboratory, Sand Hutton, UK.

Broekhuizen, P., Broekhuizen, F., Cornelissen, R., Reijnders, L. (2011) Use of nanomaterials in the European construction industry and some occupational health aspects thereof. *J Nanopart Res* 13: 447–462.

Chabas, A., Lombardo, T., Cachier, H., Pertuisot, M.H., Oikonomou, K., Falcone, R., Verita, M., Geotti-Bianchini, F. (2008) Behaviour of self-cleaning glass in urban atmosphere. *Building Environ* 43: 2124–2131.

Cleveland, D., Long, S.E., Pennington, P.L., Cooper, E., Fulton, M.H., Scott, G.I., Brewer, T., Davis, J., Petersen, E.J., Wood, L. (2012) Pilot estuarine mesocosm study on the environmental fate of Silver nanomaterials leached from consumer products. *Sci Total Environ* 421: 267–272.

DEPA (2012) Assessment of nanosilver in textiles on the Danish market. Environmental Project No. 1432. Copenhagen, Danish Environmental Protection Agency.

Dhanirama, D., Gronow, J., Voulvoulis, N. (2011) Cosmetics as a potential source of environmental contamination in the UK. *Environ Technol* 33: 1597–1608.

Elzey, S., Grassian, V.H. (2010) Agglomeration, isolation and dissolution of commercially manufactured silver nanoparticles in aqueous environments. *J Nanopart Res* 12: 1945–1958.

Farkas, J., Peter, H., Christian, P., Urrea, J.A.G., Hassellov, M., Tuoriniemi, J., Gustafsson, S., Olsson, E., Hylland, K., Thomas, K.V. (2011) Characterization of the effluent from a nanosilver producing washing machine. *Environ Int* 37: 1057–1062.

Farre, M., Sanchis, J., Barcelo, D. (2011) Analysis and assessment of the occurrence, the fate and the behavior of nanomaterials in the environment. *Trac-Trends Anal Chem* 30: 517–527.

Geranio, L., Heuberger, M., Nowack, B. (2009) Behavior of silver nano-textiles during washing. *Environ Sci Technol* 43: 8113–8118.

Gottschalk, F., Nowack, B. (2011) Release of engineered nanomaterials to the environment. *J. Environ Monitoring* 13: 1145–1155.

Gottschalk, F., Scholz, R.W., Nowack, B. (2010a) Probabilistic material flow modeling for assessing the environmental exposure to compounds: Methodology and an application to engineered nano-TiO_2 particles. *Environ Modeling Software* 25: 320–332.

Gottschalk, F., Sonderer, T., Scholz, R.W., Nowack, B. (2009a) Modeled environmental concentrations of engineered nanomaterials (TiO_2, ZnO, Ag, CNT, fullerenes) for different regions. *Environ Sci Technol* 43: 9216–9222.

Gottschalk, F., Sonderer, T., Scholz, R.W., Nowack, B. (2010b) Possibilities and Limitations of Modeling Environmental Exposure to Engineered Nanomaterials by Probabilistic Material Flow Analysis. *Environ Toxicol Chem* 29: 1036–1048.

GSM (2009) Analyser av kemikalier i varor. R 2009:8, ISSN 1401-243X, Göteborgs Stad Miljöförvaltningen.

Hagendorfer, H., Lorenz, C., Kaegi, R., Sinnet, B., Gehrig, R., Goetz, N.V., Scheringer, M., Ludwig, C., Ulrich, A. (2010) Size-fractionated characterization and quantification of nanoparticle release rates from a consumer spray product containing engineered nanoparticles. *J Nanopart Res* 12: 2481–2494.

Handy, R.D., van der Kammer, F., Lead, J.R., Hassellöv, M., Owen, R., Crane, M. (2008) The ecotoxicology and chemistry of manufactured nanoparticles. *Ecotoxicol* 17: 287–313.

Hassellov, M., Kaegi, R. (2009) Analysis and Characterization of Manufactured Nanoparticles in Aquatic Environments, Environmental and Human Health Impacts of Nanotechnology Edited by Jamie R. Lead and Emma Smith. Blackwell Publishing Ltd. ISBN: 978-1-405-17634-7.

Hauri, J.F., Niece, B.K. (2011) Leaching of Silver from Silver-Impregnated Food Storage Containers. *J Chem Educ* 88: 1407–1409.

Huang, Y.M., Chen, S.X., Bing, X., Gao, C.L., Wang, T., Yuan, B. (2011) Nanosilver migrated into food-simulating solutions from commercially available food fresh containers. *Packaging Technol Sci* 24: 291–297.

Kaegi, R., Sinnet, B., Zuleeg, S., Hagendorfer, H., Mueller, E., Vonbank, R., Boller, M., Burkhardt, M. (2010) Release of silver nanoparticles from outdoor facades. *Environ Pollut* 158: 2900–2905.

Kaegi, R., Ulrich, A., Sinnet, B., Vonbank, R., Wichser, A., Zuleeg, S., Simmler, H., Brunner, S., Vonmont, H., Burkhardt, M., Boller, M. (2008) Synthetic TiO_2 nanoparticle emission from exterior facades into the aquatic environment. *Environ Pollut* 156: 233–239.

Kaegi, R., Voegelin, A., Sinnet, B., Zuleeg, S., Hagendorfer, H., Burkhardt, M., Siegrist, H. (2011) Behavior of metallic silver nanoparticles in a pilot wastewater treatment plant. *Environ Sci Technol* 45: 3902–3908.

Kaiser, J.-P., Zuin, S., Wick, P. (2013) Is nanotechnology revolutionizing the paint and lacquer industry? A critical opinion. *Sci Total Environ* 442: 282–289.

KEMI (2012) Antibacterial substances leaking out with the washing water - analyses of silver, triclosan and triclocarban in textiles before and after washing. Sundbyberg, Swedish Chemicals Agency.

Klaine, S.J., Alvarez, P.J.J., Batley, G.E., Fernandes, T.F., Handy, R.D., Lyon, D.Y., Mahendra, S., McLaughlin, M.J., Lead, J.R. (2008) Nanomaterials in the environment: Behavior, fate, bioavailability, and effects. *Environ Toxicol Chem* 27: 1825–1851.

Klaine, S.J., Koelmans, A.A., Horne, N., Carley, S., Handy, R.D., Kapustka, L., Nowack, B., von der Kammer, F. (2012) Paradigms to assess the environmental impact of manufactured nanomaterials. *Environ Toxicol Chem* 31: 3–13.

Koponen, I.K., Jensen, K.A., Schneider, T. (2009) Sanding dust from nanoparticle-containing paints: physical characterisation. *J Phys: Conf Series* 151: 012048.

Kuhlbusch, T.A.J., Asbach, C., Fissan, H., Gohler, D., Stintz, M. (2011) Nanoparticle exposure at nanotechnology workplaces: A review. *Part Fibre Toxicol* 8: 22.

Kulthong, K., Srisung, S., Boonpavanitchakul, K., Kangwansupamonkon, W., Maniratanachote, R. (2010) Determination of silver nanoparticle release from antibacterial fabrics into artificial sweat. *Part Fibre Toxicol* 7: 8.

Labille, J., Feng, J.H., Botta, C., Borschneck, D., Sammut, M., Cabie, M., Auffan, M., Rose, J., Bottero, J.Y. (2010) Aging of TiO_2 nanocomposites used in sunscreen. Dispersion and fate of the degradation products in aqueous environment. *Environ Pollut* 158: 3482–3489.

Lee, J., Mahendra, S., Alvarez, P.J.J. (2010) Nanomaterials in the construction industry: a review of their applications and environmental health and safety considerations. *ACS Nano* 4: 3580–3590.

Liu, J.Y., Hurt, R.H. (2010) Ion Release Kinetics and Particle Persistence in Aqueous Nano-Silver Colloids. *Environ Sci Technol* 44: 2169–2175.

Lombi, E., Donner, E., Tavakkoli, E., Turney, T.W., Naidu, R., Miller, B.W., Scheckel, K.G. (2012) Fate of zinc oxide nanoparticles during anaerobic digestion of wastewater and post-treatment processing of sewage sludge. *Environ Sci Technol* 46: 9089–9096.

Lorenz, C., Hagendorfer, H., von Goetz, N., Kaegi, R., Gehrig, R., Ulrich, A., Scheringer, M., Hungerbuhler, K. (2011) Nanosized aerosols from consumer sprays: Experimental analysis and exposure modeling for four commercial products. *J Nanopart Res* 13: 3377–3391.

Lorenz, C., Windler, L., Lehmann, R.P., Schuppler, M., Von Goetz, N., Hungerbühler, K., Heuberger, M., Nowack, B. (2012) Characterization of silver release from commercially available functional (nano)textiles. *Chemosphere* 89: 817–824.

Lowry, G.V., Gregory, K.B., Apte, S.C., Lead, J.R. (2012) Transformations of nanomaterials in the environment. *Environ Sci Technol* 46: 6893–6899.

Menard, A., Drobne, D., Jemec, A. (2011) Ecotoxicity of nanosized TiO_2. Review of in vivo data. *Environ Pollut* 159: 677–684.

Mueller, N.C., Nowack, B. (2008) Exposure modeling of engineered nanoparticles in the environment. *Environ Sci Technol* 42: 4447–4453.

Norgaard, A.W., Jensen, K.A., Janfelt, C., Lauritsen, F.R., Clausen, P.A., Wolkoff, P. (2009) Release of VOCs and particles during use of nanofilm spray products. *Environ Sci Technol* 43: 7824–7830.

Nowack, B., Bucheli, T.D. (2007) Occurrence, behavior and effects of nanoparticles in the environment. *Environ Pollut* 150: 5–22.

Nowack, B., Ranville, J.F., Diamond, S., Gallego-Urrea, J.A., Metcalfe, C., Rose, J., Horne, N., Koelmans, A.A., Klaine, S.J. (2012) Potential scenarios for nanomaterial release and subsequent alteration in the environment. *Environ Toxicol Chem* 31: 50–59.

O'Brien, N., Cummins, E. (2010) Nano-Scale Pollutants: Fate in Irish surface and drinking water regulatory systems. *Hum Ecol Risk Assess* 16: 847–872.

Olabarrieta, J., Zorita, S., Pena, I., Rioja, N., Monzon, O., Benguria, P., Scifo, L. (2012) Aging of photocatalytic coatings under a water flow: Long run performance and TiO_2 nanoparticles release. *Appl Catal B: Environ* 123–124: 182–192.

Ottofuelling, S., Von der Kammer, F., Hofmann, T. (2011) Commercial Titanium Dioxide Nanoparticles in Both Natural and Synthetic Water: Comprehensive multidimensional testing and prediction of aggregation behavior. *Environ Sci Technol* 45: 10045–10052.

Park, B., Donaldson, K., Duffin, R., Tran, L., Kelly, F., Mudway, I., Morin, J.P., Guest, R., Jenkinson, P., Samaras, Z., Giannouli, M., Kouridis, H., Martin, P. (2008) Hazard and risk assessment of a nanoparticulate cerium oxide-based diesel fuel additive - A case study. *Inhal Toxicol* 20: 547–566.

Peralta-Videa, J.R., Zhao, L.J., Lopez-Moreno, M.L., de la Rosa, G., Hong, J., Gardea-Torresdey, J.L. (2011) Nanomaterials and the environment: A review for the biennium 2008–2010. *J Hazard Mater* 186: 1–15.

Quadros, M.E., Marr, L.C. (2010) Environmental and Human health risks of aerosolized silver nanoparticles. *J Air Waste Manage Assoc* 60: 770–781.

Quadros, M.E., Marr, L.C. (2011) Silver Nanoparticles and total aerosols emitted by nanotechnology-related consumer spray products. *Environ Sci Technol* 45: 10713–10719.

Scown, T.M., van Aerle, R., Tyler, C.R. (2010) Review: Do engineered nanoparticles pose a significant threat to the aquatic environment? Critical Reviews in Toxicology 40, 653–670.

Simon, P., Chaudhry, Q., Bakos, D. (2008) Migration of engineered nanoparticles from polymer packaging to food - a physicochemical view. *J Food Nutr Res* 47: 105–113.

Som, C., Nowack, B., Wick, P., Krug, H. (2010) Nanomaterialien in Textilien: Umwelt-, Gesundheits- und Sicherheits-Aspekte, Fokus: synthetische Nanopartikel. Empa und TVS Textilverband Schweiz, St. Gallen. http://www.empa.ch/plugin/template/empa/*/93326.

Song, H., Li, B., Lin, Q.B., Wu, H.J., Chen, Y. (2011) Migration of silver from nanosilverìpolyethylene composite packaging into food simulants. *Food Addit Contam: A* 28: 1758–1762.

Ulrich, A., Losert, S., Bendixen, N., Al-Kattan, A., Hagendorfer, H., Nowack, B., Adlhart, C., Ebert, J., Lattuada, M., Hungerbuhler, K. (2012) Critical aspects of sample handling for direct nanoparticle analysis and analytical challenges using asymmetric field flow fractionation in a multi-detector approach. *J Anal Atom Spectrom* 27: 1120–1130.

von der Kammer, F., Ferguson, P.L., Holden, P.A., Masion, A., Rogers, K.R., Klaine, S.J., Koelmans, A.A., Horne, N., Unrine, J.M. (2012) Analysis of engineered nanomaterials in complex matrices (environment and biota): General considerations and conceptual case studies. *Environ Toxicol Chem* 31: 32–49.

Weir, A., Westerhoff, P., Fabricius, L., Hristovski, K., von Götz, N. (2012) Titanium dioxide nanoparticles in food and personal care products. *Environ Sci Technol* 46: 2242–2250.

Wiesner, M.R., Lowry, G.V., Jones, K.L., Hochella, M.F., Di Giulio, R.T., Casman, E., Bernhardt, E.S. (2009) Decreasing Uncertainties in assessing environmental exposure, risk, and ecological implications of nanomaterials. *Environ Sci Technol* 43: 6458–6462.

Wijnhoven, S.W.P., Dekkers, S., Hagens, W.I., de Jong, W.H. (2009) Exposure to nanomaterials in consumer products. Letter report 340370001/2009. RIVM, National Institute for Public Health and the Environment, The Netherlands.

Windler, L., Height, M., Nowack, B. (2013) Comparative evaluation of antimicrobials for textile applications. *Environ Int* 53: 62–73.

Windler, L., Lorenz, C., Von Goetz, N., Hungerbuhler, H., Amberg, M., Heuberger, M., Nowack, B. (2012) Release of titanium dioxide from textiles during washing. *Environ Sci Technol* 46: 8181–8188.

Yan, Y., Yang, H., Lu, X., Li, J., Wang, C. (2012) Release Behavior of nano-silver textiles in simulated perspiration fluids. *Textile Res J* 82: 1422–1429.

Section IV

Integrating Case Studies on Methods and Materials

16 Concern-Driven Safety Assessment of Nanomaterials

An Integrated Approach Using Material Properties, Hazard, Biokinetic, and Exposure Data and Considerations on Grouping and Read-Across

Agnes Oomen, Peter Bos, and Robert Landsiedel

CONTENTS

16.1 INTRODUCTION

Although more and more nanomaterials (NMs) continue to be produced with almost infinite variations in size, shape, structure, and surface modifications (Lövestam et al. 2010), a generally applicable paradigm for NM hazard identification and risk assessment is not yet available. Not only their physicochemical characteristics may have an impact on the extent and type of biological effects they induce, additionally, unlike conventional chemicals, exposure to NMs is not to a distinct type of molecule, but to a population of primary particles, aggregates, and agglomerates of various sizes and different surface coatings that may change in characteristics with time (compare Chapter 4 and Section III). These characteristics can influence exposure and kinetic behavior, as well as the type of effect of the NM. This complicates classical risk assessment approaches and multiplies the need for efficient and effective hazard testing and exposure assessment.

In recognition of this multitude of potential variations for NMs, the *Scientific Committee on Emerging and Newly Identified Health Risks*, an advisory body to the European Commission, has recommended for the time being a case-by-case approach for NM hazard and risk assessment (SCENIHR 2009). Traditional toxicity testing calls for a substantive number of tests that are mostly restricted to the mere observation of apical (clinical or histopathological) effects (Combes et al. 2003; OECD 2012a). If based upon such traditional toxicological testing methods, one-by-one assessment of all potential hazards of NMs (just as the hazards of chemical substances as such) would result in the use of very large numbers of laboratory animals and high associated cost. This is even more true if all potential hazards associated with changes in physicochemical characteristics during a NM's lifecycle are taken into account. Yet, safety of NMs should be demonstrated before administration to the market.

This, however, would contravene the 3Rs principle to replace, reduce, and refine animal testing (Russell and Burch 1959) that has been implemented in the European Union (EU) *Directive 2010/63/EU on the protection of animals used for scientific purposes* (Anon 2010). Also the EU Chemicals Regulation No. 1907/2006 (REACH, Registration, Evaluation, Authorisation of Chemicals; Anon 2006) prescribes that animal testing should only be undertaken as a last resort.

As a consequence, there is an urgent need for new integrated approaches for the integrated testing and assessment (IATA) of NMs (George et al. 2011; Kuempel et al. 2012). An efficient and effective testing strategy should provide guidance on how to obtain the information actually relevant for risk assessment of a specific NM and its application, and include options for "non-testing" methodologies, such as "read-across" and "weight-of-evidence" (Information Box 16.1), and nonanimal test methods as part of tiered testing strategies (Combes et al. 2003; IHCP2005). The application of such nontesting methods has been included in the REACH legislation. An efficient testing strategy should exploit and combine all available information on exposure and toxicity to identify relevant concerns and to determine those data that best address these concerns (Combes and Balls 2005; van Leeuwen et al. 2007; OECD 2012a). Today there is, however, very little experience with grouping and read-across for NMs.

Integrated approaches for the testing and assessment of the safe production and use of chemicals have also become the pillar of the United States *National Research*

INFORMATION BOX 16.1 DEFINITIONS RELATED TO NONTESTING APPROACHES IN THE CONTEXT OF REACH

(Anon. 2006 (Annex XI); ECHA 2008 and 2010; German Competent Authorities 2011; ECHA 2013)

Nontesting approaches imply not generating new data, but extending and extrapolating existing information to facilitate the evaluation of the intrinsic properties of chemicals.

Grouping is the process of assigning chemical substances to a "group" (or "category"); based upon similar or regular patterns of physicochemical, biological, exposure, biokinetic, or environmental fate properties. For conventional chemicals, these similarities may be due to common functional groups, common precursors, likely common breakdown products, or a constant pattern of a given property across the group.

Read-across is the application of the grouping concept to predict end point-specific information for the target substance by using data for the same end point from other substances. Thereby, the need to test every substance for every end point is avoided. During the *analogue approach*, a very limited number of substances are compared, whereas the *structural or category approach* implies comparing larger numbers of substances.

(Q)SAR models are mathematical models (tools) that can be used for a qualitative (SAR) or quantitative (quantitative structure activity relationship [QSAR]) prediction of the physicochemical, biological, and environmental fate properties of compounds from the knowledge of their chemical structure. Computer models are also called in silico models.

Expert systems are more complex models consisting of combinations of SARs, QSARs, and databases.

Weight-of-evidence implies that the information obtained from several independent sources allows assuming or concluding that a substance has a particular property, whereas the information from each single source alone is regarded insufficient to support this notion.

Waiving of an information requirement in the context of REACH means that the submission of the standard information for the particular end point is not considered necessary in a specific case.

Council's strategy for toxicity testing in the 21st century (CTTAEA and NRC 2007). This "Tox21" initiative aims for a paradigm shift in toxicological testing that relies less on animal studies and focuses more on in vitro methods to evaluate the effects that chemicals can have on biological processes using cells, cell lines, or cellular components that are preferably of human origin. It should be noted that at present most NM risk assessments rely mostly on in vivo studies.

The development and application of integrated approaches for NM testing and assessment should be closely linked to the grouping of NMs (Damoiseaux et al. 2011). Efficient and effective testing should enable hazard and risk assessment either directly or by providing information that is also relevant for grouping, read-across, ranking, and other methodologies as a part of IATA to fill data gaps that will allow waiving of testing. In the ideal situation, the testing strategy drives a grouping approach in order to efficiently and effectively assess risk.

In the following, an overview of NM pathways from use and release to apical toxic effects is given and ongoing developments with regard to the testing and grouping of NMs are presented (Figure 16.1). Efficient and effective safety assessment of NMs should be concern-driven; that is, they should start out by identifying relevant concerns and critical lifecycle stages, based on NM exposure and use scenarios, physicochemical characterization of the NMs (related to exposure, uptake, biodistribution, and hazard), estimation of changes in physicochemical characteristics during the lifecycle, and already available toxicological information. Recognized concerns will not only influence the IATA but also drive the grouping process.

Whereas the deliberations on IATA and NM grouping are not directly related to the existing legislation, they do stand in line with EU guidance on NM safety testing (Hankin et al. 2011; ECHA 2012a). For example, alternative methods are propagated in REACH (Annex XI), if sufficiently documented, scientifically found, and applicable for risk assessment and classification and labeling. However, the material properties and exposure situation for NMs are often complex and the scientific foundation and their applicability for risk assessment are not yet established. The present deliberations are a first approach to assess the complexity for NMs and how to deal with

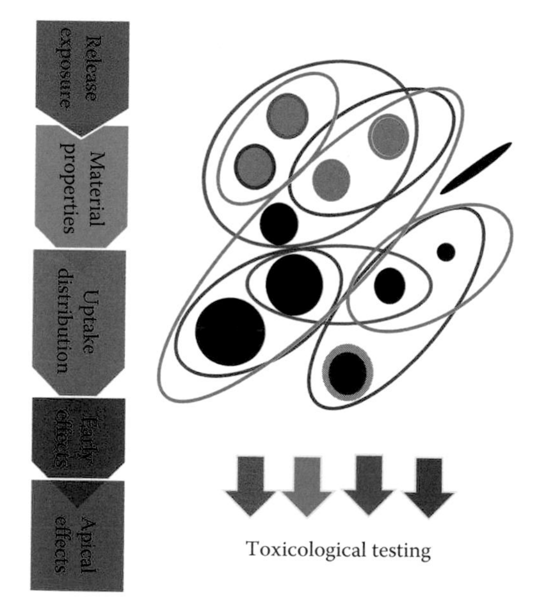

FIGURE 16.1 Using the *Source to Adverse Outcome Pathway* for grouping concepts as key elements of testing strategies. The circled figures represent potential different groups of nanomaterials, either based on material properties, exposure information, kinetic behavior, and/or effects. (Modified from Oomen AG et al. (2013), Concern-driven integrated approaches to nanomaterial testing and assessment, Report of the NanoSafety Cluster Working Group 10, *Nanotoxicol* 28 May 2013, epub ahead of print.)

that. Furthermore, they take into account the proceedings of international initiatives addressing NM hazard and risk assessment, that is, the *OECD Working Party on Manufactured Nanomaterials* (WPMN*) and the *International Standardisation Organisation Technical Committee 229.*† WPMN steering groups have addressed innovative risk assessment approaches and have compiled a list of in vitro methods that might be used for NM human hazard identification. Additionally, the WPMN has proposed a new short-term inhalation study for NM testing that provides an indication of the hazard (potential and potency for local [airway] effects and potential also for systemic effects) and includes some kinetic parameters to investigate if the NMs translocate from the respiratory tissues and thus if systemic effects should be investigated (Ma-Hock et al. 2009; OECD 2011; Klein et al. 2012).

Information on the source-to-adverse-outcome pathway can facilitate safety assessment and grouping of a NM. These pathways depend, amongst others, on the physicochemical characteristics of the NM, which may change at the various lifecycle stages (compare chapter in Section III). The different physicochemical characteristics will affect the exposure, biokinetics, and effects of a NM. Hence, these parameters can and should be considered for the safety assessment and grouping of NMs.

16.2 NANOMATERIAL TOXICITY PATHWAYS AND ADVERSE OUTCOME PATHWAYS

Aiming for an understanding of the mechanisms of toxicity that can be induced by a substance is a prerequisite to recognizing relevant concerns and selecting appropriate toxicity tests. Early biological effects of NMs (compare chapters in Section II) may be related to the particles themselves and their coatings (particle effects); ions or molecules released from the particles (chemical effects); and molecules formed by the catalytic surface of the particles (Landsiedel et al. 2010).

Different types of NMs can induce different early biological effects (Nel et al. 2013b): Titanium dioxide, copper oxide, and cobalt oxide NMs, for instance, have been observed to induce the generation of reactive oxygen species (ROS) and to increase redox activities. Metal and metal oxide NMs, such as zinc oxide and silver, can exert toxic effects by dissolution and the shedding of toxic ions. Activation of proinflammatory reactions has been reported for several NMs and persistent long-aspect ratio (fiber-like) NMs, such as carbon nanotubes (CNT), can cause sustained inflammation. Further NMs, including SiO_2 nanoparticles and Ag plates, might induce membrane damage and lysis; whereas the effects of, for example, cationic polystyrene NMs are likely related to cationic toxicity leading to lysosomal rupture and mitochondrial damage (Xia et al. 2008; Meng et al. 2009; Damoiseaux et al. 2011; Zhang et al. 2012; Nel et al. 2013b). Obviously, one NM may be able to induce more than one early biological effect.

Depending on their toxicity pathway, NMs have different modes of action (MOA) and can elicit different adverse outcome pathways (AOP) (Information Box 16.2): The proinflammatory effects caused by CNT, for instance, can result in pulmonary fibrosis, which, when progressive, might lead to severe dysfunction of the respiratory

* See: http://www.oecd.org/env/ehs/nanosafety/ note: all websites were accessed in June 2013.
† See: http://www.iso.org/iso/iso_technical_committee?commid=381983.

and cardiovascular systems (Castranova et al. 2012; Nel et al. 2013b; Shvedova et al. 2013). On the molecular level, NMs can interact with biological macromolecules, membranes, and intracellular organelles, which are oftentimes in a comparable size range as the NMs. Thereby, NMs can interfere, for example, with the native function of proteins or bind to functional sites of DNA, RNA, ribosomes, mitochondria, or the endoplasmic reticulum (Yanamala et al. 2013). The adverse outcomes of these molecular initiating events have not yet been fully described.

Figure 16.2 provides an overview of the steps of a source-to-adverse-outcome pathway for NMs taking into account their release, uptake, surface properties and states of agglomeration, biodistribution, and possible early and apical biological effects. Early biological effects can—initially—be investigated in vitro. However, in order to assess if this can result in an apical effect, information on the external and internal exposure is required. External exposure refers to the level and physicochemical form of a NM exposure outside the body and is thus related to the various lifecycle stages, used amounts, and release of the NM (referred to as dispersion in Figure 16.2). Internal

INFORMATION BOX 16.2 DEFINITIONS RELATED TO ADVERSE OUTCOME PATHWAYS

During the *molecular initiating event*, a substance interacts with a specific biological target tissue, which results in a cellular response. This is the initial point of chemical–biological interaction.

Toxicity pathways describe the sequence of the respective molecular initiating event and the resulting cellular effect.

Modes of action (MOA) link toxicity pathways to organ or organ system effects. The MOA action differs from mechanism of action, with the mode of action requiring a less detailed understanding of the molecular basis to the toxic effect.

Early biological effects summarize molecular initiating events and cellular responses.

Apical toxic effects are the clinically manifested adverse effects in an organism; they are the outcome of a cogent early biological effect.

Adverse outcome pathways (AOP) describe the steps linking the molecular initiating event to an adverse outcome at the organism—or population—level that is directly relevant for a given risk assessment context.

Source-to-[adverse-]outcome pathways relate to the comprehensive understanding of the effects of a substance from its release through to its effects on the organism/individual or population level.

Source: Ankley et al., *Env Toxicol Chem*, 29, 730–741, 2010, OECD (2012a), Proposal for a template, and guidance on developing and assessing the completeness of adverse outcome pathways, 17 pp; available at: http://www.oecd.org/env/ehs/testing/49963554.pdf, OECD. (2012b). Appendix I, Collection of working definitions, OECD, Paris, France; available at: http://www.oecd.org/env/ehs/testing/49963576.pdf.

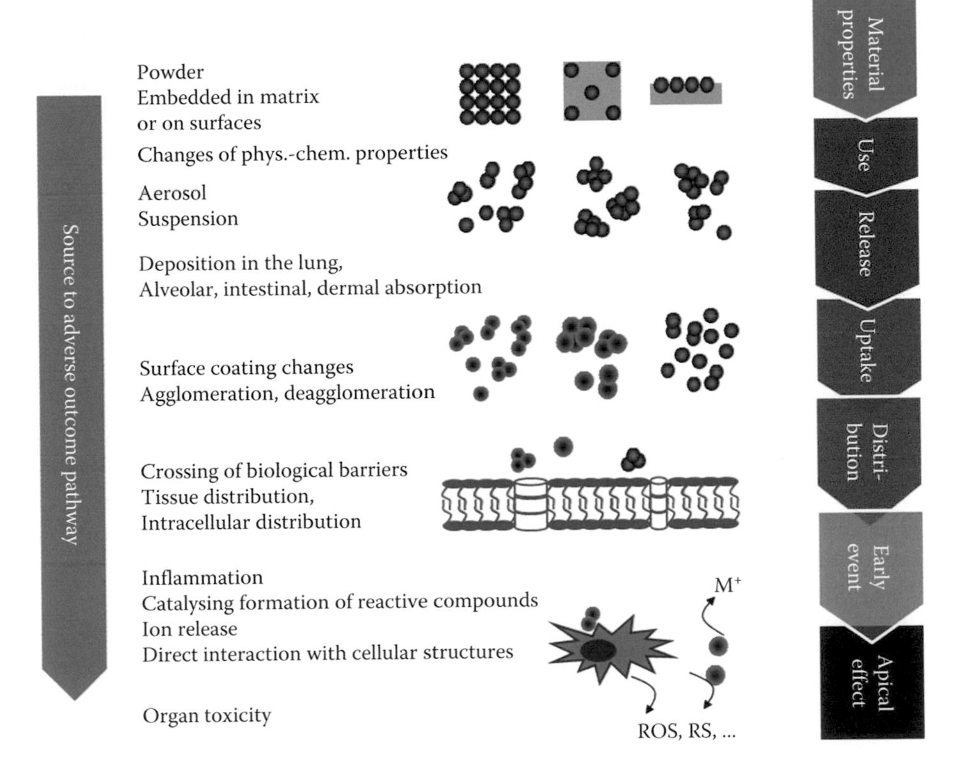

FIGURE 16.2 Source-to-adverse-outcome pathway of nanomaterials. (Landsiedel, R. et al.: Testing metal-oxide nanomaterials for human safety. *Adv. Mater.* 2010. 22. 2601–2627. Copyright Wiley-VCH Verlag GmbH & Co. KGaA. Reproduced with permission.)

exposure refers to the level and physicochemical form of a NM at the site of action in the organism. This includes biokinetics, of which transport across barriers (external barriers such as skin, lung, and gastrointestinal tract and internal barriers such as blood–brain, blood–testis, and placenta) and modification in the body are the most critical for NMs. If, for example, two NMs exert the same early biological effects at a similar degree, but internal exposure for one NM is much lower, the potency of the apical effect will be different. Furthermore, tissue distribution may be different between NMs and one NM may reach a target tissue to initiate a tissue-specific effect, whereas the other NM may not. These aspects are referred to as "uptake in the body," "modification in the body," and "distribution in the body" in Figure 16.2. Hence, information on various steps of the source-to-adverse-outcome pathway is required.

To date, no NM effects have been reported that have not yet been observed with any other substance or particle. It therefore seems generally feasible that toxicity of NMs can be investigated with standard testing methods used for conventional chemicals, adapted to NM-specific requirements for test sample preparation and characterization and test interference controls, as also recommended by the OECD's Chemicals Committee, a parent body to the WPMN (OECD 2013). For in vitro toxicity tests, in vitro doses and cellular uptake of NMs should be considered. Most NM particles >20 nm may, for example, not be able to enter Salmonella used for the Ames test.

The uptake, tissue distribution, and clearance of NMs may be different from dissolved molecules or larger particles (Landsiedel et al. 2012a), and thus needs further attention, especially as this behavior depends on physicochemical characteristics that may change during the life stage of the NM. This also implies that physicochemical characterization should be performed not only for the pristine material but also in situ at critical lifecycle stages.

16.3 GROUPING OF NM

16.3.1 Concept of Grouping

The classical grouping concept implies considering closely related materials as a group, or category, rather than as individual materials. In a category approach to hazard assessment, not every material needs to be tested for every single end point. Instead, the overall data for a given category allow estimating the hazard for the untested end points (OECD 2007). Use of information from structurally related materials to predict the end point information for another material is called read-across (Anon 2006; Information Box 16.1). In the registration dossiers submitted during the first REACH registration phase, read-across was frequently applied to avoid the generation of new data (Spielmann et al. 2011). However, according to the European Chemicals Agency (ECHA), oftentimes, it was not sufficiently detailed or adequately justified (ECHA 2012b, 2012c).

Compared to the grouping of conventional chemicals, the grouping of NMs is an even more complex challenge. So far, no single physical or chemical material property—be it surface, volume, or ROS generation—perfectly correlates with the observed biological effects for various types of NMs (Landsiedel et al. 2012b; Zuin et al. 2011; Nel et al. 2013b). More than a dozen physicochemical properties of NMs have been identified that could potentially contribute to hazardous interactions, including chemical composition, size, shape, aspect ratio, surface charge, redox activity, dissolution, crystallinity, surface coatings, and the state of agglomeration or dispersion (Thomas et al. 2011). Some of these characteristics influence also biokinetic processes and thus the fraction of a NM that reaches the target site, whereas potentially other—or another set of—material properties influence the biological effects. This may confound the correlation between one single material property and an apical effect.

NM complexity requires developing more comprehensive grouping of NMs for risk assessment. At the same time it offers the opportunity to include more information: NM grouping should not be restricted to the determination of nanostructure–activity relationships, but should take into account the entire source-to-adverse-outcome pathway, including changing of physicochemical characteristics of a NM.

Therefore, the pillars of NM grouping should be

- The lifecycle of a NM, including its production, use, and release
- The physicochemical characteristics of a NM, which can be different in different lifecycle stages (e.g., release and external exposure of organisms)

- The uptake, biodistribution, and biopersistence (biokinetics) of a NM in an organism and the physicochemical characteristics of a NM inside the organism
- The early and apical biological effects

Grouping of NMs, however, can not only be performed by different parameters but also for different purposes. A broad spectrum of incentives to perform grouping are conceivable, for example, the need to address different exposure scenarios, to prioritize potential risks, to apply read-across, to justify waiving, or to define criteria for testing strategies. Ranking of NMs can be performed within a group (e.g., high/low exposure, high/low hazard; comparable to the toxic equivalency factor approaches for dioxins or polychlorinated biphenyls; Safe 1998) and also different groups can be ranked (e.g., based on effect potency, amount of NMs released from a product, and internal doses). In this concept, a given NM may and will belong to more than one group, depending on its assignment to the respective groups (e.g., physicochemical characteristics, release, uptake and distribution—timing of—biological effects, lifecycle stages, biopersistence, susceptibility of specific population groups, and species sensitivity), all of which can govern NM toxicity. These examples show that different concerns (be they scientific or societal, see the following) may also drive the need for grouping.

Criteria need to be developed for grouping to enable justification of the application of read-across for regulatory purposes (Patlewicz et al. 2013). Clearly, the rationale of and reasons for grouping define such criteria (e.g., exposure scenarios, hazardous effects).

In principle, a grouping strategy resembles a testing strategy insofar as both may use the same criteria and identify the same concerns. Tools for grouping using physicochemical properties can be quantitative structure activity relationship models. Determination of pristine physicochemical characteristics is relatively inexpensive and reliable, too. Grouping based on exposure may sufficiently be served by use of patterns whereas in other cases actual release measurements may be necessary. Grouping based on biokinetics as well as early biological effects usually requires in vitro or in vivo testing; in the future in silico models may assist this (e.g., the transport across biological barriers). Grouping based on apical biological effects will usually require in vivo testing with the advantage of integrating biokinetics and thus providing accurate potency information.

In the beginning, grouping will be crude and incomplete (not all NMs will be assignable to groups and not all steps of the source-to-adverse-outcome pathway will be used) but may be refined with increasing knowledge. Nevertheless, one should start out with easily achievable forms of grouping. As knowledge increases, more specific grouping will become possible. Also in the context of a tiered approach, grouping becomes more detailed and complex with increasingly higher tiers, always with the main incentive to avoid having to test all NMs.

16.3.2 Grouping of NM: Ongoing Developments

Recognition of mutually common properties of different NMs is an intricate multidisciplinary task. Recently, mode-of-action integrated approaches for the testing and assessment of NMs have begun to evolve to the stage of providing quantitative estimations of the relationship of specific NM properties with mechanistic outcomes. Based on this, early attempts at NM grouping are progressing that can already be used to facilitate testing of large batches of NMs (Nel et al. 2013a), although until now this is only considered for mode-of-action.

Already in 1999, Rehn and coworkers described a multiparametric in vitro testing battery using isolated alveolar macrophages for the screening of pulmonary effects of dust aerosols. By combining the results obtained for a set of independent end points (release of glucuronidase and lactate dehydrogenase, esterase activation, release of H_2O_2, induction of tumor necrosis factor alpha, and ROS generation), assay sensitivity and specificity could be improved in comparison to single end point cytotoxicity assays, and distinct toxicity patterns of different dusts could be discerned (Rehn et al. 1999). These mentioned end points covered the four basic principles of (nano) dust toxicity: ROS generation, release of inflammatory mediators, impairment of cellular function, and cytotoxicity.

The *University of California Center for the Environmental Implications of Nanotechnology* has developed a predictive toxicological approach using in vitro mechanism-based assays for high-throughput screening (HTS) (Nel et al. 2013b). Data obtained with the HTS assays are evaluated to make predictions about those physicochemical properties of NMs that may lead to the generation of adverse effects in vivo. Thereby, multiparametric HTS platforms foster the grouping of NMs (Nel et al. 2013b).

The HTS platform described by Nel and coworkers utilizes a series of compatible fluorescent dyes to simultaneously measure a set of sublethal and lethal cellular end points in 384-well plates. The investigated end points reflect important toxicity pathways of NMs, that is, ROS production, intracellular calcium flux, mitochondrial membrane depolarization, and membrane disruption (Damoiseaux et al. 2011; George et al. 2011; Thomas et al. 2011; Xia et al. 2012; Nel et al. 2013b).

The molecular initiating events and toxicological pathways of NMs can be investigated in vitro. To be of value for risk assessment, in a predictive toxicological approach the results of the in vitro screening assays are used to quantitatively assess dose- and time-dependent molecular initiating events and toxicity pathways that are predictive of in vivo adverse outcomes (which would then have to be confirmed in limited in vitro studies) (Nel et al. 2013b).

Data from the HTS screening platforms are provided in a multivariate context (e.g., concentration, exposure times, sublethal, and lethal biological responses). Due to the complexity of the data, feature-extraction methods are applied for visual data interpretation. In this respect, so-called "heat-map clusters" provide ordered representations of data facilitating the identification of similarity patterns of large datasets based on homologous biological responses or linkage to physicochemical properties. Thereby, multiparametric mechanism-based HTS approaches combined with heat-map clustering enable NM grouping and the development of nanostructure–activity relationships (Damoiseaux et al. 2009; George et al. 2011; Liu et al. 2011; Rallo et al. 2011; Thomas et al. 2011; Nel et al. 2013b).

The *United States National Institute for Occupational Safety and Health* has presented a strategy to assign science-based hazard and risk categories for NMs that can be used for occupational exposure control. This strategy is based upon selected "benchmark" materials representing various MOA classes. New NMs are classified in accordance to the hazard and risk estimates of the respective benchmark substances that exhibit the same MOA (Kuempel et al. 2012). A key challenge in setting MOA-based classes is obtaining sufficient dose–response data to systematically evaluate the physicochemical factors influencing NM biological activity. In the absence of substance-specific information, categorization processes involve both science-based analyses and default assumptions (Kuempel et al. 2012).

Also for occupational safety assessment, the *German Federal Institute for Occupational Safety and Health* (BAuA, Bundesanstalt für Arbeitsschutz und Arbeitsmedizin) distinguishes three different groups of NMs, which combine information on specific physicochemical characteristics and effects or kinetic properties: (1) NMs that release granular biopersistent particles, (2) NMs that release fibrous dusts (for these two groups, pulmonary inflammation and lung cancer are potential health effects), and (3) NMs that have a specific "chemical" toxicity, that is, release of toxic ions, activity of chemical functional groups, and catalytic activity (Packroff 2013).

In a position paper on NMs and REACH, the German competent authorities (BAuA, together with the Federal Environment Agency [UBA; Umweltbundesamt] and the Federal Institute for Risk Assessment [BfR; Bundesinstitut für Risikobewertung]) describe three conceivable scenarios for waiving of testing during NM hazard assessment, that is, (1) use of data by referencing between the bulk and nanoform of a substance, (2) use of data by referencing between different nanoforms of a substance, and (3) read-across between substances with different chemical identities (possibly various bulk and nanoforms) (German Competent Authorities 2011). Approaches for grouping and waiving are recognized to be particularly important for substances having a large number of different nanoforms. The German position further anticipates that waiving will be rare in the beginning, but has the potential to increase as standardized tests have shown that results from substances in bulk form can be utilized for NMs (German Competent Authorities 2011).

16.4 INTEGRATED APPROACHES FOR NANOMATERIAL SAFETY TESTING AND ASSESSMENT

In the following, the different tiers of an initial framework for concern-driven integrated approach for NM testing and assessment are summarized (for further details, also covering IATA of NM effects on the environment, see Oomen et al. 2013). Further work on this framework is ongoing, as well as on ways of integration of the grouping concept.

16.4.1 Identification of "NM of Concern"

The first step of any NM risk assessment implies determining that the material under consideration is in fact a NM (Figure 16.3, step 1). This depends on the definition of NM used, but usually at least includes chemical characterization as well as some particle shape characterization. The respective EU Recommendation defines

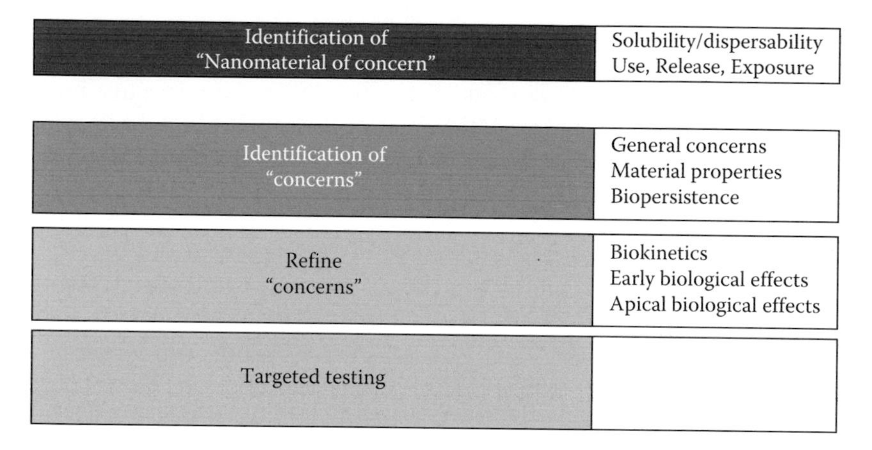

FIGURE 16.3 Identification of relevant concerns as a basis for concern-driven nanomaterials testing. (Adapted from Oomen AG et al. (2013), Concern-driven integrated approaches to nanomaterial testing and assessment, Report of the NanoSafety Cluster Working Group 10, *Nanotoxicol* 28 May 2013, epub ahead of print.)

a "nanomaterial" to be a *"natural, incidental or manufactured material containing particles, in an unbound state or as an aggregate or as an agglomerate and where, for 50% or more of the particles in the number size distribution, one or more external dimensions is in the size range 1–100 nm"* (Anon. 2011). Strategies to classify materials according to regulatory nano definitions have been outlined by the European Commission (Linsinger et al. 2012) and are described in this book (Chapter 3).

The next step of the IATA (Figure 16.3, step 2) aims at identifying "NM of concern" based on relevant exposure scenarios and the physicochemical characteristics of the specific NM. Relevant exposure scenarios should take into account the envisaged use of the NM, realistic dose levels that, for example, workers or consumers might be exposed to, its solubility, and whether the exposure is likely to be in aggregated or agglomerated states (Hristozov et al. 2013). For many NMs, inhalation is the most important route of occupational or consumer exposure, however, oral, dermal, and for specific (for example, medical) use scenarios, injection routes may also be of relevance (Oberdörster 2009; Landsiedel et al. 2012b). The physicochemical characteristics of a chemical may provide basic information not only on the exposure potential but also on the possible biokinetic fate of a chemical upon exposure and its potency to induce health effects (e.g., potential to pass specific barriers or its biochemical reactivity). Similarly, it may be possible to consider concerns with NMs or groups of NMs.

In characterizing material properties, not only the primary characterization provided by the supplier is essential but also equally important is their in situ characterization, for example, in NM stock and test substance dispersions. Test sample preparation by dispersion is a critical step determining the outcome of toxicity testing, and it must be appropriate to assess the risk under relevant real-life exposure situations (Schulze et al. 2008; Rothen-Rutishauser et al. 2010; Wang et al. 2010,

2011). Bovine serum albumin and fetal bovine serum have been suggested as dispersing agents for biological studies in mammalian tissue culture media (Thomas et al. 2011). However, dispersing agents containing lipids and lipophilic surfactant proteins reflecting lung lining fluids have also been recommended because they more closely mimic an in vivo-like situation in in vitro test systems (Sager et al. 2007; Gasser et al. 2012; Schleh et al. 2013). Over the past few years, multiple dispersion protocols have been used, and an extensive list of parameters and assays for in situ NM characterization have been proposed (Hassellov et al. 2008; Schulze et al. 2008; Card and Magnuson 2010; SCENIHR 2009; OECD 2010; Wohlleben 2012) and are discussed in Chapter 4. Further development of tools and exposure scenario libraries is required to estimate the physicochemical characteristics of a NM at various life stages, tools to identify and rank the most relevant concerns, and guidance for testing that takes the relevant physicochemical characteristics related to the concern into account.

Combining the outcome of the exposure assessment and the physicochemical characterization (material properties), taking into account all available information also from the material and other read-across approaches and physiologically based pharmacokinetic modeling, will allow determining NMs of "potential concern" (Figure 16.3, step 3). How this information should be combined needs further consideration. Many NMs will not raise specific concerns, but will, for example, readily dissolve. Other NMs are likely to possess properties raising concerns for toxic effects. These concerns may be general or restricted to specific uses and exposures (refinement of concerns, identification of "actual, relevant concerns"; Figure 16.3, step 4). For instance, NMs solely used in nonspray cosmetic sunscreen lotions may be of low concern for inhalation toxicity for consumers.

In this context, it is crucial to define which type of concern is being addressed in the course of the risk assessment: concerns can be *exposure-driven* (e.g., low exposure vs. high exposure), hazard-driven (e.g., based on known, potentially adverse effects), biokinetic-driven (e.g., biopersistent, transport across placenta, or blood–brain barrier), or *societal* (e.g., the acceptance or nonacceptance of a risk level). At an early stage, physicochemical characteristics might be considered to estimate exposure, biokinetic fate, and/or hazard potential of a NM. At some point, an open discussion is required to determine which level of certainty on the potential risks of a given NM in a specific application the society considers acceptable. Such deliberations, however, exceed the scope of the present essay.

The outcome of the four steps of identification of relevant concerns (Figure 16.3, corresponding to Tier 1 of the IATA presented in Figure 16.4) should be used to define the crucial human health end points to be tested in focused studies. It should further aid to determine the appropriate test design of these studies, including the relevant route of exposure, and all of the criteria used to identify concerns can also be considered for the grouping of NMs. If one or more concerns are identified, further data are necessary: In the case of a new NM, the next step implies generating additional data to perform Tier 2. In the case of an existing NM, appropriate tests might be chosen from Tiers 2 or 3 depending on the amount of information that is already available and depending on the information that is expected to be most effective and efficient for risk assessment.

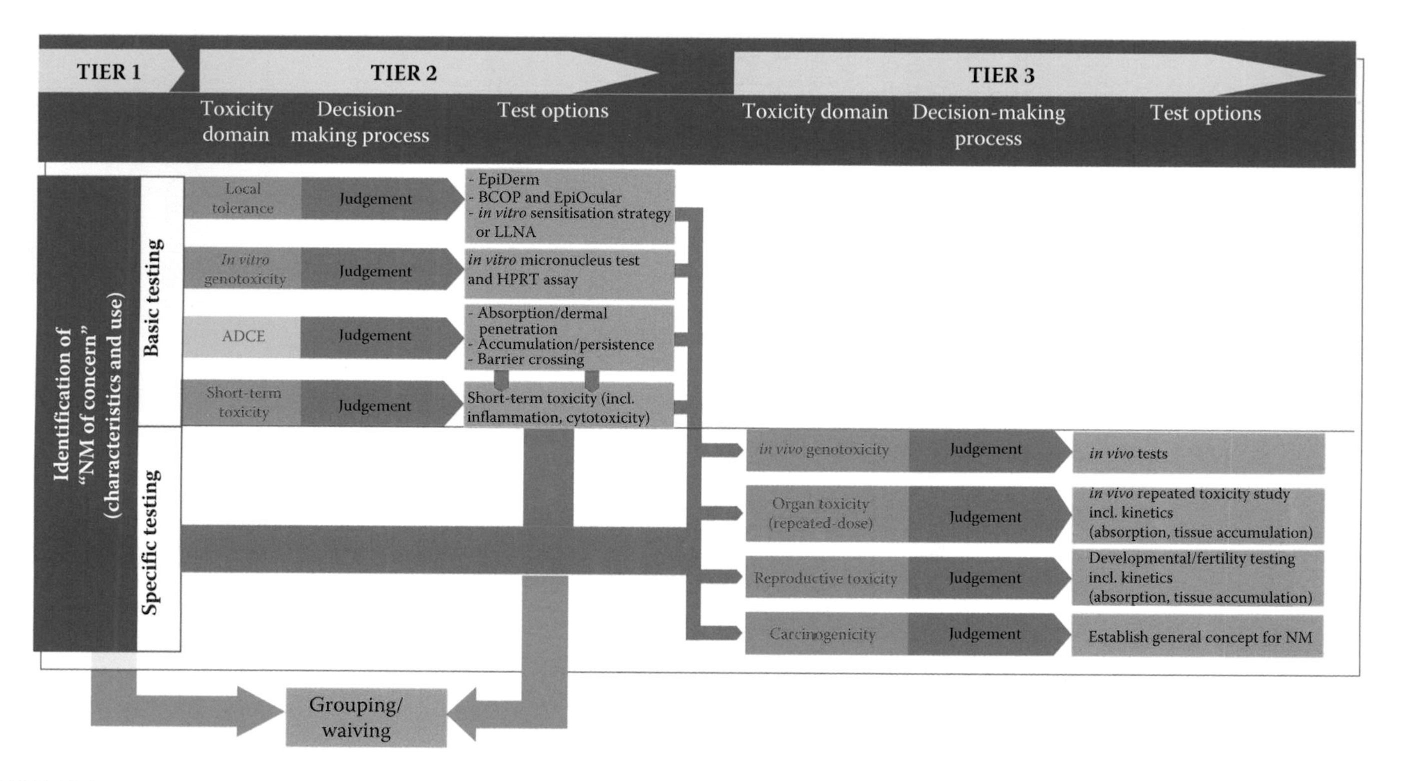

FIGURE 16.4 Integrated approach for the hazard testing and assessment of nanomaterials (NM) addressing specific concerns for an individual NM. (Reprinted with permission from Oomen AG et al. (2013), Concern-driven integrated approaches to nanomaterial testing and assessment, Report of the NanoSafety Cluster Working Group 10, *Nanotoxicol* 28 May 2013, epub ahead of print.)

16.4.2 Assessment of NM Human Health Hazards

Hazard assessment within Tier 2 of the integrated approach focuses on identifying the basic toxicological concerns related to a given NM. Tier 3 provides options to study specific end points of concern in more detail (Figure 16.4). Tiers 2 and 3 each consist of three parts, the "toxicity domain," the decision-making process, and the description of options for testing within each domain. Depending on the concerns determined for a given NM, testing can encompass basic and specific tests using noncellular and then cellular systems determining relevant MOA and toxicity pathways, and short-term and long-term in vivo testing approaches, if necessary.

At present, four toxicity domains are defined in Tier 2 for which information should become available (Figure 16.4), that is, local effects at the point of primary contact, genotoxicity, biokinetics, and short-term toxicity (including inflammation and cytotoxicity). Within each domain, a number of options for testing, grouping, or waiving will be available, from which the need for further testing, and its prioritization, can be determined. Confirmation of a concern will not necessarily lead to further testing, but may lead to the decision to stop the development of the given application or to apply specific risk management measures.

The process of decision making should facilitate an integrated combination of tests serving to obtain the maximum amount of information with the minimum resources (including animals) possible. After each tier, risk assessment is performed, that is, the data and information on hazard and exposure obtained so far are combined to characterize the risk (in a qualitative, semiquantitative, or quantitative way, in absolute or relative terms). The decision to proceed to branches of the subsequent tier(s) is based on risk considerations taking into account not only the output (i.e., acceptability or unacceptability of the risk) but also its associated uncertainty level. In each tier, the information of preceding tiers is used as guidance on how information should be obtained.

To ensure that meaningful test results are obtained when using cellular test systems, it is of paramount importance that the dosages applied in vitro encompass those applied in vivo taking into account the specific diffusion and sedimentation properties of NMs depending on their levels of dispersion and agglomeration (Teeguarden et al. 2007; Castranova et al. 2012; Sauer et al. 2013). Also the physicochemical characteristics representative of in vivo situations should be taken into consideration in in vitro tests.

16.4.3 Assessment of NM Biokinetics

Determining the biokinetic behavior of NM forms an essential part of Tier 2 of the testing strategy and might be the first domain to be addressed in this tier (Figure 16.4). This covers NM absorption/uptake, distribution, corona formation, and elimination/deposition ("ADCE," used by analogy to ADME, absorption, distribution, metabolism, and excretion): For the majority of insoluble NM, metabolism, unlike corona formation, appears to play a negligible role.

NMs may potentially cross portals of entry into the body or internal barriers. NMs with different physicochemical characteristics may differ considerably in the extent of their transportation across these barriers (in return providing indications for the choice of subsequent specific end points, such as reproduction toxicity, cardiotoxicity, or neurotoxicity). To date, little is known about the parameters that influence NM transportation (Oberdörster 2009; Landsiedel et al. 2012a). This information is, however, highly valuable for the identification of concerns and application of the grouping concept.

NMs generally tend to disappear rapidly from the blood by being taken up into tissues, mainly those containing phagocytic cells. Once filtered from the blood by macrophages, NMs are eliminated from these cells, and the body, only to a limited extent (Lankveld et al. 2010; Dan et al. 2012).

Intentional or unintentional modifications of the NM surface may have an impact on the translocation across barriers, on the tissue distribution of NMs, on the rate in which distribution occurs, and on their effects to the organism (Nel et al. 2009; Landsiedel et al. 2012a; Lundqvist et al. 2011; Liu et al. 2012). A "corona" of biological molecules will form around the NMs when they enter the body or even before, for example, when they are added to food products (Nel et al. 2009; Lundqvist et al. 2011; Johnston et al. 2012). This corona is most likely composed not only of proteins but also of lipids and carbohydrates, see Chapter 4.3 and Gasser et al. (2010). It is dynamic, because its types of molecules may change with time. A number of proteomics methods to identify the nature, composition, and dynamics of the biomolecules associated with NMs have been developed (Lai et al. 2012).

Slow elimination and persistence suggest potential bioaccumulation. Similar to conventional persistent, lipophilic chemicals that are resistant to biological degradation, NMs may have the potential to accumulate in humans (or the environment) when the frequency and magnitude of the exposure lead to an uptake exceeding the elimination in the body. Information on the potential of NMs to bioaccumulate is a criterion for higher-tier hazard and risk assessment that is more focused on effects at chronic exposure.

16.4.4 Specific Testing Based upon the Outcome of Tiers 1 and 2

The outcome of Tier 2 tests, combined with the information obtained in Tier 1, will allow identifying the need for additional information and appropriate toxicity domain(s) to be addressed in Tier 3 (Figure 16.4). So far, very limited data on long-term NM effects are available. Considering the relevance of the inhalational route of exposure, long-term inhalation studies should be performed with selected NMs, also allowing recognition of pathways by which NMs could cause lung tumors (e.g., inflammation, overload, and genotoxicity). As soon as such general principles and how these pathways can be indicated by simple tests are understood, less demanding tests can be applied in the toxicity testing strategies (e.g., appropriate in vitro assays and short-term inhalation tests). The data and tools applied at Tier 3 should allow a final conclusion on the acceptability of the risk, which infers the decision making on the risk management measures to be taken.

In summary, IATA aims at identifying and addressing concerns and thus reducing the uncertainty about potential risks by applying increasingly specific tools from tier to tier to collect or generate information. During the initial tier, as a rule, existing information is gathered and used to derive a preliminary assessment as a basis to identify relevant concerns. As necessary, the IATA proceeds through a number of tiers where additional information on exposure, kinetics, and hazard is considered to refine the assessment and derive a more comprehensive conclusion. Based on the information of the previous tiers, not all tests of a tier have to be performed but the tests targeted toward the relevant concerns can be selected. The assessment can stop at any given tier as soon as sufficient certainty on the conclusion to be drawn is obtained. Although advancing to higher tiers, the total amount of data and their complexity increase, the uncertainty is progressively reduced, and the assessment becomes more and more realistic and detailed. This difference between tiers is based not only on the amount and quality of the data used but also on the types of models and tools applied to come to a conclusion.

16.5 OUTLOOK

Comprehensive application of IATA will require less testing whenever the available information is sufficient and adequate for decision making. It will further allow obtaining more information from the selected tests, for example, on the particles' mechanism of action (Nel et al. 2013a) also by using high content screening approaches (Damoiseaux et al. 2011). Nevertheless, the decision making process might also imply requiring more information if additional specific concerns are recognized (e.g., immunotoxicity, cardiovascular toxicity, neurotoxicity, developmental toxicity, and biopersistence/bioaccumulation). Hence, IATA does not imply the generation of less or less comprehensive data but rather of more relevant information.

Future research to develop integrated approaches for testing and assessment of NMs should address the changing characteristics, biokinetics, and hazards of NMs throughout their lifecycle. In elaborating IATA that fully integrates exposure, material properties, ADCE, and primary and apical effect testing into a concern-driven risk assessment strategy aiming at an efficient and effective generation of data (including hazard testing and exposure assessment), relevant triggers for Tiers 2 and 3 of the strategy and criteria for the decision-making process should be established.

In addition, existing testing methods should be updated and amended for the specific needs of NM testing (e.g., including bronchoalveolar lavage and extended [pulmonary] histopathology in in vivo inhalation studies). Likewise, a list of NMs as performance standards for testing methods, ideally covering different toxic effects and including positive and negative controls, remains to be defined. Specific guidance is required on NM dispersion and in situ characterization, just as on methods to study NM biokinetics (ADCE) and biopersistence, and on the application of these data in the course of the IATA. General concepts for specific NM effects (carcinogenicity, cardiovascular effects, epigenetic effects, immunological effects, reproductive toxicity, and developmental toxicity) remain to be developed.

Finally, for grouping to be used as an integral part of integrated approaches for the testing and assessment of NMs, scientifically sound grouping criteria based on available data and material properties, ADCE, as well as primary and apical effects remain to be defined (Stone et al. 2010; Som et al. 2012). In this context, defining the AOP for different NMs is essential both for the elaboration of suitable in vitro methods as well as to promote the grouping of NMs. The same holds for in vitro and in silico tools that assess the changes in physicochemical characteristics and biokinetic profile. Suitable individual concepts for grouping should be integrated into comprehensive decision-tree guidance for grouping of NMs, including parameters for efficient acquisition of data supporting the grouping of NMs. Whereas such scientifically sound grouping criteria will undoubtedly form the "heart" of grouping and IATA, their definition will take time to develop and criteria will only evolve while applied. Guidance should be developed on what information is sufficient for a risk assessment of a group of NMs. Finally, grouping requires a set of information of at least one material, which serves as a reference for the whole group of NMs.

ACKNOWLEDGMENTS

This book chapter is based on the work performed at the EU projects NanoSafety Vision and MARINA. The work of the project members is greatly appreciated. The NanoSafety Vision report has been published (Oomen et al. 2013); Dr. Ursula Sauer has helped to assemble and edit the manuscript.

REFERENCES

Ankley GT, Bennett RS, Erickson RJ, Hoff DJ, Hornung MW, Johnson RD, Mount DR, Nichols JW, Russom CL, Schmieder PK, Serrano JA, Tietge JE, Villeneuve DL. (2010). Adverse outcome pathways: A conceptual framework to support ecotoxicology research and risk assessment. *Env Toxicol Chem* 29:730–741.

Anon. (2006). Regulation (EC) No 1907/2006 of the European Parliament and of the Council of 18 December 2006 concerning the Registration, Evaluation, Authorisation and Restriction of Chemicals (REACH), establishing a European Chemicals Agency, amending Directive 1999/45/EC and repealing Council Regulation (EEC) No 793/93 and Commission Regulation (EC) No 1488/94 as well as Council Directive 76/769/EEC and Commission Directives 91/155/EEC, 93/67/EEC, 93/105/EC and 2000/21/EC. O.J. L 396/1, 30 December 2006.

Anon. (2010). Directive 2010/63/EU of the European Parliament and of the Council of 22 September 2010 on the protection of animals used for scientific purposes. O.J. L 276/33, 20 October 2010.

Anon. (2011). Commission recommendation on the definition of nanomaterial. OJ L 275/38, 18 October 2011.

Card JW, Magnuson BA. (2010). A method to assess the quality of studies that examine the toxicity of engineered nanomaterials. *Int J Toxicol* 29:402–410.

Castranova V, Schulte PA, Zumwalde RD. (2012). Occupational Nanosafety Considerations for Carbon Nanotubes and Carbon Nanofibers. *AccChem Res* 5 December 2012, epub ahead of print.

Combes, R, Balls M. (2005). Intelligent testing strategies for chemicals testing–a case of more haste, less speed? *ATLA* 33:289–297.

Combes R, Barratt M, Balls M. (2003). An overall strategy for the testing of chemicals for human hazard and risk assessment under the EU REACH system. *ATLA* 31:7–19.

CTTAEA, NRC. (2007). Committee on Toxicity Testing and Assessment of Environmental Agents, US National Research Council. Toxicity Testing in the 21st Century: A Vision and a Strategy. National Academics Press, 216 pp. http://www.nap.edu/catalog.php?record_id=11970#toc.

Damoiseaux R, George S, Li M, Pokhrel S, Ji Z, France B, Xia T, Suarez E, Rallo R, Mädler L, Cohen Y, Hoek EMV, Nel AE. (2011). No time to lose—high throughput screening to assess nanomaterial safety. *Nanoscale* 3:1345–1360.

Dan M, Wu P, Grulke EA, Graham UM, Unrine JM, Yokel RA. (2012). Ceria engineered nanomaterial distribution in and clearance from blood: size matters. *Nanomed* 7:95–110.

ECHA. (2008). Guidance on information requirements and chemical safety assessment. Chapter R.6. QSARs and grouping of chemicals. European Chemicals Agency. 134 pp.; availableat:http://echa.europa.eu/documents/10162/13632/information_requirements_r6_en.pdf.

ECHA. (2010). Practical guide 4: How to report data waiving. European Chemicals Agency ECHA-10-B-09-EN, 24 March 2010. 19 pp.; available at: http://echa.europa.eu/documents/10162/13655/pg_report_data_waiving_en.pdf.

ECHA. (2012a). Guidance on information requirements and chemical safety assessment. Appendix R7-1 Recommendations for nanomaterials applicable to Chapter R7a Endpoint specific guidance. European Chemicals Agency ECHA-12-G-03-EN, April 2012. 60 pp.; available at:http://echa.europa.eu/documents/10162/13632/appendix_r7a_nanomaterials_en.pdf.

ECHA. (2012b). Background Paper. An introduction to the assessment of read across. European Chemicals Agency. 8pp.; available at:http://echa.europa.eu/documents/10162/5649897/ws_raa_20121003_background_paper_an_introduction_to_the_assessment_of_read-across_in_echa_en.pdf.

ECHA. (2012c). Assessment of read-across in REACH. European Chemicals Agency. 58pp.; available at:http://echa.europa.eu/documents/10162/5649897/ws_raa_20121003_assessment_of_read-across_in_echa_de_ raat_en.pdf.

ECHA. (2013). Grouping of substances and read-across approach. Part I. Introductory note. European Chemicals Agency ECHA-13-R-02-EN. 11 pp.; availableat:http://echa.europa.eu/documents/10162/13628/read_across_introductory_note_en.pdf.

Gasser M, Rothen-Rutishauser B, Krug HF, Gehr P, Nelle M, Yan B, Wick P. (2010). Lipids of the pulmonary surfactant and functional groups on multi-walled carbon nanotubes influence blood plasma proteins adsorption differently. *J Nanobiotech* 8:31.

Gasser M, Wick P, Clift MJ, Blank F, Diener L, Yan B, Gehr P, Krug HF, Rothen-Rutishauser B. (2012). Pulmonary surfactant coating of multi-walled carbon nanotubes (MWCNTs) influences their oxidative and pro-inflammatory potential in vitro. *Part Fibre Toxicol* 9:17.

George S, Xia T, Rallo R, Zhao Y, Ji Z, Lin S, Wang X, Zhang H, France B, Schoenfeld D, Damoiseaux R, Liu R, Lin S, Bradley KA, Cohen Y, Nel AE. (2011). Use of a high-throughput screening approach coupled with in vivo zebrafish embryo screening to develop hazard ranking for engineered nanomaterials. *ACS nano* 5:1805–1817.

German Competent Authorities, (2011). Nanomaterials and REACH. Background paper on the position of German competent authorities. 52 pages; availableat:http://www.reach-clp-helpdesk.de/de/Downloads/Hintergrundpapier%20Nano%20und%20REACH%20engl.%20Version.pdf?__blob = publicationFile.

Hankin SM, Peters SAK, Poland CA, Foss Hansen S, Holmqvist J, Ross BL, Varet J, Aitken RJ. (2011). Specific advice on fulfilling information requirements for nanomaterials under REACH (RIP-oN 2)–final project peport. REACH-NANO consultation. RNC/RIP-oN2/FPR/1/FINAL. 356 pp; available at: http://ec.europa.eu/environment/chemicals/nanotech/pdf/report_ripon2.pdf.

Hassellov M, Readman J, Rabville J, Tiede K. (2008). Nanoparticle analysis and characterization methodologies in environmental risk assessment of engineered nanoparticles. *Ecotoxicol* 17:344–361.

Hristozov DR, Gottardo S, Cinelli M, Isigonis P, Zabeo A, Critto A, Van Tongeren M, Tran L, Marcomini A. (2013). Application of a quantitative weight of evidence approach for ranking and prioritising occupational exposure scenarios for titanium dioxide and carbon nanomaterials. *Nanotoxicol* 15 Jan 2013, epub ahead of print. doi:10.3109/17435390.2012.760013.

IHCP2005. REACH and the need for intelligent testing strategies. Institute for Health and Consumer Protection, European Commission, Joint Research Centre EUR 21554 EN.

Johnston H, Brown D, Kermanizadeh A, Gubbins E, Stone V. (2012). Investigating the relationship between nanomaterial hazard and physicochemical properties: Informing the exploitation of nanomaterials within therapeutic and diagnostic applications. *J Controlled Release* 164:307–313.

Klein CL, Wiench K, Wiemann M, Ma-Hock L, van Ravenzwaay B, Landsiedel R. (2012). Hazard identification of inhaled nanomaterials: Making use of short-term inhalation studies. *Arch Toxicol* 86:1137–1151.

Kuempel ED, Castranova V, Geraci CL, Schulte PA. (2012). Development of risk-based nanomaterial groups for occupational exposure control. *J Nanopart Res* 14:1029.

Lai ZW, Yan Y, Caruso F, Nice EC. (2012). Emerging techniques in proteomics for probing nano-bio interactions. *ACS Nano* 6:10438–48.

Landsiedel R, Fabian E, Ma-Hock L, Wohlleben W, Wiench K, Oesch F, van Ravenzwaay B. (2012a). Toxico-/biokinetics of nanomaterials. *Arch Toxicol* 86(7):1021–1060.

Landsiedel R, Ma-Hock L, Haussmann HJ, van Ravenzwaay B, Kayser M, Wiench K. (2012b). Inhalation studies for the safety assessment of nanomaterials: Status quo and the way forward. *Wiley Interdiscip Rev Nanomed Nanobiotechnol* 4(4):399–413.

Landsiedel R, Ma-Hock L, Kroll A, Hahn D, Schnekenburger J, Wiench K, Wohlleben W. (2010). Testing metal-oxide nanomaterials for human safety. *Adv Mater* 22:2601–2627.

Lankveld DP, Oomen AG, Krystek P, Neigh A, Troost-de Jong A, Noorlander CW, Van Eijkeren JC, Geertsma RE, De Jong WH. (2010). The kinetics of the tissue distribution of silver nanoparticles of different sizes. *Biomat* 31:8350–8361.

Linsinger T, Roebben G, Gilliand D, Calzolai L, Rossi F, Gibson P, Klein CH. (2012). Requirements on measurements for the implementation of the European Commission definition of the term ‘nanomaterial’. JRC publication no JRC73260, ISBN 978-92-79-25603-5; available at: (http://publications.jrc.ec.europa.eu/repository/handle/111111111/26399).

Liu J, Legros S, Ma G, Veinot JG, von der Kammer F, Hofmann T. (2012). Influence of surface functionalization and particle size on the aggregation kinetics of engineered nanoparticles. *Chemosphere* 87:918–924.

Liu R, Rallo R, George S, Ji Z, Nair S, Nel AE, Cohen Y. (2011). Classification NanoSAR development for cytotoxicity of metal oxide nanoparticles. *Small* 7(8):1118–1126.

Lövestam G, Rauscher H, Roebben G, Sokull Klüttgen B, Gibson N, Putaud J-P, StammH. (2010). Considerations on a definition of nanomaterial for regulatory purposes. JRC Reference Reports. EUR 24403 EN.

Lundqvist M, Stigler J, Cedervall T, Berggård T, Flanagan MB, Lynch I, Elia G, Dawson K. (2011). The evolution of the protein corona around nanoparticles: A test study. *ACS Nano* 5:7503–7509.

Ma-Hock L, Burkhardt S, Strauss V, Gamer AO, Wiench K, van Ravenzwaay B, Landsiedel R. (2009). Development of a short-term inhalation test in the rat using nano titanium dioxide as a model substance. *Inhal Toxicol* 21:102–118.

Meng H, Xia T, George S, Nel AE. (2009). A predictive toxicological paradigm for the safety assessment of nanomaterials. *ACS nano* 3:1620–1627.

Nel AE, Mädler L, Velegol D, Xia T, Hoek EMV, Somasundaran P, Klaessig F, Castranova V, Thompson M. (2009). Understanding biophysicochemical interactions at the nano-bio interface. *Nat Mater* 8:543–557.

Nel AE, Nasser E, Godwin H, Avery D, Bahadori T, Bergeson L, Beryt E, Bonner JC, Boverhof D, Carter J, Castranova V, Deshazo JR, Hussain SM, Kane AB, Klaessig F, Kuempel E, Lafranconi M, Landsiedel R, Malloy T, Miller MB, Morris J, Moss K, Oberdorster G,

Pinkerton K, Pleus RC, Shatkin JA, Thomas R, Tolaymat T, Wang A. (2013a) A multi-stakeholder perspective on the use of alternative test strategies for nanomaterial safety assessment. *ACS Nano*: 7:6422–6433.

Nel AE, Xia T, Meng H, Wang X, Lin S, Ji Z, Zhang H. (2013b). Nanomaterial toxicity testing in the 21st century: Use of a predictive toxicological approach and high-throughput screening. *Acc Chem Res* 46:607–621.

Oberdörster G. (2009) Safety assessment for nanotechnology and nanomedicine: Concepts of nanotoxicology. *J Int Med* 267:89–105.

OECD. (2007). Guidance on grouping of chemicals. Series on testing and assessment No. 80. ENV/JM/MONO(2007)28. OECD, Paris, France, 26 September 2007.

OECD. (2010). Guidance manual for the testing of manufactured nanomaterials: OECD's sponsorship programme; first revision. ENV/JM-MONO(2009)20/REV, OECD, Paris, France, 2 June 2010.

OECD. (2011). A steering group 7 case-study for hazard identification of inhaled nanomaterials: An integrated approach with short-term inhalation studies. ENV/CHEM/NANO(2011)6/REV1. 10th Meeting of the Working Party on Manufactured Nanomaterials, OECD, Paris, France, 27–29 June 2012.

OECD. (2012a). Proposal for a template, and guidance on developing and assessing the completeness of adverse outcome pathways. 17 pp; available at: http://www.oecd.org/env/ehs/testing/49963554.pdf.

OECD. (2012b). Appendix I. Collection of working definitions. OECD, Paris, France; available at: http://www.oecd.org/env/ehs/testing/49963576.pdf.

OECD. (2013). Revised Draft Recommendation of the Council on the safety testing and assessment of manufactured nanomaterials. ENV/CHEM/NANO(2013)3/REV1 Working Party on Manufactured Nanomaterials. OECD, Paris, France, 25 February 2013.

Oomen AG, Bos PMJ, Fernandes TF, Hund-Rinke K, Boraschi D, Byrne HJ, Aschberger K, Gottardo S, van der Kammer F, Kühnel D, Hristozov D, Marcomini A, Migliore L, Scott-Fordsmand J, Wick P, Landsiedel R. (2013). Concern-driven integrated approaches to nanomaterial testing and assessment - Report of the NanoSafety Cluster Working Group 10. *Nanotoxicol* 28 May 2013, epub ahead of print.

Packroff R. (2013). The synergic dimensions of OSH strategy in the context of other EU policies: Nanomaterials ... and other advanced materials. Presentation on behalf of the German Federal Institute for Occupational Safety and Health. ETUI Conference: EU Strategy for health and safety and work 2013–2020, Brussels, 27 March 2013; available at: http://www.etui.org/Events/Conference-on-the-European-Health-and-Safety-Strategy-2013–2020.

Patlewicz G, Ball N, Booth ED, Hulzebos E, Zvinavashe E, Hennes C. (2013). Use of category approaches, read-across and (Q)SAR: General considerations. *Regul Toxicol Pharmacol* 67:1–12.

Rallo R, France B, Liu R, Nair S, George S, Damoiseaux R, Giralt F, Nel AE, Bradley K, Cohen Y. (2011). Self-organizing map analysis of toxicity-related cell signaling pathways for metal and metal oxide nanoparticles. *Env Science Technol* 45:1695–1702.

Rehn B, Rehn S, Bruch J. (1999). Ein neues in-vitro-Prüfkonzept (Vektorenmodell) zum biologischen Screening und Monitoring der Lungentoxizität von Stäuben. *Gefahrstoffe–Reinhaltung der Luft* 59:181–188.

Rothen-Rutishauser B, Brown DM, Piallier-Boyles M, Kinloch IA, Windle AH, Gehr P, Stone V. (2010). Relating the physicochemical characteristics and dispersion of multiwalled carbon nanotubes in different suspension media to their oxidative reactivity in vitro and inflammation in vivo. *Nanotoxicol* 4:331–342.

Russell WMS, Burch RL. (1959). The principles of humane experimental technique. London, UK. Methuen. Reprinted by UFAW, 1992, 238 pp., 8 Hamilton Close, South Mimms, Potters Bar, Herts EN6 3QD England.

Safe SH. (1998). Development validation and problems with the toxic equivalency factor approach for risk assessment of dioxins and related compounds. *J AnimSci* 76:134–141.

Sager TM, Porter DW, Robinson VA, Lindsley WG, Schwegler-Berry DE, Castranova V. (2007). An improved method to disperse nanoparticles for in vitro and in vivo investigation of toxicity. *Nanotoxicol* 1:118–129.

Sauer UG, Vogel S, Aumann A, Strauss V, Dammann M, Treumann S, Wohlleben W, Kolle SN, Gröters S, Wiench K, van Ravenzwaay B, Landsiedel R. (2013). Cytotoxicity, apoptosis, oxidative stress, and inflammation induced by 16 OECD reference nanomaterials in rat precision-cut lung slices. *Toxicol Appl Pharmacol* in press.

SCENIHR. (2009). The Scientific Committee on Emerging and Newly Identified Health Risks Opinion on Risk assessment of products of nanotechnologies. 19 January 2009; available at: http://ec.europa.eu/health/archive/ph_risk/committees/04_scenihr/docs/scenihr_o_023.pdf.

Schleh C, Kreyling WG, Lehr C-M. (2013). Pulmonary surfactant is indispensable in order to simulate the in vivo situation. *Part Fibre Toxicol* 10:6.

Schulze C, Kroll A, Lehr C, Schäfer UF, Becker K, Schnekenburger J, Schulze Isfort C, Landsiedel R, Wohlleben W. (2008). Not ready to use - Overcoming pitfalls when dispersing nanoparticles in physiological media. *Nanotoxicol* 2(2):51–61.

Shvedova AA, Tkach AV, Kisin ER, Khaliullin T, Stanley S, Gutkin DW, Star A, Chen Y, Shurin GV, Kagan VE, Shurin MR. (2013). Carbon nanotubes enhance metastatic growth of lung carcinoma via up-regulation of myeloid-derived suppressor cells. *Small* 9:1691–1695.

Som C, Nowack B, Krug HF, Wick P. (2012). Toward the development of decision supporting tools that can be used for safe production and use of nanomaterials. *Acc Chem Res* 46:863–872.

Spielmann H, Sauer UG, Mekenyan O. (2011). A critical evaluation of the 2011 ECHA reports on compliance with the REACH and CLP regulations and on the use of alternatives to testing on animals for compliance with the REACH regulation. *ATLA* 39, 481–493.

Stone V, Nowack B, Baun A, van den Brink N, von der Kammer F, Dusinska M, Handy R, Hankin S, Hassellöv M, Joner E, Fernandes TF. (2010). Nanomaterials for environmental studies: classification, reference material issues, and strategies for physico-chemical characterisation. *Sci Total Env* 408:1745–1754.

Teeguarden JG, Hinderliter PM, Orr G, Thrall BD, Pounds JG. (2007). Particokinetics in vitro: dosimetry considerations for in vitro nanoparticle toxicity assessments. *Toxicol Sci* 95:300–12. (erratum in: Toxicol Sci. 2007. 97:614).

Thomas CR, George S, Horst AM, Ji Z, Miller RJ, Peralta-Videa JR, Xia T, Pokhrel S, Mädler L, Gardea-Torresdey JL, Holden PA, Keller AA, Lenihan HS, Nel AE, Zink JI. (2011). Nanomaterials in the environment: From materials to high-throughput screening to organisms. *ACS nano* 5:13–20.

van Leeuwen CJ, Patlewicz GY, Worth AP. (2007). Intelligent Testing Strategies In: van Leeuwen CJ, Vermeire TG, Vermeire T, eds. Risk assessment of chemicals: An introduction, Springer, Heidelberg, Germany, 467–509.

Wang X, Xia T, Ntim SA, Ji Z, George S, Meng H, Zhang H, Castranova V, Mitra S, Nel AE. (2010). Quantitative techniques for assessing and controlling the dispersion and biological effects of multiwalled carbon nanotubes in mammalian tissue culture cells. *ACS nano* 4:7241–7252.

Wang X, Xia T, Ntim SA, Ji Z, Lin S, Meng H, Chung CH, George S, Zhang H, Wang M, Li N, Yang Y, Castranova V, Mitra S, Bonner JC, Nel AE. (2011). Dispersal state of multiwalled carbon nanotubes elicits profibrogenic cellular responses that correlate with fibrogenesis biomarkers and fibrosis in the murine lung. *ACS nano* 5:9772–9787.

Wohlleben W. (2012). Validity range of centrifuges for the regulation of nanomaterials: From classification to as-tested coronas. *J Nanopart Res* 14:1300.

Xia T, Kovochich M, Liong M, Zink JI, Nel AE. (2008). Cationic polystyrene nanosphere toxicity depends on cell-specific endocytic and mitochondrial injury pathways. *ACS Nano.* 2:85–96.

Xia T, Malasarn D, Lin S, Ji Z, Zhang H, Miller RJ, Keller AA, Nisbet RM, Harthorn BH, Godwin HA, Lenihan HS, Liu, R, Gardea-Torresdey J, Cohen Y, Mädler L, Holden PA, Zink JI, Nel AE. (2012). Implementation of a multidisciplinary approach to solve complex nano EHS problems by the UC Center for the Environmental Implications of Nanotechnology. *Small* 9:1428–1443.

Yanamala N, Kagan VE, Shvedova AA. (2013). Molecular modeling in structural nano-toxicology: Interactions of nano-particles with nano-machinery of cells. *Adv Drug Deliv Rev* 28 May 2013, epub ahead of print. doi: 10.106/j.addr.2013.05.005.

Zhang H, Ji Z, Xia T, Meng H, Low-Kam C, Liu R, Pokhrel S, Lin S, Wang X, Liao Y-P, Wang M, Li L, Rallo R, Damoiseaux R, Telesca D, Mädler L, Cohen Y, Zink JI, Nel AE. (2012). Use of metal oxide nanoparticle band gap to develop a predictive paradigm for oxidative stress and acute pulmonary inflammation. ACS nano 6(5):4349–4368.

Zuin S, Micheletti C, Critto A, Pojana G, Johnston H, Stone V, Tran L, Marcomini A. (2011). Weight of evidence approach for the relative hazard ranking of nanomaterials. *Nanotoxicol* 5:445–458.

17 Case Study
Paints and Lacquers with Silica Nanoparticles

Keld A. Jensen and Anne T. Saber

CONTENTS

17.1 INTRODUCTION

Several different manufactured nanomaterials (MN) are already widely used in commercial protective surface coatings. Even the types of MN used in paints and lacquers may be exhaustive and this sector may be one of the largest industrial users of MN by volume. However, it is often difficult to know the exact composition of paints and lacquers and which MN might be applied therein. Experimental paints and lacquers have been made containing MN of, for example, silica (Carneiro et al. 2012), titania (Marolt et al. 2011), nanoclays/organoclays (Kowalczyk and Spychaj 2009), CuO and Ag (Cioffi et al. 2005), and nanodiamond and nanoalumina (Sajjadi et al. 2013), as well as carbon nanofibers and multiwalled carbon nanotubes (Il'darkhanova et al.

2012). Kaiser et al. (2013) have recently discussed the possibilities of application of several of these MN in paints and lacquers.

Due to the nature of the paint and lacquer products, consisting of a mixture of passive and functional fillers in a liquid matrix, it is evident that they may be associated with great risk of exposure not only during production and downstream use but also during renovation and removal of coated surfaces. However, there is currently limited knowledge available on the specific hazards of the MN used in such surface coatings and the exposure risks at the different lifecycle stages. Moreover, surface modification of the MN is often necessary to enable good dispersion and binding to the product matrix. In other cases doping of the primary MN is done to achieve antimicrobial properties. Due to these surface modifications and doping, the potential hazards associated with paint MN may not be limited to the core material alone. Consequently, the MN producers as well as the paint and lacquer manufacturers using MN have a difficult task of ensuring the safety of workers, consumers, and the environment.

In this chapter, we give an overview of the typical silica nanomaterials used in experimental and produced paints and lacquers, emission characteristics of dusts generated by abrasion testing of such surface coatings, and their associated potential human hazard.

17.2 EXAMPLES OF SILICA NANOMATERIALS USED IN PAINTS AND LACQUERS

The use of silica MN may improve several different mechanical paint or lacquer properties, including increased strength and scratch resistance, hydrophobicity, high gloss, thermal stability/fire resistance, and resistance to ultraviolet (UV) light (Kaiser et al. 2013; Mizutani et al. 2006; Zhu et al. 2008). In other applications, silica nanoparticles are used or proposed as a carrier or doped composite nanomaterial to achieve well-controlled biocide release properties (Chmielewska et al. 2006; Edge et al. 2001; Kristensen et al. 2010; Lukasiewicz et al. 2003; Sørensen et al. 2010). In the following sections, we highlight some of the silica MN that have been used in experimental and commercial paints and lacquer products.

17.2.1 Examples of Nanosilica Used to Improve Mechanical Properties

Nanosilica (primary particle diameter 15–25 nm) was used to produce a superhydrophobic low-friction epoxy paint (Karmouch and Ross 2010). The optimum performance was observed at 2–3 wt% silica nanoparticles.

Another example of silica use is in a nanocomposite emulsion for wall paint proposed as a replacement for traditional solvent-type paints. In this case, 17 wt% of ca. 60 nm-size polyacrylate-coated silica (silica core diameter ca. 30 nm) was added to an acrylic paint replacing mainly the acrylic resin (Mizutani et al. 2006). The performance of this product was compared to the performance of a commercial acrylic paint and a conventional solvent-type paint. The pigment and filler content in the solvent-type paint was 30 wt%, whereas it was only 21 wt% in the commercial paint. In the nanocomposite paint, the total pigment and filler content was 38 wt%, including the 17 wt% nanosilica. Several advantageous phenomena including reduced total environmental pollution load, high water and flame resistance, and surface hardness were observed for the nanocomposite wall paint. The adhesion performance was, however, lowered, but could be managed by use of appropriate primers. The higher particle content and lower adhesion could be factors that possibly increase the risk of exposure during abrasion of such paints.

In a third example, a hybrid material consisting of polystyrene–butyl acrylate–acrylic acid-grafted silica was made from trimethoxysilane-functionalized fume silica with an average primary particle size of ca. 25 nm. It was used to form a nanosilica-doped latex nanocomposite for emulsion-type paints (Zhu et al. 2008). The addition of 1.5 wt% of the functionalized hybrid silica nanoparticles significantly improved the interfacial adhesion properties, UV shielding, water resistance, and thermoresistance of the latex nanocomposite.

From 1 to 7 wt% of ca. 40 nm-size *o*-phenylenediamine-coated silica was added to a nitrocellulose paint intended for corrosion-resistant coating on steel (Hegazy et al. 2013). The starting material of the silica nanocomposite was not disclosed and the functionalization was completed as part of the study and included further grinding of the silica. It was found that the protection efficiency of the paint increased significantly as well as its adhesion to the steel surface increased with an increase in content of SiO_2/*o*-phenylenediamine.

For application in both paints and lacquers, Bindzil® CC30 (Akzo Nobel Chemicals), NANOCRYL® XP 21/0768, and Axilat™ Ultrafine LS5000 (Hexion Specialty Chemicals B.V.) are examples of industrial products used in the Danish NanoKem study (Mikkelsen et al. 2013; Saber et al. 2012a). Bindzil CC30 is a 7 nm-size organo-modified colloidal nanosilica (Figure 17.1a) intended for improving the hardness, adhesion, abrasive resistance, drying time, and sanding properties of latex coatings and water-based lacquers. In the work of Saber et al. (2012c), 11.63 wt% of the Bindzil CC30 was introduced into an experimental acrylic paint produced by the Danish Coatings and Adhesives Association. NANOCRYL XP 21/0768 (HanseChemie; now EvonikHanse GmbH with trade name NANOCRYL A 210) is a product belonging to a series of nanoparticle-modified acrylate monomers. The NANOCRYL XP 21/0768 contains 50 wt% of 20 nm-size nanosilica particles dispersed in 50 wt% 1,6-hexanediol diacrylate and is used a.o. for formation of transparent tear and abrasion resistant coatings (epoxy paints, lacquers, gel coat formulations, and styrene-free vinyl ester resin systems) and stereolithography formulations (Evonik, 2008). In the NanoKem study, ca. 5 wt% of the products was applied in a nanosilica-enhanced UV-hard coat lacquer (Saber et al. 2012c). Axilat Ultrafine LS5000 is an acrylate-modified colloidal nanosilica product with a reported particle size smaller than 50 nm. Diluted in water, the hydrodynamic size was 6.5 nm as determined by dynamic light scattering. Axilat Ultrafine LS5000 is intended for latex-based coatings and water-based lacquers and improves impregnation of the substrate, adhesion, and translucence of primers and paints. In the work of Saber et al. (2012c), 28.5 wt% of the Axilat was used in a binder, also produced by the Danish Coatings and Adhesives Association. In all cases, the formulations were made for testing the potential hazards of paint- and lacquer-relevant nanomaterials and sanding dust from surface coatings.

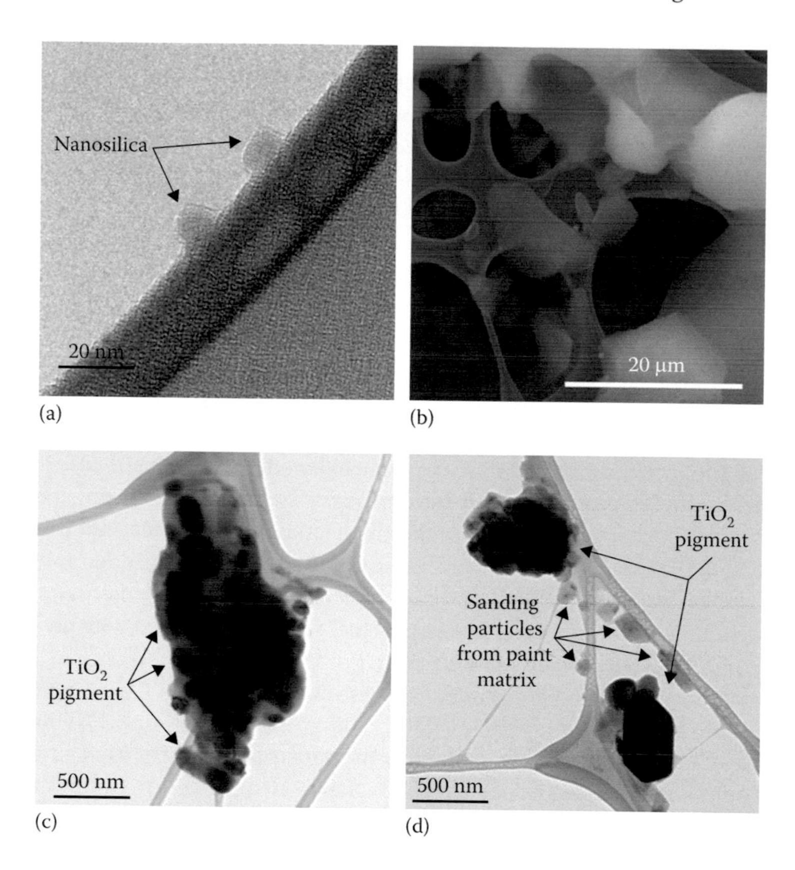

FIGURE 17.1 Electron microscopy images of nanosilica and different paint dusts. (a) Transmission electron microscopy (TEM) images of Bindzil® CC30 nanoparticles adhered to the edge of a hole in the carbon film of the TEM grid. The sample was prepared directly from dispersion in in vivo test medium. (b) Scanning electron microscopy image of sanding dust of UV-hard coat lacquer added NANOCRYL® XP 21/0768. The sanding dust has a wide size distribution ranging into the nanorange, but free nanosilica particles were not observed in the sanding dust. (c) TEM image of a paint dust particle from sanding a model reference indoor acryl paint. TiO_2 pigments are visible inside the paint dust particle. Some occur at the surface of the dust particle. (d) TEM image of free sub-µm-size TiO_2 pigment particles and nano- to fine-carbonaceous particles generated during sanding the reference indoor acryl paint.

17.2.2 Examples of Nanosilica Used to Achieve Biocidal Properties

Silica and nanosilica are not antimicrobial by themselves. However, nanocomposite materials have been developed where nanospheres or nanoporous silica are doped with antimicrobial or antifungal agents or even smaller nanoparticles thereof. The potential applicability of such colloidal silica as a carrier and controlled release of biocides such as isothiazolinone (Edge et al. 2001) as well as quaternary ammonium (benzalkonium) chloride and combined silver–quaternary ammonium chloride (Chmielewska et al. 2006; Lukasiewicz et al. 2003) have been demonstrated at least a decade back. In both cases, the used silica materials were not described in any detail, but they had nanomaterial characteristics such as high nanoporosity and high specific surface areas (Edge et al. 2001) or by being described as colloidal silica (Chmielewska et al. 2006). More detailed information on this type of nanosilica-based biocidal nanocomposites was found in recent references.

In the first example, silica nanospheres doped (modified) with either 3.5 wt% Ag or 4.5 wt% Cu nanoparticles were used in silicone acrylic emulsion paint intended for architectural paints and impregnates (Zielecka et al. 2011). Only 0.5 and 0.1 wt% of the Cu-doped silica nanoparticles (average dynamic light scattering zeta size = 132 nm) and 0.1 wt% of the Ag-doped silica nanoparticles (average dynamic light scattering zeta size = 40 nm) were added to the different test paints. In the Ag-doped nanosilica, the Ag is distributed as a homogeneous coating on the nanosilica surfaces. In the case of Cu-doped silica nanoparticles, small metallic Cu nanoparticles were well-dispersed on the nanosilica surfaces. Cu-doped nanosilica was observed to have greater antifungal capability than Ag-doped nanosilica, but both types had highly efficient biocidal behavior.

In the second example, mesoporous microsilica was produced with 3-ioprop-2-ynyl *N*-butylcarbamate biocide and used in an experimental water-based wood paint (Sørensen et al. 2010). The results from this study indicated that the mesoporous silica was able to prolong the release of the organic biocide. Efficient protection was achieved at 0.05 wt% 3-ioprop-2-ynyl *N*-butylcarbamate biocide in the paint corresponding to addition of up to ca. 0.08 wt% doped mesoporous silica. At the same time increased UV resistance was observed. Both observations indicate that addition of the 3-ioprop-2-ynyl *N*-butylcarbamate biocide-doped mesoporous silica would extend the in-service lifetime of the treated wood paint.

17.3 EXPOSURE TO NANOSILICA FROM MN PAINT INGREDIENTS AND SANDING OF PAINTS AND LACQUERS

The manufacturing of paints and lacquers usually includes handling powders, dispersions, and slurries, whereas application of the coatings may be done by, for example, brush, roller, dip coating, or spray. The direct risk of inhalation exposure can generally be considered to be reduced during handling of MN dispersed in liquids rather than powders. Dermal exposure, however, is expected to be the same as for conventional paint formulations and dispersed additives. Finishing and renovation may involve the use of polishing, sanding, and even high-pressure propellants (sand blasting) of which the latter two processes are known to have high dust emission potential consisting of paint/lacquer nanocomposite fragments and more or less liberated fillers (Daniels et al. 2001; Koponen et al. 2011; Saber et al. 2012c). Figure 17.1 shows examples of such airborne particles generated during sanding of various paint and lacquer products.

17.3.1 Exposure to Nanosilica in the Paint and Lacquer Manufacturing

We did not find any published measurements on the occupational exposure to silica MN in the paint and lacquer industry. One of our own studies submitted for publication investigates the airborne concentrations and exposure levels to dust at different paint mixers in two different paint manufacturing companies (Koponen and Jensen 2014). Silica (SIPERNAT® produced by Evonik Degussa GmbH) was an ingredient in one of the paints. The product was not specified as a MN, but online monitoring with a Fast Mobility Particle Sizer (TSI Inc.) showed that the dust was dominated by ca. 200 nm-size particles, which we often observe in dust from powder MN.

The near-field particle concentration around the worker increased by ca. 20,000 cm^{-3} during pouring and wet mixing seven 25 kg bags of the SIPERNAT product. The daily personal PM_1 (mass concentration of particles ≤1 μm aerodynamic diameter) exposure level of the worker was 0.65 mg/m^3. The specific contribution from the SIPERNAT to the mass concentration was not quantified, but from the online exposure data and electron microscopy, the exposure was dominated by dust from handling 280 kg talc and 2,675 kg pigment TiO_2 (Koponen and Jensen 2014).

17.3.2 Powder Handling and Dustiness

In addition to workplace measurements, assessment of the exposure potential can be made from powder dustiness data on silica powders. A dustiness standard, EN15051, has been established and contains two different methods: a rotating drum and a continuous drop dustiness tester. However, a range of other different methods for dustiness testing exist and have been developed recently for nanopowders (Evans et al. 2013; O'Shaughnessy et al. 2012; Tsai et al. 2012). Unfortunately, at this point in time, it is not possible to compare the quantitative levels of dust release between these different methods.

Six different silica MN powders have been studied using a miniaturized EN15051 rotating drum dustiness tester (Schneider et al. 2008). The results show a wide variation depending on the product. The least dusty nanosilica was the OECD WPMNM sample NM-202 (91 ± 11 mg/kg respirable dust) and the most NM-204 (2,058 mg/kg respirable dust) (Rasmussen et al. 2013). Following the dustiness categorization in the EN15051 standard, all the products had high dustiness levels.

17.3.3 Abrasion

A small number of groups are currently investigating the potential release of nanomaterials from surface coatings such as paints and lacquers during various abrasion processes (e.g., sanding and Taber abrasion). Only a few such studies have been carried out on surface coatings with silica nanoparticles. These were conducted in the Danish NanoKem study, which was conducted in collaboration with the Danish Coatings and Adhesives Association (Koponen et al. 2011). Two nanosilica types were investigated, namely silica (Bindzil CC30, code: Bindzil) and nanosilica-reinforced acrylate (NANOCRYL XP 21/0768, code: Nanocryl). A binder containing silica-acrylate (Axilat™ Ultrafine LS500, code: Axilat) was also available, but without reference binder and is not further discussed in this section. Other test materials included powders of a pigment size TiO_2 (RDI-S, code: FineTiO_2) and seven NM (UV-Titan L181, code: NanoTiO_2), photocatalytic TiO_2 (W2730X, code: PhotocatTiO_2), kaolinite (ASP-G90, code: Kaolin), and a carbon black (Flammruss 101, code: FineCB). The paint and lacquer matrices tested in NanoKem were: polyvinyl acetate paint (PVA), indoor (Indoor acryl), outdoor acrylic (Outdoor acryl) paint and UV hard coat lacquer (Lacquer), and a binder (Binder). Sanding was performed using a handheld electrically powered orbital sander (Metabo Model FSR 200 Intec) mounted with grit 240 sanding paper. The airborne dust was drawn from the sanded surface through the sander fan and collected at 100 m^3/h from the in-built filter chamber in the sander (Koponen et al. 2011). The dust was then passed through a 0.03 m^3 mixing chamber where it was monitored using a TSI Fast Mobility Particle Sizer (FMPS) Model 3091 and a TSI Aerodynamic Particle Sizer Model 3321 mounted with a TSI Diluter Model 3302A and finally collected in a modified commercial electrostatic precipitator. The sampled dust was used for subsequent characterization and toxicological studies (Saber et al. 2012b; Saber et al. 2012c). The work was performed in a HEPA-filtered exposure chamber and the mixing air was additionally HEPA filtered to reduce the risk of contamination from the chamber air during sampling.

It was found that all the airborne sanding dust was complex and had a wide size distribution, which could be described in five size modes (Figures 17.1c, 17.1d, and 17.2). The two finest size modes occurred at ca. 10 and 12–23 nm and were dominated by emissions from the electrically powered sander. The third size mode was 0.05 μm in lacquer and 0.13–0.18 μm for paints. The two coarser size modes were located at 0.9–1.2 μm and 1.6–2.0 μm, respectively, again with lacquer dust at the lower ranges. The three upper size modes were highly dominated by paint- and lacquer-generated sanding dust. It appeared that these three dust size modes were specific for each product and that they generally only changed to a minor degree after addition of nanofillers, whereas the number of particles in each size mode usually increased (Koponen et al. 2011).

In the specific experiments with the nanosilica-doped acrylic paint (ca. 10 wt% Bindzil CC30) and UV hard coat lacquer (ca. 5 wt% Nanocryl), the three sanding dust size modes were also practically unchanged. However, the addition of nanosilica appeared to have opposite effects on the number concentrations of the sanding dust. In the case of the acrylic paint, the number of mode 3 particles significantly increased after addition of Bindzil 30CC (Figure 17.2a). In the case of UV hard coat lacquer the number of particles significantly decreased in all three sanding dust size modes (3, 4, and 5) after addition of Nanocryl (Figure 17.2a). However, normalized to the total number concentrations in each of the sanding dust size spectra, a ca. 1.4 increase was observed for mode 3 particle concentration of the acryl paint dust and a major decrease in the coarser size modes 4 and 5, after addition of Bindzil 30CC (Figure 17.2b). As before, the relative changes in the UV hard coat dust are different with negligible to minor change in the normalized concentrations of mode 3 and mode 5 particles, and an almost doubling of size mode 4 particles (1.8 times) after addition of Nanocryl (Figure 17.2b). This demonstrates highly different consequences in the emission characteristics, which may be due to both the type of matrix and the type and amount of nanofiller added.

17.4 HUMAN SAFETY

The scientific literature on toxicological tests of silica nanoparticles intended for use in nanomaterial-based paints and lacquers appears to be rather scarce. Similarly, only a few studies have looked into the potential hazards of dusts generated by sanding paints and

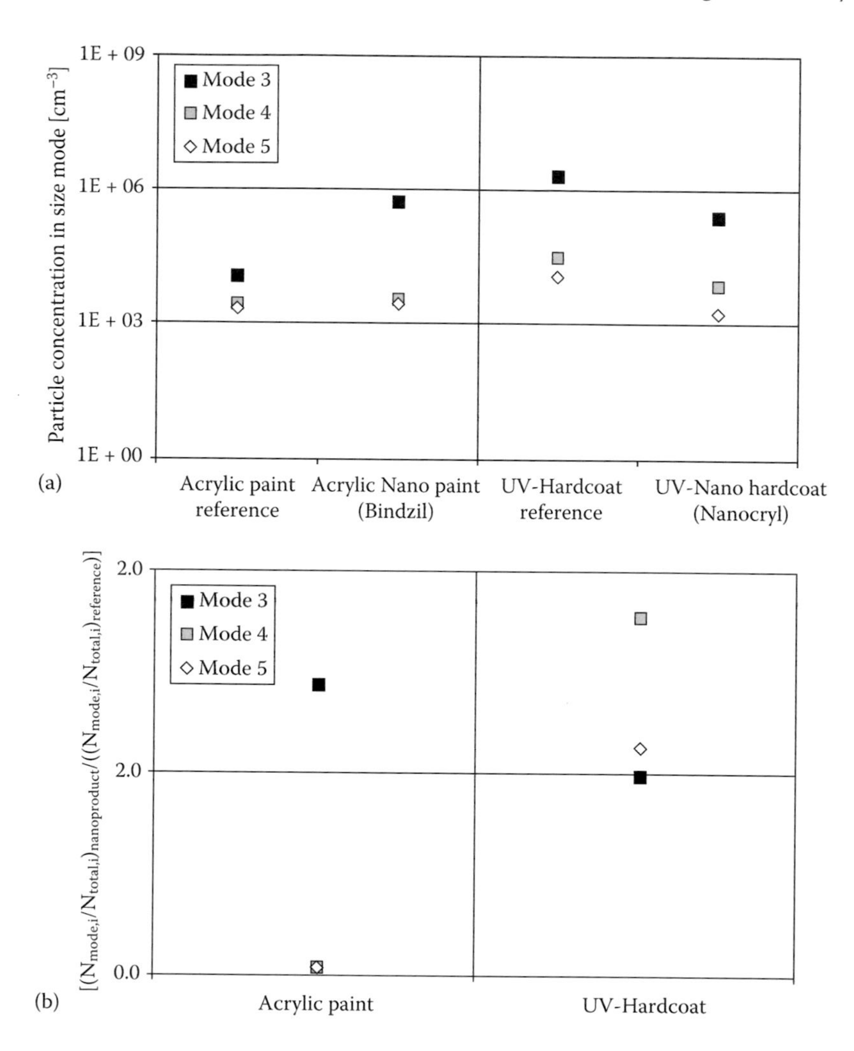

FIGURE 17.2 (a) Average number concentration for each size mode of sanding dust particles in a 0.03 m^3 size mixing chamber during sanding of acrylic paint and UV-hard coat lacquer with and without nanofillers (Bindzil® and NANOCRYL®). (b) Ratios between the normalized modal number concentrations of sanding dust particles in the nanofiller-doped paint/lacquer and the corresponding size mode in the reference paint/lacquer. Normalization was made using the total particle number concentration in the three size modes discussed for each size spectra. (Data obtained from Koponen, I. K. et al., *J. Phys. Conf. Ser.,* 151, 012048, 2009; and Koponen, I. K. et al., *J. Expo. Sci. Env. Epid.,* 21, 408–418, 2011.)

lacquers with added MN. Currently, synthetic amorphous silica registered as "Silicon dioxide" is listed with Ref.-No. 86240 and CAS-No.7631-86-9 as an additive without any restrictions and/or specification in EU directive 2000/72/EC. Pure synthetic silica nanoparticles are therefore thought by many to be a low toxic material. In line with this opinion silica MN are heavily investigated for their application in bioanalysis and drug delivery (Knopp et al. 2009; Mamaeva et al. 2013; Tang and Cheng 2013). However, the hazard of MN may be different from that of the comparable bulk material. In the following sections, we summarize some findings from in vivo studies on the toxicity of silica MN relevant for paints and lacquers as well as the few studies available on paint dust particles generated by sanding of paints and lacquers added silica MN.

17.4.1 In Vivo Toxicity of Manufactured Nanosilica

As listed earlier, the types of silica used in paints and lacquers comprise fumed and precipitated synthetic amorphous silica, colloidal silica, surface-treated derivates, and doped silica. Several in vivo toxicological studies of pure amorphous silica nanoparticles have been recently published (reviewed in Fruijtier-Poelloth 2012). Based on this review it can be concluded that silica has inflammatory properties but none of the reported studies detected genotoxic or carcinogenic effects in animal studies. To exemplify the current developments, a few recent publications on the toxicity of silica are described in greater detail as follows.

Brown et al. (2014) studied the effects 24 h after intratracheal instillation of neutral and positively (NH_2) charged 50 and 200 nm-size silica in Male Sprague Dawley rats (30 μg/rat). The silica was dispersed and dosed in different media, including saline, 0.2% BSA water, and 0.2% rat lung lining fluids. There was a statistically significant higher pulmonary neutrophil influx in rats instilled with 50 nm-size silica compared with 200 nm-size silica when dispersed in saline and BSA, but not when dispersed in lung lining fluids (see Table 17.2 later in this chapter).

Saber et al. (2012a) tested the toxicity of two specific paint and binder-relevant silica nanoparticles (Bindzil CC30, Akzo Nobel Chemicals and Axilat LS5000, and Hexion Specialty Chemicals B.V.). DNA damaging activity and inflammogenicity (pulmonary cell composition and mRNAs) were determined 24 h after intratracheal instillation of a single dose of 54 μg in mice (see Table 17.2 later in this chapter). Instillation of Axilat increased the pulmonary influx of neutrophils compared to control animals. Bindzil CC30 did not induce inflammation at the tested dose and none of the nanosilicas were DNA damaging tested by the comet assay 24 h after exposure.

As shown in Table 17.1, the MN often have more complex chemical composition than silica alone due to different surface treatments. In the studies of Saber and colleagues, the surface modifications did not appear to add toxicity to the nanosilica. Similar observations have been made as part of the German nanoGEM project from which first results are presented in Chapter 8 (Hahn et al. ibid). In this study, three pure nanosilica and three surface-modified silica denoted SiO_2.naked (SiO_2 Levasil®200 nanoparticles), SiO_2.PEG, SiO_2.amino, or SiO_2.phosphate, respectively, were tested in vivo for inflammation, genotoxicity, and adjuvant effects. Immediately after, as well as three weeks after, five consecutive days of 6 h exposure to 50 mg/m^3 electrosprayed nanosilica using a male Wistar rat inhalation model, significant inflammatory effects (neutrophil influx) were observed only for the SiO_2.naked. The allergy tests, however, showed significant adjuvant effects of both SiO_2.naked and SiO_2.PEG, but no significant effect of the SiO_2.amino or SiO_2.phosphate, after single dose intratracheal instillation of 50 μg in mice and subsequent OVA-albumin inhalation challenge. In other cases, however, surface modifications may increase the toxicity; for example,

TABLE 17.1
Examples of Silica Nanomaterials Used in Experimental Paint Formulations

Silica Type	Primary Particle Size	Optimal/Used Percent Nanofiller	Surface Modification	Paint/Lacquer Type	Achieved/Improved Properties	Reference
Fume silica	80 nm	6 wt%	Methacryloxypropyltrimethoxysilane	Acrylic paint	Dirt repellent, self-cleaning, UV-resistance	Carneiro et al. (2012)
Not reported	15–25 nm	2–3 wt%	NA	Epoxy paint	Superhydrophobicity, low friction	Karmouch and Ross (2010)
Ag-doped silica	40[b] nm	0.1 wt%	Ag-doped/silicone acrylic	Acrylic paint	Antimicrobial activity	Zielecka et al. (2011)
Cu-doped silica	132[b] nm	0.1 and 0.5 wt%	CuO-coated/silicone acrylic	Acrylic paint	Antimicrobial activity	Zielecka et al. (2011)
Colloidal silica	30 nm	17 wt%	Polyacrylate	Acrylic paint	Reduced pollution load, high water and flame resistance, and surface hardness	Mizutani et al. (2006)
Silica type not specified	40 nm	7 wt%	*o*-Phenylenediamine	Nitrocellulose	Adhesion, corrosion resistance	Hegazy et al. (2013)
Mesosilica	ca. [a]1,000 nm	0.08 wt%	IPBC-doped	Not reported	UV-resistance, antimicrobial activity	Sørensen et al. (2010)
Fume silica	25 nm	1.5 wt%	Polystyrene–butyl acrylate–acrylic acid grafted/trimethoxysilane	Latex emulsion	Interfacial adhesion, UV-resistance, water-resistance thermoresistance	Zhu et al. (2008)

Colloidal silica Bindzil® CC30 (Akzo Nobel Chemicals)	[c]7 nm	11.63 wt%	[c]Organo-modified	Acrylic outdoor paint, but especially made for latex coatings, water-based lacquers	[c]Hardness, abrasive resistance, friction, adhesion, drying time, sanding properties	Saber et al. (2012c)
NANOCRYL® A 210 (Evonik Hanse GmbH)	ca. 20 nm	>40 wt% (>20 wt% nanosilica) may be applicable, testing is required 5 wt% applied in present study	[c]Acrylate (1,6-hexanediol diacrylate)	UV-cured adhesives; epoxy paints, lacquers,	Transparency, tear and abrasion resistance	Saber et al. (2012c)
Axilat Ultrafine LS5000 (Hexion Specialty Chemicals B.V.)	[c]<50 nm	28.5 wt%	[c]Acrylate (1,6-hexanediol diacrylate)	[c]Latex coatings (water-based lacquers)	[c]Impregnation of material, adhesion, and transluscence of primers and paints	Saber et al. (2012c)

Abbreviation: IPBC: 3-iodoprop-2-ynyl N-butylcarbamate.

[a] Mesoporous silica.

[b] Hydrodynamic zeta-average diameter of nanoporous aggregate.

[c] Data from supplier.

by the release of toxicants from the surface. In a recent study, it was found that silica nanoparticles doped with cadmium expectedly caused more oxidative stress than uncoated silica in rats after intratracheal instillation (Coccini et al. 2012). This shows that increased hazard can occur when silica is coated with a more toxic material.

The aforementioned review and selected studies on relevant silica nanoparticles show that silica MN in some cases may be inflammogenic in animal models for human toxicity and that doping of silica MN with, for example, toxic metals may cause increased toxicity compared to pure silica MN. Other surface modifications used for improving dispersibility may, but not always, reduce the toxicity of silica NM when compared to the unmodified analogues. As a new end point, it may be necessary to investigate the potential adjuvant effects of exposure to silica MN. No DNA damage was observed or reported in any of the studies.

17.4.2 In Vitro Toxicity of Nanosilica

The recent review (Fruijtier-Poelloth 2012) concludes that there is also no evidence for mutagenicity of silica nanoparticles in vitro. Genotoxicity was observed in some of the reviewed in vitro studies, but this most often occurred at doses where cytotoxicity was also observed. As an example in a recent in vitro test of nanosilica, four synthetic amorphous silica samples (NM-200, NM-201, NM-202, and NM-203) from the OECD Working Party on Manufactured Nanomaterials (WPMNM) also did not show any genotoxicity using the human lymphocyte micronucleus assay associated with any of these industrial grade MN. No effects were observed even after exposure to 1.25 mg/ml (Tavares et al. 2014).

Even though limited hazardous effects have been observed for silica MN, recent studies including new toxicological end points suggest that both traditional and doped silica MN may disturb the biochemical function on the intracellular level. Christen and Fent (Christen and Fent 2012) showed that fumed silica nanoparticles (Sigma Aldrich) and Ag-doped silica (1 and 5 wt% Ag) (produced by flame spray pyrolysis) affected human liver cells (Huh7) by affecting the CYP1A enzyme activity as well as perturbation of the endoplasmic reticulum (ER), which lead to stress response. By number, 50% of all three silica particles were smaller than 100 nm and mainly occurred in aggregates. ER stress affects the cellular protein homeostasis. It is worth noting that the effects were observed for both silica and Ag-doped silica. The level of toxicity increased systematically as function increased Ag-doping and was to some degree related to the amount of dissolved Ag. The authors state that the ER stress response appears to be a new mechanism of MN toxicity. The observed effect remains to be demonstrated in cells relevant to the respiratory tract and in vivo.

17.4.3 Toxicity of Sanding Dust Particles from Paints and Lacquers

In contrast to the presence of several studies on the biological effects of pure silica nanoparticles, data on the toxicity of silica included in a paint or binder matrix are currently only found for three nanosilica materials (Table 17.2). These were all included in the Danish NanoKem study previously mentioned. The toxicological tests were completed on the Axilat™ Ultrafine LS5000 (Axilat) and Bindzil

TABLE 17.2
Examples of Toxicity of Silica Manufactured Nanomaterials in Experimental Animals

Silica Type	Size	Surface Modification	Dose	Test System	Main Effect	Reference
Silica nanoparticles	50 nm	–	30 μg/animal	Intratracheal instillation of particles in different dispersion media; male Sprague Dawley rats	50 nm particles resulted in increased neutrophil influx when particles were dispersed in BSA and saline. >200 nm particles gave no response	Brown et al. (2014)
	50 nm	NH_2				
	200 nm	–				
	200 nm	NH_2				
Colloidal silica Bindzil® CC30 (Akzo Nobel Chemicals)	[a]7 nm; 6.5 nm diluted 1:9 in Nanopure water	Organo-modified	54 μg/mouse (2.8 mg/kg)	Intratracheal instillation; female C57BL/6	No inflammation and DNA damage 24 h after instillation	Saber et al. (2012a)
Axilat Ultrafine LS5000 (Hexion Specialty Chemicals B.V.)	[a]<50 nm; 5.6 nm diluted 1:9 in Nanopure water	Acrylate (1,6-hexanediol diacrylate)	54 μg/mouse (2.8 mg/kg)	Intratracheal instillation; female C57BL/6	Increased pulmonary influx of neutrophils and inflammation but no DNA damage 24 h after instillation	(Saber et al. 2012a)
Pure HiSilTM T700 nanosilica (SiNP) (Degussa GmbH) and Cd-coated SiNP (SiNP-Cd)	Original SiNPs: 20 nm pore size	32.5 wt% Cd	SiNPs: (600 μg/rat)	Intratracheal instillation; male Sprague–Dawley rats	Oxidative lung tissue injury and inflammatory effects at 30 days post- exposure as observed from F2-IsoPs SOD1, iNOS, COX-2, granuloma, and stromal fibrogenic reaction	Coccin et al. (2012)
	Cd-treated SiNPs: 20–80 nm primary particle size (200 m^2/g)		SiNPs-Cd: (1 mg/rat, corresponding to about 250 μg Cd/rat)		Pure silica SiNPs did not influence any of the evaluated end points	

[a] Data according to the supplier.

CC30 (Akzo Nobel Chemicals) (Bindzil). Because it was not possible to provide NANOCRYL XP 21/0768 as a pure material, it was only tested as part of the lacquer matrix. As described previously, Axilat silica induced inflammatory responses in mice 24 h after intratracheal instillation of 54 μg/mouse, whereas no inflammation was observed for Bindzil in the same test setup (Saber et al. 2012a). No toxicity was detected when Axilat was encapsulated in a binder matrix. Only sanding dust particles from the lacquers with and without Bindzil as well as the outdoor acrylic-based reference paint resulted in statistically significant increased level of DNA damage in the bronchoalveolar lavage cells when compared to the vehicle. This indicates that the acute toxicity of total emitted sanding dust of these paints and lacquer types is controlled by the paint matrix rather than the added fillers and nanomaterials, at least when low toxic nanofillers are used.

This hypothesis of matrix effects was further supported by an in vivo study specifically investigating this question using the inflammogenic and DNA-damaging nanofiller (UV Titan L181) (Saber et al. 2012a) and sanding dusts from the corresponding paints with 10 wt% UV-Titan L181 (Indoor-NanoTiO_2) and a reference paint (Indoor-R) (Saber et al. 2012b). The instilled doses were adjusted so that the doses of the UV Titan L181 nanopaint sanding dust (54, 162, and 486 μg/mouse) would contain approximately the same amount of UV Titan L181 as the pure MN reference test (18, 54, and 162 μg/mouse). As shown in Figure 17.3, the results showed no additive effect when UV-Titan L181 was added to the paint when compared to the reference paint for any of the measured toxicological endpoints at 1, 3, and 28 days after instillation (Saber et al. 2012b). However, for paints with more less-durable

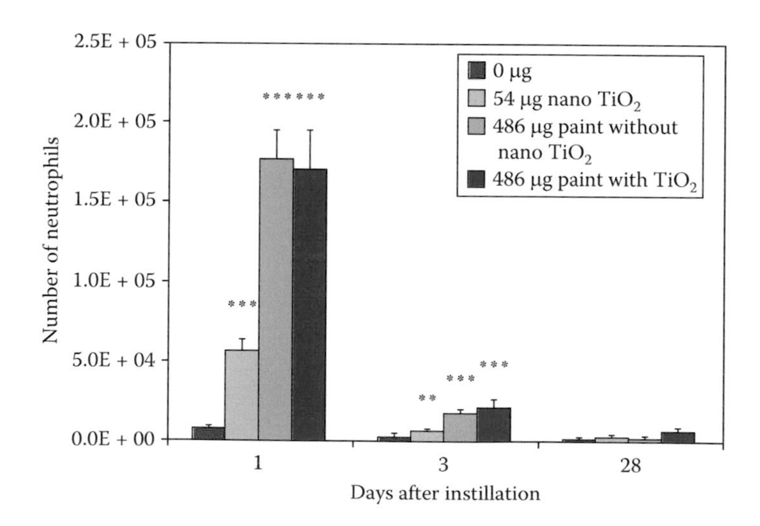

FIGURE 17.3 Neutrophil influx in the lungs of mice after exposure to a low and high dose of UV-Titan L181 (nanoTiO_2) as well as sanding dusts from an acryl reference paint and acryl paint with added nanoTiO_2. The symbols *, **, and *** illustrate that the results are statistically significantly different when compared to effect levels in control mice at the 0.5, 0.01, and 0.001 level, respectively. (Data originates from Saber, A.T. et al., *Nanotoxicology*, 6, 776–788, 2012c.)

matrices (mechanically and/or water-soluble) and/or more toxic fillers with higher hydrous solubility, this observation may not be true.

The complete selection of NanoKem nanoparticles and sanding dusts were also tested in vitro in primary human umbilical vein endothelial cells. The results were in agreement with the in vivo results showing that sanding dusts were not more toxic than sanding dusts from conventional paints (Mikkelsen et al. 2013).

17.4.4 Lessons from Toxicological Studies on Abrasion Particles from Other Nanocomposites

To the best of our knowledge the aforementioned publications are the only ones testing the toxic effects of sanding dusts from nanoparticle-doped paints and lacquers. However, two other publications focusing on the toxic effects of other kinds of nanocomposites have been published. Wohlleben and coworkers have published two studies within this research area: (1) a study in rats testing toxic properties of sanding dusts from cement and plastic with and without carbon nanotubes (Wohlleben et al. 2011) and (2) an in vitro study using Precision Cut Lung Slices of sanding dust from thermoplastic polyurethane with and without carbon nanotubes (Wohlleben et al. 2013). The conclusions based on these studies are in agreement with the results from the NanoKem study: No additional toxicity was observed for the nanoparticle-containing products compared to the corresponding products without nanoparticles.

17.5 OUTLOOK

To summarize, a very limited number of published studies on the toxicological effects of sanding dusts from nanocomposites, including the NanoKem study specifically testing silica in paint, lacquer, and binder, are in good agreement with each other: No additional toxicity has been detected for any of the nanocomposites compared to the corresponding composites without MN. Noteworthy, more results on the emission characteristics and hazards of various process-generated dust emissions from nanocomposites and reference materials, including paint dusts from a recently completed FP7 project NanoSustain (www.nanosustain.eu), are currently under preparation for publication. In this project, we investigated the effects of adding nano-TiO_2, CNTs, nano-ZnO, and nanocellulose to a range of products, including paints, epoxy, paper, and a glass treatment product. As these MN and products have a wider range of potential hazard and physicochemical matrix properties (e.g., hydrous solubility and matrix integrity), the results from NanoSustain may add additional valuable understanding of the emission characteristics and potential hazards associated with process-generated emissions from a wider range of nanocomposites.

For future work, it should be considered that all current studies have analyzed and tested the total airborne dust generated during processing (e.g., sanding, grinding, and cutting). It is observed that the composition of the dust varies with particle size. Coarser particles are mainly paint fragments, whereas the smaller particles increasingly consist of smaller paint aggregate fragments, fillers, and nanomaterials (Saber et al. 2012c). It would be of high interest to investigate in greater detail the size-fractioned dust from mechanical treatment processes. Additionally, it is necessary to complement the previous studies with studies on emissions/exposures observed during use of engineered and local exhaust ventilation control and to conduct toxicological study on these dust materials as well. This will be increasingly important when adding more complex MN, such as the doped biocidal silica reported in Table 17.1, to the products. Potentially, the effects on mass-dose basis of finer size particles would be different from that of the total generated dust due to the inhibitory effect of the matrix reported so far. In addition, products containing MN doped with toxic compounds should be tested in greater detail for longer exposure times to ensure that sudden breakdown of the matrix will not cause a delayed hazardous impact.

ACKNOWLEDGMENTS

We gratefully acknowledge the funding to the Danish Centre for Nanosafety' (grant #20110092173/3) from the Danish Working Environment Research Foundation, which funded this case study. We also thank the Danish Working Environment Research Foundation and the European Community's Seventh Framework Programme (FP7/2007-2013) for funding the NanoKem project (grant #20060068816) and the NanoSustain (grant agreement no. 247989), respectively, which enabled us to generate the key data discussed in this publication.

REFERENCES

Brown, D. M., N. Kanase, B. Gaiser, H. Johnston, and V. Stone. (2014) Inflammation and gene expression in the rat lung after instillation of silica nanoparticles: Effect of size, dispersion medium and particle surface charge. *Toxicol Lett* 224 (1): 147–156.

Carneiro, C., R. Vieira, A. M. Mendes, and F. D. Magalhes. (2012). Nanocomposite acrylic paint with self-cleaning action. *J Coat Technol Res* 9 (6): 687–693.

Chmielewska, D. K., A. Lukasiewicz, J. Michalik, and B. Sartowska. (2006). Silica materials with biocidal activity. *Nukleonika* 51 (Supplement 1): S69–S72.

Christen, V. and K. Fent. 2012. Silica nanoparticles and silver-doped silica nanoparticles induce endoplasmatic reticulum stress response and alter cytochrome P4501A activity. *Chemosphere* 87 (4): 423–434.

Cioffi, N., N. Ditaranto, L. Torsi, R. A. Picca, E. De Giglio, L. Sabbatini, L. Novello, G. Tantillo, T. Bleve-Zacheo, and P. G. Zambonin. (2005). Synthesis, analytical characterization and bioactivity of Ag and Cu nanoparticles embedded in poly-vinyl-methyl-ketone films. *Anal Bioanal Chem* 382 (8): 1912–1918.

Coccini, T., E. Roda, S. Barni, C. Signorini, and L. Manzo. (2012). Long-lasting oxidative pulmonary insult in rat after intratracheal instillation of silica nanoparticles doped with cadmium. *Toxicology* 302 (2–3): 203–211.

Daniels, A. E., J. R. Kominsky, and P. J. Clark. (2001). Evaluation of two lead-based paint removal and waste stabilization technology combinations on typical exterior surfaces. *J Hazard Mater* 87 (1–3): 117–126.

Edge, M., N. S. Allen, D. Turner, J. Robinson, and K. Seal. (2001). The enhanced performance of biocidal additives in paints and coatings. *Prog Org Coat* 43 (1–3): 10–17.

Evans, D. E., L. A. Turkevich, C. T. Roettgers, G. J. Deye, and P. A. Baron. (2013). Dustiness of fine and nanoscale powders. *Ann Occup Hyg* 57 (2): 261–277.

Evonik (2008). Adhesive applications. product portfolio, http://hanse.evonik.com/sites/hanse/Documents/hanse-adhesives-en-web.pdf, date accessed 04.09.2014.

Fruijtier-Poelloth, C. (2012). The toxicological mode of action and the safety of synthetic amorphous silica-A nanostructured material. *Toxicology* 294 (2–3): 61–79.

Hegazy, M. A., M. M. Hefny, A. M. Badawi, and M. Y. Ahmed. (2013). Nanosilicon dioxide/o-phenylenediamine hybrid composite as a modifier for steel paints. *Prog Org Coat* 76 (5): 827–834.

Il'darkhanova, F. I., G. A. Mironova, K. G. Bogoslovsky, V. V. Men'shikov, and E. D. Bykov. (2012). Development of paint coatings with superhydrophobic properties. *Prot Met Phys Chem Surf* 48 (7): 796–802.

Kaiser, J. P., S. Zuin, and P. Wick. (2013). Is nanotechnology revolutionizing the paint and lacquer industry? A critical opinion. *Sci Total Environ* 442 282–289.

Karmouch, R. and G. G. Ross. (2010). Superhydrophobic wind turbine blade surfaces obtained by a simple deposition of silica nanoparticles embedded in epoxy. *Appl Surf Sci* 257 (3): 665–669.

Knopp, D., D. P. Tang, and R. Niessner. (2009). Bioanalytical applications of biomolecule-functionalized nanometer-sized doped silica particles. *Anal Chim Acta* 647 (1): 14–30.

Koponen, I. K. and K. A. Jensen. (2014). Worker exposure and high time-resolution analyses of nanoparticle and pigment dust release at different mixing stations in two paint factories. *submitted.*

Koponen, I. K., K. A. Jensen, and T. Schneider. (2009). Sanding dust from nanoparticle-containing paints: Physical characterisation. *J Phys Conf Ser* 151: 012048.

Koponen, I. K., K. A. Jensen, and T. Schneider. (2011). Comparison of dust released from sanding conventional and nanoparticle-doped wall and wood coatings. *J Expo Sci Env Epid* 21 (4): 408–418.

Kowalczyk, K. and T. Spychaj. (2009). Protective epoxy dispersion coating materials modified a posteriori with organophilized montmorillonites. *Surf Coat Tech* 204 (5): 635–641.

Kristensen, J. B., R. L. Meyer, C. H. Poulsen, K. M. Kragh, F. Besenbacher, and B. S. Laursen. (2010). Biomimetic silica encapsulation of enzymes for replacement of biocides in antifouling coatings. *Green Chem* 12 (3): 387–394.

Lukasiewicz, A., D. K. Chmielewska, L. Walis, and L. Rowinska. (2003). New silica materials with biocidal active surface. *Pol J Chem Tech* 5 (4): 20–22.

Mamaeva, V., C. Sahlgren, and M. Linden. (2013). Mesoporous silica nanoparticles in medicine-Recent advances. *Adv Drug Deliver Rev* 65 (5): 689–702.

Marolt, T., A. S. Skapin, J. Bernard, P. Zivec, and M. Gaberscek. (2011). Photocatalytic activity of anatase-containing facade coatings. *Surf Coat Tech* 206 (6): 1355–1361.

Mikkelsen, L., K. A. Jensen, I. K. Koponen, A. T. Saber, H. Wallin, S. Loft, U. Vogel, and P. Moller. (2013). Cytotoxicity, oxidative stress and expression of adhesion molecules in human umbilical vein endothelial cells exposed to dust from paints with or without nanoparticles. *Nanotoxicology* 7 (2): 117–134.

Mizutani, T., K. Arai, M. Miyamoto, and Y. Kimura. (2006). Application of silica-containing nano-composite emulsion to wall paint: A new environmentally safe paint of high performance. *Prog Org Coat* 55 (3): 276–283.

O'Shaughnessy, P. T., M. Kang, and D. Ellickson. (2012). A novel device for measuring respirable dustiness using low-mass powder samples. *J Occup Environ Hyg* 9 (3): 129–139.

Rasmussen, K., Mech. A., J. Mast, P.-J. De Temmermann, N. V. S. F. Waeggeners, J. C. Pizzolon, L. De Temmermann, E. Van Doren, K. A. Jensen, R. Birkedal, M. Levin, S. H. Nielsen, I. K. Koponen, P. A. Clausen, Y. Kembouche, N. Thieret, O. Spalla, C. Giuot, D. Rousset, O. Witschger, S. Bau, B. Bianchi, B. Shivachev, D. Gilliland, F. Pianella, G. Ceccono, G. Cotogno, H. Rauscher, N. Gibson, and H. Stamm. (2013). Synthetic amorphous silicon dioxide (NM-200, NM-201, NM202, NM-203, NM-204): Characterisation and physico-chemical Properties. EUR 26046 EN 1-201. (Luxembourgh, European Union).

Saber, A. T., N. R. Jacobsen, A. Mortensen, J. Szarek, P. Jackson, A. M. Madsen, K. A. Jensen, I. K. Koponen, G. Brunborg, K. B. Gutzkow, U. Vogel, and H. Wallin. (2012b). Nanotitanium dioxide toxicity in mouse lung is reduced in sanding dust from paint. *Part Fibre Toxicol* 9: 4.

Saber, A. T., K. A. Jensen, N. R. Jacobsen, R. Birkedal, L. Mikkelsen, P. Møller, S. Loft, H. Wallin, and U. Vogel. (2012a). Inflammatory and genotoxic effects of nanoparticles designed for inclusion in paints and lacquers. *Nanotoxicology* 6 (5): 453–471.

Saber, A. T., I. K. Koponen, K. A. Jensen, N. R. Jacobsen, L. Mikkelsen, P. Moller, S. Loft, U. Vogel, and H. Wallin. (2012c). Inflammatory and genotoxic effects of sanding dust generated from nanoparticle-containing paints and lacquers. *Nanotoxicology* 6 (7): 776–788.

Sajjadi, S., M. Avazkonandeh-Gharavol, S. Zebarjad, M. Mohammadtaheri, M. Abbasi, and K. Mossaddegh. (2013). A comparative study on the effect of type of reinforcement on the scratch behavior of a polyacrylic-based nanocomposite coating. *J Coat Technol Res* 10 (2): 255–261.

Schneider, T. and K. A. Jensen. (2008). Combined single-drop and rotating drum dustiness test of fine to nanosize powders using a small drum. *Ann Occup Hyg* 52 (1): 23–34.

Sørensen, G., A. L. Nielsen, M. M. l. Pedersen, S. Poulsen, H. Nissen, M. Poulsen, and S. D. Nygaard. (2010). Controlled release of biocide from silica microparticles in wood paint. *Prog Org Coat* 68 (4): 299–306.

Tang, L. and J. Cheng. (2013). Nonporous silica nanoparticles for nanomedicine application. *Nano Today* 8 (3): 290–312.

Tavares, A. M., H. Louro, S. Antunes, S. Quarré, S. Simar, P. J. De Temmerman, E. Verleysen, J. Mast, K. A. Jensen, H. Norppa, F. Nesslany, and M. J. Silva. (2014). Genotoxicity evaluation of nanosized titanium dioxide, synthetic amorphous silica and multi-walled carbon nanotubes in human lymphocytes. *Toxicol* in vitro 28 (1): 60–69.

Tsai, C. J., G. Y. Lin, C. N. Liu, C. E. He, and C. W. Chen. (2012). Characteristic of nanoparticles generated from different nano-powders by using different dispersion methods. *J Nanopart Res* 14 (4): 777.

Wohlleben, W., S. Brill, M. W. Meier, M. Mertler, G. Cox, S. Hirth, B. von Vacano, V. Strauss, S. Treumann, K. Wiench, L. Ma-Hock, and R. Landsiedel. (2011). On the life cycle of nanocomposites: Comparing released fragments and their in-vivo hazards from three release mechanisms and four nanocomposites. *Small* 7 (16): 2384–2395.

Wohlleben, W., M. W. Meier, S. Vogel, R. Landsiedel, G. Cox, S. Hirth, and Z. Tomovic. (2013). Elastic CNT-polyurethane nanocomposite: Synthesis, performance and assessment of fragments released during use. *Nanoscale* 5 (1): 369–380.

Zhu, A., A. Cai, Z. Yu, and W. Zhou. (2008). Film characterization of poly(styrene-butylacrylate-acrylic acid)–silica nanocomposite. *J Colloid Interf Sci* 322 (1): 51–58.

Zielecka, M., E. Bujnowska, B. Kepska, M. Wenda, and M. Piotrowska. (2011). Antimicrobial additives for architectural paints and impregnates. *Prog Org Coat* 72 (1–2): 193–201.

18 Case Study
The Lifecycle of Conductive Plastics Based on Carbon Nanotubes

Richard Canady and Thomas A.J. Kuhlbusch

CONTENTS

18.1 INTRODUCTION

Conductive plastics using nanometer-scale fillers such as carbon nanotubes (CNTs), graphene, and carbon black are being used as a light-weight replacement for metals in many applications, as well as in novel materials that extend or create functionality in many others (De Volder et al. 2013; Future Markets 2012; Han and Fina 2011; Kingston et al. 2014; Ma and Zhang 2014; Sahoo et al. 2010; Sun et al. 2013). For example, conductive plastics are being used in antistatic plastic films and trays to provide low-cost, secure packaging for computer chips and electronic components, spark-resistant fuel lines in automotive and aerospace applications, lightning strike protection coatings for aerospace, and electromagnetic interference shielding. The possibilities and benefits are tremendous and therefore identification of safe development pathways is critical for these uses.

For these reasons, leading experts shared data, knowledge, and participated in reviews of materials used in commerce, lifecycle release scenarios, methods to measure nanomaterial release, and studies of the released materials for CNT–polymer composites as part of the multistakeholder and international NanoRelease project (Canady et al. 2013; Kingston et al. 2014; NanoRelease Consumer Products 2014; Nowack et al. 2013). Projects such as NanoRelease and nanoGEM worked on methods to measure the release of nanomaterials and characterize released nanomaterials so that a reliable understanding of exposures can be built. CNT–polymers were chosen as a case study for method requirements in the NanoRelease project. The approaches chosen in these projects give insight into the relative likelihood of human health risk for nanoscale particles by an "understand exposure first" approach to the risk management that is necessitated by the tremendous variation in possible nanomaterial characteristics. Toxicity studies of the "released" nanomaterials are needed to allow full assessment of the potential risk; however, in the absence of those studies, we are forced to rely on studies of "pristine" materials, but only after we have a better understanding of the materials to which actual exposures occur.

18.2 RELEASE OF CNTs DURING THE LIFECYCLE

As discussed by the NanoRelease Project Steering Committee (NanoRelease Consumer Products 2014) and reviewed by Nowack et al. (2013), it is important to segregate the question of "Are uses of a nanomaterial safe?" into materials, uses, and stages of the life of the material (lifecycle) so that statements of risk can be properly focused on particular populations and possible risk management points. In so doing, it is clear that the potential for release of CNTs will vary from essentially zero for parts of the lifecycle in which contact does not occur with potential exposure media, to measureable levels under conditions in which disaggregation of the matrix holding the CNT may occur. The releases will depend on the material, use, and environmental conditions for the particular points in the lifecycle. Nowack et al. (2013) illustrate this point and identify release and exposure points of greatest interest across anticipated uses for CNTs in a polymer matrix. No specific lifecycle descriptions for conductive plastics have been published to our knowledge so the approach provides an approximation that aids in bounding discussion of likelihoods of release and exposure. Nowack et al. considered scenarios with respect to the following:

- Mechanisms of release (e.g., abrasion, weathering by ultra violet (UV) radiation or water)
- Form of the release (e.g., free CNTs or CNT embedded in particles)
- Magnitude of release
- Estimates of frequency/duration of release
- Properties of the composite affecting release (e.g., UV stabilizers or polymer type)
- Site of release (e.g., to air following abrasion/cutting in a manufacturing facility, dermal/air exposure for general population/consumer, workplace exposures to release from products, and air/water for environmental and workplace exposures during disposal)
- Environmental conditions affecting release (e.g., UV intensity, humidity, or abrasion conditions)
- Populations potentially exposed (e.g., for manufacturing: workers; for product life/usage: workers, the general public [if the product is in a public building/space], or specific consumers [who buy the product])

The general structure of the stages considered is represented in Figure 18.1.

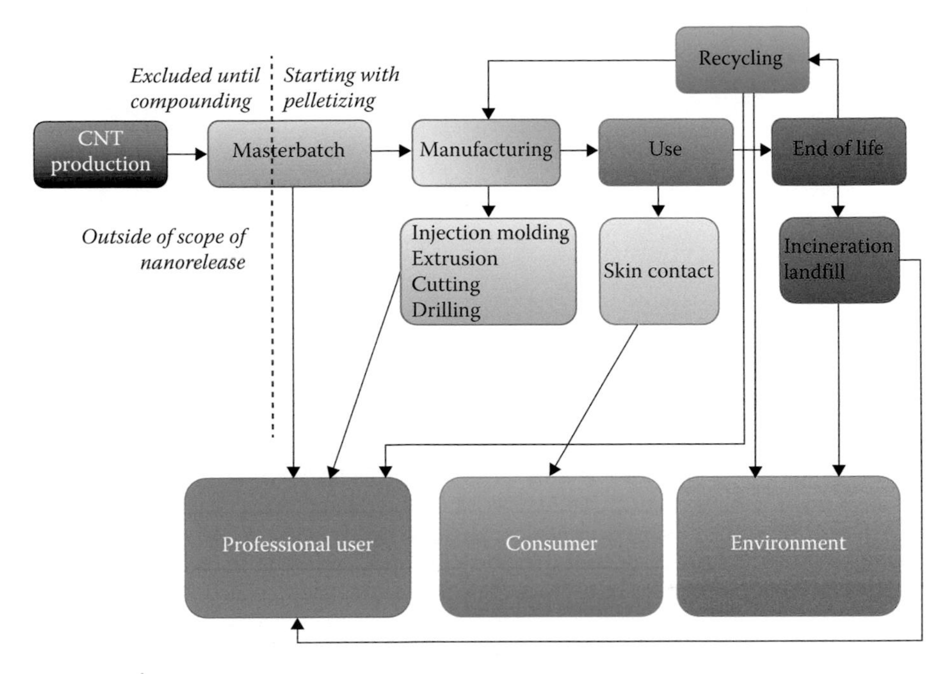

FIGURE 18.1 Lifecycle map for carbon nanotube composites used in electronics. The thickness of the lines corresponds to the likelihood and magnitude of release based on the general release scenario with all possible lifecycle steps and stressors. (Adapted from *Environ. Int.*, Nowack, B. et al., Potential release scenarios for carbon nanotubes used in composites, 1–11, Copyright 2013, with permission from Elsevier.)

For conductive polymer applications such as consumer electronics and automotive fuel system components, releases to consumers are essentially zero simply because of lack of access to the materials in normal use and lack of off-gassing for the materials (e.g., in comparison to volatile organic chemical constituents of the materials). However, professional users (automobile mechanics, electronics repair technicians) will have a higher contact with the materials and possibly under conditions of abrasion, and thus exposure to released materials is possible. The nature of the released materials is dependent on physical–chemical factors of the CNT, the matrix material, as well as the actual mechanisms involved in the release process.

For example, Figures 18.2 and 18.3 illustrate outcomes for mechanical stress processes such as abrasion affecting the attachment of CNTs to the polymer. Initially, the following steps may occur: the CNT is fixed in the matrix (a) and shear forces (b) may loosen at one end from the matrix, rupture the fiber into two attached strands (c), "pull-out" inner sheets of the CNT (d), or elongate connecting across cracks in the matrix (e). A fifth option of course is the release of free CNT fibers, resulting from simultaneous release of the CNT from all matrix surfaces. This release of free CNTs seems less likely in cases in which the matrix is not completely degraded due to the need for simultaneous disengagement along the entire length of the fiber. In fact, observation of released fragments independently across laboratories rarely shows

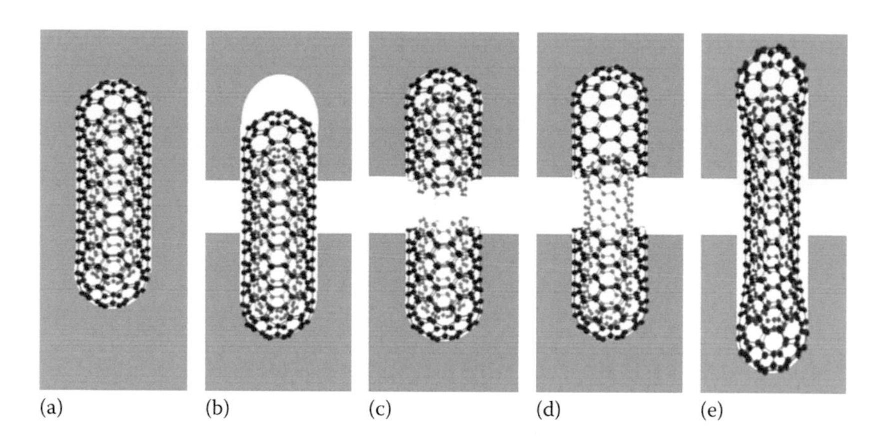

FIGURE 18.2 (a) Initial state of carbon nanotubes (CNTs) in matrix. (b) Pull-out resulting from weak interfacial adhesion. (c) Rupture of CNTs resulting from strong interfacial adhesion plus extensive/fast local deformation. (d) Telescopic pull-out due to stronger interfacial bonding than van der Waals between the tube layers. (e) Crack bridging and partial debonding of the interface. (Reprinted from *Compos. Sci. Technol.*, Gojny, F. et al., Influence of different carbon nanotubes on the mechanical properties of epoxy matrix composites: A comparative study, 2300–2313, Copyright 2005, with permission from Elsevier.)

free CNT fiber (Table 18.1). This finding may suggest less of a "fiber" toxicology basis for polymer–CNT composites under abrasion scenarios. Furthermore, adding to the complexity of predicting release, the bonding between the CNT and the matrix may change significantly if surface modifications such as –OH or –NO groups are added to the CNT (Kingston et al. 2014).

Stressors in addition to mechanical forces can also be thermal or chemical reactions influencing the matrix material as well as the CNT. Thermal stress in particular may occur along with mechanical stress during sanding as mentioned by Kuhlbusch and Kaminski (2014). Friction forces lead to heating of the sanded material as is sometimes seen during wood sanding and the resulting smoke. The effect of heating by mechanical forces on the likeliness of release has not been studied in detail thus far.

Examples of further release processes during the lifecycle of conductive plastics are the accidental combustion, incineration, and shredding at the end of the lifecycle as well as weathering if left in dumps or in the environment. Table 18.1 presents a broad overview on release scenarios and study results.

Three kinds of information are essential to correctly address the potential risk of added CNTs in polymer composite for a particular release scenario: (1) quantitative characterization of particles released, including minimum detection levels, relevant to the added CNT; (2a) transformations between release and exposure, allowing quantitative characterization of the CNT-related particles at exposure points, or (2b) measurement of CNT-related particles at the point of exposure; and (3) dose response for the CNT-related particles reaching individuals at the point of exposure. The release rate and the transformation and transport in the environment determine the exposure, whereas the type of particles (e.g., CNTs still embedded in the matrix

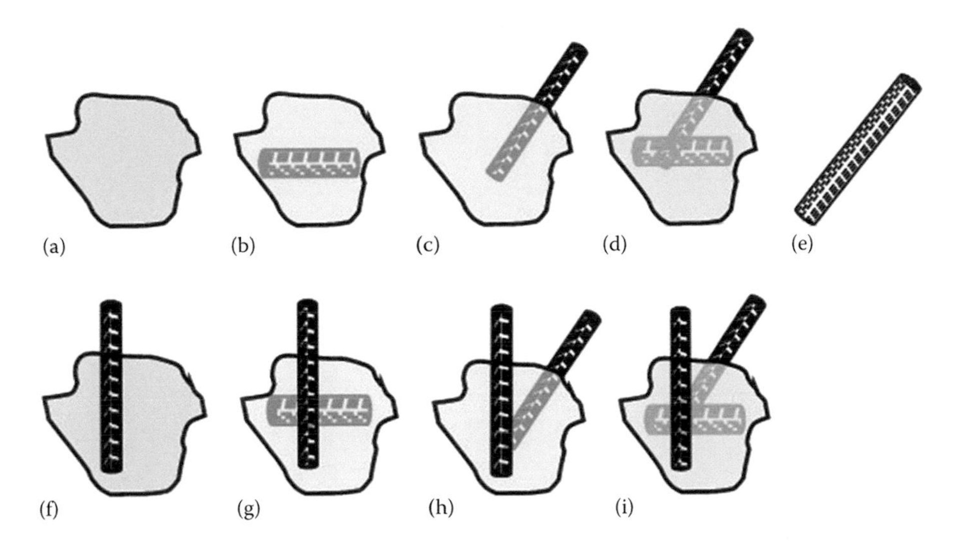

FIGURE 18.3 Various forms of material from an abrasive release event: polymer fragments containing the following: (a) no multiwalled carbon nanotubes (MWCNTs); (b) and (c) MWCNTs bonded to the polymer matrix fully encased (b) or protruding from the polymer surface (c); (e) idealized unbound MWCNTs; (f) MWCNTs loosely adhered to the composite surface; (d), (g), (h), and (i): combinations of (b), (c), and (f). For simplicity, MWCNTs are portrayed as straight tubes, rather than as curved, twisted, and intertwined tubes typically observed in actual MWCNT–polymer composites. (Adapted from Kaiser, D. et al. (2014), Methods for the Measurement of Release of MWCNTs from MWCNT-Polymer Composites, A Report of Task Group 1 of the NanoRelease Consumer Products Project, Submitted for Steering Committee use in project monograph, personal communication.)

material Figure 18.3) and environmental transformation determine the potential hazard. As mentioned for abrasion, it is relatively rare (although not quantified) to observe release of free CNTs at least in initial release processes (Table 18.1).

18.3 ENVIRONMENTAL BEHAVIOR OF CNTs

The possibility of environmental persistence of CNTs coupled with matrix degradation may, however, result in as yet unrecognized exposure pathways to free CNTs. Very little information is available on the behavior of CNTs in the environment due to the difficulties in detecting these materials in natural environments. The methods most commonly used in CNT studies include radioactive labeling (Simon et al. 2010), detection of the catalysts used for CNT production (Maynard et al. 2004), or thermal analysis (Lee et al. 2010), next to the commonly used electron microscopy. CNTs are believed to be stable in the environment but some studies with enzymes show a significant degradability (Zhao et al. 2011). Before this enzymatic degradation may occur, release of fibers from the matrix has to occur. Although this release is possibly rare during the many initial release processes, it may take place by selective degradation of the matrix. This leads ultimately to the release of the CNT fibers.

TABLE 18.1
Areas for Detection and Identification before Selective Quantification Can Begin

Release Scenario (Anticipated Degradation Mechanism)	Method to Simulate the Lifecycle to Induce, Detect, and Quantify Release	Morphology of Typical Released Fragments		Key Findings and Research Needs
Outdoor use (UV degradation, hydrolysis and weak mechanical forces)	Accelerated dry or wet weathering (ISO 4892; International Organization for Standardization 2011), followed by immersion to induce release into water, detected by AUC, TEM, EDX, LD, XPS	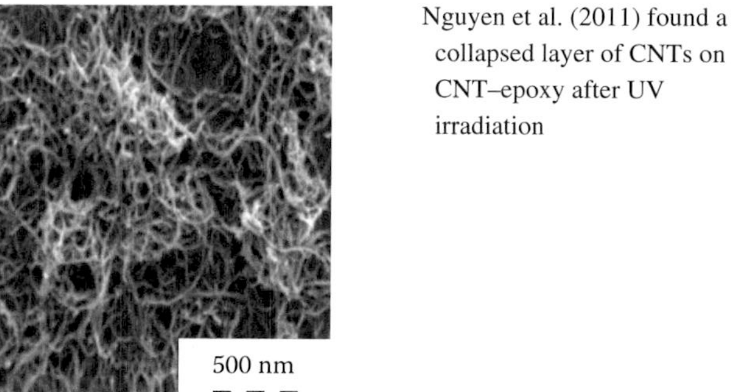	Nguyen et al. (2011) found a collapsed layer of CNTs on CNT–epoxy after UV irradiation	Weathering tests are highly standardized for plastics and coatings, but the CNT network formed on the composite essentially eliminates the release of CNT (Nguyen et al. 2011) Methods to induce a release during or after degradation are exploratory, and miss selective quantification
		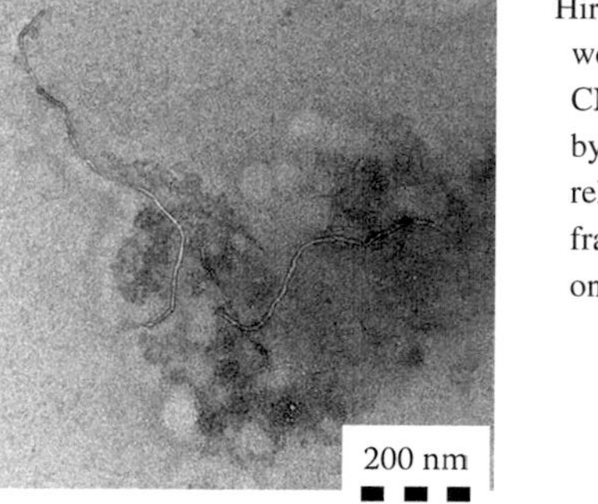 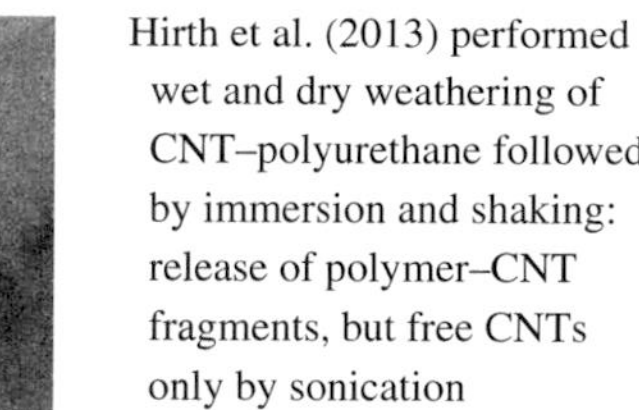	Hirth et al. (2013) performed wet and dry weathering of CNT–polyurethane followed by immersion and shaking: release of polymer–CNT fragments, but free CNTs only by sonication	

Machining by sanding, band-saw, abrasion, rotary cutting wheel, wet saw cutting, wet drilling, grinding. (moderate temperature + moderate to strong mechanical stresses: energy input per test by sanding estimated is 10-fold higher than by sawing) (LeBihan et al. 2013)	No standard simulation machines: monitoring by APS, CPC, SMPS, ESP, FMPS, LDPSA, NSAM, OPC, and sampling for characterization by SEM/TEM, EDX, AUC, LD, XPS	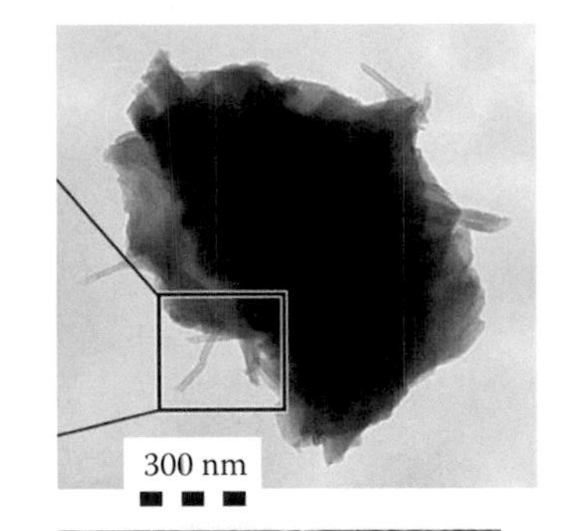	Fragments from CNT–epoxy were submicrons in diameter and carried protrusions of CNTs that were pulled out from the matrix (Cena and Peters 2011)	Many laboratory studies on different CNT–polymer and nanoparticle–polymer composites found no release of free nanofillers (Cena and Peters 2011; Fleury et al. 2013; Göhler et al. 2010, 2013; Hirth et al. 2013; Koponen et al. 2010; Wohlleben et al. 2011, 2013) with exceptions linked to agglomerates in the polymer (Golanski et al. 2012; Huang et al. 2012)
			Fragments from CNT–polyoxymethylene measured microns in diameter with no protrusions (Wohlleben et al. 2011)	Protrusions of CNTs were not observed for tough thermoplastics (Wohlleben et al. 2011, 2013) which can elongate during machining and keep CNTs embedded (Hirth et al. 2013; Schlagenhauf et al. 2012)
Machining by dry drilling or milling (moderate mechanical stress at elevated temperatures)	No standard simulation machines. Detect aerosols around machines by APS, CPC, Dust TrakTM, ESP, FMPS, TP, TSI		Polymorph aerosol sampled during dry drilling of a graphite-epoxy pre-preg with aligned carbon fibers and aligned CNTs (Bello et al. 2010)	By synergy of degradation and stress, machining with local overheating is a key release scenario

(*Continued*)

TABLE 18.1 (*Continued*)
Areas for Detection and Identification before Selective Quantification Can Begin

Release Scenario (Anticipated Degradation Mechanism)	Method to Simulate the Lifecycle to Induce, Detect, and Quantify Release	Morphology of Typical Released Fragments		Key Findings and Research Needs
Compounding by hot melt extrusion and Combustion (thermal degradation)	Commercial twin-screw extruders, monitored by CPC, SMPS, sampled for TEM/SEM, or TGA (ISO 11358) with off-gases monitored and sampled	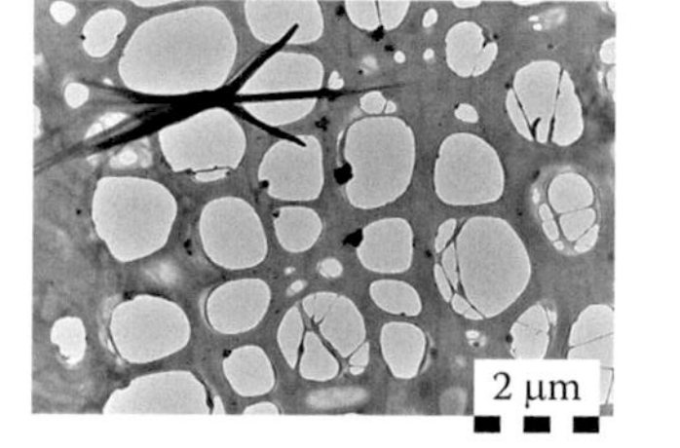	Fleury et al. (2013) conducted extrusion of CNT–acrylonitrile–butadiene–styrene: complex polymer fumes, no free CNT	At elevated temperatures, polymer vaporizes and re-condenses. Release of free CNTs was observed for thermal decomposition by TGA (Bouillard et al. 2013) and for dry-core drilling of CNT–epoxy composites (Bello et al. 2010) but not for extrusion of CNT–polymer (Fleury et al. 2013) Capacity to simulate real-world scenarios needs to be validated
		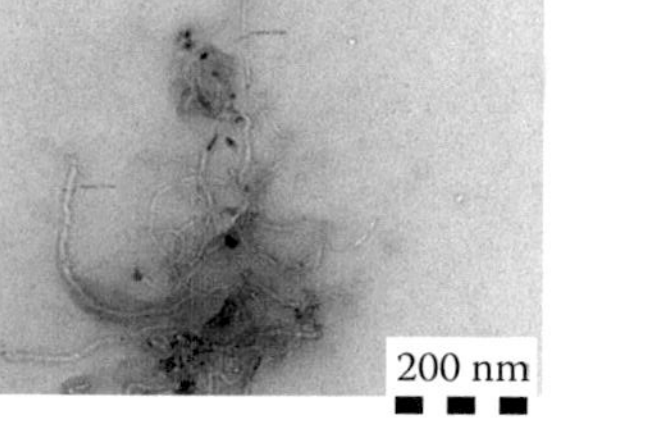	Bouillard et al. (2013) found that low-temperature (400°C) combustion of the same composite can release CNT in flue gases	

Waste incineration (thermal degradation at controlled oxygen and temperatures above 1,000°C)	Cone calorimeter ISO 5659-2 with off-gases monitored by CPC and sampled for TEM/SEM	 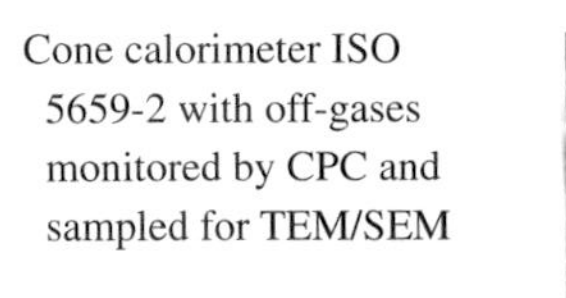	Fire smoke from carbon-nanofiber-polyurethanes is soot with no remaining fibers (Petersen et al. 2011; Uddin and Nyden 2011; Stahlmecke et al. 2014)	Existing reports on carbon nanofiber-polymer find no free fibers in flue gas (Petersen et al. 2011; Uddin and Nyden 2011) Release rate needs to be studied for end-of-life scenarios of CNT–polymer

Abbreviations: APS, aerodynamic particle sizer; AUC, analytical ultracentrifugation; CNT, carbon nanotubes; CPC, condensation particle counter; DLS, dynamic light scattering; EDX, energy dispersive X-ray analysis; ELPI, electrical low pressure impactor; ESP, electrostatic precipitator; FE-SEM, field emission scanning electron microscope; FMPS, fast mobility particle sizer; FTIR, Fourier transform infrared spectroscopy; ICP-MS, inductively coupled plasma mass spectrometry; LDPSA, laser diffraction particle size analyzer; LIBS, laser-induced breakdown spectroscopy; NAS, nano-aerosol sampler; OPC, optical particle counter; PAS, photoelectric aerosol sensor; PSSD, particle surface sensitive device (e.g., NSAM, DiscMini, Nanocheck); SEM, scanning electron microscopy; SMPS, scanning mobility particle sizer; TEM, transmission electron microscopy; TGA, thermogravimetric analyzer; TOF-SIMS, time-of-flight secondary ion mass spectrometry; TP, thermophoretic precipitator; UNPA, universal nano particle analyzer; WRASS, wide-range aerosol particle sampling system; XPS, X-ray photoelectron spectroscopy.

For example, the matrix degradation may occur due to weathering or heat stress during the use and transport in water or storage in soils and sediments.

Although the transport of CNTs in different matrices can be simulated like any other particle in the environment, transport of CNTs either as single fibers or in bundles may behave differently due to their chemical and physical nature. One determining factor for the transport of CNTs is their chemical composition. Basically, a CNT consists of one or several rolled-up layers of graphene and behaves chemically very similar to carbon black and particles of elemental carbon. The CNT is hydrophobic and hence is not easy to disperse in water. It readily quenches radicals and is black in color.

This behavior can significantly change when groups such as –OH, –COOH, –NO, and others are added to the surface of CNTs (Balasubramanian and Burghard 2005). These groups may be introduced in the CNTs directly during production to give them specific features needed for dispersion or embedding into a matrix or may occur during the lifecycle through, for example, thermal stress or UV weathering. Such surface modifications may also render the CNT readily dispersible in aquatic media. Another mechanism leading to higher dispersion of CNT in the environment is concentration-dependent adsorption to particular types of organic matter (Hyung and Kim 2008).

The physical parameters of CNTs (i.e., their specific morphology and state of agglomeration) may be of lesser importance for the particle mobility in the environment, whereas the chemical nature of the CNT significantly determines its dispersibility in water and hence its mobility. Fibers and tubes often align themselves according to the flow, with the tube diameter as the determining particle diameter. The morphology, that is, the length of the fibers increases this mobility diameter slightly but is of some importance when the flow direction, in liquids or gases, changes such as when the long fibers "attach" more rapidly to surfaces. Therefore, the transport of CNTs in air can be described in first approximation being similar to spherical particles with a diameter as that of the CNT-tube. Free CNTs will agglomerate readily according to the particle number concentration with other particles and form agglomerates. Depending on the size and agglomeration, some CNT fibers can stay airborne up to a few days in outdoor air and can be transported over long distances in analogy to spherical particles (Pruppacher and Klett 2010).

If released and transported in water, agglomeration can happen quite rapidly. Sedimentation tests using radiolabeled ^{14}C-CNTs showed high sediment partitioning coefficients (>95%) for both the addition of CNTs to the systems with the sediment as well as CNTs dispersed in water (A. Schaeffer, personal communication). Significant stabilization was observed in some cases, such as the addition of natural organic matter (Schwyzer et al. 2013).

Once the CNTs have been transported in air or water, they will ultimately deposit in either sediments or soils. The large aspect ratio of CNTs can lead to relatively higher deposition rates compared to colloids or other nanomaterials (Jaisi and Elimelech 2009; Wang et al. 2012), possibly because the CNTs can coil around soil and sediment particles (Sedlmair et al. 2012). This may explain the observed relative low mobility of CNTs in soil column tests and the possible accumulation in the upper soil layer (Cornelis et al. 2014). Details on the soil–nanoparticle interaction for CNTs and other nanoparticles are well summarized by Cornelis et al. (2014).

Thus far, no definite predictions on the transport and fate of CNTs in water and soils can be made to better describe the exposure in the environment. Previous studies showed that CNTs can be taken up by animals such as *Daphnia magna* (Petersen et al. 2009). Uptake by plants is realistic taking the current general knowledge of nanoparticle into account (Jacob et al. 2013; Ma et al. 2010) even though it was not tested for CNTs to our knowledge.

18.4 HAZARD AND RISK POTENTIALS OF PURE CNTs

It appears from initial findings regarding the nanometer-scale materials that are released from actual uses that risk evaluation for CNTs in conductive plastics should be based on exposure and toxicity studies of the released particles, and not on extrapolation from studies of free CNT fibers. Furthermore, there is as yet no evidence to assert that the released material (e.g., Figure 18.3 on primary particles with embedded CNTs following abrasion/cutting or Table 18.1 on degradation products following weathering or thermal processes) would be more hazardous than the material released from the same polymer without added CNTs (Wohlleben et al. 2011, 2013). More studies of such differential toxicity between material with and without embedded CNTs are therefore needed.

Alternatively, risk evaluation could be based on the assumption that exposure to free CNTs could occur as a small fraction of the released material or at some later point in the environmental fate of the released material in which CNTs are released from the polymer matrix as free fibers. In the first case, the fraction may need to be determined on a case-by-case basis (e.g., for a type of CNT, composite, and release scenario) and the difference between the released fibers and the added fibers would need to be determined. In the second case, research is needed to identify fate and transport of the released particles, as well as points of eventual exposure to free CNTs.

However, unfortunately, in lieu of such studies, the most relevant specific toxicity information with which to consider hazard and risk for such released nanometer-scale material comes from studies of free CNT fibers prior to their addition to any composite (see Bonner 2014). It should be noted that application of toxicity evaluation of these "prior to use" CNT fibers would be expected to greatly overestimate "fiber-related" health risk for the release scenarios that have not yet shown release of free CNT fibers. Furthermore, basing risk evaluation on assumed free CNTs would provide poor discriminating utility across different use categories, because the primary exposure entity is not being evaluated.

Taking into account these caveats to use of toxicity information from the study of free CNT fibers, the first challenge of drawing inference is again variation in the possible entities to which exposure may occur. Many of the potentially risk-relevant CNT characteristics, including thickness, length, and surface characteristics, vary substantially from one CNT manufacturing process to the next. Furthermore, research has shown that such characteristics do in fact affect toxicity (Aschberger et al. 2010; Coccini et al. 2013; Donaldson et al. 2006, 2013; Liu et al. 2012).

Accordingly, risk evaluation should consider such characteristics in evaluating the potential for adverse health effects.

For example, a study by Poland et al. (2008) compared the toxic responses of thin and shorter CNTs to thicker and longer CNTs. The thin and short fibers were described as forming "tightly packed spherical agglomerates" and "tangled agglomerates," and the thicker fibers were described as forming "dispersed bundles and singlets" or "regular bundles and ropes." The toxic effects noted by Poland et al. were caused by the thick and longer CNTs. Toxicity has also been observed for these thicker and longer CNTs in other studies (Liu et al. 2008; Sakamoto et al. 2009; Sargent et al. 2014; Takagi et al. 2008). The CNTs that did not show toxic effects in the study by Poland et al. (2008) were those that were described as tangled agglomerates.

Thinner and shorter CNTs also did not show cancer effects in a 2-year bioassay (Muller et al. 2009). However, studies of some thinner CNTs have shown potential for other effects (Liu et al. 2012; Ma Hock et al. 2009; Mitchell et al. 2007). Adverse effects have also been shown for high-dose exposures (high relative to dust levels in occupational settings) to the thin and flexible CNT agglomerates (Ma Hock et al. 2009; Pauluhn 2009). The dose–response range for these effects relative to possible exposure levels has not been fully evaluated. Therefore, more information is needed before broad generalizations about the biological interactions of CNTs are understood well enough to make risk management decisions without individual studies of toxicity. At this time, a generalized model for risk management of CNTs cannot be generated. In lieu of such a generalized model, it is essential to generate information specific to each CNT, if there is demonstration or expectation of fiber release to an exposure pathway.

18.5 CONCLUSIONS AND RECOMMENDATIONS

Specific combinations of factors such as aging, polymer additives, CNT surface modification, media contact, UV exposure, and abrasion will dictate where matrix degradation is more likely to occur for composites containing CNTs and the nature of the released material. Therefore, an understanding of these release-related characteristics through the lifecycle (product manufacture, use, and disposal/reuse) is necessary to evaluate the likelihood and nature of materials entering into exposure pathways. Through this kind of structured review of data regarding environmental conditions and the underlying material properties of CNT–polymer composites, some generalizations and therefore some exposure likelihoods can be bounded through evaluation of release likelihood. Such evaluations indicate that release of free CNTs is expected to be rare and below likely toxicity levels for many of the specific release and exposure scenarios occurring during the lifecycle, and a consequent low likelihood of human health risk is expected based on the added CNT for those exposure scenarios (Kingston et al. 2014; Nowack et al. 2013).

For other release scenarios, studies of the actual released material are needed before assessment of risk can be undertaken. Hazard inference for "pure" CNTs varies greatly, from essentially no evidence of toxicity for some CNT forms (with several well-conducted studies) to initial reports of tumor promotion for other forms. Although the nature and amount of free CNTs from actual uses in relation to these purer forms is not clear, the potential for exposure to a form of CNT that may be toxic should be addressed cautiously (e.g., through appropriately validated dust control and personal protection equipment in occupational settings where dust is released) as methods and data are developed to characterize the exposure and toxicology of the released material. In particular, in the short term, evaluation of release likelihoods and the persistence of released material as polymer bound-CNT fragments is needed to understand the relationship between released materials and toxicology studies of pure CNTs.

Before such studies can be undertaken, funding must be applied to the development of methods to routinely and reliably characterize the "added and persistent nanoscale properties" of the released materials. Evaluations of methods presented and discussed by experts of the NanoRelease project at workshops in 2012 and 2013 concluded that technology and standard approaches are not available to more than qualitatively detect (e.g., electron microscopic evaluation of nonrepresentative dust samples) the presence of CNTs in material released to exposure pathways. Quantitative methods needed to ascertain dose in relationship with measures of toxicity may not even be possible with current technology, and are certainly not in standard practice. Without such methods, the risk assessment of CNTs in polymer composites can only be done in a binary or bounded fashion, either demonstrating that there is no exposure and no risk or that there is some exposure bounded by crude measures of mass of added CNTs. Differentiation within the bounding is possible, given the differences in toxicity expressed in the literature for different CNT types, with greater attention for materials that are clearly asbestos-like in initial form and release, and a greater expectation of safety for materials that do not have this initial form.

Considering the likelihood of exposure to free CNT fiber, the US National Institute of Occupational Safety and Health (2013) has issued a "Current Information Bulletin" with a "recommended exposure limit (REL)" of 1.0 μg/m^3 for CNT fibers based on evaluation of a number of studies on various forms of CNTs and carbon nanofibers. Based on consideration of the material chemistry and the rare observance of free CNTs in studies of released particles from abrasion, it would appear that this safety level is not likely to be exceeded in direct exposure scenarios for users of conductive CNT–polymer composites. However, this qualitative comparison should be confirmed with quantitative methods sufficient to show the upper boundary level for free fiber release. Furthermore the lack of evidence to expect that CNTs embedded in polymer particles would be inherently more toxic than particles without embedded CNTs indicates that, again, the use scenarios with CNTs are no more hazardous than those without CNTs. Other routes of exposure (e.g., dermal or oral) have not been evaluated for CNT exposure, with most of the research community focused on the inhalation route of exposure for CNTs.

Risk evaluation for direct exposure pathways through disposal and recycling is less well understood, although for each there would be a similar expectation of low likelihood of pure CNT exposure. Risk evaluation for subsequent environmental transport, degradation of polymer matrix, and exposure to free CNTs is in need of further research. In particular, persistence and accumulation potential for the materials should be evaluated so that places of higher concentration of free CNT fiber can be identified and studied.

ACKNOWLEDGMENT

We thank Wendel Wohlleben for his very helpful comments and for the contribution of Table 18.1 to this chapter.

REFERENCES

Aschberger, K., Johnston, H. J., Stone, V., Aitken, R. J., Hankin, S. M., Peters, S. A., Tran, C. L., Christensen, F. M. (2010). Review of carbon nanotubes toxicity and exposure-appraisal of human health risk assessment based on open literature. *Crit. Rev. Toxicol.* 40: 759–790.

Balasubramanian K., Burghard, M. (2005). Chemically functionalized carbon nanotubes, *Small* 1: 180–192.

Bello D., Wardle, B. L., Zhang, J., Yamamoto, N., Santeufemio, C., Hallock, M. et al. (2010). Characterization of exposures to nanoscale particles and fibers during solid core drilling of hybrid carbon nanotube advanced composites. *Int. J. Occup. Environ. Health.* 16: 434–450.

Bonner J.C. (2014). Uptake and Effects of Carbon Nanotubes. In Wohlleben, Kuhlbusch, Schnekenburger, Lehr, eds, *Safety of Nanomaterials along Their Life cycle: Release, Exposure and Human Hazard*, Taylor & Francis.

Bouillard J. X., R'Mili, B., Moranviller, D., Vignes, A., Le Bihan, O., Ustache, A. et al. (2013). Nanosafety by design: Risks from nanocomposite/nanowaste combustion. *J. Nanopart. Res.* 15: 1–11.

Canady, R., Kuhlbusch, T., Renker, M., Lee, E., Tsytsikova, L. (2013). Comparison of Existing Studies of Release Measurement for CNT-Polymer Composites. http://www.ilsi.org/ResearchFoundation/RSIA/Documents/NanoRelease%20Consumer%20Products%20Phase%202.5%20Report.pdf.

Cena L. G., Peters, T. M. (2011). Characterization and control of airborne particles emitted during production of epoxy/carbon nanotube nanocomposites. *J. Occup. Environ. Hyg.* 8: 86–92.

Coccini, T., Manzo, L., Roda, E. (2013). Safety evaluation of engineered nanomaterials for health risk assessment: An experimental tiered testing approach using pristine and functionalized carbon nanotubes. *ISRN Toxicol.* 2013: 825427.

Cornelis G., Hund-Rinke, K., Kuhlbusch, T., Van den Brink, N., Nickel, C. (2014). Fate and bioavailability of engineered nanoparticles in soils: A review. *Crit. Rev. Environ. Sci. Technol.* In press.

De Volder, M. F., Tawfick, S. H., Baughman, R. H., Hart, A. J. (2013). Carbon nanotubes: Present and future commercial applications. *Science* 339: 535–539.

Donaldson, K., Aitken, R., Tran, L., Stone, V., Duffin, R., Forrest, G., Alexander, A. (2006). Carbon nanotubes: A review of their properties in relation to pulmonary toxicology and workplace safety. *Toxicol. Sci.* 92: 5–22.

Donaldson, K., Poland, C. A., Murphy, F. A., MacFarlane, M., Chernova, T., Schinwald, A. (2013). Pulmonary toxicity of carbon nanotubes and asbestos—Similarities and differences. *Advanced drug delivery reviews* 65: 2078–2086.

Fleury, D., Bomfim, J. A. S., Vignes, A., Girard, C., Metz, S., Muñoz, F., Romili, B., Ustache, A., Guiot, A., and Bouillard, J.X. (2013). Identification of the main exposure scenarios in the production of CNT-polymer nanocomposites by melt-moulding process. *J. Cleaner Prod.* 53: 22–36.

Future Markets. (2012). Nanomaterials in plastics and advanced polymers. Market Report #52.

Göhler, D., Nogowski, A., Fiala, P., Stintz, M. (2013). Nanoparticle release from nanocomposites due to mechanical treatment at two stages of the life-cycle. *J. Phys. Conf. Ser.* 429: 012045.

Göhler, D., Stintz, M., Hillemann, L., Vorbau, M. (2010). Characterization of nanoparticle release from surface coatings by the simulation of a sanding process. *Ann. Occup. Hyg.* 54: 615–624.

Gojny, F., Wichmann, M. H. G., Fiedler, B., Schulte, K. (2005). Influence of different carbon nanotubes on the mechanical properties of epoxy matrix composites–A comparative study, *Compos. Sci. Technol.* 65: 2300–2313.

Golanski, L., Guiot, A., Pras, M., Malarde, M., Tardif, F. (2012). Release-ability of nano fillers from different nanomaterials (toward the acceptability of nanoproduct). *J. Nanopart. Res.* 14: 1–9.

Han, Z., Fina, A. (2011). Thermal conductivity of carbon nanotubes and their polymer nanocomposites: A review. *Prog. Polym. Sci.* 36: 914–944.

Hirth, S., Cena, L., Cox, G., Tomovic, Z., Peters, T., Wohlleben, W. (2013). Scenarios and methods that induce protruding or released CNTs after degradation of composite materials. *J. Nanopart. Res.* 15: 1504.

Huang, G., Park, J., Cena, L., Shelton, B., Peters, T. (2012). Evaluation of airborne particle emissions from commercial products containing carbon nanotubes. *J. Nanopart. Res.* 14: 1–13.

Hyung, H., Kim, J. H. (2008). Natural organic matter (NOM) adsorption to multi-walled carbon nanotubes: Effects of NOM characteristics and water quality parameter. *Environ. Sci. Technol.* 42: 4416–4421.

International Organization for Standardization (2011). ISO 4892: Plastics - Methods of exposure to laboratory light sources - Part 2: Xenon-arc lamps (ISO/DIS 4892-2:2011).

Jacob, D. L., Borchardt, J. D., Navarathnam, L., Otte, M. L., Bezbaruah, A. N. (2013). Uptake and translocation of Ti from nanoparticles in crops and wetland plants. *Int. J. Phytoremediation* 15: 142–153.

Jaisi, D. P., Elimelech, M. (2009). Single-walled carbon nanotubes exhibit limited transport in soil columns. *Environ. Sci. Tech.* 43: 9161–9166.

Kaiser, D., Stefaniak, A., Scott, K., Nguyen, T., Jurg Schutz, J. (2014). Methods for the Measurement of Release of MWCNTs from MWCNT-Polymer Composites. A Report of Task Group 1 of the NanoRelease Consumer Products Project. Submitted for Steering Committee use in project monograph.

Kingston, C., Zepp, R., Andrady, A., Boverhof, D., Fehir, R., Hawkins, D., Roberts, J., Sayre, P., Shelton, B., Sultani, Y., Vejinsj, V., Wohlleben, W. (2014). Release characteristics of selected carbon nanotube polymer composites. *Carbon* 68: 33–57.

Koponen, I. K., Jensen, K. A., Schneider, T. (2010). Comparison of dust released from sanding conventional and nanoparticle-doped wall and wood coatings. *J. Exposure Sci. Environ. Epidemiol.* 21: 408–418.

Kuhlbusch, T. A. J., Kaminski, H. (2014). Release from composites by mechanical and thermal treatment: Test methods. In Wohlleben, Kuhlbusch, Schnekenburger, Lehr, eds, *Safety of Nanomaterials along Their Life cycle: Release, Exposure and Human Hazard*, Taylor & Francis.

Le Bihan, O., Shandilya, N., Gheerardyn, L., Guillon, O., Dore, E., and Morgeneyer, M. (2013) Investigation of the Release of Particles from a Nanocoated Product. *Adv. Nanopart.* 2: 39–44.

Lee, J. H., Lee, S. B., Bae, G. N., Jeon, K. S., Yoon, J. U., Ji, J. H., Sung, J. H., Lee, B. G., Lee, J. H., Yang, J. S., Kim, H. Y., Kang, C. S., Yu, I. J. (2010). Exposure assessment of carbon nanotube manufacturing workplaces. *Inhalation Toxicol.* 22: 369–381.

Liu, A., Sun, K., Yang, J., Zhao, D. (2008). Toxicological effects of multi-wall carbon nanotubes in rats. *J. Nanopart. Res.* 10: 1303–1307.

Liu, Y., Zhao, Y., Sun, B., & Chen, C. (2012). Understanding the toxicity of carbon nanotubes. *Acc. Chem. Res.* 46: 702–713.

Ma, P. C., Zhang, Y. (2014). Perspectives of carbon nanotubes/polymer nanocomposites for wind blade materials. *Renewable Sustainable Energy Rev.* 30: 651–660.

Ma, X., Geiser-Lee, J., Deng, Y., Kolmakov, A. (2010). Interactions between engineered nanoparticles (ENPs) and plants: Phytotoxicity, uptake and accumulation. *Sci. Total Environ.* 408: 3053–3061.

Ma-Hock, L., Treumann, S., Strauss, V., Brill, S., Luizi, F., Mertler, M., Wiench, K., Gamer, A. O., van Ravenzwaay, B., and Landsiedel, R. (2009). Inhalation toxicity of multiwall carbon nanotubes in rats exposed for 3 months. *Toxicol. Sci.* 112: 468–481.

Maynard, A. D., Baron, P. A., Foley, M., Shvedova, A. A., Kisin, E. R., Castranova, V. (2004). Exposure to carbon nanotube material: Aerosol release during the handling of unrefined single walled carbon nanotube material. *J. Toxicol. Environ. Health. A* 67: 87–107.

Mitchell, L. A., Gao, J., Van der Wal, R., Gigliotti, A., Burchiel, S. W., Jacob D. McDonald, J. D. (2007). Pulmonary and systemic immune response to inhaled multiwalled carbon nanotubes. *Toxicol. Sci.* 100: 203–214.

Muller, J., Delos, M., Panin, N., Rabolli, V., Huaux, F., Lison, D. (2009). Absence of carcinogenic response to multiwall carbon nanotubes in a 2-year bioassay in the peritoneal cavity of the rat. *Toxicol. Sci.* 110: 442–448.

NanoRelease Consumer Products. (2014). NanoRelease Consumer Products website. http://www.ilsi.org/ResearchFoundation/RSIA/pages/NanoRelease1.aspx. Accessed March 3, 2014.

Nguyen, T., Pellegrin, B., Bernard, C., Gu, X., Gorham, J. M., Stutzman, P., Stanley, D., Shapiro, A., Byrd, E., Hettenhouser, R., Chin, J. (2011). Fate of nanoparticles during life cycle of polymer nanocomposites. *J. Phys. Conf. Ser.* 304: 012060.

Nowack, B., David, R. M., Fissan, H., Morris, H., Shatkin, J. A., Stintz, M., Zepp, R., Brouwer, D. (2013). Potential release scenarios for carbon nanotubes used in composites. *Environ. Int.* 59: 1–11.

Pauluhn, J. (2009). Subchronic 13-week inhalation exposure of rats to multiwalled carbon nanotubes: Toxic effects are determined by density of agglomerate structures, not fibrillar structures. *Toxicol. Sci.* 113: 226–242.

Petersen, E. J., Akkanen J., Kikkonen J. V. K., Weber W. J. (2009). Biological uptake and depuration of carbon nanotubes by Daphnia magna. *Environ. Sci. Technol.* 43: 2969–2975.

Petersen E. J., Zhang, L., Mattison, N. T., O'Carroll, D. M., Whelton, A. J., Uddin, N., Nguyen, T. L., Huang, Q., Henry, T. B., Holbrook, R. D., Chen, K. L. (2011) Potential release pathways, environmental fate, and ecological risks of carbon nanotubes. *Environ. Sci. Technol.* 45: 9837–9856.

Poland, C. A., Duffin R., Kinloch I., Maynard A., Wallace W. A., Seaton A., Stone V., Brown S., Macnee W., Donaldson K. (2008). Carbon nanotubes introduced into the abdominal cavity of mice show asbestos-like pathology in a pilot study. *Nat Nanotechnol.* 3: 423–428.

Pruppacher H. R., Klett J. D. (2010). Microphysics of Cloud and Precipitation, Springer, New York, ISBN 978-0-7923-4211-3, p. 250.

Sahoo, N. G., Rana, S., Cho, J. W., Li, L., Chan, S. H. (2010). Polymer nanocomposites based on functionalized carbon nanotubes. *Prog. Polym. Sci.* 35: 837–867.

Sakamoto, Y., Nakae, D., Fukumori, N., Tayama, K., Maekawa, A., Imai, K., Hirose, A., Nishimura, T., Ohashi, N., Ogata, A. (2009). Induction of mesothelioma by a single intrascrotal administration of multi-wall carbon nanotube in intact male Fischer 344 rats. *J. Toxicol. Sci.* 34: 65–76.

Sargent, L. M., Porter, D. W., Staska, L. M., Hubbs, A. F., Lowry, D. T., Battelli, L., Siegrist, K. J., Kashon, M. L., Mercer, R. R., Bauer, A. K. (2014). Promotion of lung adenocarcinoma following inhalation exposure to multi-walled carbon nanotubes. *Part. Fiber Toxicol.* 11: 3.

Schlagenhauf L., Chu, B. T., Buha, J., Nuesch, F. Wang, J. (2012). Release of carbon nanotubes from an epoxy-based nanocomposite during an abrasion process. *Environ. Sci. Technol.* 46: 7366–7372.

Schwyzer, I., Kaegi, R., Nowack, B. (2013). Colloidal stability of suspended and agglomerate structures of settled carbon nanotubes in different aqueous matrices, *Water Res.* 47: 3910–3920.

Sedlmair, J., Gleber, S.-C., Wirick, S., Guttmann, P., Thieme, J. (2012). Interaction between carbon nanotubes and soil colloids studied with X-ray spectromicroscopy. *Chem. Geol.* 329: 32–41.

Simon, A., Zielke, H., Schmidt, B., Seiler, T.-B., Schäffer, A., Kukkonen, J. V. K., Hollert, H. (2010). Chemical and radioactive analyses of the partitioning of organic chemicals in sediment-water-organism-systems. Proceedings, 20th SETAC Annual Meeting, Sevilla, Spain.

Stahlmecke B., Asbach C., Todea A. M., Kaminski H., Kuhlbusch T. A. J. (2014). Investigations on CNT release from composite materials during end of life, In *Handbook of nanosafety measurements, exposure and toxicology*, Vogel U., Savolainen K, Wu Q., van Tongeren M., Brower D., Berges M., eds., Elsevier, ISBN: 978-0-12-416604-2 pp.242-255,

Sun, D. M., Liu, C., Ren, W. C., Cheng, H. M. (2013). A review of carbon nanotube and graphene based flexible thin film transistors. *Small* 9: 1188–1205.

Takagi, A., Hirose, A., Nishimura, T., Fukumori, N., Ogata, A., Ohashi, N., Kitajima, S., Kanno, J. (2008). Induction of mesothelioma in p53+/- mouse by intraperitoneal application of multi-wall carbon nanotube. *J. Toxicol. Sci.* 33: 105–116.

Toyokuni, S. (2013). Genotoxicity and carcinogenicity risk of carbon nanotubes. *Adv. Drug Deliv. Rev.* 65: 2098–2110.

Uddin, N. M., M. R. Nyden. (2011) Characterization of nanoparticle release from polymer nanocomposites due to fire. Nanotechnology 2011: Advances Materials, CNTs, Particles, Films and Composites, Nano Science Technology Institute, Danville, CA, pp. 523–526.

US National Institute of Occupational Safety and Health. (2013). NIOSH current intelligence bulletin 65: occupational exposure to carbon nanotubes and nanofibers. http://www.cdc.gov/niosh/docs/2013–145/.

Wang, Y., Kim, J.-H., Baek, J.-B., Miller, G. W., Pennell, K. D. (2012). Transport behavior of functionalized multi-wall carbon nanotubes in water-saturated quartz sand as a function of tube length. *Water Res.* 46: 4521–4531.

Wohlleben, W., Brill, S., Meier, M. W., Mertler, M., Cox, G., Hirth, S., von Vacano, B., Strauss, V., Treumann, S., Wiench, K., Ma-Hock, L., Landsiedel, R. (2011) On the life cycle of nanocomposites: Comparing released fragments and their in-vivo hazards from three release mechanisms and four nanocomposites. *Small* 7: 2384–2395.

Wohlleben W., Meier, M. W., Vogel, S., Landsiedel, R., Cox, G., Hirth, S., Tomovic, Z. (2013). Elastic CNT-polyurethane nanocomposite: Synthesis, performance and assessment of fragments released during use. *Nanoscale* 5: 369–380.

Zhao, Y., Allen, B. L., Star, A. (2011), Enzymatic degradation of multiwalled carbon nanotubes. *J. Phys. Chem. A* 115: 9536–9544.

19 Case Study
Challenges in Human Health Hazard and Risk Assessment of Nanoscale Silver

Christian Riebeling and Carsten Kneuer

CONTENTS

19.1 BACKGROUND

Several types of nanomaterials are currently subject to discussion of potential health hazards and risks associated with their use, and corresponding testing programs have been initiated (e.g., OECD 2010). Silver nanoparticles might have found their way into the largest number of practical applications including consumer products. For the 1,628 products claiming to employ nanotechnology that are listed in the online inventory of the Woodrow Wilson International Center for Scholars (WWICS) Project on Emerging Nanotechnologies (PEN) (WWICS 2013), the most often identified nanomaterial is silver. Twenty-four percent, or a total of 383 products supposedly contained nanosilver (WWICS 2013a). Furthermore, often employed materials include titanium/titanium dioxide in 179 products, carbon in 87 products, silicon dioxide in 52 products, zinc/zinc oxide in 36 products, and gold in 19 products. This inventory is the most comprehensive, freely accessible database to date. It lists products available to consumers, which claim to employ nanotechnology, many of which do not identify a nanomaterial. On the German site www.nanowatch.de of the nongovernmental organization BUND,* 99 products are itemized under the category "contains nanomaterial: silver" out of a total of 1007 products claiming to contain nanomaterials listed on this site: http://www.bund.net/nc/themen_und_projekte/nanotechnologie/nanoproduktdatenbank/produktsuche/ (accessed on 10/25/2013). Searching the global trade website www.alibaba.com for the terms "nano-material, nano-technology, nano-particles" yields >70,000 products, whereas a specific search for "nanosilver, nano-silver" yields >7400 products, confirming that a multitude of products claiming to contain nanosilver have found their way into the global market (Alibaba 2013).

19.2 INTRODUCTION

Consensus has been reached that risks to human health from manufacture and use of nanomaterials including nanosilver can, in principle, be characterized using the established risk assessment paradigm. This approach to risk assessment is based on four fundamental steps: identification of the hazard potential of a product and its components, characterization of the relevant hazards namely their dose–response relationships, assessment of exposures to the product and/or its critical components, and finally, a characterization of the likely risk by integration of the information on hazard and exposure characteristics (Figure 19.1). A more in-depth description of this approach, as well as recommendations for additional considerations aiming to improve its utility for risk assessment and the technical analyses supporting risk assessment, has been provided by the National Research Council (NRC) (NRC 2009). With regard to application of the risk assessment paradigm to nanomaterials, a detailed analysis of the applicability of this approach and its (default) assumptions by the Organization for Economic Co-operation and Development (OECD) Working

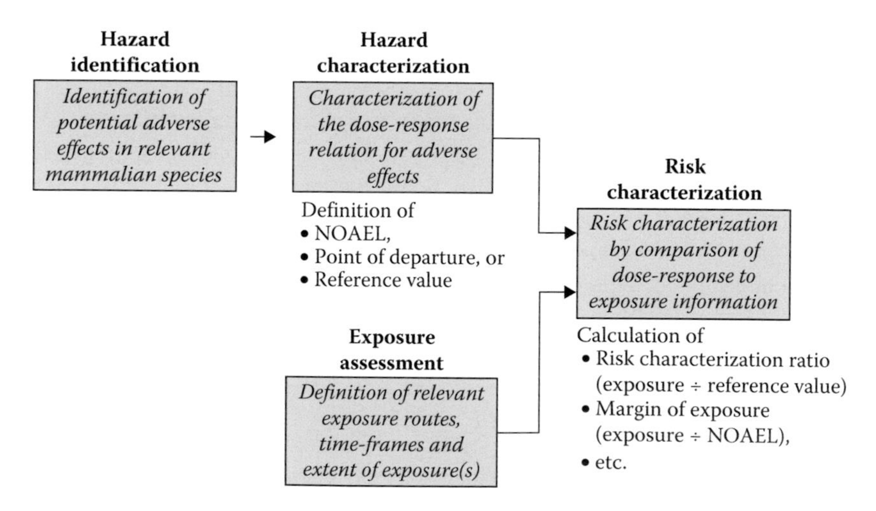

FIGURE 19.1 The basic four steps of the risk assessment paradigm.

* Bund für Umwelt und Naturschutz.

Party on Manufactured Nanomaterials (WPMN) revealed a number of nanomaterial-specific challenges requiring special consideration (OECD 2012). In this chapter of the book, some of those challenges will be highlighted using nanosilver as an example.

19.3 USE(S) AND POTENTIAL SOURCES OF NANOSILVER EXPOSURE

According to the WWICS database, nanosilver may be found in nearly every type of consumer product. Silver nanoparticle powders and solutions are available for general antibacterial use on surfaces, for self-medication purposes, or as food and water supplement. It is also found in wound dressings. The inventory lists cosmetic products including soap, cream, gel, foam, skin cleanser, toothpaste, mouthwash, facemasks, whitening masks, and hair care spray. Moreover, cosmetic appliances that utilize silver are wipes, toothbrushes, toothbrush sterilizers, shavers, epilators, hair brushes, hair dryers, curling irons, hair straighteners, and make-up instruments. Textiles are a large category including underwear and every kind of clothing, bed linen, blankets, towels, pillows, and also laundry detergents, fabric softener, and washing machines releasing silver. The database also lists shoes, shoe deodorizers, and sanitizing shoe cabinets presumably containing nanosilver. Products for babies in addition to textiles include supplemented water, pacifiers, baby teeth developers, baby mugs, baby bottles, baby bottle brushes, mats for diaper changing, carriages, and stuffed toys. Home appliances include kitchenware, tableware, coffee makers, food storage containers, food storage bags, beverage containers, refrigerators, vacuum cleaners, air purifiers, air sanitizers, air conditioners, humidifiers, water taps, water filters, water purifiers, water energizers, locks, and foot massagers. Consumer electronics include computer keyboards, computer mice, notebooks, and mobile phones. Further uses of silver are in watchbands, ear protectors, toilet seats, yoga mats, rubber gloves, condoms, inks, and wall paint.

Many of the products found at www.alibaba.com when searching for nanosilver are the same or similar to the aforementioned products, or basic raw materials for the manufacture of the said products such as fabrics, yarns, pigments, plastic master batches, and ceramic balls. In addition, products not found in the database of the WWICS are wallets, purses, hand bags, textile/leather finishings, insoles, razors, hair clippers, dental floss, sanitary pads, diapers, mist facial moisturizers, facial steamers, acupressure suction cups, breast milk pumps, breast milk bags, baby rattles, computer/mobile phone cases, paints/paint additives, spray paint, glass coatings, wall coating, formaldehyde elimination (mould, mildew) solutions (photocatalyst), coating additive, flooring tiles, jewellery, car wax, thermal paste, ice packs, bath mats, boots keepers, backpacks, blenders, microwave ovens, lamp reflectors, ultrasonic vegetable and fruit cleaners, and ultrasonic denture cleaners.

In addition to its antibacterial activity, which is the basis of almost all of its consumer-related use, nanoparticles of silver are also used in electronic components

(Lee et al. 2006) and biosensors (Lee et al. 2013a). Nanoscale silver has also been traditionally used to produce yellow glass for windows (Solomon et al. 2007).

To date, little quantitative information about potential exposure resulting from such nanosilver applications has been published. To test a setup for measuring the release of nanoparticles in aerosols generated by sprays intended for the consumer market, an aqueous solution containing silver nanoparticles was used with two types of spray dispensers provided by the producer (Hagendorfer et al. 2010). It was shown that a pump spray dispenser did not produce any measurable droplets in the setup, that is, the generated droplets were presumably too big and heavy to reach the detector placed 80 cm apart. In contrast, a propellant-based dispenser produced droplets in the range of 10–300 nm measured by SMPS (Scanning Mobility Particle Sizer, see Chapter 2) with a range of 10–500 nm. The droplets contained silver nanoparticles (Hagendorfer et al. 2010). Using the same setup, a second study investigated the release of nanomaterials from four commercially available sprays, two of which contained silver (Lorenz et al. 2011). One of the silver-containing sprays was a plant strengthening spray using a pump spray dispenser. As in the earlier study, the solution contained silver nanoparticles and no droplets were detected at the detector after pump spraying due to the droplet size. A propellant-based antiperspirant spray generated nanometer-size droplets and released very low concentrations of silver likely in its ionic form.

In a different setup using a mannequin head for sampling, amongst 11 tested sprays, a silver nanoparticle containing spray intended for topical or internal antibacterial use was compared to a spray labeled to contain silver without reference to nanotechnology intended for topical and nasal application (Nazarenko et al. 2011). All sprays were propellant-based and generated droplets in the nanometer range and reaching into the micrometer range. In both silver-containing sprays nanoscale silver particles were found. Although the primary particles of the nanosilver spray were of a relatively defined size of 3–65 nm and exhibited some agglomeration, the silver-containing spray particles were of a broad size distribution around <3–435 nm. Three commercially available sprays, an antiodor spray for hunters, a disinfectant spray, and a throat spray, were investigated for nanoparticle content and release in another study (Quadros and Marr 2011). Nanoparticulate silver was found in the antiodor spray and in the throat spray. Although nanosize droplets were observed in the aerosols, only low amounts of silver were released, and were in micrometer scale agglomerates. In all studies, the propellant-based sprays generated nanosize droplets that have the ability to carry nanoscale materials into the alveolar region without the necessity of prior evaporation of the carrier liquid. Thus, the likelihood of exposure and therefore the level of potential risk from nanosilver sprays were linked to the mode of aerosol generation.

Silver content and release have also been investigated for textiles (see Chapter 15 for details). The emphasis here was mostly on environmental exposure (Benn and Westerhoff 2008; Lorenz et al. 2012; Pasricha et al. 2012) and lifecycle analysis (Meyer et al. 2011) rather than human exposure. The release of particles was apparently dependent on the type of fabric and the particle integration into the fabric because both near complete release in the first three washing cycles and very low release in several washing cycles have been observed (Benn and Westerhoff 2008;

Lorenz et al. 2012). In addition, a washing machine releasing silver nanoparticle has been investigated (Farkas et al. 2011). It uses a patented mechanism (Patent WO/2005/056908) to generate and release silver nanoparticles during the washing cycle. Release of nanoparticulate silver was found in the wastewater as well as deposited on the laundry (Farkas et al. 2011).

Food contact materials have been investigated in the form of food storage bags (Huang et al. 2011) and food storage containers (von Goetz et al. 2013). Using four different food simulating solutions on bags, silver was released in nanoparticulate form in a time- and temperature-dependent manner (Huang et al. 2011). Silver was detected in two of four tested bodies of plastic food storage containers (von Goetz et al. 2013). Both containers released silver into food-simulating solution with a comparable rate to the bags.

Human exposure may also occur at the workplace, during production and handling of nanosilver, its formulation into products, or the fabrication of goods from nanosilver-treated material.

A study monitoring workplace air in nanosilver production facilities measured silver metal concentrations from 0.00002 to 0.00118 mg/m^3, depending on the type of workplace, the operation state, and the manufacturing method. When silver nanoparticles were generated using induced coupled plasma, particle numbers reached 535–25,022 per cm^3 at the workplace outside the reactor. A broad range of particle sizes was observed, presumably as a result of agglomeration or aggregation. When a wet method was used for manufacturing, particle numbers in the workplace air were significantly lower with 393–3,526 particles per cm^3 (Lee et al. 2011). Particle numbers during production, postproduction handling, and processing in a test facility producing a silica–nanosilver microcomposite powder were monitored using SMPS by Demou et al. (2008). In the size range from 6 to 673 nm an increase over background of approx. 9,000 particles per cm^3 to a maximum of 50,000 particles per cm^3 was observed during production, and a maximum of 15,000 particles per cm^3 was determined during processing and handling. Although measurements were limited to particle size and numbers, not discriminating between silver and other materials, these results were interpreted as to indicate the potential for breakaway of silver nanoparticles from the microcomposite (US EPA 2011). In addition, information from health surveillance of two male individuals with a 7-year history of work in nanosilver manufacturing was reported. For the first individual, a level of 0.34 μg/L silver in blood and 0.43 μg/L in urine were associated with an estimated exposure at the workplace of 0.35 μg/m^3. For the other individual, blood and urine silver levels were 0.30 μg/L and not detectable (<0.1 μg/L), respectively, at an estimated occupational exposure of 1.35 μg/m^3 (Lee et al. 2012). For comparison, blood silver levels in an unexposed control group ranged from not detectable (<0.1 μg/L) in 11 of 15 individuals to 0.2 μg/L in the remaining four workers (Armitage et al. 1996). In conclusion, occupational exposures during nanosilver production and handling may result in significant inhalation exposures that are reflected by increases in blood silver levels. Finally, spraying or roll-on application of paints or other coatings, manipulation, and handling of treated materials, for example, cutting and sewing nanosilver-treated textiles, as well as secondary tasks including cleaning and others may provide potential sources of nanosilver exposure.

TABLE 19.1
Potential Routes of Exposure to Nanosilver

Exposure	Oral	Dermal	Inhalation
Professional (direct)	Negligible	Potential	Potential
Professional (indirect)	Negligible	Potential	Potential
Consumer (direct)	Potential	Potential	Potential
Consumer (indirect)	Potential	Negligible	Negligible

BOX 19.1 KEY PRODUCT CHARACTERISTICS OF DUMMY PRODUCTS AND INITIAL EXPOSURE CHARACTERIZATION

Name	DP1 (dummy product 1)	DP2 (dummy product 2)
Description	Liquid at 10 mg/L nanosilver, packaged in 500 mL ready-to-use containers	Powdered nanosilver (100%), packaged in containers at 10 kg
Production	Wet chemical synthesis (fluid phase)	Evaporation–condensation method (gas phase)
Application	Pump spray, direct application to household surfaces	Textile fiber modification (polymer doping)

Overall, the available information suggests a variety of potential exposure scenarios either directly through use of the material or the product by the consumer and the worker (professional), or indirectly through the environment or the food chain as summarized in Table 19.1. For a specific product and use, the sources and routes of exposure to be taken into consideration for risk assessment will be more limited. This is exemplified in Box 19.1 for two biocidal dummy products: a liquid intended for spray application and a solid product for production of nanosilver containing textiles. To quantify potential risks, actual exposures may need to be measured experimentally or estimated using exposure modeling. In some cases, worst-case estimates may be sufficient for a first tier risk assessment. Further details on exposure are provided in Section III of this book, especially Chapter 15.

19.4 NANOMATERIAL IDENTIFICATION

Any risk assessment requires the unambiguous identification of the substance or the material under evaluation. For "conventional" chemicals, technical guidance has been developed on this issue such as, for example, the *Guidance for identification and naming of substances under REACH and CLP* (ECHA 2013). Accordingly, identification of the material usually includes information on the main constituents as well as on impurities and additives (Box 19.1). For nanosilver, there are a significant number of potentially relevant impurities and additives, including citrate, various amines (e.g., polyethylenimine), Tween 20, polyvinylpyrrolidone (PVP), cetyl trimethylammonium bromide (CTAB), sodium dodecylsulfate (SDS), (poly)saccharides, dextran, (poly)ethylene glycol, polyacrylic acid, functional thiols, inorganic coatings, and others. Although impurities resulting from the production process may in principle be removed from the nanomaterial, others have been purposely added to, for example, prevent aggregation of the nanosilver particles. The minimum set of information that should be provided for constituents, impurities, and additives includes the chemical names, molecular and structural formulas, and content or typical concentration ranges. Unfortunately, such information is currently rarely provided on stabilizers and residual reducing agents in study reports published in the open literature. It is widely accepted that some substances including many inorganic minerals and also nanomaterials require additional identifiers such as crystallinity to allow for unequivocal description. Currently, there is no full consensus on the type of additional information that should be part of the substance identification of a nanomaterial. In many cases the debate is whether a certain parameter such as morphology, size, and size distribution should be regarded as a characterizer of the substance rather than an identifier (JRC 2011). Significant research activity (compare Section II of this book, especially Chapter 9) is ongoing to elucidate those nanomaterial characteristics that have a major impact on the hazard potential and should thus be relevant identifiers in the context of risk assessment.

For nanosilver, different particle morphologies can be obtained depending on the production protocol: In addition to spheres that dominate in the literature, pyramids, cubes, plates, and rods can be generated (Wiley et al. 2007). It has been demonstrated that chemical reactivity differs between nanosilver shapes as morphology influences accessibility of the different crystal faces of the silver (Xu et al. 2006). Therefore, nanosilver shape may also influence toxicity and may deserve consideration as a substance identifier in risk assessment. In addition, the crystallite size may also vary for particles of comparable dimensions. Depending on the number of nucleation sites, monocrystalline, twinned, and multiple twinned crystalline particles have been described for nanosilver (Wiley et al. 2007).

Nanosilver particles can be synthesized with narrow monodisperse but nonoverlapping size distributions of, for example, 25 ± 3 and 70 ± 4 nm, as demonstrated by Li et al. (2012a). For most other nanomaterials, reaction products contain particles of a wider size range or show a polydisperse distribution resulting from aggregation during synthesis. In in vitro experiments, an inverse correlation of nanosilver particle size and cytotoxicity has been shown; whereas, the situation in vivo including the effect of particle size on toxicokinetics is less clear (Carlson et al. 2008; Lankveld et al. 2010; Li et al. 2012b).

The OECD (2012) currently recommends considering at least chemical composition, size and size range, crystallinity, surface coatings, and morphology to identify the nanoscaled constituents (see Box 19.2 as an example). However, the OECD document also states that where further information on potentially relevant properties is available, such as zeta potential, redox potential, specific surface area, and others,

BOX 19.2 NANOMATERIAL IDENTIFICATION FOR DUMMY PRODUCTS

Name	DP1(dummy product 1)	DP2(dummy product 2)
Active substance	Ag (silver, metallic; CAS 7440-22-4) 0.09–0.11 g/L	Ag (silver, metallic; CAS 7440-22-4) >99.9%
Other constituents	Water	None above 0.1%
Impurities	Ammonium carbonate, sodium carbonate, ammonia, tannic acid, all at <1%	None above 0.1%
Further identifiers of active substance	Polycrystalline (TEM) particles with an average size of 25 nm (range 10–100 nm, measured by DLS), negatively charged surface (zeta potential −37 mV), coated with PVP and carbonate	Average size 5 nm, range 2–10 nm (TEM), spherical shape
Method of production	Wet chemistry (silver salt reduction)	Laser ablation

"risk assessors are encouraged to consider this information as part of a regulatory risk assessment and include a justification for how this information contributes to a nanomaterial risk assessment (e.g., determining if the dose-response relationship is influenced by these parameters)." (OECD 2012)

When there is uncertainty about the appropriate degree of specificity in substance identification as it is the case for nanomaterials, definition of identity could be either broader or narrower. Although broadening the definition—for example with regard to particle size and size distribution or surface coating—can be expected to improve the database available for risk assessment quantitatively, it will inevitably cause additional uncertainty. This uncertainty results from a currently limited understanding of the impact of physicochemical properties on the hazard properties of nanomaterials (OECD 2013). When choosing a narrower specification of the nanosilver material there is an increased likelihood that toxicity data are not available for certain endpoints. In these cases, read-across of information obtained with a similar type of nanosilver will be considered, and the degree of uncertainty resulting from the extrapolation between materials can be described. A recent analysis by the OECD WPMN in principle supports the latter approach and clarified that read-across of data between nanomaterials or a nanomaterial and its corresponding non-nanoform should be applied cautiously and with clear scientific justification (OECD 2012). In the absence of repeated-dose toxicity data on the nanosilver–silica composite AGS-20, for example, the US Environmental Protection Agency (EPA) decided to use data from an oral subacute toxicity study by Park et al. (2010) on presumably uncoated size-fractionated 42 nm particles to characterize the hazard from nanosilver breaking away from the composite. Both nanoforms of silver were regarded analogous, and the resulting uncertainties were considered to be covered by inclusion of an additional uncertainty factor of 10 in quantitative risk assessment to account for the quality of the database (US EPA 2011).

19.5 TRANSFORMATION (SPECIATION) OF NANOSILVER MATERIAL ALONG ITS LIFECYCLE

Nanosilver particles can undergo transformation reactions, most notably oxidative dissolution with release of silver ions. Ion release from coated silver nanoparticles has been shown to be dependent primarily on available surface area, thus increasing with decreasing particle size (Ma et al. 2012). In principle, the presence of surface modifiers could be expected to control nanosilver degradation. Unfortunately, the current literature seems to be contradictive in this respect, which may be due to an exchange and/or release of the surface modifier depending on the particular environment and interaction between modifier and particle core. Modification of 5 nm sized silver particles by addition of 0.4–10 mM citrate, 0.4 mM sodium sulfide, or 4 mM 11-mercaptoundecanoic acid was shown to slow the time-dependent silver release (Liu et al. 2010). Although the effect of citrate was comparatively weak (~50% reduction), the sulfur compounds had a major impact with >90% reduction in silver release. However, the surface modifiers used during wet chemistry synthesis of nanosilver are usually noncovalently bound such as those evaluated by Ma et al. (2012). These authors measured silver release from particles synthesized using different methods, including wet chemistry and gas phase condensation, and which were either unmodified or coated with PVP or gum arabic. Notably, the method of synthesis and the different coatings only had a small effect on ion release in this study, whereas the effect of surface area and size was confirmed (Ma et al. 2012). This apparent discrepancy might be explained by the oxygen-mediated formation of a surface layer of Ag_2S on silver nanoparticles, termed sulfidation (Liu et al. 2011). This process generates a highly insoluble Ag_2S shell, which reduces the rate of silver oxidation (Levard et al. 2011). It has been proposed that the sulfide-rich, anoxic environment of sewage treatment plants also facilitates rapid sulfidation of silver nanoparticles (Kim et al. 2010). Importantly, Ag_2S is less toxic to microorganisms than elemental silver nanoparticles and silver ions (Levard et al. 2012). Similarly, a modeling approach predicted that sulfidation will reduce oxidation and ion release of silver nanoparticles in sediments (Dale et al. 2013). Depending on redox conditions, sulfidized silver nanoparticles might persist in the environment with a half-life ranging from a few years to over a century (Dale et al. 2013). The seasonally variable availability of organic carbon and dissolved oxygen dictates silver speciation and persistence (Dale et al. 2013). In contrast, desulfurization of proteins and amino acids by silver nanomaterials and concomitant formation of Ag_2S appears to be negligible (Chen et al. 2013). Nevertheless, in the case of argyria Ag_2S and Ag_2Se are formed in the skin upon sun exposure (Liu et al. 2012). Some oxidation to Ag_2O occurs as well in solution, possibly generating a shell around the particle (Li et al. 2012b). AgCl nanoparticles can be formed from elemental silver nanoparticles in the presence of bleach, for instance by textile washing (Impellitteri et al. 2009). They are also used in functional textiles as part of nanocomposites (Lorenz et al. 2012) (compare the detailed review in Chapter 15). However, under conditions of a wastewater treatment plant also AgCl nanoparticles are transformed into Ag_2S (Lombi et al. 2013). As has been postulated by the Smoluchowski coagulation theory, heteroaggregation with natural colloids in surface waters leads to rapid sedimentation of nanoparticles. This has also been demonstrated for silver nanoparticles (Quik et al. 2013; Zhou et al. 2012), whereas humic acids appear to have little influence on aggregation and sedimentation (Piccapietra et al. 2012).

During textile washing the surfactants can coat the particles and the different molecules have an influence on the aggregation, both positive and negative depending on various factors including charge (Hedberg et al. 2012; Skoglund et al. 2013). Upon exposure of organisms, nanosilver particles may come into contact with a variety of bodily fluids, prominently the lung surfactant and blood. Especially, the contained proteins, lipids, and polysaccharides add to the formation of a so-called corona, a layer of such molecules covering the surface of the particle (see Chapter 4). It has been shown that the composition of the protein corona varies with the material, size, and surface modification of the nanoparticles (Monopoli et al. 2012). Initial formation of the protein corona is rapid and is followed by maturation over time (Tenzer et al. 2013). Interestingly, in the investigated set of nanoparticles, the majority of proteins in the corona exhibited an overall negative charge independent of the charge of the surface modification (Tenzer et al. 2013). Moreover, changes in the environment will dynamically alter the corona (Lundqvist et al. 2011). Importantly, the composition of the corona influences the cellular uptake of nanoparticles (Treuel et al. 2013). Apparently, more proteins are shared amongst silver nanoparticles of the same size with different surface modifications compared to the corona on different sized particles with the same surface modification (Shannahan et al. 2013). Lipids that are found in the mucus of the lung can form a bilayer around particles, which does not significantly affect ion release. Silver ion release from these particles was dependent on the pH value and varied from 10% at pH 3 to 2% at pH 5, with negligible dissolution at pH 7 over 14 days (Leo et al. 2013). Similarly, the calculated half-life of silver nanoparticles of different sizes after inhalation exposure of rats for 28 days was in the range of 24–260 days for the different tissues with an average of 80 days (Lee et al. 2013b). In fact, deposition of silver in human skin is considered permanent in argyria after oral or intravenous application of silver (Drake and Hazelwood 2005). Thus, it is apparent that silver nanoparticles exhibit a high tendency to accumulate in organisms and the environment with a slow release of silver ions. Moreover, silver ions themselves show a high propensity to bioaccumulate and persist in the environment (Fabrega et al. 2011).

In conclusion, there are a multitude of potential transformation reactions that any given nanosilver particle can undergo during its lifecycle, depending on both the initial product properties and the environments encountered. These reactions determine the form of silver to which exposure occurs, ultimately affecting the resulting risk.

19.6 HAZARD IDENTIFICATION AND CHARACTERIZATION

For the purpose of risk assessment, the characterization of human health hazards of nanosilver should address the routes of exposure and time frames that are of relevance to the particular use. As shown in Table 19.1, this could—in principle—include all routes, that is, dermal, oral, and inhalation, as well as all different time frames from acute to long-term exposure. Hazard and risk from exposure via different routes can be described with either route-specific (i.e., oral, dermal, and inhalation) toxicity studies or by route-to-route extrapolation from one representative mode of exposure. Such route-to-route extrapolation can significantly reduce animal experiments, but requires a good understanding of the toxicokinetic behavior. Moreover, route-to-route extrapolation can usually not be applied for local effects such as inflammatory and functional changes in the lung as observed following nanosilver inhalation (see the following). Likewise, extrapolation from short-term studies to long-term exposures may be possible, and extrapolation factors have been recommended for chemicals. However, a reasonable understanding of the relevant mechanisms of toxicity and kinetic aspects (e.g., accumulation) should be considered a prerequisite of such time extrapolation. OECD WPMN has concluded on this issue: *"Ideally the use of long(er) term or chronic data is recommended over extrapolation from acute or subchronic to chronic. If these data are not available and default assessment factors are used, this source of additional uncertainty to the RA should be noted"* (OECD 2012). In addition, information on hazard may need to cover not only the particular nanoform of nanosilver as manufactured but also all relevant transformation products to which humans may become exposed along the materials lifecycle (also refer to the earlier section).

There has been a series of reviews of the publicly available database on mammalian toxicity of nanosilver over recent years. In 2009, Wijnhoven et al. identified various data gaps, namely an insufficient understanding of the toxicokinetic behavior including transformation of nanosilver; the influence of particle properties on toxicity and the contribution of silver ion release; the potential for developmental and reproductive toxicity as well as neurotoxicity and carcinogenicity; and finally, the impact of long-term exposures (Wijnhoven et al. 2009).

Since then, the amount of information available on mammalian toxicity of nanoscale silver has continuously increased. Nevertheless, the EPA has concluded in December 2011 that the studies available then did not cover all life stages and did not address all relevant endpoints. In particular, no studies on chronic toxicity, reproductive and developmental toxicity, and carcinogenicity could be identified. Dose-dependent changes in cytokine levels (IL-1, IL-4, IL-6, IL-10, IL-12, and TGFβ), an increase in serum IgE levels, and a decrease in the CD4/CD8 T-lymphocyte ratio in an oral 28-day study in mice reported by Park et al. (2010) raised questions about potential immunotoxicity of the tested nanosilver. Although no increase in micronucleated polychromatic erythrocytes was detected in an earlier study by Kim et al. (2008), the relevance of this finding for an overall conclusion on mutagenicity was disputed in the absence of information on distribution of the tested nanosilver to the bone marrow (US EPA 2011). The Agency responded to this situation by applying an uncertainty factor of 10 to account for the overall quality of the database and choosing the effects on immune parameters observed in mice as point of departure for assessment of oral, dermal, and inhalation exposures. In addition, authorization of the product under review occurred under the condition that the manufacturer makes available—depending on refinement of exposure assessment—further toxicity data, namely a subchronic inhalation and/or oral study with a bone marrow micronucleus assay and a detailed assessment of neurotoxic potential, an in vitro micronucleus assay, a screening test for reproductive and developmental toxicity in rats, and a dermal toxicity study.

Recently, the nanosilver toxicity database was revisited in a conference report by Schäfer et al. (2013) and a provisional margin of safety calculation by Hadrup and Lam (2013). With regard to human health risk assessment it was concluded at a conference on state of the art in risk assessment of silver compounds in consumer products held at the German Federal Institute for Risk Assessment (BfR) in 2012 that development of consumer exposure scenarios for nanosilver products is still at an early stage and reliable exposure measurement data are lacking. Major gaps in the toxicological database were identified with regard to toxicokinetic behavior of different types of nanosilver, genotoxicity, reproductive toxicity, and further repeated dose toxicity studies designed according to relevant exposure settings. It was pointed out that a better mechanistic understanding is required to allow linking differences in physicochemical properties between nanoforms of silver to their toxic potential. Participants discussed that in such a data-deficient situation, approaches to risk assessment and management can include law enforcement and/or communication measures to limit nanosilver use to essential applications, or mandatory provisions for labeling of nanosilver containing products and the obligation for manufacturers to generate additional safety data. It was pointed out that in any case uncertainties in current safety assessment would need to be clearly pointed out and adequately accounted for (Schäfer et al. 2013). This is contrasted by considerations of Hadrup and Lam (2013), who proposed a tolerable daily intake (TDI) of 2.5 µg/kg bw/d based on the subacute study in mice (Park et al. 2011) and the default safety factor of 100. When compared to estimates for dietary oral exposure, the authors concluded that "*a Margin of Safety calculation indicates at least a factor of five before a level of concern to the general population is reached.*"

Different approaches have been discussed to compensate for the current lack of chronic toxicity data on nanoforms of silver. In its assessment of AGS-20, the EPA has chosen to apply an additional uncertainty factor of 3 to the assessment of long-term scenarios to account for extrapolations from a 28-day study in the case of oral and dermal exposure, and a 90-day study in the case of inhalation exposure (US EPA 2011). In contrast, Hadrup and Lam derived their TDI value from a no observed adverse effect level (NOAEL) of a 28-day study using the default uncertainty factor of 100 without consideration of time extrapolation (Hadrup and Lam 2013). For comparison, European Chemicals Agency's (ECHA) guidance on chemical safety assessment chapter on the characterization of dose–response for human (R.8) recommends an assessment factor of 6 for extrapolation from subacute to chronic exposure to chemicals (ECHA 2013).

A more generic concept for derivation of chronic reference values was developed in the NanoGEM research project: Experimental observations in the rat led to the concept of dust overloading of the lung as an underlying mechanism for development of an adverse effect. It was later formulated as the overload hypothesis by Morrow (1988). The overload hypothesis postulates that loading of an alveolar macrophage with inert biopersistent material occupying 6% of the macrophages volume causes a reduced mobility of the cell. Chronic exposure at this level would lead to lung inflammation and might ultimately result in tumorigenesis (Morrow 1988). Although there is no direct evidence that such a mechanism is also causing adverse effects in humans, the hypothesis itself was translated to humans for theoretical purposes by considering their larger macrophage volumes (Oberdörster 1995). In the NanoGEM project, the level at which 6% of the macrophage volume is occupied was considered as a theoretical lowest observed effect dose (t-LOEL). This LOEL was converted into a theoretical no observed adverse effect level (t-NOAEL) using two factors: a factor of 10 for the intraspecies variability and another factor of 10 for extrapolation from LOEL to NOAEL (which is considered very conservative). This calculation leads to a theoretical $NOAEL_{rat} = 0.6\ \mu m^3$ loading with inert biopersistent material per alveolar macrophage. Using data on the diameter of the particles and the density of the material, the number of particles occupying the $t\text{-}NOAEL_{rat}$ volume can be calculated. The density can be adjusted to account for spaces between individual particles. In a worst case scenario that would have those spaces with no density, the adjusted density equals the difference in volume between a solid particle and the same weight of any size of smaller particles $\rho^{Vcube}/_{Vsphere} = {}^{1}/_{6}\pi\rho_{material}$. Assuming a particle diameter of 18 nm, 3.3 pg of silver nanoparticles (corresponding to 1×10^5 particles) would occupy a volume of $t\text{-}NOAEL_{rat} = 0.6\ \mu m^3$ using the adjusted density. Hence, disregarding any specific properties of silver nanoparticles an exposure to 1×10^5 18 nm-nanosilver particles per rat alveolar macrophage should remain without adverse effects according to this concept. Data on the loading of individual macrophages can be modeled using the algorithms included in the multiple-path particle dosimetry software (MPPD), which includes nanoparticle specific modeling in version 2.11. This software may therefore be used to compare the t-NOAEL of $0.6\ \mu m^3$ (or 1×10^5) nanosilver particles per rat macrophage to the results of inhalation toxicity studies: In a subchronic inhalation toxicity study on rats, a dose of $0.133\ mg/m^3$ 18 nm silver nanoparticle aerosol with a geometric standard deviation of 1.5 remained without adverse effect in the lung and was identified as no observed adverse effect concentration (NOAEC), whereas indications for alveolar inflammatory and lung function changes were observed at $0.515\ mg/m^3$, identified as lowest observed adverse effect concentration (LOAEC) (Sung et al. 2009). For these doses, MPPD calculates a mass deposition of such particles per macrophage per breath of 2.7 and 7.8 attograms, corresponding to a number of particles per macrophage of 0.084 and 0.21 per breath, or 3102 and 7601 particles per macrophage per 6-hour exposure day, respectively. Thus, the cumulative doses at the experimental NOAEC and LOAEC would reach the theoretical NOAEL (t-NOAEL) of 1×10^5 particles per macrophage on day 3 of the seventh of 13 weeks or already on day 4 of the third week of exposure, respectively. However, this calculation did not adjust for particle clearance from the lung and macrophage cell turnover. When clearance is taken into account by using the default parameters for clearance of MPPD software, the cumulative dose at the experimental NOAEC would reach the t-NOAEL just before termination of the study on day 4 of week 12. In contrast, the same loading is predicted for day 2 of week 4 at the experimental LOAEC. Overall, there was a surprising level of agreement between the theoretical approach based on macrophage loading and the experimental data even though aspects like dissolution and specific toxicity of (nano)silver in the lung were not accounted for, warranting further development and validation of this model.

19.7 CHALLENGES CAUSED BY LIMITATIONS IN NANOPARTICLE MEASUREMENT TECHNIQUES

Methods to characterize nanomaterials and their limitations are discussed in detail in Section I, Chapters 1–4 of this book. Airborne particles can be measured with a variety of instruments (Kuhlbusch et al. 2011). The instruments generally provide qualitatively comparable measurements, each with its specific limitations (Asbach et al. 2012; Kaminski et al. 2013). Most methods are useful to characterize the virgin materials because background and contaminations cannot be differentiated and nanosilver particles in a mixed exposure atmosphere cannot be distinguished from other nanomaterials (Kuhlbusch et al. 2011). Currently, the aerosol mass spectrometer is the only instrument sizing and chemically analyzing nanoscale particles online. However, this instrument does not detect metals such as nanosilver or metal oxides. Similarly, light scattering methods to characterize nanoparticles in aqueous dispersions exhibit clear strengths and limitations, including that only simple dispersions can be measured—subpopulations of different material or size cannot be, or at least reliably, detected (Landsiedel et al. 2012; Roebben et al. 2011). Noble metal particles such as silver nanoparticles exhibit plasmon resonance and can be detected using characteristic UV-Vis spectra, in which the shift in the adsorption maximum acts as an indicator of particle size—a property that still remains to be exploited for routing measurement of nanosilver (Hagendorfer et al. 2012). An important consideration for lifecycle evaluations is also the limited range of detected size for the different methods. In the case of partial homo- or heteroagglomeration of nanosilver, this might lead to incorrect measurement of total particle numbers. Because of the size to mass relationship this can lead to severe over- or underestimations in the dose metrics. Hence, a combination of methods is recommended to characterize the material (Linsinger et al. 2012). For the detection of the material in the environment, or bodies or organs, measurement of total silver is usually possible. However, detection of the physical state, that is, the presence of nanoparticles is extremely difficult. The method of choice for the identification of nanomaterials *in situ*, for example, in organs of animals from a toxicokinetic study or products presumably containing nanosilver, is electron microscopy coupled with single particle chemical analysis such as EDX. However, taking into account the aforementioned plasticity of silver within organisms and the environment, detected electron-dense particles do not necessarily represent the exogenously introduced nanoparticles. Silver ions form nanoscale Ag_2S and Ag_2Se deposits, and might be derived from other sources and/or the introduced nanoparticles (Liu et al. 2012). Moreover, electron microscopy requires extensive sample preparation, which is not always compatible with the specimen, although some methods in development allow for minimally processed samples (Dudkiewicz et al. 2011). Often, the distribution or dilution is so high that the likelihood of finding a nanoparticle for instance in an organ using electron microscopy after ultrathin slicing is extremely small and defies quantification. Conversely, due to these limitations current technology may not allow for a definitive proof of absence of nanomaterials, even if automated particle identification software is used (Kuhlbusch et al. 2011).

REFERENCES

Alibaba product search, http://www.alibaba.com/trade/advancesearch?advancedSearchText, accessed 10/25/2013.

Armitage, S. A., White, M. A., Wilson, H. K. (1996). The determination of silver in whole blood and its application to biological monitoring of occupationally exposed groups. *Ann Occup Hyg* 40: 331–338.

Asbach, C., Kaminski, H., von Barany, D., Kuhlbusch, T. A., Monz, C., Dziurowitz, N., Pelzer, J., Vossen, K., Berlin, K., Dietrich, S., Götz, U., Kiesling, H. J., Schierl, R., Dahmann, D. (2012). Comparability of portable nanoparticle exposure monitors. *Ann Occup Hyg* 56: 606–621.

Benn, T. M., Westerhoff, P. (2008). Nanoparticle silver released into water from commercially available sock fabrics. *Environ Sci Technol* 42: 4133–4139.

Carlson, C., Hussain, S. M., Schrand, A. M., Braydich-Stolle, L. K., Hess, K. L., Jones, R. L., Schlager, J. J. (2008). Unique cellular interaction of silver nanoparticles: size-dependent generation of reactive oxygen species. *J Phys Chem B* 112: 13608–13619.

Chen, S., Theodorou, I. G., Goode, A., Gow, A., Schwander, S., Zhang, J., Chung, K. F., Tetley, T., Shaffer, M. S., Ryan, M. P., Porter, A. E. (2013). High resolution analytical electron microscopy reveals cell culture media induced changes to the chemistry of silver nanowires. *Environ Sci Technol* 47:13813–13821.

Dale, A. L., Lowry, G. V., Casman, E. A. (2013). Modeling nanosilver transformations in freshwater sediments. *Environ Sci Technol* 47: 12920–12928.

Demou, E., Peter, P., Hellweg, S. (2008). Exposure to manufactured nanostructured particles in an industrial pilot plant. *Ann Occup Hyg* 52: 695–706.

Drake, P. L., Hazelwood, K. J. (2005). Exposure-related health effects of silver and silver compounds: a review. *Ann Occup Hyg* 49: 575–585.

Dudkiewicz, A., Tiede, K., Loeschner, K., Jensen, L. H. S., Jensen, E., Wierzbicki, R., Boxall, A. B. A., Molhave, K. (2011). Characterization of nanomaterials in food by electron microscopy. *TrAC - Trends Anal Chem* 30: 28–43.

ECHA (2013). Guidance for identification and naming of substances under REACH and CLP, ECHA-11-G-10.1-EN. Helsinki, European Chemicals Agency. http://echa.europa.eu/documents/10162/13643/substance_id_en.pdf, date accessed 3-12-2013.

Fabrega, J., Luoma, S. N., Tyler, C. R., Galloway, T. S., Lead, J. R. (2011). Silver nanoparticles: behaviour and effects in the aquatic environment. *Environ Int* 37: 517–531.

Farkas, J., Peter, H., Christian, P., Gallego Urrea, J. A., Hassellov, M., Tuoriniemi, J., Gustafsson, S., Olsson, E., Hylland, K., Thomas, K. V. (2011). Characterization of the effluent from a nanosilver producing washing machine. *Environ Int* 37: 1057–1062.

Hadrup, N., Lam, H. R. (2013). Oral toxicity of silver ions, silver nanoparticles and colloidal silver - A review. *Regul Toxicol Pharmacol* 68: 1–7.

Hagendorfer, H., Kaegi, R., Parlinska, M., Sinnet, B., Ludwig, C., Ulrich, A. (2012). Characterization of silver nanoparticle products using asymmetric flow field flow fractionation with a multidetector approach—A comparison to transmission electron microscopy and batch dynamic light scattering. *Anal Chem* 84: 2678–2685.

Hagendorfer, H., Lorenz, C., Kaegi, R., Sinnet, B., Gehrig, R., von Goetz, N., Scheringer, M., Ludwig, C., Ulrich, A. (2010). Size-fractionated characterization and quantification of nanoparticle release rates from a consumer spray product containing engineered nanoparticles. *J Nanopart Res* 12: 2481–2494.

Hedberg, J., Lundin, M., Lowe, T., Blomberg, E., Wold, S., Wallinder, I. O. (2012). Interactions between surfactants and silver nanoparticles of varying charge. *J Colloid Interface Sci* 369: 193–201.

Huang, Y., Chen, S., Bing, X., Gao, C., Wang, T., Yuan, B. (2011). Nanosilver migrated into food-simulating solutions from commercially available food fresh containers. *Packag Technol Sci* 24: 291–297.

Impellitteri, C. A., Tolaymat, T. M., Scheckel, K. G. (2009). The speciation of silver nanoparticles in antimicrobial fabric before and after exposure to a hypochlorite/detergent solution. *J Environ Qual* 38: 1528–1530.

JRC (2011). REACH Implementation Project Substance Identification of Nanomaterials (RIP-oN 1) AA No.070307/2009/D1/534733 between DG ENV and JRC - Advisory Report. European Commission, Joint Research Centre, Institute for Health and Consumer Protection. http://ec.europa.eu/environment/chemicals/nanotech/pdf/report_ripon1.pdf, date accessed 3-12-2013.

Kaminski, H., Kuhlbusch, T. A. J., Rath, S., Götz, U., Sprenger, M., Wels, D., Polloczek, J., Bachmann, V., Dziurowitz, N., Kiesling, H. J., Schwiegelshohn, A., Monz, C., Dahmann, D., Asbach, C. (2013). Comparability of mobility particle sizers and diffusion chargers. *J Aerosol Sci* 57: 156–178.

Kim, B., Park, C. S., Murayama, M., Hochella, M. F. (2010). Discovery and characterization of silver sulfide nanoparticles in final sewage sludge products. *Environ Sci Technol* 44: 7509–7514.

Kim, Y. S., Kim, J. S., Cho, H. S., Rha, D. S., Kim, J. M., Park, J. D., Choi, B. S., Lim, R., Chang, H. K., Chung, Y. H., Kwon, I. H., Jeong, J., Han, B. S., Yu, I. J. (2008). Twenty-eight-day oral toxicity, genotoxicity, and gender-related tissue distribution of silver nanoparticles in Sprague-Dawley rats. *Inhal Toxicol* 20: 575–583.

Kuhlbusch, T. A., Asbach, C., Fissan, H., Gohler, D., Stintz, M. (2011). Nanoparticle exposure at nanotechnology workplaces: A review. *Part Fibre Toxicol* 8: 22.

Landsiedel, R., Fabian, E., Ma-Hock, L., van Ravenzwaay, B., Wohlleben, W., Wiench, K., Oesch, F. (2012). Toxico-/biokinetics of nanomaterials. *Arch Toxicol* 86: 1021–1060.

Lankveld, D. P., Oomen, A. G., Krystek, P., Neigh, A., Troost-de Jong, A., Noorlander, C. W., Van Eijkeren, J. C., Geertsma, R. E., De Jong, W. H. (2010). The kinetics of the tissue distribution of silver nanoparticles of different sizes. *Biomaterials* 31: 8350–8361.

Lee, D. Y., Hwang, E. S., Yu, T. U., Kim, Y. J., Hwang, J. (2006). Structuring of micro line conductor using electro-hydrodynamic printing of a silver nanoparticle suspension. *Appl Phys A: Mater Sci Process* 82: 671–674.

Lee, J. H., Hwang, J. H., Nam, J. M. (2013a). DNA-tailored plasmonic nanoparticles for biosensing applications. *Wiley Interdiscip Rev Nanomed Nanobiotechnol* 5: 96–109.

Lee, J. H., Kim, Y. S., Song, K. S., Ryu, H. R., Sung, J. H., Park, J. D., Park, H. M., Song, N. W., Shin, B. S., Marshak, D., Ahn, K., Lee, J. E., Yu, I. J. (2013b). Biopersistence of silver nanoparticles in tissues from Sprague-Dawley rats. *Part Fibre Toxicol* 10: 36.

Lee, J. H., Kwon, M., Ji, J. H., Kang, C. S., Ahn, K. H., Han, J. H., Yu, I. J. (2011). Exposure assessment of workplaces manufacturing nanosized TiO_2 and silver. *Inhal Toxicol* 23: 226–236.

Lee, J. H., Mun, J., Park, J. D., Yu, I. J. (2012). A health surveillance case study on workers who manufacture silver nanomaterials. *Nanotoxicology* 6: 667–669.

Leo, B. F., Chen, S., Kyo, Y., Herpoldt, K. L., Terrill, N. J., Dunlop, I. E., McPhail, D. S., Shaffer, M. S., Schwander, S., Gow, A., Zhang, J., Chung, K. F., Tetley, T. D., Porter, A. E., Ryan, M. P. (2013). The stability of silver nanoparticles in a model of pulmonary surfactant. *Environ Sci Technol* 47: 11232–11240.

Levard, C., Hotze, E. M., Lowry, G. V., Brown, G. E., Jr. (2012). Environmental transformations of silver nanoparticles: Impact on stability and toxicity. *Environ Sci Technol* 46: 6900–6914.

Levard, C., Reinsch, B. C., Michel, F. M., Oumahi, C., Lowry, G. V., Brown, G. E. (2011). Sulfidation processes of PVP-coated silver nanoparticles in aqueous solution: Impact on dissolution rate. *Environ Sci Technol* 45: 5260–5266.

Li, L., Sun, J., Li, X., Zhang, Y., Wang, Z., Wang, C., Dai, J., Wang, Q. (2012a). Controllable synthesis of monodispersed silver nanoparticles as standards for quantitative assessment of their cytotoxicity. *Biomaterials* 33: 1714–1721.

Li, X., Lenhart, J. J., Walker, H. W. (2012b). Aggregation kinetics and dissolution of coated silver nanoparticles. *Langmuir* 28: 1095–1104.

Linsinger, T. P. J., Roebben G., Gilliland, D., Calzolai, L., Rossi, F., Gibson, N., Klein, C. (2012) Requirements on measurements for the implementation of the European Commission definition of the term nanomaterial. JRC Reference Report. Luxembourg July 2012.

Liu, J., Pennell, K. G., Hurt, R. H. (2011). Kinetics and mechanisms of nanosilver oxysulfidation. *Environ Sci Technol* 45: 7345–7353.

Liu, J., Sonshine, D. A., Shervani, S., Hurt, R. H. (2010). Controlled release of biologically active silver from nanosilver surfaces. *ACS Nano* 4: 6903–6913.

Liu, J., Wang, Z., Liu, F. D., Kane, A. B., Hurt, R. H. (2012). Chemical transformations of nanosilver in biological environments. *ACS Nano* 6: 9887–9899.

Lombi, E., Donner, E., Taheri, S., Tavakkoli, E., Jamting, A. K., McClure, S., Naidu, R., Miller, B. W., Scheckel, K. G., Vasilev, K. (2013). Transformation of four silver/silver chloride nanoparticles during anaerobic treatment of wastewater and post-processing of sewage sludge. *Environ Pollut* 176: 193–197.

Lorenz, C., Hagendorfer, H., von Goetz, N., Kaegi, R., Gehrig, R., Ulrich, A., Scheringer, M., Hungerbühler, K. (2011). Nanosized aerosols from consumer sprays: Experimental analysis and exposure modeling for four commercial products. *J Nanopart Res* 13: 3377–3391.

Lorenz, C., Windler, L., von Goetz, N., Lehmann, R. P., Schuppler, M., Hungerbühler, K., Heuberger, M., Nowack, B. (2012). Characterization of silver release from commercially available functional (nano)textiles. *Chemosphere* 89: 817–824.

Lundqvist, M., Stigler, J., Cedervall, T., Berggard, T., Flanagan, M. B., Lynch, I., Elia, G., Dawson, K. (2011). The evolution of the protein corona around nanoparticles: A test study. *ACS Nano* 5: 7503–7509.

Ma, R., Levard, C., Marinakos, S. M., Cheng, Y., Liu, J., Michel, F. M., Brown, G. E., Lowry, G. V. (2012). Size-controlled dissolution of organic-coated silver nanoparticles. *Environ Sci Technol* 46: 752–759.

Meyer, D. E., Curran, M. A., Gonzalez, M. A. (2011). An examination of silver nanoparticles in socks using screening-level life cycle assessment. *J Nanopart Res* 13: 147–156.

Monopoli, M. P., Aberg, C., Salvati, A., Dawson, K. A. (2012). Biomolecular coronas provide the biological identity of nanosized materials. *Nat Nanotechnol* 7: 779–786.

Morrow, P. E. (1988). Possible mechanisms to explain dust overloading of the lungs. *Fundam Appl Toxicol* 10: 369–384.

Nazarenko, Y., Han, T. W., Lioy, P. J., Mainelis, G. (2011). Potential for exposure to engineered nanoparticles from nanotechnology-based consumer spray products. *J Expo Sci Environ Epidemiol* 21: 515–528.

NRC (2009). Science and Decisions: Advancing Risk Assessment. Committee on Improving Risk Analysis Approaches Used by the U.S. EPA, Board on Environmental Studies and Toxicology, Division on Earth and Life Studies, National Research Council of the National Academies., p. 424. The National Academies Press, Washington, D.C.

Oberdörster, G. (1995). Lung particle overload: Implications for occupational exposures to particles. *Regul Toxicol Pharmacol* 21: 123–135.

OECD (1-12-2010). List of Manufactured Nanomaterials and List of Endpoints for Phase One of the Sponsorship Programme for the Testing of Manufactured Nanomaterials: Revision, Series on the Safety of Manufactured Nanomaterials No. 27, ENV/JM/MONO(2010)46. Paris, OECD. http://search.oecd.org/officialdocuments/displaydocumentpdf/?cote=env/jm/mono%282010%2946&doclanguage=en, 2-12-2013.

OECD (28-3-2012). Important Issues on Risk Assessment of Manufactured Nanomaterials, Series on the Safety of Manufactured Nanomaterials No. 33, ENV/JM/MONO(2012)8. Paris, OECD. http://search.oecd.org/officialdocuments/displaydocumentpdf/?cote=env/jm/mono%282012%298&doclanguage=en, date accessed 2-12-2013.

OECD (21-8-2013). Co-Operation on Risk Assessment: Prioritisation of Important Issues on Risk Assessment of Manufactured Nanomaterials - Final Report, Series on the Safety of Manufactured Nanomaterials No. 38, ENV/JM/MONO(2013)18. Paris, OECD. http://search.oecd.org/officialdocuments/displaydocumentpdf/?cote=env/jm/mono%282013%2918&doclanguage=en, date accessed 2-12-2013.

Park, E. J., Bae, E., Yi, J., Kim, Y., Choi, K., Lee, S. H., Yoon, J., Lee, B. C., Park, K. (2010). Repeated-dose toxicity and inflammatory responses in mice by oral administration of silver nanoparticles. *Environ Toxicol Pharmacol* 30: 162–168.

Pasricha, A., Jangra, S. L., Singh, N., Dilbaghi, N., Sood, K. N., Arora, K., Pasricha, R. (2012). Comparative study of leaching of silver nanoparticles from fabric and effective effluent treatment. *J Environ Sci (China)* 24: 852–859.

Piccapietra, F., Sigg, L., Behra, R. (2012). Colloidal stability of carbonate-coated silver nanoparticles in synthetic and natural freshwater. *Environ Sci Technol* 46: 818–825.

Quadros, M. E., Marr, L. C. (2011). Silver nanoparticles and total aerosols emitted by nanotechnology-related consumer spray products. *Environ Sci Technol* 45: 10713–10719.

Quik, J. T., Velzeboer, I., Wouterse, M., Koelmans, A. A., van de Meent, D. (2013). Heteroaggregation and sedimentation rates for nanomaterials in natural waters. *Water Res* 48: 269–279.

Roebben, G., Ramirez-Garcia, S., Hackley, V. A., Roesslein, M., Klaessig, F., Kestens, V., Lynch, I., Garner, C. M., Rawle, A., Elder, A., Colvin, V. L., Kreyling, W., Krug, H. F., Lewicka, Z. A., McNeil, S., Nel, A., Patri, A., Wick, P., Wiesner, M., Xia, T., Oberdörster, G., Dawson, K. A. (2011). Interlaboratory comparison of size and surface charge measurements on nanoparticles prior to biological impact assessment. *J Nanopart Res* 13: 2675–2687.

Schäfer, B., Brocke, J. V., Epp, A., Götz, M., Herzberg, F., Kneuer, C., Sommer, Y., Tentschert, J., Noll, M., Günther, I., Banasiak, U., Bol, G. F., Lampen, A., Luch, A., Hensel, A. (2013). State of the art in human risk assessment of silver compounds in consumer products: A conference report on silver and nanosilver held at the BfR in 2012. *Arch Toxicol* 87: 2249–2262.

Shannahan, J. H., Lai, X., Ke, P. C., Podila, R., Brown, J. M., Witzmann, F. A. (2013). Silver nanoparticle protein corona composition in cell culture media. *PLoS One* 8: e74001.

Skoglund, S., Lowe, T. A., Hedberg, J., Blomberg, E., Wallinder, I. O., Wold, S., Lundin, M. (2013). Effect of laundry surfactants on surface charge and colloidal stability of silver nanoparticles. *Langmuir* 29: 8882–8891.

Solomon, S. D., Bahadory, M., Jeyarajasingam, A. V., Rutkowsky, S. A., Boritz, C., Mulfinger, L. (2007). Synthesis and study of silver nanoparticles. *J Chem Educ* 84: 322–325.

Sung, J. H., Ji, J. H., Park, J. D., Yoon, J. U., Kim, D. S., Jeon, K. S., Song, M. Y., Jeong, J., Han, B. S., Han, J. H., Chung, Y. H., Chang, H. K., Lee, J. H., Cho, M. H., Kelman, B. J., Yu, I. J. (2009). Subchronic inhalation toxicity of silver nanoparticles. *Toxicol Sci* 108: 452–461.

Tenzer, S., Docter, D., Kuharev, J., Musyanovych, A., Fetz, V., Hecht, R., Schlenk, F., Fischer, D., Kiouptsi, K., Reinhardt, C., Landfester, K., Schild, H., Maskos, M., Knauer, S. K., Stauber, R. H. (2013). Rapid formation of plasma protein corona critically affects nanoparticle pathophysiology. *Nat Nanotechnol* 8: 772–781.

Treuel, L., Jiang, X., Nienhaus, G. U. (2013). New views on cellular uptake and trafficking of manufactured nanoparticles. *J R Soc Interface* 10: 20120939.

US EPA (1-12-2011). Decision Document: Conditional Registration of HeiQ AGS-20 as a Materials Preservative in Textiles, EPA-HQ-OPP-2009-1012-0064. Environmental Protection Agency, Office of Pesticide Programs, Antimicrobials Division. http://www.regulations.gov/#!documentDetail;D=EPA-HQ-OPP-2009-1012-0064, date accessed 3-12-2013.

von Goetz, N., Fabricius, L., Glaus, R., Weitbrecht, V., Gunther, D., Hungerbühler, K. (2013). Migration of silver from commercial plastic food containers and implications for consumer exposure assessment. *Food Addit Contam Part A Chem Anal Control Expo Risk Assess* 30: 612–620.

Wijnhoven, S. W. P., Peijnenburg, W. J. G. M., Herberts, C. A., Hagens, W. I., Oomen, A. G., Heugens, E. H. W., Roszek, B., Bisschops, J., Gosens, I., van de Meent, D., Dekkers, S., De Jong, W. H., Van Zijverden, M., Sips, A. J. A. M., Geertsma, R. E. (2009). Nanosilver - A review of available data and knowledge gaps in human and environmental risk assessment. *Nanotoxicology* 3: 109–138.

Wiley, B., Sun, Y., Xia, Y. (2007). Synthesis of silver nanostructures with controlled shapes and properties. *Acc Chem Res* 40: 1067–1076.

WWICS (2013). The Project on Emerging Nanotechnologies: Consumer Products: An inventory of Nanotechnology-Based Consumer Products Currently on the Market: Analysis. http://www.nanotechproject.org/cpi/about/analysis/, date accessed 25-10-2013.

Xu, R., Wang, D., Zhang, J., Li, Y. (2006). Shape-dependent catalytic activity of silver nanoparticles for the oxidation of styrene. *Chem Asian J* 1: 888–893.

Zhou, D., Abdel-Fattah, A. I., Keller, A. A. (2012). Clay particles destabilize engineered nanoparticles in aqueous environments. *Environ Sci Technol* 46: 7520–7526.

Index

O

P

R

S

T

U

V

W

X

Z